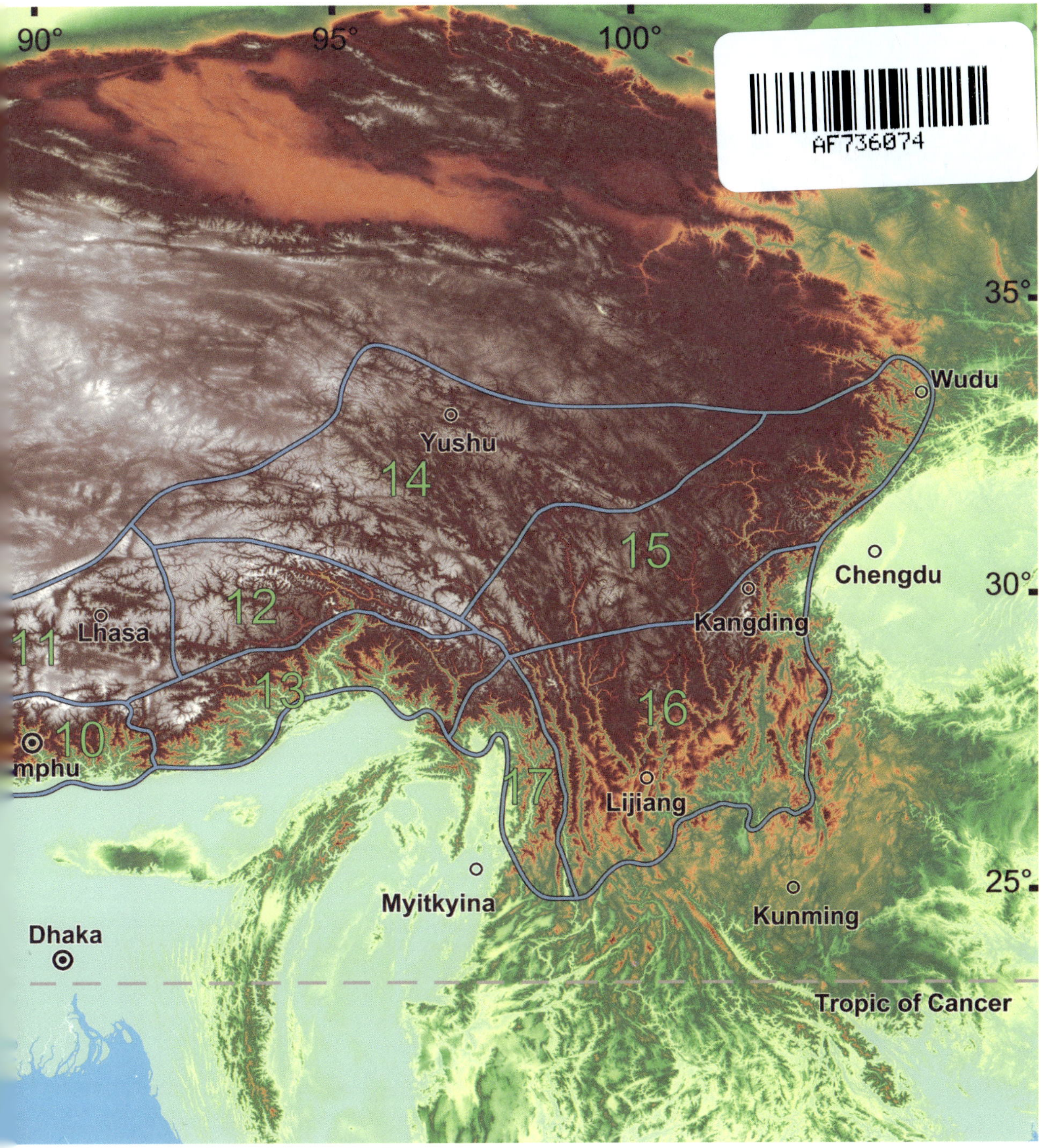

Pan-Himalaya

ng Zangbo; 6. W Nepal; 7. C Nepal; 8. E Nepal; 9. Sikkim & Darjiling;
ıhmaputra; 14. Tangut; 15. N Hengduan; 16. S Hengduan; 17. U Irrawaddy.

Supported by the National Fund for Academic Publication in Science and Technology

Flora *of* Pan-Himalaya

Volume 47

Aquifoliaceae, Helwingiaceae, Campanulaceae, Lobeliaceae, Menyanthaceae

De-Yuan HONG

State Key Laboratory of Systematic and Evolutionary Botany,
Institute of Botany, Chinese Academy of Sciences
Email: hongdy@ibcas.ac.cn

CAMBRIDGE UNIVERSITY PRESS

Science Press Beijing

Published by Science Press (Beijing) under ISBN 9787030466266

and by

CAMBRIDGE
UNIVERSITY PRESS

University Printing House, Cambridge CB2 8BS, United Kingdom

Cambridge University Press is part of the University of Cambridge.

It furthers the University's mission by disseminating knowledge in the pursuit of education, learning, and research at the highest international levels of excellence.

www.cambridge.org

Information on this title: www.cambridge.org/9781107158979

First published 2016

Printed in the People's Republic of China by Science Press

A catalogue record for this publication is available from the British Library.
Library of Congress Cataloguing in Publication Data

ISBN 978-1-107-15897-9 Hardback

Flora *of* Pan-Himalaya

Office: Ren SA
Tel: 86-10-62836084
Fax: 86-10-62590843

Email: flph@ibcas.ac.cn
Website: www.flph.org

It is with great pleasure that we acknowledge the generous financial support of the following:

China Ministry of Science and Technology
National Natural Science Foundation of China
Chinese Academy of Sciences

Foreword

We are delighted to welcome the appearance of this highly significant regional flora, and congratulate De-Yuan HONG and all of the others that worked together to make it possible. The Himalaya is one of the most beautiful, extensive, and revered ranges on earth, containing nine of our ten highest peaks, and its plants deserve the kind of inclusive and collaborative treatment that we have been presented in these fine volumes.

The mighty Himalaya were formed as the basaltic Indian plate plowed into the underbelly of Asia. Carrying the lighter rocks that would become the mountain ranges we see now, it broke out of the gigantic southern continent of Gondwanaland approximately 90 million years ago and started its journey northward to its current position. In the south, any part of India that rose above the ocean would have been covered with an archaic southern vegetation similar to that which survives today in southern Australia and South America, New Zealand, and New Caledonia. As the fragment that includes India journeyed northward, it passed from the south across the equator and started crowding into the Asian plate perhaps 55-50 million years ago, continuing to thrust up the Himalaya from that time onward. The same movement folded what is now China into a series of accordion-like pleats that became the current mountain ranges of southern and southwestern China and its neighboring countries.

Crossing so many latitudes seems to have led to the extinction of virtually all of the plants and animals that occupied India when its position lay south of the equator, but when it came within reach of Asia and under the influence of a tropical and subtropical climate from the early Eocene Period onward. The plants and animals of southern Asia colonized the new lands, and more migrated from the north as the mountains of the region were thrust ever upward. Importantly, there was a great opportunity for the evolution of endemic plants and animals in the new mountains, and the mixture of all of these rich biological strands have combined to form the incredible biological richness of the area as we see it today.

Both because of the political subdivisions of the region and its complex floristic composition, it took very courageous and forward-looking leadership for De-Yuan HONG and his colleagues to undertake this monumental flora. Except for China, there are relatively few endemics in most of the countries of the Himalayan ranges and their bordering lowlands, but many in the region. This lack of national endemics, however, stems mainly from the artificial presence of countries in the area, which taken as a whole is an extremely rich one for endemics in virtually all groups of plants, animals, and fungi that occur there. The plants of some of the countries are reasonably well known, those of others poorly known, and our historical efforts to bring the available information together and improve its quality by trans-national comparisons have been inadequate.

That situation is all the more unfortunate because of the severe threats that the ecosystems of the regions face from rapid human population growth and the even more rapidly rising expectations for increased consumption and especially from the climate change that threatens all of the ecosystems of the area, including the possible destruction of its glaciers and the disruptions of the monsoons on which so many depend for the food and livelihood. A sound knowledge of the plants and ecosystems of the region is the only way to win sustainability, even if population, consumption levels, and the runaway use of malign technologies like burning coal are brought under control.

For all of these reasons, we declare once more our delight in the appearance of this flora and congratulate those who have worked together to make it a reality. It is a critically important first step in improving our knowledge of the plants of the area on a regional basis, and one of great and general importance.

Peter H. Raven,
President Emeritus,
Missouri Botanical Garden

Introduction

—General Guideline of the *Flora of Pan-Himalaya*

The Himalaya and adjacent regions (Pan-Himalaya) form a natural phytogeographical unit, from the Vakhan Corridor eastwards to the Hengduan Mountains via the Karakorum and the Himalaya. This region covers the northeastern corner of Afghanistan, northern Pakistan, northern India, Nepal, Bhutan, northern Myanmar, and southwestern China (S Tibet, SE Qinghai, SE Gansu, W Sichuan and NW Yunnan). This region is divided into 17 subregions: Vakhan, N Pakistan, Jammu & Kashmir, U Ganga & Indus, U Yarlung Zangbo, W Nepal, C Nepal, E Nepal, Sikkim & Darjiling, Bhutan, M Yarlung Zangbo, L Yarlung Zangbo, Yarlung Zangbo-Brahmaputra, Tangut, N Hengduan, S Hengduan, and U Irrawaddy (see the maps).

The *Flora of Pan-Himalaya* (FLPH) will be published in English in two editions: printed and online. It treats all the native and naturalized vascular plants of this region.

The Flora will be published in 50 volumes with ca. 80 books, with the larger families divided into two to seven parts. The classification systems used in this Flora will reflect our current understanding of phylogenetic or phylogenomic relationships of the relevant groups. The third edition of the Angiosperm Phylogeny Group system (APG III) will be adopted for the angiosperms, and up-to-date phylogenetic or phylogenomic systems for the gymnosperms, ferns, and lycophytes will be followed for the treatments of those groups.

The natural and rational delimitation of species is the critical task of taxonomy, and is the most important criterion for judging the scientific value and standard of a Flora. Population concepts should be employed in determining the taxonomy; character analysis should be considered as the basis for a rational taxonomic treatment. Therefore, field observations at the population level, and statistic analysis based on broad population sampling and collections are greatly encouraged to investigate the variation of characters and evaluate their taxonomic significance for the delimitation of taxa. Herbarium specimens kept at the following herbaria must be examined and identified: PE, KUN, K, BM, E, CDBI, and SZ. The authors are also encouraged to visit the herbaria of A, B, CAL, CAS, G, KATH, LE, P, TI, W, and HNWP.

Delimitation of closely related species should be based on correlated discontinuous (including statistically discontinuous) variation of at least two characters. As to apomicts (such as *Taraxacum* spp., *Sorbus* spp.) and vegetatively reproducing plants, "macrospecies" should be adopted instead of "microspecies".

The use of subspecies is encouraged in infraspecific classifications, but varieties only in special cases. Subspecies are geographical races and horizontally or vertically vicarious, whilst varieties are prominent ecotypes. Genotype variants within populations should not be recognized as different taxa. The description of a species should cover all its elements. Variants which do not merit formal taxonomic recognition may be noted below the description of the species.

Larger families and genera may be concisely subdivided. These systems should reflect the current understanding of phylogenetic or phylogenomic relationships, especially those generated from solid molecular and genomic data.

The concept of the genus should reflect the current understanding of phylogenies. Generic concepts strongly supported by both molecular/genomic and morphological evidence should be adopted, but those with weak support in molecular/genomic analysis and have no sound morphological evidence should not be accepted. The treatment of more poorly known groups should be conservative and any perceived problems indicated.

Identification keys are one of the most important products of classification, and the best reflection of its usefulness. Authors should construct a key based on examination of all relevant specimens.Where possible, keys should not be based on a single character, and must avoid vague words, such as "relatively large" vs. "relatively small", "longer" vs. "shorter". It is important to use easily visible characters in keys. Dichotomous indented keys such as those in the *Flora of China* are to be used in this Flora.

Description of species should be concise with emphasis on diagnostic characters, and should include habitat information and the geographical distribution, plus chromosome numbers, phenology, and other biological features, as well as reliable economic uses and conservation information where available.

Literature citation for accepted names and their basionyms must be provided. Important references, such as monographs and Floras, should be cited, but not more than three titles (at most five if absolutely necessary). All synonyms based on types from within the Flora area should be listed along with other synonyms that have been widely used in relevant publications.

Detailed type information should be indicated after the nomenclatural citation of taxa. Designation of types is highly encouraged in this Flora.

A distribution map is given for each species. This will be a dot map based on the specimens seen for the Flora. The vouchers are to be cited to record geographical distribution for each species, subspecies (or variety). At least one representative specimen is needed for each distribution unit (county in China, district in the other countries). For the more widely distributed taxa (across nine or more subdivisions), a larger distribution unit can be used, no citation of vouchers is needed, and the distribution outlines are also acceptable.

The nomenclatural novelties (new taxa, new combinations, and *nomina nova*) may be published in the Flora or elsewhere.

Sterile hybrids, and those hybrids not forming populations, are to be briefly discussed after the descriptions of their parents. Plants only konwn from cultivation may be indicated at the end of the treatment of the genus, but should not be included in the identification keys.

All genera, and at least one third of species, should be illustrated with good quality line drawings with scale bars to indicate magnifications. The illustrations can be drawn from herbarium materials, living materials or images available, selected by the authors, or taken from publications with permission. Wherever possible the material used should be indicated. Diagnostic characters of the species should be clearly illustrated. One plate may hold illustrations of one, two, or three species. For larger genera composite plates of diagnostic characters are encouraged.

The web edition of the Flora will follow the corresponding printed edition. Colour photographs, detailed citation of vouchers, and other sources of information are to be included only in the web edition.

The physical geography, history of studies of vascular plants from this region, biogeography, biodiversity conservation, general references, and key to families are to be included in the first (introductory) volume. A comprehensive index to taxa and other subjects are to be published in the last volume.

More detailed guidelines and a model treatment for preparation of Flora accounts will be provided by the Editorial Committee.

The Editorial Committee of *Flora of Pan-Himalaya*

Preface

Volume 47 of the *Flora of Pan-Himalaya* is the first one of the 50 volumes with about 80 books in the series. It includes five families (Aquifoliaceae, Helwingiaceae, Campanulaceae, Lobeliaceae, and Menyanthaceae), 18 genera, and 192 species, among which one genus (*Himalacodon*) and 80 species (42%) are endemic to the Pan-Himalaya.

The Lobeliaceae is included in the Campanulaceae in the APG III system, but the Campanulaceae in sense of APG III system consists of five monophyletic groups, of which the Campanulaceae s. str. and Lobeliaceae are independent from each other, and they are distinct from each other not only in morphology, but also in palynology and phytochemistry (see p. 254 for further explanation). Therefore, the author of the present volume considers that it may be better to treat them as separate family than as subfamily.

The nomenclatural novelties in the volume are four new species, one new subspecies, three new combinations, one new status, 10 new synonyms, and 48 lectotypes are newly designated.

During the work on this volume, the author made eight expeditions to the Pan-Himalaya for field observations and collection of materials: three to S Tibet (U, M and L Yarlung Zangbo, and Yarlung Zangbo-Brahmaputra), one to N Pakistan (Hazara), three to N and S Hengduan, and one to U Irrawaddy (Tengchong in Yunnan). The author and his team determined three new genera, *Pseudocodon* D. Y. Hong & H. Sun, *Himalacodon* D. Y. Hong & Q. Wang, and *Pankycodon* D. Y. Hong & X. T. Ma, merged two genera, *Campanumoea* and *Leptocodon*, into *Codonopsis*, reduced 15 species as synonymy, and also excluded records of five species of Aquifoliaceae in the Pan-Himalaya due to misidentification by a previous author. In addition to the five new taxa published in this volume, the author also described other eight new species (one in *Ilex*, seven in *Codonopsis*). There are 329 scientific (specific and infraspecific) names in this volume, among which 208 had holotypes already designated, while 121 needed to have lectotypes. Of these 121 names, the present author designated lectotypes for 84 (80%) since 2010. Unfortunately, we still have 13 names, i.e. 4% for the total, to be lectotypified.

Mr. M. Gilbert reviewed the volume, checking the nomenclature, improving the English, and making the descriptions better. His contribution to the present volume is significant.

The editor of figures is Miss Ai-Li LI, who is a plant artist. The distribution maps were automatically produced based on cited vouchers presented by the author. The automatic procedure was designed by Mr. Min LI and Mr. Jing XUAN.

Acknowledgements

Partial funding of the *Flora of Pan-Himalaya* has been provided by China Ministry of Science and Technology (grant 2013FY112100), the National Natural Science Foundation of China (grant 31110103911), the Chinese Academy of Sciences (grant KSCX2-EW-Z-1), and the State Key Laboratory of Systematic and Evolutionary Botany, Institute of Botany, Chinese Academy of Sciences.

The directors, curators, and librarians of the following institutions are warmly thanked for their assistance in sending loans and images of specimens, and/or providing the author with herbarium and library facilities: A, B, BM, CAL, CAS, CDBI, E, G, GXMI, HAST, HNWP, IFP, ISL, K, KATH, KUH, KUN, LD, LE, P, PE, PMNH, SM, SZ, TI, W, WU, and XZ. Special thanks are due to Dr. D. Simpson at Kew, who provided Mr. M. Gilbert with long-term working facilities in the herbarium.

Author's field trips to Tibet in 2010, 2011, and 2012 were assisted by Mr. Xin-Tang MA and Dr. Qiang WANG from author's laboratory, to NW Yunnan in 2010 by Prof. Hang SUN from Kunming Institute of Botany, Chinese Academy of Sciences, and his students, Huan-Chong WANG, Yang YANG, and Ji-Pei YUE, and in 2014 by Prof. Hong YU from Yunnan University and Dr. Qiang WANG; to SW Sichuan in 2011 by Prof. Xing-Jin HE from Sichuan University, and his student, Xiang Guang MA; to W Sichuan in 2012 by Dr. Qiang WANG and Prof. Kai-Yu PAN; to NE Qinghai by Prof. Shang-Wu LIU from Northwest Plateau Institute of Biology and his student Mr. Qing-Bo GAO, and to North Pakistan in 2013 by Prof. Zhi-Duan CHEN from the author's laboratory.

The present volume was financially supported also by the National Natural Science Foundation of China (grant no. 31170175).

Contents

AQUIFOLIACEAE

Trees or shrubs, evergreen or deciduous, dioecious. Leaves alternate, rarely opposite; stipules minute, persistent or caducous; petiole present or lacking; leaf blade leathery to membranous, margin entire, serrate, crenulate, or spinose. Inflorescences axillary, solitary on current year's shoots, or fasciculate on previous year's growth, cymose, sometimes umbelliform, simple to much branched. Flowers small, regular, heteromerous, unisexual by abortion, (2–) 4–8 (–16)-merous; calyx persistent, shallowly dish-shaped; petals imbricate, usually connate at base. Male flowers: stamens isomerous, alternate with petals, epipetalous, anthers oblong-ovoid, introrse, longitudinally dehiscent; rudimentary ovary subglobose or pulvinate, rostrate. Female flowers: staminodes sagittate or cordate; ovary superior, ovoid, 4–8(–10)-loculed; style very rarely developed; stigma capitate, discoid, or columnar. Fruit a drupe, red, rarely brown or black, usually globose, less frequently ellipsoid or oblate; exocarp membranous or papery; mesocarp fleshy. Pyrenes (1–) 4–6 (–23), smooth, striate, striate-sulcate, or rugose and/or pitted; endocarp leathery, woody, or stony.

Recent references: Basu, 2000. Aquifoliaceae. In: Singh *et al.* (eds), Fl. India 5: 50–74. Calcutta: Government of India; Chen *et al.*, 2008. Aquifoliaceae. In: Wu, Raven & Hong (eds), Fl. China 11: 359–438. Beijing & St. Louis: Science Press & Missouri Botanic Garden Press.

One genus, *Ilex*, with 400–500 species; tropical, subtropical to temperate regions of the Northern and Southern Hemispheres, mainly in tropical regions of Central and South America, and East Asia; 41 species in the Pan-Himalaya (12 endemic) , mainly in the eastern part.

ILEX L.

Linnaeus, Sp. Pl. 1: 125. 1753.

According to Loesner's system (Loesner, 1901, Nova Acta Acad. Caes. Leop.-Carol. Nat. Cur.: 78), *Ilex* consists of three subgenera, subg. *Ilex*, subg. *Byronia*, and subg. *Prinos*. We here recognize two of these subgenera including five sections for the Pan-Himalaya based on Hu's work (Hu, 1949, J. Arnod Arbor. 30: 258).

Recent reference: Hong, 2015. A taxonomic revision of *Ilex* (Aquifoliaceae) in the Pan-Himalaya and unraveling its distribution patterns. *Phytotaxa* 230(2):151-171.

1a. Deciduous trees or shrubs, often with spur branchlets; current year's branchlets often with lenticels; leaf blade membranous, papery, thinly leathery, rarely subleathery [subg. **1. Prinos** (L.) Loes. , 3 spp.]

2a. Fruit black, 10–14 mm in diameter; pyrenes cylindrical, 3-striate, 2-sulcate; petiole puberulent .... 3. *I. macrocarpa*

2b. Fruit red, 3–6 mm in diameter; pyrenes ellipsoid, striate or rugose, esulcate or 1-sulcate; petiole glabrous.

3a. Spur branchlets lacking; peduncle 4–12 mm long; female inflorescences a pseudo-umbell of over 5 flowers........ 2. *I. micrococca*

3b. Spur branchlets present; peduncle absent or less than 1.5 mm long; female inflorescences a 1–3-flowered cyme .... 1. *I. fragilis*

1b. Evergreen trees or shrubs, without spur branchlets; current year's branchlets without lenticels; leaf blade leathery, rarely papery (subg. **2. Ilex**, 38 spp.).

4a. Female and male inflorescences both solitary, axillary on current year's shoots; pyrenes abaxially unicanaliculate or 3-striate and 2-sulcate [sect.1. **Lioprinus** (Loes.) S. Y. Hu, 6 spp.].

5a. Leaf blade usually crenate, serrate, rarely entire; inflorescences cymose; pyrenes unicanaliculate on dorsal side.

6a. Leaf blade usually less than 10 cm long, glabrous; fruit ellipsoid; branchlets glabrous .... 4. *I. chinensis*

6b. Leaf blade over 12 cm long, usually hairy; fruit subglobose; branchlets pubescent or glabrous.

7a. Cymes (umbels) simple, without secondary axis; lateral veins 14–18 pairs .... 5. *I. atrata*

7b. Cymes (umbels) compound, with secondary axis; lateral veins 11–13 pairs .... 6. *I. shweliensis*

5b. Leaf blade entire; inflorescences usually umbelliform; pyrenes abaxially 3-striate and 2-sulcate.

8a. Leaf blade 5–8(–10) cm long; inflorescences mostly in axils of normal leaves .... 7. *I. rotunda*

8b. Leaf blade 9–15 cm long; inflorescences mostly in axils of scales, less often in axils of normal leaves.

9a. Current year's branchlets glabrous; fruit ca. 6 mm in diameter .... 9. *I. umbellulata*

9b. Current year's branchlets puberulent; fruit 3.5–4 mm in diameter .... 8. *I. godajam*

4b. Male inflorescences fasciculate, on previous year's growth, female ones fasciculate or solitary, on previous year's or current year's growth; pyrenes smooth, rugose and pitted, or striate.

10a. Female inflorescences solitary, axillary on current year's shoots, male inflorescences fasciculate on previous year's growth or rarely solitary on current year's shoots; pyrenes smooth, or striate and esulcate with insculpt striae or slightly roughened [sect. 2. **Paltoria** (Ruiz & Pavon) Maxim., 5 spp.].

11a. Leaf blade abaxially not punctate, shorter than 3 cm or longer than 7 cm.

12a. Leaf blade longer than 7 cm; fruiting peduncle plus pedicel 3–4 cm long; young shoots glabrous .... 14. *I. gansuensis*

12b. Leaf blade shorter than 3 cm; fruiting peduncle plus pedicel 0.3–1.4 cm long; young shoots densely yellow-villous .... 13. *I. yunnanensis*

11b. Leaf blade abaxially punctate, 1.9–10 cm long.

13a. Leaf blade 1.9–3.4 cm long, 1–1.7 cm broad; shrubs less than 3 m tall .... 10. *I. crenata*

13b. Leaf blade 3.5–10 cm long, 1.5–4 cm broad; small trees or shrubs 3–10 m tall.

14a. Leaf blade oblong or broadly elliptic, 7–10 cm long, 3–4.5 cm broad; fruiting cymes fasciculate or solitary; fruiting peduncle plus pedicel 4–7 mm long .... 12. *I. dicarpa*

14b. Leaf blade elliptic, mostly 3.5–5 cm long, 1.5–2.0 cm broad, rarely up to 7.5 cm long, 3.5 cm broad; fruiting cymes solitary; fruiting peduncle plus pedicel 8–10 mm long .... 11. *I. szechwanensis*

10b. Female and male inflorescences both fasciculate, axillary, on previous year's growth; pyrenes rugose and pitted, or striate with elevated striae.

15a. Individual branches of female inflorescences fasciculate or pseudo-racemose, consisting of 3-flowered cymes, or nearly umbellate, or 1-flowered.

16a. Individual branches of both male and female inflorescences in compound cymes or subumbels, each with over 10 flowers; leaf blade 10–20 cm long, 4–7 cm broad (sect. 4. **Lauroilex** S. Y. Hu, 1 sp.) .... 35. *I. venulosa*

16b. Individual branches of male and female inflorescences 1-, 3-, or sometimes 5- flowered; leaf blade up to 10 cm long, 3.5 cm broad. (sect. 5. **Pseudoaquifolium** S. Y. Hu, 6 spp.)

17a. Leaf blade serrate, crenate or nearly entire, apex short-acuminate or short- caudate; pyrenes 6 or 7.

18a. Leaf blade 4.5–10 cm long, 1.5–3.5 cm broad, caudate-acuminate at apex with acumen 1–1.5 cm long; infructescence axis up to 8 mm long, puberulent; pedicel puberulent .... 40. *I. forrestii*

18b. Leaf blade 3.5–5.5 cm long, 1.5–2 cm broad, nearly without acumen; infructescences nearly sessile; pedicel glabrous .... 41. *I. tsiangiana*

17b. Leaf blade entire, apex usually caudate; pyrenes 4, rarely 5.

19a. Leaf blade punctate on abaxial surface.

20a. Leaf blade with acumen 1–2 cm long; female flowers often on short and leafless shoots, appearing pseudo-paniculate; fruit 2.5–3.5 mm in diameter .... 36. *I. longecaudata*

20b. Leaf blade with acumen 0.6–1cm long; female cymes only in axils, not appearing pseudo-paniculate; fruit 4 mm in diameter ..... 37. *I. wilsonii*

19b. Leaf blade not punctate.

21a. Fruiting pedicel glabrous, 2.5–4.5 mm long; fruit globose, 3.5–4 mm long and in diameter .... 8. *I. sublongecaudata*

21b. Fruiting pedicel puberulent, 6–8 mm long; fruit oblate, 4 mm long, 5 mm in diameter .... 39. *I. wuana*

15b. Individual branches of female inflorescences 1-flowered; pyrenes 4, rarely less; endocarp woody or stony (sect. 3. **Ilex**, 20 spp.).

22a. Leaf margins spinose or entire but apex ending in a spine (ser. 1. **Dipyrenae** S. Y. Hu, 7 spp.).

23a. Leaf blade over 3.5 cm long, ovate or wide-elliptic, with 1–15 spines on each side, but sometimes entire; trees 4–20 m tall, very rarely shrubs.

24a. Leaf blade 10–20 cm long, 4–7 cm broad, with 13–15 pairs of lateral veinlets; fruit globose, 8–10 mm in diameter ....1. *I. kingiana*

24b. Leaf blade 2.5–10 cm long, 1.5–4 cm broad, with 4–9 pairs of lateral veinlets; fruit globose, 6–9 mm in diameter or ellipsoid, 6–10 mm long, 5–7 mm wide.

25a. Fruit cylindrical, 7–9 mm long, 5–7 mm in diameter; shrubs, usually 1–3 m tall, or sometimes small trees, up to 5 m tall; leaf blade always spinose, 3–6.5 cm long.. 16. *I. bioritsensis*

25b. Fruit globose, 6–9 mm in diameter; trees up to 20 m tall; leaf blade often entire, 5–12 cm long .... 15. *I. dipyrena*

23b. Leaf blade shorter than 3.5 cm (up to 4 cm, but narrow, lanceolate in *I. georgei*), 1.5 cm wide, elliptic, lanceolate, or rectangular, with usually 1 or 2 (–3) spines on each side (up to 5–7 only in *I. miguensis* and *I. georgei*); shrubs usually 1–2 m, very rarely up to 6 m tall.

26a. Leaf blade elliptic, with 5, less often 4, spines on each side .... 20. *I. miguensis*

26b. Leaf blade mostly lanceolate or rectangular, with only 1 or 2, less often 3, very rarely up to 7 spines on each side.

27a. Leaf blade lanceolate, with length/width ca. 3; fruit obovoid or cylindrical, 5 mm long, 3.5–4 mm wide .... 18. *I. georgei*

27b. Leaf blade more or less rectangular at lower (spinose) part, upper part much narrower.

28a. Prostrate shrubs, less than 0.5 m tall; leaf blade nearly square in lower part (below uppermost lateral spine), upper part less than 1/2 as long as lower part .... 19. *I. perryana*

28b. Erect shrubs or small trees up to 8 m tall; leaf blade rectangular or nearly square in lower part, upper part longer than, or nearly equal to, lower part .... 17. *I. pernyi*

22b. Leaf blade entire, serrate or crenate, never spinose.

29a. Leaf blade chartaceous or subleathery; veinlets impressed on adaxial surface; fruiting pedicel 4–7 mm long (ser. 2. **Hookerianae** S. Y. Hu, 6 spp.).

30a. Branchlets verruculose; leaf blade 1–5 cm, rarely up to 8 cm long, apex acute or obtuse.

31a. Leaf blade elliptic, mostly 4–5 cm, sometimes up to 8 cm long; fruit fasciculate .... 24. *I. delavayi*

31b. Leaf blade oblong or obovate-elliptic, 1–3.2 cm long; fruit usually solitary or paired.

32a. Small trees, up to 6 m tall; leaf blade broadly oblong, cuspidate at apex; petiole 4–5 mm long .... 25. *I. nothofagifolia*

32b. Prostrate shrubs, less than 1 m tall; leaf blade obovate-elliptic, rounded or acute at apex; petiole 1–2 mm long .... 26. *I. intricata*

30b. Branchlets smooth; leaf blade 5–18 cm long, apex more or less acuminate or caudate.

33a. Leaf blade 12–18 cm long, 3.5–6 cm broad; trees 5–15 m tall .... 27. *I. sikkimensis*

33b. Leaf blade 5–12 cm long, less than 4 cm broad; shrubs or small trees up to 5 m tall.

34a. Infructescences umbelate; fruiting pedicel longer than diameter of fruit, glabrous .... 23. *I. hookeri*

34b. Infructescences pseudo-racemose; fruiting pedicel shorter or longer than diameter of fruit, glabrous or hispidulous .... 22. *I. fargesii*

29b. Leaf blade leather, rarely subleathery; veinlets never impressed on adaxial surface [ser.3. **Repandae** (Loes.) S. Y. Hu, 7 spp.].

35a. Current year's shoots puberulent; infructescence axis and fruiting pedicel puberulent or glabrous; leaf blade acuminate or shortly caudate-acuminate with acumen not over 8 mm long.

36a. Fruiting pedicel 3–5 mm long, puberulent; leaf blade 3–9.5 cm long, 1.5–3.2 cm broad .... 33. *I. odorata*

36b. Fruiting pedicel 15 mm long, mostly glabrous, less often puberulent; leaf blade 6–13 cm long, 2.5–4.5 cm broad .... 28. *I. intermedia*

35b. Current year's shoots glabrous; infructescence axis glabrous; leaf blade caudate-acuminate with acumen 10–25 mm long, or acuminate with shorter acumen.

37a. Leaf blade with acumen less than 1 cm long; infructescence with axis ca. 8 mm long, or sessile; fruiting pedicel 4–5 mm long or shorter.

38a. Infructescence axis ca. 8 mm long; fruiting pedicel 4–5 mm long; fruit 3–3.5 mm in diameter; petiole 1–2 cm long; leaf blade 8–21 cm long, 2.5–7.5 cm broad .... 32. *I. pseudo-odorata*

38b. Infructescences sessile; fruiting pedicel less than 2 mm long; fruit ca. 3 mm in diameter; petiole less than 1 cm long; leaf blade 5–13 cm long, 1.5–5 cm broad .... 30. *I. corallina*

37b. Leaf blade with acumen 1–2.5 cm long; infructescence axis less than1 mm long; fruiting pedicel 1–3 mm long.

39a. Petiole puberulent; pyrenes esulcate; leaf blade with lateral veins 7 or 8 pairs .... 29. *I. cyrtura*

39b. Petiole glabrous; pyrenes sulcate; leaf blade with lateral veins 8–11 pairs.

40a. Leaf blade 10–13 cm long, 4–6 cm broad, with lateral veins 10 or 11; fruit 6–7 mm in diameter .... 31. *I wattii*

40b. Leaf blade 6–11 cm long, 2–3.5 cm broad, with lateral veins 8–10 pairs; fruit 4–5 mm in diameter .... 34. *I. medogensis*

Subg. 1. Prinos (L.) Loes., Vorst. Monog. Aquif. - Diss. 25: 26. 1890. ≡ *Prinos* L., Sp. Pl. 330. 1753.

Deciduous trees or shrubs, often with both elongated shoots and spurs; lenticels usually conspicuous on current year's shoots; leaf blade membranous or papery, rarely subleathery; female inflorescences solitary.

Three species in the Pan-Himalaya.

1. Ilex fragilis Hook. f., Fl. Brit. India 1: 602. 1875; Hu, J. Arnold Arbor. 30: 259. 1949; Chen *et al.* in Wu, Raven & Hong, Fl. China 11: 434. 2008. Type: Bhutan, Aug. 1837, *Griffith 1098* (**lectotype here designated**: K; isolectotype: E).

Ilex burmanica Merr., Brittonia 4: 102. 1941. Type: Upper Burma, Adung Valley, 28°20′N, 97°47′E, 28 May 1931, *F. Kingdon-Ward 9559* (holotype: A; isotypes: BM); loc. eodem, 31 May 1931. *F. Kingdon-Ward 9583* (paratype: A).

Deciduous shrubs or small trees, 1.5–15 m tall. Spur branchlets 1–1.5 cm long, with persistent leaf scars and bud scales. Petiole 5–15 mm long, glabrous; leaf blade ovate or elliptic, 4–8 cm long, 3–5 cm broad, adaxially glabrous, abaxially glabrous or pubescent only along veins, or entirely pubescent, base rounded to obtuse, margin serrate, apex acuminate, midrib impressed adaxially, lateral veins 8–10 pairs. Inflorescences on currect year's growth, fasciculate or solitary, 1-flowered cymes; male flowers 6–8-merous; pedicel 3–6 mm long, glabrous; calyx ca. 4 mm in diameter; corolla yellow, becoming red or white, petals ob-

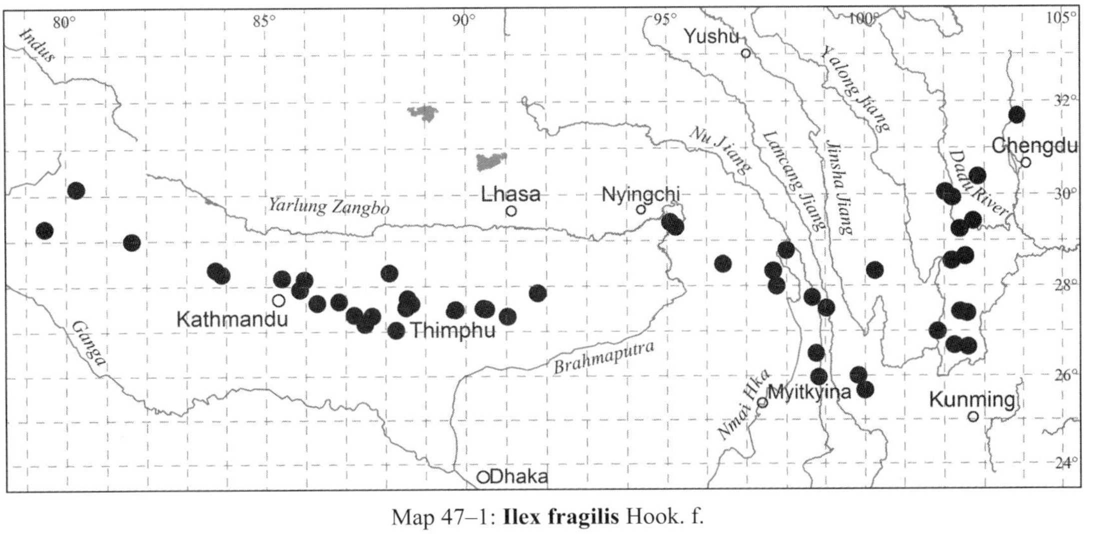

Map 47–1: **Ilex fragilis** Hook. f.

long, 2–2.5 mm long, ca. 1.5 mm broad. Female flowers 6–8 (–16)-merous; pedicel 2–3 mm long, glabrous; calyx and corolla as in male flowers; style 0.5–1.5 mm long; stigma capitate or crested. Fruiting pedicel ca. 5 mm long; fruit red, purple-red, black or brownish, oblate, 4–6 mm in diameter. Pyrenes 4–8 (–13), ellipsoid, 2–2.5 mm long, ca. 1.5 mm in diameter, abaxially longitudinally striate; endocarp woody. **Chromosomes**: 2n=36. Figure 47–1: a-d.

Phenology: flowering from May to July; fruiting from August to November.

Habitats: forests or thickets; 2000– 3200 m.

Distribution: U Ganga & Indus, U Yarlung Zangbo, W, C, and E Nepal, Sikkim & Darjiling, Bhutan, Yarlung Zangbo-Brahmaputra, N and S Hengduan, and U Irrawaddy (Map 47-1); also in Meghalaya of India.

2. **Ilex micrococca** Maxim., Mém. Acad. Imp. Sci. Saint Pétersbourg, ser. 7, 29(3): 39. pl. 1, fig. 6. 1881; Hu, J. Arnold Arbor. 30: 261. 1949; Hong, Phytotaxa 230(2): 161, 2015. Type: Japan (not seen).

Ilex micrococca Maxim. var. *polyneura* Hand.-Mazz., Symb. Sin. 7: 654. 1933. ≡ *Ilex polyneura*(Hand.-Mazz.) S. Y. Hu, J. Arnold Arbor. 30: 263. 1949. Type: China, NW Yunnan "gegenüber Tjiontson- lumba", 28. Jun. 1916, *Handel-Mazetti 9121* (holotype: W; isotype: W).

Trees deciduous, 4–20 m tall. Petiole 1.5–3.2 cm long, glabrous; leaf blade ovate, ovate-elliptic, or ovate-oblong, 7–13 cm long, 3–5 cm broad, both surfaces glabrous or abaxially pubescent, base rounded or broadly cuneate, often oblique, margin nearly entire or serrate, apex long-acuminate, midrib slightly impressed adaxially, lateral veins 5–12 pairs. Inflorescences on current year's growth, solitary, compound cymes with 3 or 4 orders of branching, glabrous; peduncle 9–12 mm long, secondary axis 2–7 mm long, bracts absent; tertiary axis with broadly deltoid bracts; pedicel 2–3 mm long, bracteole 1, basal, deltoid. Male flowers 5- or 6-merous; calyx-lobes obtuse, glabrous or ciliate, or abaxially pilose; corolla rotate, petals oblong, 1.2–1.5 mm long. Female inflorescence a pseudo-umbell of more than 5 flowers; flowers 6–8-merous; calyx-lobes obtuse, ciliate; petals oblong, ca. 1 mm long, stigma discoid. Fruit bright red, globose, ca. 3 mm in diameter; pyrenes 6–8, ellipsoid, ca. 2 mm long, 1 mm wide, ends obtuse, abaxially slightly rugose, longitudinally 1-sulcate, laterally smooth; endocarp leathery. Figure 47–1: e-h.

Phenology: flowering in May and June; fruiting from July to November.

Habitats: forests, rarely thickets; 1600–2600 m.

Distribution: N and S Hengduan, and U Irrawaddy (Map 47–2); also widely distributed in Japan, Vietnam, and E China (Anhui, Fujian, Guangdong, Guangxi, Guizhou, Hainan, Hubei, and S Yunnan).

N HENGDUAN: **Songpan**, 1928, *W. P. Fang 6029* (PE). **S HENGDUAN**: **Bijiang**, Yueliang- tian, 1800 m, 1964, *S. K. Wu 8523* (KUN); **Fugong**, Guqian, 1600 m, 1982, *Qinghai-Xizang Exped. 7291* (KUN, PE); **Gongshan**, Binzhong-luo, 1982, *Qinghia-Xizang Exped. 7895* (KUN, PE); Salween Valley, *T. T. Yü 22918* (holotype: A); **Lushui**, Pianma to Pass, 2100 m, 1978, *Bijiang Exped. 1601* (KUN); **Miyi**, Malong, 2300 m, 1983, *Qinghai-Xizang Exped. 11756* (KUN, PE); **Tianquan**, Sijing, 1953, *W. K. Hu & Z. He 11955* (PE); **Weixi**, Yezhi, 2100– 2200 m, 1982, *Qinghai-Xizang Exped. 6288* (KUN, PE); **Yan-**

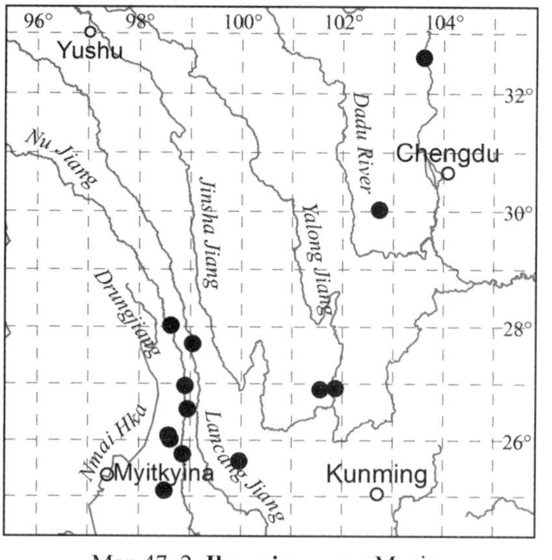

Map 47–2: **Ilex micrococca** Maxim.

Figure 47–1: **a–d**, **Ilex fragilis** Hook. f.: **a**, a section of female flowering branch; **b**, male flower; **c**, fruit; **d**, pyrene. **e–h**, **Ilex micrococca** Maxim.: **e**, a section of male flowering branch; **f**, male flower; **g**, fruit; **h**, pyrene. From *Fl. Reipubl. Popularis Sin.* 45(2), 1999.

bian, Zhemulong, 2250 m, 2002, *X. F. Gao et al. 3505* (PE); **Yangbi**, Huajiazhuang, 1940, *C. R. Ching 25401* (KUN, PE); Mekong-Salween Divide, Aiwa Shan, 2000 m, 1938. *T. T. Yü 22879* (paratype of *I. polyneura*: KUN). **U IRRAWADDY**: **Hpimaw**, 2743 m, 1924, *G. Forrest 24998* (E); **Tengchong**, 2440 m, 1924, *G. Forrest 26159* (PE).

3. Ilex macrocarpa Oliv., Hooker's Icon. Pl. 8: t. 1787. 1888; Hu, J. Arnold Arbor. 30: 271. 1949. Type: Oliver's plate cited in protolog.

Trees deciduous, 5–17 m tall. Spur branchlets 4–10 cm long. Petiole 1–1.2 cm long, sparsely puberulent; leaf blade ovate or ovate- elliptic, 3–15 cm long, 2–6 cm broad, base rounded or obtuse, margin shallowly serrate, apex acuminate to acute, both surfaces glabrous or sparsely puberulent when young, midrib flat or impressed adaxially, raised abaxially, glabrous or sparsely puberulent, lateral veins 8–10 pairs. Male inflorescences on current or previous year's branchlets, solitary or fasciculate, cymes of 1–5 flowers; peduncle 2–3 mm long; pedicel 3–7 mm long. Flowers 5- or 6-merous; calyx lobes deltoid-ovate; corolla white, ca. 7 mm in diameter; petals obovate-oblong, ca. 3 mm long, 1.5–2 mm broad, slightly connate at base. Female inflorescence a solitary, 1-flowered cyme; flowers 7–9-merous; pedicel 6–18 mm long, glabrous or pubescent, bracteoles 2, basal; calyx shallowly lobed, lobes ovate-deltoid; corolla 1–1.2 mm in diameter; ovary conical-ovoid; style present, stigma columnar. Fruiting pedicel 1–3.3 cm long; fruit black, globose, 1–1.4 cm in diameter; pyrenes 7–9, oblong, laterally compressed, abaxially 3-striate, 2-sulcate, laterally reticulately striate and sulcate, endocarp stony.

Phenology: flowering in April and May; fruiting in October and November.

Habitats: forests in ravine; 750–3000 m.

Distribution: N and S Hengduan (Map 47–3); also widely distributed in E China, from S Shaanxi southwards to Guangdong and SE Yunnan.

N HENGDUAN: **Wenxian**, Bifenggou, 750 m, 2000, *X. Y. Zhu et al. 20252* (PE). **S HENGDUAN**, **Baoxing**, 1933, *T. T. Yü 2011* (PE); **Dali**, 1946, *T. N. Liou 14131* (KUN); **Miyi**, Baipo Shan, 2800–2900 m, 1983, *Qinghai-Xizang K. Hu & Z. He 12053* (PE); **Yanbian**, Yashi to Jiudaozhulin, 2600 m, 1983, *Qinghai-Xizang Exped. 11882* (PE); **Shimian**, 1955, *C. C. Hsieh 39537* (PE); **Tianquan**, Shiyang, 1951, *W. Exped. 11473* (PE); **Yangbi**, Ha-lamuda Village, 1946, *T. N. Liou 15991* (KUN).

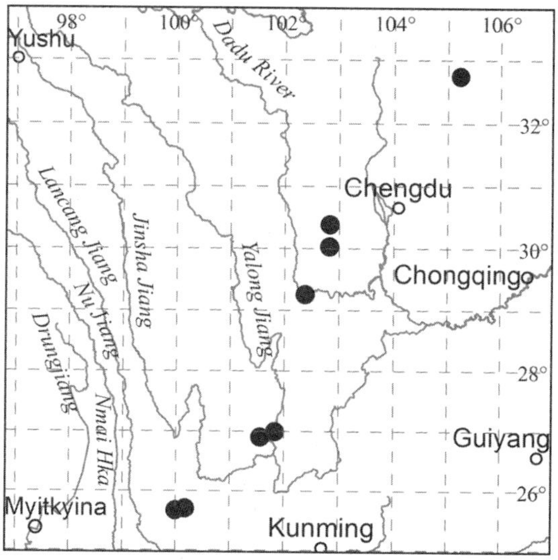

Map 47–3: **Ilex macrocarpa** Oliv.

Subg. 2. Ilex

Evergreen trees or shrubs; branchlets all elongated, spur shoots absent; lenticels usually lacking on current year's shoots; leaf blade leathery, rarely papery; inflorescences solitary or fasciculate.

Five sections with 38 species in the Pan-Himalaya.

Sect. 1. Lioprinus (Loes.) S. Y. Hu, J. Arnold Arbor. 30: 284. 1949. ≡ *Ilex* subg. *Euilex* ser. *Lioprinus* Loes., Verhandl. Bot. Ver. Prov. Brandenb. 33: 26. 1891.

Trees, rarely shrubs, 2–25 m tall. Leaf margin entire, crenate or serrate. Inflorescences on current year's shoots, solitary in axils of leaves or scales. Flowers 4- or 5-merous. Pyrenes 4–6 (–7); endocarp leathery.

Six species, all in the Pan-Himalaya (1 endemic).

4. Ilex chinensis Sims, Bot. Mag. 46: pl. 2043. 1819; Hu, J. Arnold Arbor. 30: 299. 1949; Chen *et al.*, in Wu, Raven & Hong, Fl. China 11: 377. 2008. Type: Sims' plate cited in protolog.

Evergreen trees, rarely shrubs, 2–25 m tall. Bark gray, smooth, current year's shoots glabrous. Petiole 8–10 mm long, glabrous, adaxially flat, sometimes slightly sulcate; leaf blade thinly leathery, elliptic, lanceolate or rarely ovate, 5–11 cm long, 2–4 cm broad, base obtuse or cuneate, margin crenate or rarely serrate, apex acuminate, acumen 5–10 mm long, midrib adaxially flat, abaxially elevated, lateral veins 6–9 pairs. Inflorescences on current year's shoots, in axils of leaves, solitary, cymose, glabrous. Male cymes 3 or 4 times dichotomous or sometimes irregularly dichotomous, 7–15- flowered; peduncle 4–14 mm long; secondary axis well developed, 2–5 mm long. Pedicel ca. 2 mm long, glabrous; calyx ca. 2.5 mm across, glabrous, lobes ciliate; corolla purple, petals ca. 2.5 mm long, 1.8 mm broad; stamens shorter than petals. Female cymes once or twice dichotomous, 3–7-flowered; peduncle 3–10 mm long; pedicel 6–10 mm long. Fruit red, ellipsoid, 10–12 mm long, 6–8 mm in diameter. Pyrenes 4 or 5, trigonous in cross-section, 9–11 mm long, 2.5–3 mm wide. Figure 47–2: a-d.

Phenology: flowering in May and June; fruiting from September to November.

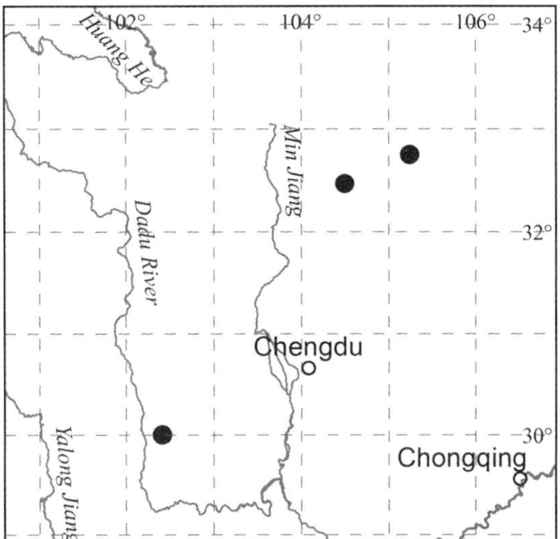

Map 47–4: **Ilex chinensis** Sims

Habitats: forests or bamboo thickets; 800–2100 m.

Distribution: N and S Hengduan (Map 47–4); also in E and S China, and Japan.

N HENGDUAN: **Pingwu**, Tucheng, 2120 m, 1958, *H. L. Tsiang 11145* (PE); **Wenxian**, Bikou, 900 m, 1992, *X. Wang 556* (PE). **S HENGDUAN**: **Tianquan**, Meiziling, 1380 m, 1959, *S. Jiang & C. L. Jin 1747* (PE).

5. Ilex atrata W. W. Smith, Notes Roy. Bot. Gard. Edinburgh 10: 40. 1917. Type: China, Yunnan, NE Tengyueh (Tengchong), 25°12'N, 2130–2440 m , Dec. 1912, *G. Forrest 9414* [**lectotype here designnated**: E (E00265728); isolectotype: E (E00078816), K].

Ilex atrata W. W. Smith var. *wangii* S. Y. Hu, J. Arnold Arbor. 30: 305. 1949. Type: China, Yunnan, Chiu-Kiang (Drungjiang), W of Champutung, *C. W. Wang 67359* (holotype: A; isotype: PE).

Small evergreen trees, 3–10 m tall. Current year's shoots glabrous to densely ferrugineous or fulvous-pubescent, terminal buds large, bud-scales ovate, outer ones 8 mm long, 5 mm broad, margin entire, densely ciliate. Petiole 1.5–2.5 cm long, flat or slightly grooved adaxially, glabrous or fulvous-pubescent; leaf blade thickly papery or thinly leathery, elliptic, 12–16 cm long, 4–6 cm broad, base rounded, obtuse or cuneate, margin serrate, crenate-serrate, rarely nearly entire, apex acuminate; glabrous except for villous midrib abaxially, or villous on both surfaces, sparsely so adaxially, densely so abaxially, lateral veins 14–19-paired, prominent on both surfaces. Inflorescences a cyme on current year's shoots, solitary, 1–3-flowered, glabrous to yellowish tomentose; fruiting peduncle 3–15 mm long; pedicel 5–8 mm long. Fruit red, globose, ca. 6 mm in diameter. Pyrenes 5, oblong in outline, trigonous in cross section, ca. 5.5 mm long, 2.5 mm wide, smooth, striate, esulcate; endocarp leathery.

Phenology: flowering in June and July; fruiting from October to December.

Habitats: evergreen broad-leaved and mixed forests or shrubs: 2000–2700 m.

Distribution: Yarlung Zhangbo-Brahmapu-

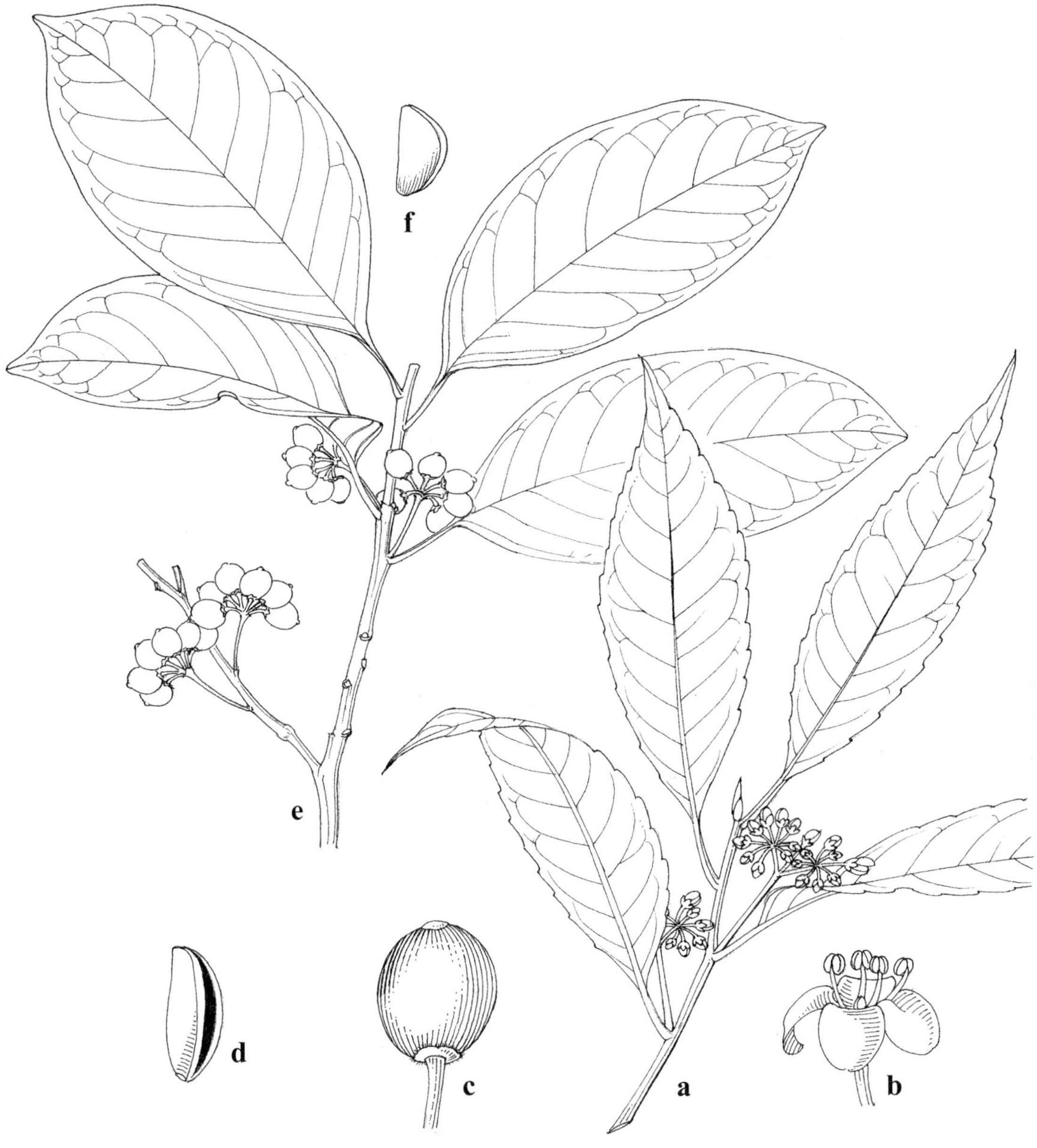

Figure 47–2: **a–d**, **Ilex chinensis** Sims: **a**, male flowering branch; **b**, male flower; **c**, fruit; **d**, pyrene. **e** & **f**, **Ilex umbellulata** (Wall.) Loes.: **e**, a section of fruiting branch; **f**, pyrene. Form *Fl. Reipubl. Popularis Sin.* 45(2), 1999.

tra, S Hengduan, and U Irrawaddy (Map 47–5); also in SW Yunnan of China.

YARLUNG ZANGBO-BRAHMAPUTRA: **Bangxing**, 2100 m, 1982, *S. Z. Cheng & B. S. Li 3326* (PE); **Gutang**, 2200 m, 1993, *H. Sun et al. 5022* (KUN). **S HENGDUAN**: **Bijiang**, Nongzhong, 2550 m, 1978, *Bijiang Exped. 0113* (KUN); **Gongshan**, Cikai, 1940, *K. M. Feng 8636* (KUN, PE); **Lushui**, 2500 m, 1964, *S. K. Wu 8291* (KUN). **U IRRAWADDY**: **Adung Valley**, 2130– 2440 m, 1931, *F. Kingdon-Ward 9253* (BM); **Nam Tamai Valley**, 2440–2743 m, 1937, *F. Kingdon-Ward 13228* (BM) ; **Tengchong**, Shweli- Salween Divide, 1917, *G. Forrest 16009* (K, P).

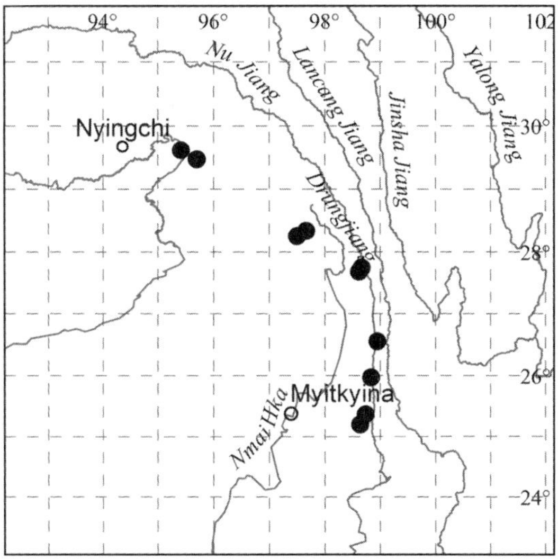

Map 47–5: **Ilex atrata** W. W. Smith

6. Ilex shweliensis H. F. Comber, Notes Roy. Bot. Gard. Edinburgh 18: 57. 1933; Hu, J. Arnold Arbor. 30: 303, 1949, pro syn; Hong, Phytotaxa 230(2): 165. 2015. Type: China, Yunnan, Shweli Valley, 25°20′N, 1830 m, Oct. 1917, *G. Forrest 15947* (**lectotype here designated**: E; isolectotype: K).

Small evergreen trees, 6–10 m tall. Branchlets longitudinally rugose. Petiole 6–10 mm long, sparsely puberulent, grooved adaxially; leaf blade subleathery, oblong-elliptic, 7–12 cm long, 4–4.5 cm broad, base rounded or broadly cuneate, margin entire or remotely serrulate, apex obtuse, adaxially glabrous or nearly glabrous, midrib abaxially elevated and yellow lannate, lateral veins 11–13 pairs. Flowers not seen. Infructescences solitary and axillary, a compound cyme of 3–7 drupes, on current shoots; peduncle 9–10 mm long, puberulent, secondary axis 2–4 mm long; pedicel 4–7 mm long, puberulent; persistent calyx 4-lobed, ca. 6 mm across. Fruit subglobose, ca. 8 mm in diameter. Pyrenes 4, abaxially sulcate.

Phenology: fruiting in October.

Habitats: thickets; ca. 1800 m.

Distribution: **endemic**; confined to U Irrawaddy (Tengchong) (Map 47–6).

Notes: *Ilex shweliensis* was reduced by Hu (1949) to a synonym of *I. kwangtungensis* Merr. This treatment was followed by Chen *et al.* in *Fl. China* (2008). *Ilex shweliensis* is indeed rather similar to *I. kwangtungensis* Merr. in habit, leaf shape, and the development of secondary inflorescence axes. However, it differs from that species by having leaves with the midrib abaxially yellow-lanate (vs. mostly glabrous, less often golden-hirsute), a shorter petiole (6–10 mm long vs. 13–20 mm), and more lateral veins (11–13 pairs vs. 9–11). It is also similar to *I. atrata* W. W. Smith (with its type from the same locality, Tengchong) in habit, leaf shape, and yellow-lanate hairs on the midrib, but differs from it by the presence of secondary inflorescence axes (vs. absence), fewer lateral veins (11–13 vs. 14–18 pairs), and a much shorter petiole (6–10 vs. 20–

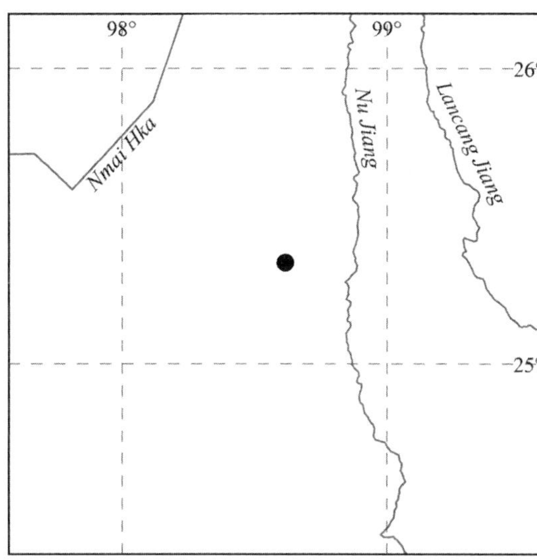

Map 47–6: **Ilex shweliensis** H. F. Comber

30 mm). I have seen only the type collection with two duplicates, and thus I think that the identity of the collection and its systematic position need further study.

7. Ilex rotunda Thunb. subsp. **excelsa** (Wall.) D. Y. Hong, Phytotaxa 230(2): 163. 2015. ≡ *Cassine excelsa* Wall. in Roxburgh, Fl. India 2: 376. 1824. ≡ *Ilex excelsa* (Wall.) Hook. f., Fl. Brit. India 1: 603. 1875; Hu, J. Arnold Arbor. 30: 306. 1949; Basu in Singh *et al.*, Fl. India 5: 55. 2000. Type: Nepal, *Wallich 4328* (holotype: K).

Ilex excelsa Wall. var. *hypotricha* (Loes.) S. Y. Hu, J. Arnold Arbor. 30: 308. 1949. ≡ *Ilex hypotricha* Loes., Nova Acta Acad. Caes. Leop.–Carol. Nat. Cur. 78: 103. 1901. Type: Bhutan, Aug. 1837, *Griffith 2008* (**lectotype here designated**: K); Kumaon, Binsar, 1524 m, *R. Strachy & J. E. Winterbottom No. 1* (paratype: K).

Evergreen trees up to 20 m tall, trunk up to 1 m d. b. h. Young branchlets pale red, glabrous, old branchlets dark brown. Petiole 1–2 cm long, grooved adaxially, glabrous; leaf blade thinly leathery or papery, ovate, obovate or elliptic, 5–10 cm long, 2–4 cm broad, base cuneate or rounded, margin entire, apex shortly acuminate, acumen 5–8 mm long, both surfaces glabrous, midrib slightly impressed adaxially, elevated abaxially, lateral veins 6–9 pairs. Inflorescences on current year's growth, solitary, umbelliform or cymose, 3–13-flowered. Male inflorescence: peduncle 3–10 mm long, puberulent; pedicel 4–5 mm long, puberulent; calyx ca. 2 mm across, shallowly 4-lobed, glabrous; corolla rotate, ca. 6 mm across, petals oblong, ca. 2.5 mm long, 1.5 mm broad, connate at base; stamens longer than petals. Female inflorescence: peduncle 5–13 mm long, puberulent or glabrous; pedicel 3–8 mm long, puberulent or glabrous. Fruit red, subglobose, rarely ellipsoid, 3–5 mm across. Pyrenes 5–7, ellipsoid, abaxially nearly smooth, esulcate.

Phenology: flowering in May and June; fruiting from October to December.

Habitats: mixed forests; 1220–1850 m.

Distribution: U Ganga & Indus, W, C and E Nepal, Bhutan, Yarlung Zangbo-Brahmaputra, S Hengduan, and U Irrawaddy (Map 47–7); also in other parts of India (Meghalaya) and Myanmar (Mogok).

Notes: 1) Loesener enumerated three collections in the protologue of *Ilex hypotricha*, i. e. *Griffith 2008* (K), *Strachey & Wintebottom* (K), and *Brandis* (K). I have seen the former two collections, so that I designate here *Griffith 2008* as lectotype. I can not understand why Hu (1949: 308) indicated the collection Yunnan, s. loc. *Forrest 27200*, not mentioned in the protologue, as type (A). 2) The typical subspecies is very similar, but its pyrenes are 3-striate, obviously 2-sulcate, and it is widely distributed from SE Yunnan and N Vietnam to Japan (Hong, 2015).

U GANGA & INDUS: **Kangra** Distr., Dharmsala, 1524 m, 1934, *A. N. Garker 3317* (K); **Tehri Garhwal**, Lunagadb, 1150 m, 1900, *U. Kanjilal 764* (K); Tehri-Garhwal, 1200 m, 1891, *J. S. Gamble 25148* (K). **W NEPAL**: **Karnali**, Jajarkot, 1370 m, 1952, *O. Polunin et al. 5727* (E). **C NEPAL**: **Dhawalagiri**, Baglung, 2134 m, 1954, *Stainton, Sykes & Williams 521* (E); **Gandaki**, Pohara, ca. 1950 m, 1983, *H. Ohba et al. 836282* (E); **Kathmandu**, 1967, *Williams & Stainton 8136* (K); Kathmandu, Godavari, 1676 m, 1966, *A. D. Schilling 797* (K); **E NEPAL**: **Kosi**, Sankhuwasabha, 1520 m, 1998, *S. Noshiro et al. 9840027* (E); **Mechi**, Sitranti, above W side of Tamur Valley, 1753 m, 1981, *A. D. Schilling 2561* (K); Taplejung, 1380 m, 1992, *M. Suzuki et al. 9263036* (E). **BHUTAN**: **Sarbhang** Distr., 590 m, 1982, *A. J. C. Grierson & D. G. Long 3554* (E); **Sankosh**, 200 m, 1982, *A. J. C. Grierson & D. G. Long 3855* (E); **Shali**, 1830m, 1915, *R. E. Cooper* s. n. (E). **YARLUNG ZANGBO-BRAHMAPUTRA**: **Dirang** Dzong, 1524–2130 m, 1938, *F. Kingdon-Ward 13766* (BM); **Zakhoma**, 1830 m, 1949, *F. Kingdon-Ward 19042* (BM); between Ukhrul and Tusom Khulen, 1830 m, 1948, *F. Kingdon-Ward 18368* (BM). **S HENGDUAN**: **Huili**: Waibei, 1900m, 1965, *Acad. Sin. Southwest China Integrative Exped. 841* (PE); **Lushui**, 1924, *G. Forrest 26015* (K); **Mianning**; Haha, 2800 m, 1976. *Southwest Normal Univ. 74 Grade 11940* (PE); **Yongping**; Beidou, Yuntai Shan, 1900–2000 m, 1981. *Qinghai-Xizang Exped. 74* (PE). **U IRRAWADDY**:

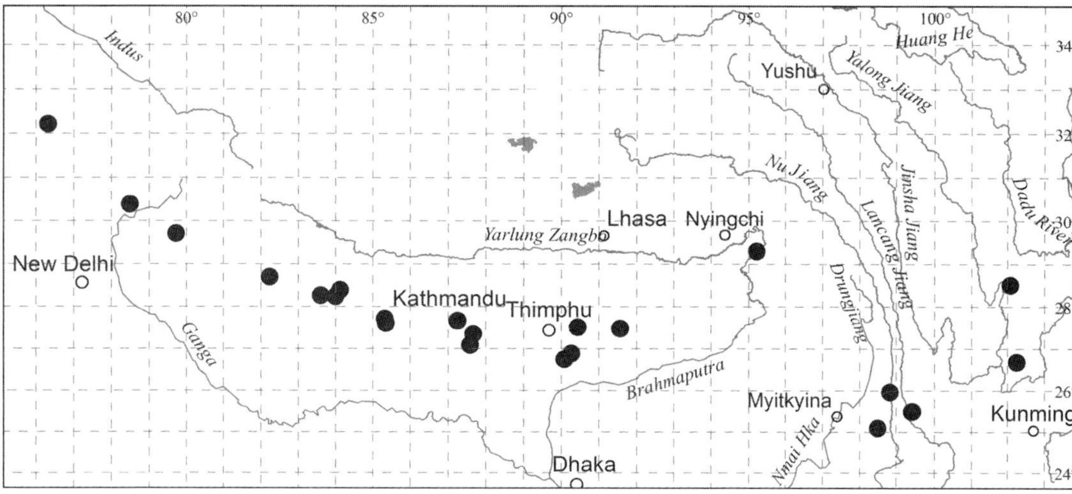

Map 47–7: **Ilex rotunda** Thunb. subsp. **excelsa** (Wall.) D. Y. Hong

Tengchong, hills NW of Tengchong (Tengyueh), 1830 m, 1924, *G. Forrest 26103* (PE); W Tengchong, 1524 m, 1912, *G. Forrest 7627* (K); Tengchong, by Guyong River, 1680 m, 1964, *S. K. Wu 6828* (KUN).

8. Ilex godajam (Colebr. ex Wall.) Wall., Cat. 4329. 1839; Hooker, Fl. Brit. India 1: 604, 1875, p. p. excl. syn. ≡ *Prinos godajam* Colebr. ex Wall., Pl. Asiat. Rar. 3: 38, pl. 261. 1832. Type: Wallich's pl. cited in protolog.

Evergreen trees or shrubs, 5–13 m tall. Bark gray-white. Branchlets glabrescent; current year's branchlets densely puberulent. Petiole 1–1.5 cm long, puberulent, grooved adaxially; leaf blade thinly leathery, ovate, elliptic, or oblong, 9–15 cm long, 4–6 cm broad, base cuneate, margin entire, apex short-acuminate, both surfaces glabrous, midrib impressed adaxially, prominent abaxially, lateral veins 8–10 pairs. Inflorescences on current year's growth, mostly in axils of scale-like bracts, subumbelliform. Flowers 4- or 5-merous. Male inflorescences 8–30-flowered; peduncle 1.4–1.8 cm long; pedicel 2–4 mm long; calyx ca. 2.5 mm across, pubescent, lobes ovate, ciliate; corolla white, petals obovate or ovate-oblong, ca. 2.5 mm long, connate at base; stamens as long as petals. Female inflorescences 4–15 flowered; fruiting peduncle 1–1.3 cm long; pedicel 2–5 cm long. Fruit globose, 3.5–4 mm across. Pyrenes 4–6, ca. 2 mm long, 1.5 mm wide, 3-striate, 2-sulcate. Figure 47–3.

Phenology: flowering from March to May; fruiting in late August.

Habitats: forests; 200–600 m.

Distribution: E Nepal, Sikkim & Darjiling, and Bhutan (Map 47–8); also in Bangladesh, Cambodgia, Laos, W Myanmar, Thailand, Vietnam, and other parts of India (Assam, Mizoram, Meghalaya, Nagaland, and Tripura) and S China (Guangxi, Hainan, and S Yunnan).

Figure 47–3: **Ilex godajam** (Colebr. ex Wall.) Wall.: a section of fruiting branch, showing two fruiting umbels. Drawn by Zhi-Juan CHEN.

E NEPAL: **Kosi**, Dharan, 610 m, 1967, *J. D. A. Stainton 5681* (BM); **Mechi**, Jhapa 250 m, 1967, *D. H. Nicolson 3087* (BM); Ilam, Ganjbari, 305 m, 1967, *J. D. A. Stainton 5741* (BM). **SIKKIM & DARJILING**: **Darjiling**, 610 m, 1923, *J. M. Cowan* s. n. (E). **BHUTAN**: **Sankosh** Distr., 200 m, 1982, *A. J. C. Grierson & D. G. Long 3855* (K); **Sarbhang** Distr., 590 m, 1982, *A. J. C. Grierson & D. G. Long 3554* (K).

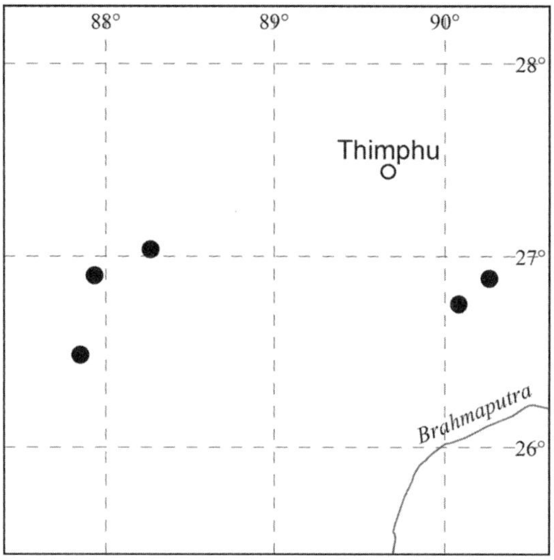

Map 47–8: **Ilex godajam** (Colebr. ex Wall.) Wall.

9. Ilex umbellulata (Wall.) Loes., Nova Acta Acad. Caes. Leop.-Carol. Nat. Cur. 78: 99. 1901; Hu, J. Arnold Arbor. 30: 315. 1949. ≡ *Ehretia umbellulata* Wall. in Roxburgh, Fl. India 2: 344. 1824; Hooker, Fl. Brit. India 1: 604. 1875, pro syn. Type: s. loc., *Wallich 4329* [**lectotype here designated**: K (K000 669423); isolectotype: K (K000669422)].

Evergreen trees 4–18 m tall. Bark gray-brown, smooth or with wrinkles. Branchlets glabrous. Petiole 1–1.5 cm long, glabrous, adaxially grooved; leaf blade papery or membranous, elliptic, elliptic-oblong or ovate, 9–15 cm long, 4–6 cm broad, base rounded or obtuse, margin entire, apex abruptly acuminate or acute, acumen ca. 5 mm long, both surfaces glabrous, midrib impressed adaxially, strongly prominent abaxially, lateral veins 9–12 pairs. Inflorescences on current year's shoots, mostly in axils of scale-like bracts, axillary, subumbelliform. Flowers 4 or 5-merous. Male pseudoumbels 6–16 (–30)-flowered; peduncle 1.8–3 cm long, puberulent; pedicel 3–5 mm long, puberulent; calyx 2–3 mm across, puberulent; petals oblong, ca. 2 mm long, 1.5 mm broad, connate at base; stamens as long as or slightly longer than petals. Female flowers: ovary ovoid, style present, stigma 4- or 5-lobed. Peduncle of infrutescence 1–2 cm long, puberulent; pedicel 3–5 mm long, puberulent; fruit red when mature, globose, 5–7 mm in diameter. Pyrenes 6–10, oblong, ca. 4 mm long, 1–1.5 mm wide, abaxially longitudinally 2- or 3-striate, 1- or 2-sulcate. Figure 47–2: e & f.

Phenology: flowering in April and May; fruiting from late August to November.

Habitats: mixed forests; 780–1350 m.

Distribution: Sikkim & Darjiling (Map 47–9); also in other parts of Myanmar, Bangladesh, Thailand, Vietnam, and other parts of China (S Yunnan) and India (Assam, Meghalaya, Andaman & Nicobar Islands).

Notes: The distribution of this species in the Pan-Himalaya is poorly known. I visited 12 major herbaria, such as BM, E, K,KUN, P, PE, but I have seen only some specimens from Myanmar and S Yunnan of China, both out of our region. In the Pan-Himalaya, I have seen only the type specimen, *Wallich 4329* without locality, and *J. D. Hooker* **s.** n. from Sikkim. Basu (2000, Fl. India , 5) recorded this species from Assam, Yarlung Zangbo-Brahmaputra, Meghalaya and Anderson & Nicobar Islands but without the citation of vouchers.

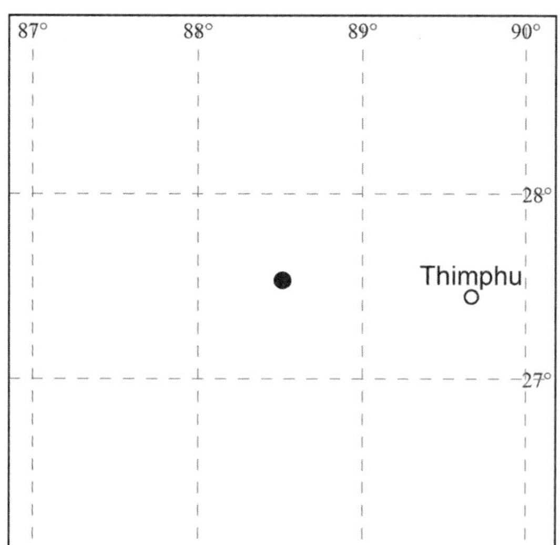

Map 47–9: **Ilex umbellulata** (Wall.) Loes.

SIKKIM & DARJILING: Sikkim, *J. D. Hooker* s. n. (K).

Sect. 2. Paltoria (Ruiz & Pavon) Maxim., Mém. Acad. Sci. St. Pétersb. Ser. 7. 29(3): 20. 1881. ≡ *Paltoria* Ruiz & Pavon, Fl. Peruv. Prodr. 2, pl. 33. fig. 1-9. 1797.

Shrubs creeping or erect, up to 5 m tall. Leaf blade glandular-punctate or epunctate abaxially. Male inflorescences fasciculate on previous year's growth or solitary on current year's shoots; female inflorescences on current year's shoots, solitary cymes with 1–3 flowers. Pyrenes 4, rarely 5 or 6, smooth or rarely slightly rugose abaxially, striate, esulcate.

Four species in the Pan-Himalaya (2 endemic).

10. Ilex crenata Thunb.in Murray, Syst. Veg. ed. 14: 168. 1784 & Fl. Jap. 78. 1784; Hu, J. Arnold Arbor. 30: 321. 1949; Hong, Phytotaxa 230(2): 154. 2015. Type: Japan (not seen).

Ilex thomsonii Hook. f., Fl. Brit. India 1: 602. 1875. ≡ *Ilex crenata* Thunb. var. *thomsonii* (Hook. f.) Loes., Nova Acta Acad. Caes. Leop.-Carol. Nat. Cur. 78: 202. t. 42C. 1901. Type: Bhutan, Tongsa, *Griffith 2003* (holotype: K).

Ilex yuana ('yuiana') S. Y. Hu, J. Arnold Arbor. 32: 396. 1951. Type: China, Yunnan, Taron in Kiu-Kiang (Drungjiang) Valley, Chünwangtum, 1350 m, 7 Sept. 1938, *T. T. Yü 20181* (holotype: A; isotype: PE).

Small evergreen trees or shrubs up to 4 m tall. Branchlets puberulent. Petiole 2–5 mm long, puberulent entirely or only adaxially; leaf blade elliptic, 1.2–3 cm long, 0.5–1.5 cm broad, base cuneate or rounded, margin serrate, apex rounded or acute, puberulent only along midrib adaxially, punctate abaxially, midrib impressed adaxially, lateral veins obscure. Male inflorescences on current year's shoots, axillary, solitary, umbellate, 1–5-flowered; peduncle 2–9 mm long, puberulent; pedicel 2–3 mm long, glabrous; flowers 4-merous, sepals and petals glabrous. Female inflorescences also on current year's shoot, axillary, solitary, 1-flowered; peduncle 2–5 mm long, puberulent; pedicel 1–3 mm long, glabrous or sparsely puberulent; flowers 4-merous, stigma 4-lobed. Fruit slightly oblate-globose, 4–5 mm long, 5–6 mm across. Pyrenes 3 or 4, ca. 2.5 mm long, 4 mm wide, smooth, slightly grooved, endocarp leathery. Figure 47–4: a-d.

Phenology: flowering from April to June; fruiting in October and November.

Habitats: forests or thickets; 1350–2750 m.

Distribution: E Nepal, Sikkim & Darjiling, Bhutan, Yarlung Zangbo-Brahmaputra, S Hengduan, and U Irrawaddy (Map 47–10); also in Manipur and Meghalaya of India, and disjunctly in Japan, Korea Peninsula, and E China (Anhui, Fujian, Guangdong, NE Guangxi, Hainan, Hubei, Hunan, Jiangxi, Taiwan, and Zhejiang).

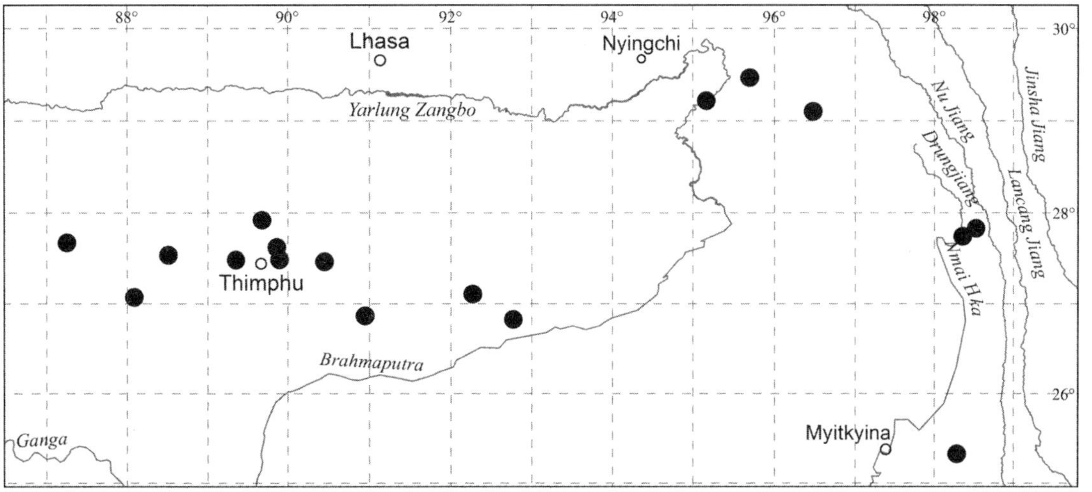

Map 47–10: **Ilex crenata** Thunb.

Figure 47–4: **a–d, Ilex crenata** Thunb.: **a**, female flowering branch; **b**, fruiting branch; **c**, pyrene; **d**, abaxial view of leaf blade, showing glandular puncta. **e–g, Ilex yunnanensis** Franch.: **e**, male flowering branch; **f**, fruit; **g**, pyrene. From *Fl. Reipubl. Popularis Sin*. 45(2), 1999.

E NEPAL: Kosi, Sankhuwasabha, 2720 m, 1998, *S. Noshiro et al. 9840138* (BM). **SIKKIM & DARJILING: Darjiling**, Palmajua 2200 m – Batasi 2100 m, 1972, *H. Kanai 721564* (BM); **Sikkim**, s. loc., 1857, *Thomson* s. n. (K). **BHUTAN: Gasa** Dzong, MoChu, 2440 m, 1949; *F. Ludlow et al. 16543* (BM, E); **Paro**, below Taksang, 2590 m, 1987, *K. Rushforth 1065* (E): **Punakha**, 2740 m, 1914, *R. E. Cooper 3149* (E); **Tashiling**, 1915, *R. E. Cooper 3975* (E); **Thimphu**, 2300 m, 1989, *J. R. I. Wood 6962* (E); **Tongsa**, Changkha, 2400 m, 1979, *A. J. C. Grierson & D. G. Long 1146* (E. K). **YARLUNG ZANGBO-BRAHMAPUTRA: Baibung**, Gelin, 1690 m, 1983, *B. S. Li et al. 3686* (PE); **Gutang**, Xingkai to Jiangbietang, 2200–2400 m, 1982, *B. S. Li & S. Z. Cheng 1069* (PE); **Shangzayü**, near Buzhong, 2100 m, 1980, *C. C. Ni et al. 0711* (PE); **Shergaon**, 2440–2743 m, 1935, F. *Kingdon-Ward 12445* (BM); **Talung** Dzong, Balipara Frontier, 2134 m, 1935, F. *Kingdon-Ward 11420* (BM). **S HENGDUAN: Salween-Irrawaddy Divide**, 3400 m, 1916, *Handel-Mazzetti 9214* (WU). **U IRRAWADDY**: **Drungjiang**, 1600 m, 1982, *Qinghai- Xizang Exped. 9483* (KUN, PE) ; **Tengchong**, Houqiao, 1680–1750 m, 1964, *S. K. Wu 6863* (KUN).

11. Ilex szechwanensis Loes., Nova Acta. Acad. Caes. Leop. –Carol. Nat. Cur. 78: 347 (Monog. Aquif. 1: 347). 1901; Hu, J. Arnold Arbor. 30: 325. 1949. Type: China, Szechwan (Sichuan), Wushan, *A. Henry 5716* (**lectotype here designated**: K; isolectotype: P); Hupeh (Hubei), Chien Shih (Jianshi CO), *A. Henry 5808* (paratype: K).

Evergreen shrubs or small trees 1–6 m tall. Branchlets subquadrangular, current year's shoots ridged and grooved, puberulent, especially in grooves. Petiole 4–6 mm long, puberulent; leaf blade subleathery or leathery, elliptic or oblong, 3.5–7 cm long, 2–4 cm broad, base obtuse, rarely cuneate, margin serrate, rarely crenate, apex acuminate, abaxially punctate and sometimes puberulent particularly along midrib, lateral veins 6 or 7 pairs. Both male and female inflorescences on current year's shoots. Male inflorescences solitary, rarely fasciculate, cymes of 1–7 flowers; peduncle 4–15 mm long; pedicel 2–5 mm long; flowers 4–7-merous; calyx 2–2.5 mm across, glabrous or sparsely puberulent, deeply lobed; corolla white, rotate, petals ovate, glabrous, connate at base; stamens shorter than petals. Female inflorescences solitary, 1-flowered; peduncle 3–10 mm long; pedicel 3–8 mm long; flowers 4-merous. Fruit globose or oblate-globose, ca. 6 mm long, 7–8 mm wide. Pyrenes 4, ovoid or suborbicular, trigonous in cross-section, 4.5–5 mm long, ca. 4 mm wide.

Phenology: flowering from May to July; fruiting from October to December.

Habitats: forests, mainly evergreen forests; 850–3350 m.

Distribution: Yarlung Zangbo-Brahmaputra, N and S Hengduan, and U Irrawaddy (Map 47–11); also in other parts of China (Chongqing, SW Guizhou, W Hubei, S Sichuan, and S Yunnan).

YARLUNG ZANGBOBRAHMAPUTRA: Xiazayü, 1950 m, 2012, *D. Y. Hong et al. H12032* **(PE). N HENGDUAN**: **Wenchuan**, Sanjiang, 1600 m, 1975, *Sichuan Plant Surv. 8401* (PE). **S HENGDUAN**: **Binchuan**, Jizu Shan, 1940, *T. N. Liou 17847* (PE); **Fugong**, Lumadeng, 2700 m, 1982, *Qinghai-Xizang Exped. 6971* (KUN, PE); **Hanyuan**, Xiangling Shan, 2100 m, 1938, *T. P. Wang 8782* (KUN, PE); **Huidong**, Lunan, 2800 m, 1959, *S. K. Wu 1039* (KUN); **Tianquan**,

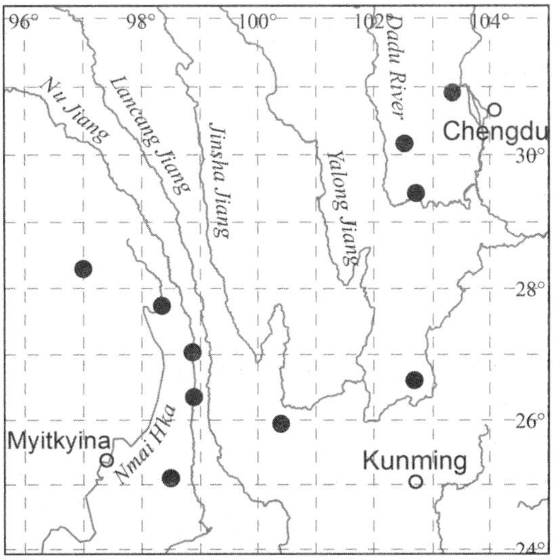

Map 47–11: **Ilex szechwanensis** Loes.

Erlang Shan, 1300 m, 1953, *H. L. Tsiang 35148* (PE). **U IRRAWADDY**: **Drungjiang**, 1250 m, 1959, *K. M. Feng 24284* (KUN); W flank of N' Maikha - Salween Divide, 3350 m, 1925, *G. Forrest 27557* (E) ; **Tengchong**, Salween - Tengchong; 2400 m, *C. Y. Wu et al. 903* (PE) & *909* (PE); Tengchong (Tengyueh), 2134 m. 1912, *G. Forrest 8113* (syntypes of *I. crenata* var. *scoriarum* W. W. Smith: E, K); W of Tengchong, 1524 m, 1912, *G. Forrest 7867* (syntype of *I. crenata* var. *scoriarum*: K).

12. Ilex dicarpa Y. R. Li, Acta Bot. Yunnan. 3: 352. fig. 2. 1981; Hong, Phytotaxa 230(2): 157. 2015. Type: China, Xizang [Tibet], Mêdog, De'ergong, 1600 m, 20 Aug. 1974, *Qinghai- Xizang Exped. 74–4435* (holotype: KUN; isotype: PE).

Small evergreen trees, ca. 5 m tall. Bran- chlets gray-green, striate, glabrous or puberulent. Petiole 6–7 mm long; leaf blade leathery or papery, broadly oblong to elliptic, 8–10 cm long, 3–4 cm broad, base broadly cuneate or rounded, margin obscurely crenate or nearly entire, apex abruptly acuminate with acumen ca. 1 cm long, glabrous, abaxially punctate, midrib impressed adaxially, elevated abaxially, lateral veins ca. 10-paired. Flowers not seen. Infructescences on previous year's growth, fasciculate, umbelliform, 1–4-flowered; peduncle less than 1 mm long; pedicel 4–12 mm long; bracts at base. Fruit globose, ca. 5 mm across; persistent calyx 4-lobed, discoid, or more or less reflexed, leathery, margin mem- branaceous and white. Pyrenes 4. Figure 47–5: d-f.

Phenology: fruiting from September to December.

Habitats: forests; 900–1600 m.

Distribution: **endemic**; confined to Yarlung Zangbo- Brahmaputra (Map 47–12).

YARLUNG ZANGBO-BRAHMAPUTRA: **Baibung**, 910 m, 1992, *H. Sun et al. 0714* (KUN); Baibung, Gelin, 1600 m, 1974, *Qinghai-Xizang Exped. Veget. Group 2906* (PE); **Mêdog**, near the city, 1150 m, 2012, *D. Y. Hong et al. H12002* (PE).

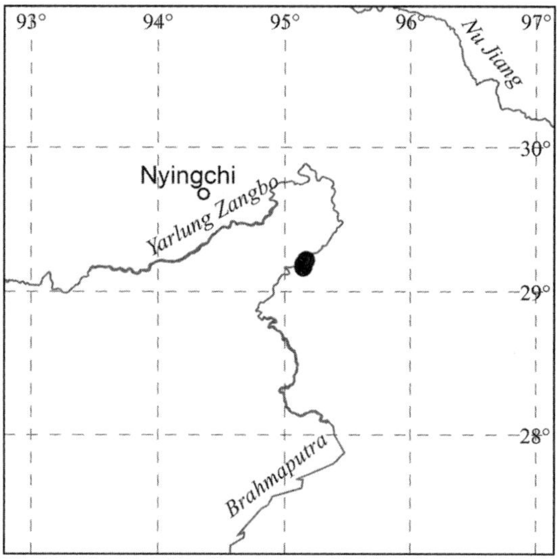

Map 47–12: **Ilex dicarpa** Y. R. Li

13. Ilex yunnanensis Franch., Pl. Delav. 2: 128. 1889; Hu, J. Arnold Arbor. 30: 338. 1949; Hong, Phytotaxa 230(2): 165. 2015. Type: China, Yunnan, "Kou-toui, in silvis supra Mo So Yn", 3000 m, 17 Jun. 1887, *Delavay 2673* (holotype: P; isotype: K).

Ilex yunnanensis Franch. var. *paucidentata* S. Y. Hu, J. Arnold Arbor. 30: 340. 1949. Type: China, Yunnan, Weixi, *C. W. Wang 67855* (holotype: A).

Ilex rockii S. Y. Hu, J. Arnold Arbor. 30: 336. 1949. Type: China, Yunnan, Kun-Shan, Solo-la, 3810 m, May–Jun. 1932, *J. F. Rock 22299* (holotype: A; isotype: K).

Evergreen erect shrubs, rarely small trees, up to 4 m tall. Bark light-gray. Branchlets densely ferrugineous-villose. Petiole 2–5 mm long, pubescent; leaf blade leathery, ovate-lanceolate, lanceolate, or elliptic, 1.5–5.2 cm long, 0.8–2.5 cm broad, base rounded or cuneate, margin crenulate-serrate, teeth often aristate, apex acute, mucronate or rounded, surfaces villose especially near base and along margin, sometimes glabrescent, lateral veins obscure. Inflorescences on current year's shoots, in axils of leaves or scales, solitary, cymose, pubescent. Male cymes (1–) 3-flowered; peduncle 8–14 mm long; pedicel 2–4 mm long. Flowers 4-merous; calyx ca. 2 mm across, deeply

4- lobed, lobes ciliate; corolla white, green, or pink to red, petals ovate, ca. 2 mm long, 1.5 mm broad, connate at base. Female cymes 1(–3)- flowered; fruiting pedicel 4–14 mm long. Fruit red, globose, 4.5–6 mm across. Pyrenes 4, oblong-ovoid, subtriangular in crosssection, ca. 5 mm long, 3 mm wide. Figure 47– 4: e-g.

Phenology: flowering from May to July; fruiting from August to October.

Habitats: forests or thickets often of *Rhododendron* and bamboos; 1600–4300 m.

Distribution: L Yarlung Zangbo, Yarlung Zangbo-Brahmaputra, N and S Hengduan, and U Irrawaddy (Map 47–13); also in W Sichuan of China (Emei and Ebian).

L YARLUNG ZANGBO: **Yi'ong** Farm, 2500 m, 1983, *B. S. Li et al. 6739* (PE). **YARLUNG ZANGBO –BRAHMAPUTRA**: **Mêdog**, 2800–3000 m, 1982, *B. S. Li & S. Z. Chen 1867* (PE); **Nage**, 3200 m, 1974, *Qinghai-Xizang Exped. 74–3853* (KUN); **Gyigang**, 2800 m, 1973, *Qinghai-Xizang Exped. 73–598* (KUN, PE); **Walong**, 3050–3350 m, 1950, *F. Kingdon-Ward 20256* (BM). **N HENGDUAN**: **Pingwu**, Heijia, 2120 m, 1958, *H. L. Tsiang 10464* (PE); **Wenchuan**, Wolong, 2980 m, 1975, *s. coll. 8160* (PE); **Wenxian**, Tiexiu, 1800 m, 1958, *Y. Q. He 1004* (PE). **S HENGDUAN**: **Baoxing**, Dengcigou, 1680 m, 1933, *T. T. Yü 4874* (PE); **Butuo**, Wuyi, 2400 m, 1976, *Acad. Sin. Chengdu Biol. Inst. Veget. Group. 14153* (PE); **Cawarong**, Saleen-Kui- chiang Divide, *G. Forrest 19074* (K). **Dali**, Zhonghesi, 1940, *R. C. Ching 22836* (KUN, PE); **Deqên**, Cizhong, 2900–3200 m, 1940, *K. M. Feng 5029* (KUN, PE); N of Deqên, 4267 m, *G. Forrest 14051* (E, K); **Gongshan**, Sijitung, 2600–2800 m, 1940, *K. M. Feng 7945* (KUN, PE); **Lanping**, 3000 m, 1981, *Bot. Inst. Hengduan Mts Exped. 919* (PE); **Luding**, Erlang Shan, 2300 m, 1953, *W. K. Hu 37910* (PE); **Lushui**, W of Pianma Pass, 2900 m, 1981, *Bot. Inst. Hengduan Mts Exped. 287* (PE); **Meigu**, 1959, *Sichuan Econ. Plant Surv. (Liang 59) 1767* (PE); **Mianning**, Tuowu, 2600 m, 1959, *S. K. Wu 1741* (PE); **Shangri-la** (Zhongdian), by Yangtze River, 2800–2900 m, 1939, *K. M. Feng 3254* (KUN); **Shimian**, 1955, *C. C. Hsien 41939* (PE); **Tianquan**, Lancizi, 1800 m, 1951, *Y. W. Tsui 4102* (PE); **Weixi**, Biluoxue Shan, 3250 m, 1981, *Inst. Bot. Hengduan Mts Exped. 1407* (PE); **Xichang**, Luoji shan, 2600 m, 1965, *Acad. Sin. Southwest China Comprehensive Exped. 1029* (PE); **Yangbi**, Meiweng, 3000 m, 1963, *NW Yunnan Yangtze Exped. 9043* (KUN, PE); **Yulong** (Lijiang), 3350 m, 1913, *G. Forrest 10247* (K, PE). **U IRRAWADDY**: **Adung Valley**, sources of the Irrawaddy, 2440 m, 1931, *F. Kingdon-Ward 9587* (BM); **Nam Tamai Valley**, 2743–3350 m, 1937, *F. Kingdon-Ward 13296* (BM) ; **Tengchong**, Guyong, 2400 m, 1964, *S. K. Wu 6944* (KUN).

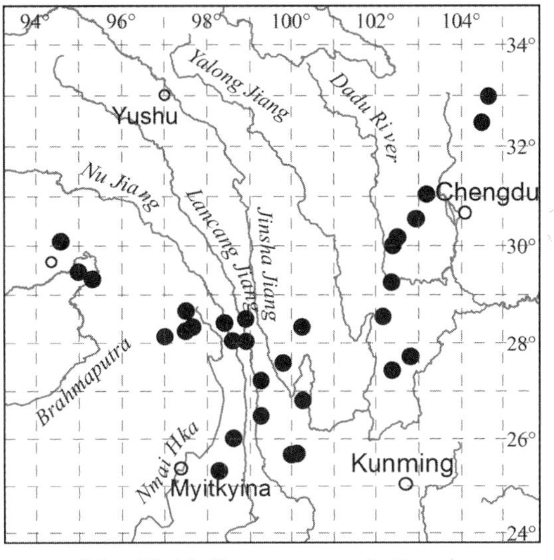

Map 47–13: **Ilex yunnanensis** Franch.

14. Ilex gansuensis D. Y. Hong, Phytotaxa 230(2): 159. 2015. Type: China, Gansu Prov., Wenxian County, Bikou, Bifenggou, 1500 m, forests, 30 Aug. 1992, *X. Wang 432* (holotype: PE).

Evergreen trees, glabrous throughout except for calyx. Branchlets pale brown; current year's shoots powdery white. Petiole 1.5–2 cm long, powdery white; leaf blade broadly elliptic or lanceolate, 7–9 cm long, 2.5–3.5 cm broad, base rounded, margin sparsely serrate, or serrate only in distal part, apex acuminate or acute, adaxially powdery white, abaxially brown. Flowers not seen. Infructescences on previous and current year's shoots, solitary, a cyme with 1 fruit; peduncle 1–3 cm long; bracts in lower, middle or upper part of infructescence axis. Pedicel

Figure 47–5: **a–c, Ilex gansuensis** D. Y. Hong: **a**, a section of fruiting branch; **b**, fruit with long pedicel; **c**, pyrene. From *Phytotaxa* 230(2): 160. 2015, with permission. **d–f, Ilex dicarpa** Y. R. Li: **d**, a section of fruiting branch; **e**, fruit; **f**, pyrene. From *Fl. Reipubl. Popularis Sin*. 45(2), 1999.

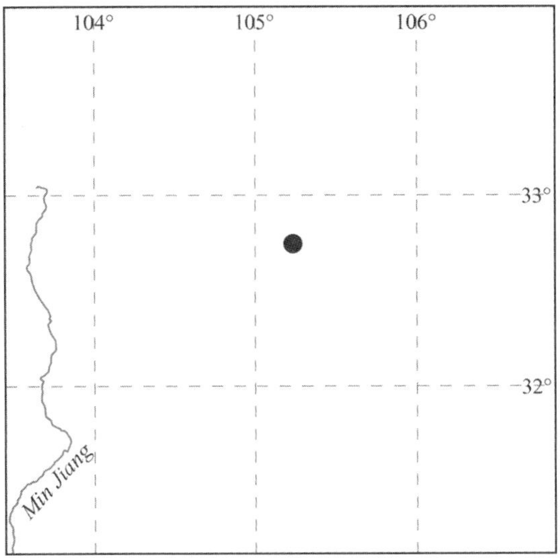

Map 47–14: **Ilex gansuensis** D. Y. Hong

1–1.5 cm long; persistent calyx ca. 4 mm across, 4-lobed, sparsely hispidulous-ciliate; fruit globose, 5–7 mm in diameter; persistent stigma sessile, 2-lobed. Pyrenes 3, smooth, trigonous in cross- section, ca. 5 mm long, 2 mm wide. Figure 47–5: a-c.

Phenology: fruiting in late August.

Habitats: forests; ca. 1500 m.

Distribution: **endemic**; recorded only in N Hengduan (Wenxian County in SE Gansu) (Map 47–14).

Notes: The new species should have its position in sect. *Paltoria* in having solitary and 4-merous female flowers, which are axillary on both the current and previous year's growth. In the section it is very distinct by the long peduncle plus pedicel and glabrous shoots and leaves.

Sect. 3. Ilex

Trees or shrubs. 0.5–12(–25) m tall. Leaf margin spiny or serrate, rarely sinuate or entire. Inflorescences fasciculate, in axils of leaves of previous year's or older growth. Pyrenes 4, rarely fewer, striate and sulcate, or rugose and pitted; endocarp stony or woody.

Twenty-two species in the Pan-Himalaya (7 endemic).

15. Ilex dipyrena Wall. in Roxburgh, Fl. India 1: 473. 1820 & Pl. Asiat. Rar. 3: 68. pl. 292. 1832; Hu, J. Arnold Arbor. 30: 358. 1949. Type: Nepal, *Wallich 4327* (**lectotype here designated**: K; isolectotypes: E, P).

Evergreen trees or less often shrubs, 3–20 m tall. Lenticels obscure; leaf scars semicircular. Petiole 3–6 mm long; leaf blade thickly leathery, elliptic-oblong, elliptic or ovate-elliptic, 4–10 cm long, 2–4 cm broad, base broadly cuneate to rounded, margin entire or with 3–14 spines each side, apex acuminate, always with a sharp spine, both surfaces glabrous; lateral veins 6–9 pairs. Inflorescences on previous year's branchlets, fasciculate, cymes one-flowered; flowers

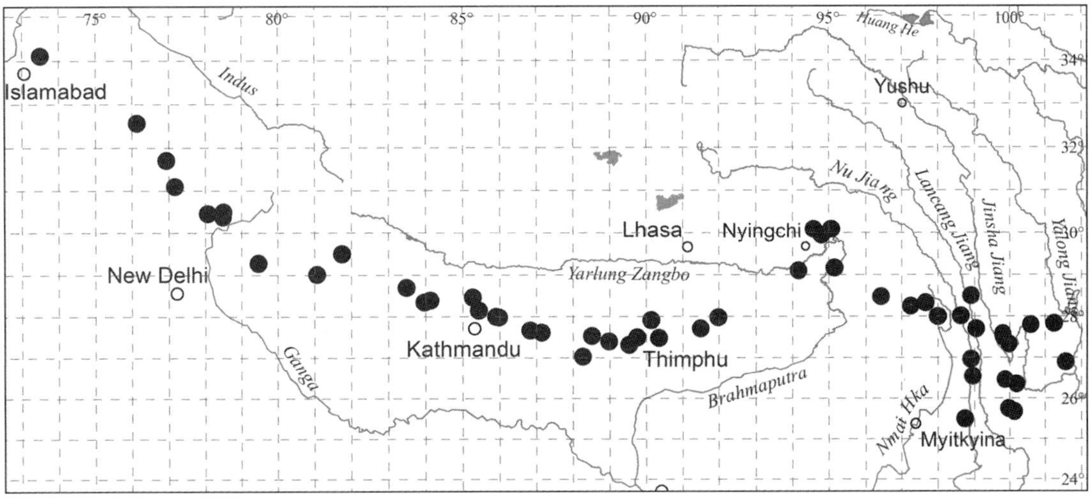

Map 47–15: **Ilex dipyrena** Wall.

green, 4-merous. Male flowers: pedicel 2–3 mm long; bracteoles basal, lanceolate; calyx discoid, ca. 3 mm across; corolla rotate, ca. 7 mm across. Female flowers: pedicel 1–3 mm long, sparsely puberulent or glabrescent; calyx and corolla as in male flowers; ovary ovoid, stigma discoid, lobed. Fruit red, globose, 6–9 mm in diameter. Pyrenes usually 2, rarely 1, 3 or 4, oblong-ellipsoid or nearly globose, 5–7 mm long, ca. 5 mm wide, abaxially palmately striate and sulcate, adaxially striate-sulcate; endocarp woody. **Chromosomes**: 2n=40. Figure 47-6: a-c

Phenology: flo-wering from April to July; fruiting from June to December.

Habitats: forests or thickets; 1700– 3500 m.

Distribution: widely distributed in the Pan-Himalaya, absent only from Vakhan, Tangut, and N Hengduan (Map 47–15); also in S Yunnan of China.

Notes: The species usually grows in forests: evergreen broad-leaved forests, deciduous broad-leaved forests or broad-leaved and conifers mixed forests, but may be found in thickets at edges of forests. The leaf margin is mostly entire or with 1–3 short and weak spines plus an apical spine. However several strong spines may be found, but our observations indicate that individuals of this form are usually shrubs or small trees and occur in thickets or deforested areas. The present author supposes that the latter form is a kind of modification or the juvenile growth, but not an ecotype.

16. Ilex bioritsensis Hayata, J. Coll. Sci. Tokyo 30(1): 53. 1911; Hu, J. Arnold Arbor. 30: 366. 1949; Hong, Phytotaxa 230(2): 153. 2015. Type: China, Taiwan (Formosa), Biōritsu: Taizan, Oct. 1908, *T. Kawakami & U. Mori 7185* (holotype: TI, n. v., photo: K).

Ilex ciliospinosa Loes. in Sargent, Pl. Wils. 1: 78. 1911. ≡ *Ilex bioritsensis* Hayata var. *ciliospinosa* (Loes.) H. F. Comber, Notes Roy. Bot. Gard. Edinburgh 18: 43. 1933. Type: China, Sichuan, Chien-shi (Hanyuan), 1500 m, 15 Sept. 1908, *E. H. Wilson 996* (holotype: A; isotype: K).

Evergreen shrubs or small trees up to 5 m tall. Branchlets sparsely puberulent or glabrescent. Petiole 1–3 mm long, pubescent; leaf blade thickly leathery, ovate-quadrangular or ovate, 2.5–6 cm long, 1.5–3.5 cm broad, puberulent along midrib above, base rounded or truncate, margin sinuate, with (1–) 2–4 (–6) strong spines on each side, apex acuminate, acumen 5–15 mm long, terminated by a spine 3 mm long, lateral veins 4–6 pairs. Inflorescences on previous year's growth, fasciculate, individual cymes one-flowered; flowers 2–4-me-rous. Male flowers: pedicel ca. 2 mm long, glabrous; calyx ca. 3 mm across, lobes sparsely ciliate; corolla rotate, yellow-green; petals broadly elliptic, 3 mm long, connate at base. Female flowers: pedicel 1–2 mm long, glabrous. Fruit ellipsoid, 6–10 mm long, 5–7 mm wide. Pyrenes 2, dorsiventrally compressed, 5–6 mm long, 4–5 mm wide, slightly convex, abaxially palmately striate and sulcate, adaxially striate. Figure 47–6: d & e.

Phenology: flowering in May and June; fruiting from September to November.

Habitats: various forests or thickets; 1500–3300 (3800) m.

Distribution: S Hengduan (Map 47–16); also in E and C China: Chongqing, NE Guangxi, Guizhou, Hubei, Hunan, other parts of Sichuan, Taiwan, and NE Yunnan.

S HENGDUAN: **Hanyuan**, 3800 m, 1938, *T.*

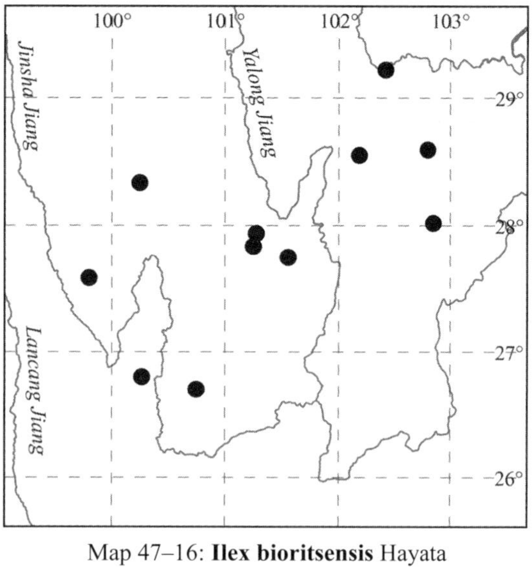

Map 47–16: **Ilex bioritsensis** Hayata

Figure 47–6: **a–c, Ilex dipyrena** Wall.: **a**, fruiting branch; **b**, a different form of leaf blade: **c**, pyrene. **d** & **e**, **Ilex bioritsensis** Hayata: **d**, fruiting branch; **e**, pyrene. **f** & **g**, **Ilex georgei** H. F. Comber: **f**, fruiting branch; **g**, a different form of leaf blade. From *Fl. Reipubl. Popularis Sin.* 45(2), 1999.

P. Wang 9445 (PE); **Mianning**, Maidi to Huilong, 2700 m, 1960, *T. S. Ying 4719* (PE); **Muli**, Bo'ao, 2400 m, 1983, *Qinghai-Xizang Exped. 13510* (PE); Muli, Wachang to Muli Town, 2400 m, 1984, *W. L. Chen et al. 6030* (PE); **Meigu**, 1959, *Sichuan Econ. Plant Surv. (Liang 59) 1789* (PE); **Shangri-la** (Zhongdian), Anangu Valley, 1939, *K. M. Feng 966* (PE); **Yanyuan**, Yuanbao, Dalin, 3200 m, 1983, *Qinghai-Xizang Exped. 12059* (PE); **Yongshen**, 2200 m, 1932, *H. T. Tsai 50962* (PE); **Yuexi**, Lajiba, 2700 m, 1976, *Sichuan Plant Surv. 14048* (PE); **Yulong** (Lijiang), s. loc., 1939, *K. M. Feng 622* (PE); **Zhaojue**, Suosuogou, 2500 m, 1960, *S. Jiang 7643* (PE).

17. Ilex pernyi Franch., Nouv. Arch. Mus. His. Nat. Ser. 2, 5: 221. 1883; Hu, J. Arnold Arbor. 30: 361. 1949; Chen *et al.* in Wu, Raven & Hong, Fl. China 11: 391. 2008. Type: China, Chensi (Shaanxi), Tsin-ling (Qinling Range), March 1873, *A. David* s. n. [**lectotype here designated**: P (P03274228); isolectotypes: K (fragments), P (P0327- 4226)].

Evergreen shrubs or small trees up to 6 m tall. Branchlets pubescent. Petiole ca. 2 mm long, pubescent; leaf blade leathery, ovate or ovate- lanceolate, 1.3–3 cm long, 0.5–1.5 cm broad, broadest between two largest marginal spines, base rounded or truncate, margin sinuate- dentate, with 1 or 2, rarely 3, spines on each side, apex triangular- acuminate, acumen 1.2–1.4 cm long, terminated by a strong spine ca. 3 mm long, midrib slightly impressed adaxially, puberulent near base, elevated abaxially, lateral veins 1–3 pairs. Inflorescences on previous year's growth, fasciculate, individual cymes one-flowered; flowers all 4-merous. Male flowers: pedicel ca. 1 mm long, glabrous; calyx ca. 2 mm across, deeply lobed, lobes broadly triangular or semi- orbicular, ciliate; corolla ca. 7 mm across, yellow, petals ciliate; stamens slightly longer than petals. Female flowers: pedicel 1–2 mm long. Fruit globose or subglobose, red, 7–8 mm long, 6–7 mm wide. Pyrenes 4, obovoid or cylindrical, 4.5–5.5 mm long, 3–3.5 mm wide, palmately striate or sulcate on back. Figure 47–7: e-h.

Phenology: flowering from April to July; fruiting from September to November.

Habitats: forests or thickets; 800–2600 m.

Distribution: N and S Hengduan (Map 47–17); also in other parts of China: Guizhou, W Hubei, S Shaanxi, and other parts of SE Gansu.

N HENGDUAN: **Jiuzhaigou** (Nanping), 1624 m, 2003, *As Sich No 2391* (K); **Maoxian**, Dagou, 2010 m, 1995, *RBG Kew Exped. to Sichuan Sich1676* (K); **Pingwu**, *H. L. Tsiang 10623* (PE); **Wenchuan**, Wolong, 2000–2100 m, 1982, *Lang & Li 1362* (KUN, PE); **Wenxian**, Bijiang, Bifenggou, 1300 m, 1992, *X. Wang 427* (PE); **Zhugqu**, Jiao'erqiao, 2100 m, 1958, *Y. Q. He 627* (PE). **S HENGDUAN**: **Baoxing**, 1954, *T. P. Soong 38863* (KUN, PE); **Ganluo**, Pingba, 2200 m, 1959, *Sichuan Econ. Plant Surv. (Liang 59) 4342* (PE); **Kangding**, 1930, *Z. P. Huang et al. 1865* (KUN, PE); **Luding**, Gongga Shan, 2120 m, 1991, *RBG Kew Exped. to Sichuan Sich 760* (K); **Mianning**, Tuowu, 2400–2500 m, *S. K. Wu 1774* (KUN, PE); **Muli**, Bo'ao, 2800–3350 m, 1959, *S. K. Wu 3243* (KUN); **Shimian**, Ke'erma Shan, 2600 m, 1993, *Z. C. Zhao 1597* (PE); **Tianquan**, Erlang Shan, 2300 m, 1953, *H. L. Tsiang 37740* (PE); **Yuexi**, 1930, *Z. P. Huang 2681* (KUN).

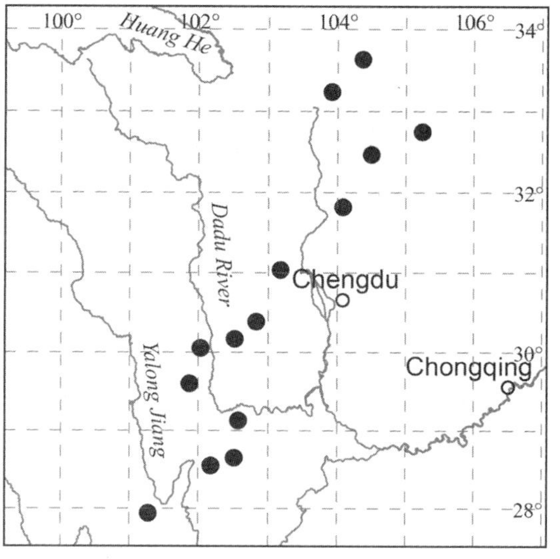

Map 47–17: **Ilex pernyi** Franch.

18. Ilex georgei H. F. Comber, Notes Roy. Bot. Gard. Edinburgh 18: 50. 1933, p. p. excl. var. *rugosa* H. F. Comber; Hu, J. Arnold Arbor. 30: 363. 1949; Chen *et al.* in Wu, Raven & Hong, Fl. China 11: 392. 2008. Type: China,

Figure 47–7: **a–d**, **Ilex perryana** S. Y. Hu: **a**, male flowering branch; **b**, male flower; **c**, fruit; **d**, pyrene. **e–h**, **Ilex pernyi** Franch.: **e**, fruiting branch; **f**, fruit; **g**, pyrene; **h**, cross section of pyrene. From *Fl. Reipubl. Popularis Sin.* 45(2),1999.

Yunnan, Oct. 1917, *G. Forrest 16061* (**lectotype here designated**: E; isolectotype: A, K); China, Yunnan, Tengchong (Tengyueh), 1840 m, 1925, *G. Forrest 26254* [paratypes (as type of ♂ flower): A, E, K, PE].

Evergreen shrubs or small trees, up to 8 m tall. Branchlets pubescent. Petiole 1–2 mm long, pubescent; leaf blade thickly leathery, lanceolate, ovate-lanceolate, 3–5 cm long, 0.7–1.5 cm broad, base rounded, margin with up to 7 spines on each side, rarely entire, apex acuminate, acumen ca. 1 cm long, terminated by a spine ca. 3 mm long, midrib pubescent, lateral veins 5–7 pairs. Inflorescences on previous year's growth, fasciculate; male inflorescences with individual cymes 1–3-flowered, uniflorous pedicel 2–4 mm long, 3-flowered peduncle 1 mm long, puberulent; flowers 4-merous. Calyx 1.5–2 mm across, 4-lobed, lobes ovate, ciliate; corolla rotate, 4–5 mm across, petals ca. 2 mm long, sparsely puberulent-ciliate, connate at base; stamens longer than petals. Fruit usually paired; pedicel ca. 2 mm long, puberulent; fruit obovoid-ellipsoid, 4–6 mm long, 3–4 mm wide. Pyrenes 1 or 2, obovoid-oblong, ca. 4.5 mm long, 2.5–3 mm wide, longitudinally 7–9 striate, shallowly sulcate abaxially. Figure 47–6: f & g.

Phenology: flowering from February to May; fruiting in October

Habitats: thickets; 1500–2700 m.

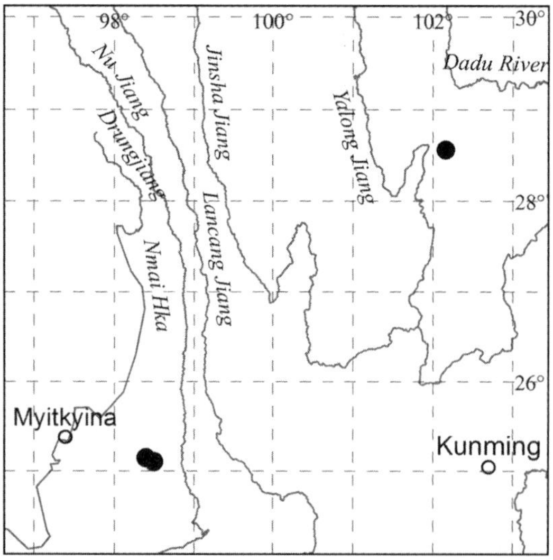

Map 47–18: **Ilex georgei** H. F. Comber

Distribution: S Hengduan and U Irrawaddy (Map 47–18); also in SW Yunnan of China (Lincang).

S HENGDUAN: **Mianning**: Chengxiang Town, 1950 m, 1995, *M. Ogisu 95099* (K). **U IRRAWADDY**: *F. Kingdon-Ward 463* (A); **Tengchong**, Tengchong, NW of Tengchong, 1924, *G. Forrest 24826* (paratype: PE); Tengchong, 1500 m, *G. Forrest 9712* (paratype: K); loc. eodem, *G. Forrest 7577 & 9608* (paratypes: K).

19. Ilex perryana S. Y. Hu, J. Arnold Arbor. 30: 367. 1949; Chen *et al.* in Wu, Raven & Hong, Fl. China 11: 390. 2008. Type: China, Yunnan, Gongshan, Taron Taru Divide, Barcuhwang, 2800 m, Nov. 1938, *T. T. Yü 20942* (holotype: A (photo, K); isotypes: E, PE).

Ilex georgei H. F. Comber var. *rugosa* H. F. Comber, Notes Roy. Bot. Gard. Edinburgh 18: 51. 1933. Type: N'Maikha-Salween divide, 26°23′N, 98°48′E, 3050–3350 m, Aug. 1924, *G. Forrest 24860* (holotype: E).

Prostrate evergreen shrubs up to 1.5 m tall, branches growing more or less horizontally. Branchlets stout, puberulent. Petiole 1–2 mm long, glabrous or puberulent; leaf blade thickly leathery, hexangular or nearly rhomboid, rarely elliptic, 1–3.5 cm long, 0.7–1.5 cm broad, base rounded or obtuse, margin with 2, less frequently 1 or 3, spines on each side, apex broadly deltoid, terminated by a spine ca. 2 mm long, midrib adaxially deeply impresssed, glabrous, abaxially slightly elevated, lateral veins 3 or 4 pairs. Inflorescences on previous year's growth, fasciculate, fascicles small, each with 2–4 one-flowered cymes; flowers all 4-merous. Male flowers: pedicel 2–3.5 mm long, glabrous; calyx ca. 3 mm across, deeply 4-lobed, lobes glabrous with minutely ciliate margins; corolla white or green-white, rotate, 4–5 mm across, petals ovate, ca. 1.5 mm long, glabrous; stamens equal to or slightly longer than petals. Fruiting pedicel 3–4 mm; fruit red, subglobose-ellipsoid, 4.5–6 mm long, 3–4.5 mm wide. Pyrene 1, subglobose, slightly dorsiventrally compressed, palmately striate, canaliculate adaxially. Figure 47–7: a-d.

Phenology: flowering in May and June;

fruiting in October and November.

Habitats: forests, mainly conifer forests, sometimes forming a compact cushion on rocks; 2800–3800 m.

Distribution: **endemic**; L Yarlung Zangbo, Yarlung Zangbo-Brahmaputra, S Hengduan, and U Irrawaddy (Map 47–19).

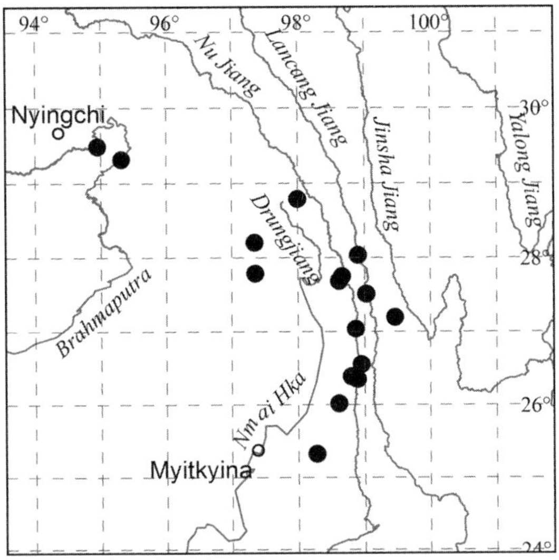

Map 47–19: **Ilex perryana** S. Y. Hu

L YARLUNG ZANGBO: **Mainling**, Doshong (Doxong) La, 3350–3660 m, 1924, *F. Kingdon-Ward 6240* (K). **YARLUNG ZANGBO-BRAHMAPUTRA**: **Mêdog**, 3350 m, 1982, *B. S. Li & S. Z. Cheng 1856* (PE). **S HENGDUAN**: **Bijiang**, Lahe Valley, 1978, *Bijiang Exped. 1293* (KUN); **Deqên**, Cizhong, 3100–3300 m, 1940, *K. M. Feng 5477* (KUN, PE); **Fugong**, Lumadeng, Biluoxue Shan, 2800 m, 1982, *Qinghai-Xizang Exped. 6926* (KUN, PE); **Gongshan**, Cikai, Heipu Shan, 1940, *K. M. Feng 8381* (KUN, PE); **Lushui**, Pianma, 3102 m, 2008, *X. H. Jin & T. Zhang 279* (PE); **Weixi**, Yanwaqiao, Biluoxue Shan, 3100–3300 m, 1981. *Inst. Bot. Hengduan Mts Exped 1371* (PE); **Yulong** (Lijiang), Ludian, 3300 m, 1956, *P. I. Mao 322* (KUN, PE). **U IRRAWADDY**: **Dichu Valley**, 3050–3350 m, 1926, *F. Kingdon-Ward 7635* (K); **N'Maikha-Salween Divide**, 1924, *G. Forrest 25429* (paratype of *I. georgei* var *rugosa*: PE); **Ridong**, Shizhudu, 3000 m, 1982, *Qinghai-Xizang Exped. 10021* (KUN, PE); **Seinghku Valley**, 3350–3660 m, 1926, *F. Kingdon-Ward 6925* (K); **Tengchong**, Guyong, 3500 m, 1980, *L. S. Xie 884* (KUN).

20. Ilex miguensis S. Y. Hu, J. Arnold Arbor. 32: 396. 1951. Type: China, Tibet, Zayul [Zayü], Rong Tö Valley, 3050–3350 m, 28 Nov. 1933, *F. Kingdon-Ward 11019* (holotype: BM).

Evergreen shrubs. Branchlets gray, glabrous, current year's branchlets 2–2.5 mm in diameter; terminal bud conical, glabrous. Petiole ca. 3 mm long, glabrous; leaf blade leathery, elliptic, 1.5–3 cm long, 0.7–1.5 cm broad, base acute, rarely obtuse, margin more or less sparsely spinose- serrate, apex acute, with a spine, midrib impressed adaxially, lateral veins 4 or 5 pairs, impressed adaxially, raised abaxially. Flowers not seen. Infructescences in fascicles, often 2-fruited; individual cymes 1-fruited. Fruiting pedicel ca. 3 mm long, glabrous; bracteoles 2, basal; persistent calyx 4-angled, ca. 2 mm in diameter, glabrous, persistent stigma discoid; fruit bright red, globose, ca. 4 mm in diameter. Pyrenes 4, ca. 4 mm long, 3 mm wide, abaxially palmately striate, not sulcate.

Phenology: fruiting in November.

Habitats: forests on rocks; 3050–3350 m.

Distribution: **endemic**; confined to Yarlung Zangbo-Brahmaputra (Map 47–20).

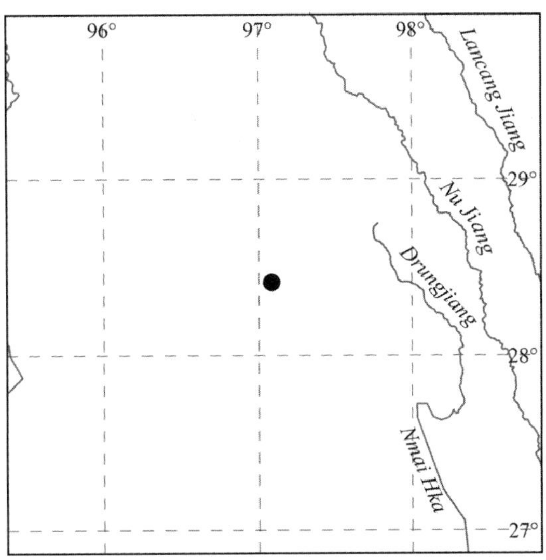

Map 47–20: **Ilex miguensis** S. Y. Hu

Notes: Kingdon-Ward, the collector of the type specimen, left his handwriting on the label of the type sheet, which runs: "Rong Tö Valley: above Migu: Zayul, Tibet, 10000–11000, 28.XI. 33, an undershrub in the upper forest on rocks. Berries scarlet. Has the appearance of being a hybrid (*I. pernyi* × *I. hookeri* ?), only one or two plants seen in the spot."

21. Ilex kingiana Cockerell, Torreya 11: 264. 1911; Chen *et al.* in Wu, Raven & Hong, Fl. China 11: 359, 2008, pro syn. ≡ *Ilex insignis* Hook. f., Fl. Brit. India 1: 599. 1875, non Heer (1869); Basu in Singh *et al.*, Fl. India 5: 59. 2000. Type; India, Sikkim, *J. D. Hooker* s. n. (**lectotype here designated**: K); Darjiling, 2133 m, 1875, *C. B. Clarke 26705* (paratype: K); Sikkim, *s. coll.* s. n. (paratype: K).

Small evergreen trees or shrubs, up to 6 m tall, glabrous throughout. Bark gray, branchlets as thick as a little finger. Petiole 1.5–2.5 cm long; leaf blade elliptic, narrowly elliptic, 10–20 cm long, 4–7 cm broad, base cuneate or rounded, margin mostly serrate, rarely spinose-toothed, or sparsely verrucose-serrate, apex acute to shortly acuminate, veins impressed adaxially, lateral veins 13–15-paired. Both male and female flowers sessile, in subglobose cluster, 4-merous, green. Male flowers: calyx 3–4 mm across; petals orbicular-obovate, 4–5 mm long, 2–3 mm broad; stamens as long as petals. Female flowers: ovary 4-celled, 1–1.7 mm in diameter, stigma obscurely 4-lobed. Fruit scarlet, globose, 8–10 mm in diameter. Pyrenes 4, abaxially striate.

Phenology: flowering from April to June; fruiting from July to November.

Habitats: fo-rests; 2100–2800 m.

Distribution: **endemic**; E Nepal, Sikkim & Darjiling, Bhutan, and U Irrawaddy (Map 47–21).

E NEPAL: **Kosi**, Morang, Meklajam, 1830 m, 1969, *J. D. A. Stainton 6530* (BM); **Mechi**, Ilam, 2130 m, 1967, *J. D. A. Stainton 5776* (BM). **BHUTAN**: **Chukka** Distr., 2000 m, 1982, *A. J. C. Grierson & D. G. Long 3117* (E). **U IRRAWADDY**: **Tengchong**, hills around Lungfan, 2438 m, 1925, *G. Forrest 27767* (E); east of Ta-Chu-Pa, 2740 m, 1924, *G. Forrest 25400* (E).

22. Ilex fargesii Franch., J. Bot. (Morot) 12: 255. 1898; Hong, Phytotaxa 230(2): 157. 2015. Type: China, "Se-tchuen (Sichuan), district de Tchen Kéou tin (Chengkou)", 2000 m, *R. P. Farges 763* [holotype: P (photo, K); isotypes: E, KUN, P].

Evergreen shrubs or small trees up to 7 m, rarely up to 25 m tall, glabrous throughout except sometimes for puberulent peduncle and pedicel. Petiole 1–1.8 cm long; leaf blade subleathery, oblanceolate, linear- oblanceolate or elliptic, 5–13 cm long, 1.2–3.5 cm broad, base cuneate, margin entire in lower half or 2/3, serrate in upper 1/3 or half, apex acuminate with acumen 5–10 mm

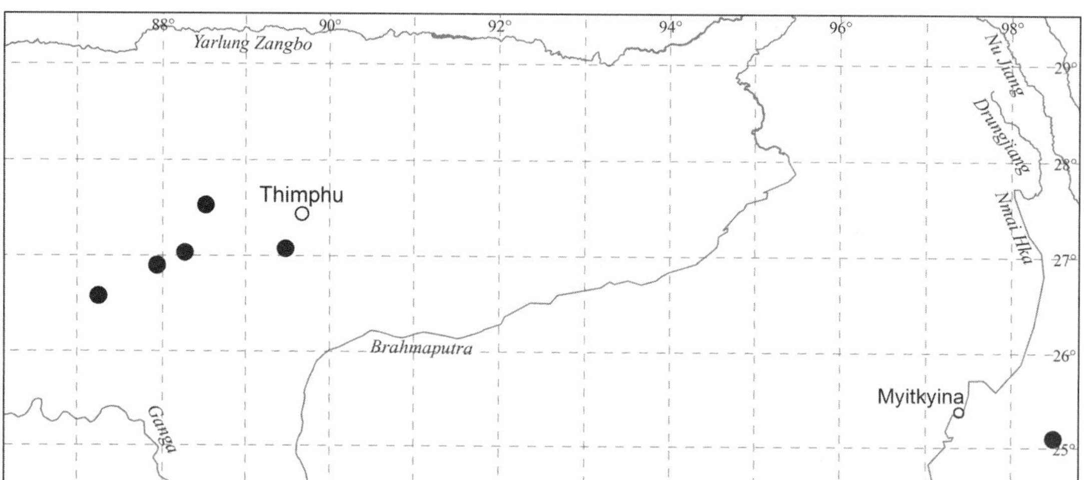

Map 47–21: **Ilex kingiana** Cockerell

long, rarely acute, midrib impressed adaxially, elevated abaxially, lateral veins 8–10 pairs. Male inflorescences on previous year's growth, fascicles of 3-flowered cymes. Flowers 4-merous; peduncle ca. 1 mm long, glabrous or puberulent; pedicel ca. 2 mm long, glabrous or puberulent; calyx ca. 2 mm across, deeply lobed, lobes very sparsely ciliate; corolla ca. 5 mm across; petals obovate-oblong, ca. 2 mm long, 1.25 mm broad; stamens shorter than petals. Infructescences fasciculate, infructescence axis from nearly absent to 16 mm long; individual cymes 1-flowered. Fruiting pedicel 5–7 mm long; fruit red or dull red, globose or oblate-globose, 4–5 mm long, 5–6 mm wide. Pyrenes 4, oblong, ca. 4 mm long, 3.5 mm wide, back palmately striate-sulcate.

Distribution: N and S Hengduan, and U Irrawaddy (Map 47–22); also in other parts of China.

1a. Fruiting pedicel glabrous, shorter than or equal to fruit diameter; infructescence axis nearly absent .... 22a. subsp. *fargesii*

1b. Fruiting pedicel mostly more or less puberulent, rarely glabrous, longer than or equal to fruit diameter; infructescence axis 1–16 mm long .... 22b. subsp. *melanotricha*

22a. Ilex fargesii subsp. fargesii

Ilex franchetiana Loes. in Sargent, Pl. Wils. 1: 77. 1911. Type: Western Szech'uan (W Sichuan), SE of Tachienlu (Kangding), 2400 m, 1908. *E. H. Wilson 1257* (lectotype designated by Hu, 1949: A; isolectotype: K); China, W Hupeh (Western Hubei), Changyang, 1200–1800 m, 1907, *E. H. Wilson 148* (paratypes: A, K).

Fruiting pedicel glabrous, 4–5 mm long, shorter than or equal to fruit diameter. Figure 47–8: a-d.

Phenology: flowering in May; fruiting in September and October.

Habitats: forests; 1500–3200 m.

Distribution: N and S Hengduan; also in other parts of China: Chongqing, other parts of SE Gansu, Guizhou, Hubei, Hunan, S Shaanxi, E Sichuan, and NE Yunnan.

N HENGDUAN: **Lixian**, Bipengou, 2765 m, 21. September 2005, *M. F. Hou 18835* (PE); **Pingwu**, Wangba, 2030 m, 1958, *s. coll. 10736* (PE); **Wenchuan**, Wolong, 2400 m, 1959, *Sichuan Econ. Plant Surv. (A 59) 2131* (KUN, PE); **Wenxian**, Tiexiu, 1900 m, 1958, *Y. Q. He 1009* (PE). **S HENGDUAN**: **Baoxing**, Ganyanggou, 3000 m, 1936, *K. L. Chu 3259* (PE); **Hanyuan**, Xiangling Shan, 3200 m, 1938, *T. P. Wang 9478* (PE); **Jiulong**, Hongba, *K. Y. Yao 3817* (PE); **Kangding**, mountain summit, 1939, *McLaren's Collector 166* (P); Kangding, 1930, *Z. P. Huang et al. 1713* (PE); **Mianning**, Lianhe Township, 3000 m, 1959, *S. K. Wu 2222* (KUN, PE); **Shimian**, 1955, *C. C. Hsieh 40291* (PE); **Tianquan**, Baisanhe, 1940, *C. L. Wu 12374* (PE); **Yuexi**, Bao'an, Dalongtang, 2000 m, *Sichuan Econ. Plant Surv. (Liang 59) 3860* (KUN, PE).

22b. Ilex fargesii subsp. **melanotricha** (Merr.) S. Andrews, Kew Mag. 3(3): 134. 1986; Hong, Phytotaxa 230(2): 158. 2015. ≡ *Ilex melanotricha* Merr., Brittonia 4: 101. 1941. Type: Upper Burma (Myanmar), Adung Valley, 2440 m, 31 Mar. 1932, *F. Kingdon-Ward 9331* [lectotype designated by Hu, 1950: A (photo, K); isolectotypes: BM, E, K].

Ilex subodorata S. Y. Hu, J. Arnold Arbor. 31: 74. 1950. Type: China, Yunnan, Tengchong, Shweli-Salween Divide, 25°20′N, 98°50′E,

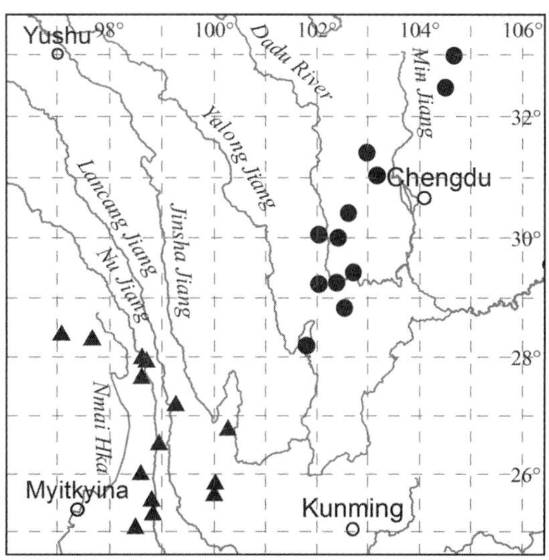

Map 47–22: **Ilex fargesii** Franch.

● subsp. **fargesii,** ▲ subsp. **melanotricha**

Figure 47–8: **a–d**, **Ilex fargesii** Franch. subsp. **fargesii**: **a**, male flowering branch; **b**, fruiting branch; **c**, fruit; **d**, pyrene. **e**, **Ilex fargesii** subsp. **melanotricha** (Merr.) S. Andrews: fruiting branch. **f**, **Ilex hookeri** King: fruiting branch. From *Fl. Reipubl. Popularis Sin.* 45(2), 1999.

9,000 ft., Nov. 1925, *G. Forrest 27726* (**lectotype here designated:** A; isolectotypes: E, K).

Ilex chartacifolia C. Y. Wu ex Y. R. Li, Bull. Bot. Res. Harbin 5(1): 11. pl 3: 2. 1985. Type: China, Yunnan, Yangbi, Mubangzuzu, 29 Oct. 1940, *R. C. Ching 25348* (holotype: KUN; isotype: PE).

Ilex chartacifolia C. Y. Wu ex Y. R. Li var. *glabra* C. Y. Wu ex Y. R. Li, Bull. Bot. Res. Harbin 5(1): 12. pl. 3: 3. 1985. Type: China, Yunnan, Gongshan, Bingzhongluo, 1900 m, 10 Oct. 1956, *P. I. Mao 549* (holotype: KUN; isotype: PE).

Ilex micropyrena C. Y. Wu ex Y. R. Li, Bull. Bot. Res. Harbin 51(1): 20. pl. 6: 3. 1985. Type: China, Yunnan, Gongshan, Changputung, 1700 m, 28 Sept. 1940. *K. M. Feng 8070* (holotype: KUN).

Ilex punctatilimba C. Y. Wu ex Y. R. Li, Bull. Bot. Res. Harbin 5(1): 21. pl. 6: 4. 1985. Type: China, Yunnan, Jingdong, Wuliang Shan, 2500 m, 1 May 1959, *S. G. Xu 4780* (holotype: KUN).

Fruiting pedicel puberulent, rarely glabrous, 5–7 mm long, longer than or equal to fruit diameter. Figure 47–8: e.

Phenology: flowering in May; fruiting in October and November.

Habitats: forests or open thickets; 1700–3100 m.

Distribution: Yarlung Zangbo Brahmaputra, S Hengduan, and U Irrawaddy; also in S Yunnan of China.

YARLUNG ZANGBO -BRAHMAPUTRA: **Zayü**, Dongqiong, 2900 m, 1973, *Qinghai- Xizang Exped. 73–778* (KUN). **S HENGDUAN**: **Bijiang**, 2800 m, 1933, *H. T. Tsai 58668* (PE); **Dali**, Canshan, Huadianba, 2900 m, 1981, *S. B. E. C. 1012* (KUN); **Gongshan**, Baihaluo, 1959, *s. coll. 24682* (KUN); **Lushui**, Shweli-Salween Divide, 3050 m, 1924–1925, *G. Forrest 26070* (PE); **Weixi**, *Qinghai-Xizang Exped. 6385* (KUN); **Yangbi**, Shizhong, 3080 m, 1981, *S. B. E. C. 0294* (KUN); **Yulong** (Lijiang), Luoxu Shan, 1939, *R. C. Ching 21932* (KUN, PE). **U IRRAWADDY**: **Adung Valley**, 2440 m, 20. 1931, *F. Kingdon-Ward 9507* (BM); **Hpimaw** Pass, 3050–3350 m, 1924, *G. Forrest 25069* (E, K); **Tengchong**, 2440 m, 1924, *G. Forrest 25995* (K, PE); Tengchong, by Guyong River, 1680 m, 1964, *S. K. Wu 6826* (KUN).

23. Ilex hookeri King, J. Asiat. Soc. Bengal, Nat. Hist. 55(2): 266. pl. 14. 1886; Hu, J. Arnold Arbor. 31: 40. 1950; Chen *et al.* in Wu, Raven & Hong, Fl. China 11: 403. 2008. Type: India, Sikkim, Tongloo, 3050 m, Jul. 1882, *J. S. Gamble 10408* [**lectotype here designated**: K (K000669435); isolectotype: K (K000669434)].

Evergreen trees or shrubs, 5–18 m tall, glabrous throughout. Branchlets stout, current year's shoots brown. Leaves persisting even on third year's growth. Petiole 1.2–2 cm long; leaf blade thickly leathery, elliptic or obovate-elliptic, 5–11 cm long, 2–4.5 cm broad, base obtuse or rounded, margin finely serrate, apex acute or shortly acuminate with acumen ca. 5 mm long, midrib impressed adaxially, elevated abaxially. Inflorescences axillary on previous year's growth, fasciculate; flowers 4-merous, white. Individual cymes of male fascicles 1–3-flowered; peduncle ca. 1 mm long; pedicel 1–3 mm long; calyx deeply 4-lobed; corolla rotate, 6 mm across, connate at base; stamens slightly shorter than petals. Female inflorescences umbellate, nearly sessile, with 3–5 fruit; pedicel 6–8 mm long; calyx and corolla as in male flowers. Fruit globose, red, 4–6 mm long and wide. Pyrenes 4, ca. 5 mm long, 2 mm wide, palmately striate and deeply sulcate on back and on sides, endocarp leathery. Figure 47–8: f.

Phenology: flowering in May and June; fruiting from August to October.

Habitats: mixed forests, mossy *Rhododendron* forests, or dense thickets; 2500–3100 m.

Distribution: E Nepal, Sikkim & Darjiling, M Yarlung Zangbo, Bhutan, S Hengduan, and U Irrawaddy (Map 47–23), also in Assam of India.

E NEPAL: **Kosi**, Sankhuwasabha, upper Kashwa Khola, 2800 m, 1981, *Grey-Wilson et al. 4395* (K); **Mechi**, Taplejung, Milke Denda, 2895 m, 1967, *Williams & Stainton 8427* (K).

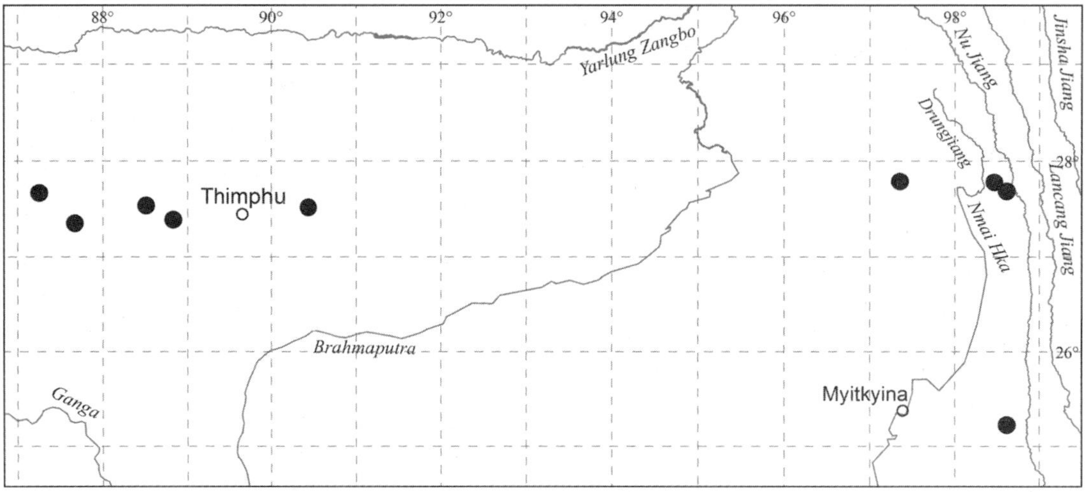

Map 47–23 **Ilex hookeri** King

M YARLUNG ZANGBO: **Yadong**, Chumbi, 28 Jul. 1884. *King's collector 466* [paratype: K (K000669433)]. **BHUTAN**: central Bhutan, 3200 m, 1988, *A. D. Schilling* s. n. (K). **S HENGDUAN**: Gongshan to Drungjiang, 2800 m, 1956, *P. I. Mao 444* (KUN, PE); **U IRRAWADDY**: **Drungjiang**, Taron-Taru Divide, Bucahwang Valley, 2500 m, 1938, *T. T. Yü 20120* (KUN, PE); **Seinghku Valley**, 2895 m, 1926, *F. Kingdon-Ward 6860* (K) ; **Tengchong**, Langya Shan, 2800 m, 1964, *S. K. Wu 7146* (KUN).

24. Ilex delavayi Franch., J. Bot. (Morot) 12: 255. 1898; Hong, Phytotaxa 230(2): 156. 2015. Type: "Chine Occidentale, Yunnan, dans les bois de Kou toui", 20 Apr. 1888, *Delavay 4414* (holotype: P).

Ilex delavayi Franch. var *exalata* H. F. Comber, Notes Roy. Bot. Gard. Edinburgh 18: 44. 1933. Type: China, Yunnan: N'Maikha- Salween Divide, 26°10′N, 2440 m, Jan. 1919, *G. Forrest 18093* (holotype: E; isotype: K).

Ilex delavayi Franch. var. *comberiana* S. Y. Hu, J. Arnold Arbor. 31: 48. 1950. Type: China, Yunnan, ShiKu. *J. F. Rock 9575* (holotype: A).

Ilex delavayi Franch. var. *linearifolia* S. Y. Hu, J. Arnold Arbor. 31: 48. 1950. Type: China, Yunnan: Li-Kiang (Lijiang), *R. C. Ching 21989* (holotype: A; isotype: KUN).

Ilex delavayi Franch. var. *muliensis* W. P. Fang & Z. M. Tan, J. Sichuan Univ. Nat. Sci. Ed. 1983 (2): 81. pl. 6. fig. 2. 1983. Type: China, Sichuan: Muli, Mahuanggou, 3520 m, 25 Jul. 1978, *Q. S. Zhao & Y. B. Yang 6817* (holotype: SZ).

Evergreen shrubs or small trees up to 6 m tall. Branchlets verrucose, glabrous to densely puberlulent. Petiole 1–1.5 cm long; leaf blade subleathery, elliptic or elliptic-lanceolate, (2.5–) 4–5 (–8.5) cm long, (0.7–) 1–2 (2.7) cm broad, base acute or cuneate, margin crenulate-serrate, teeth callose and nigrescent, apex obtuse or acute, midrib impressed adaxially, prorninent abaxially, lateral veins 5 or 6 pairs. Inflorescences on previous year's growth, fasciculate; flowers always 4-merous. Male cymes 1–3-flowered, the fascicles often appearing umbel-like; peduncle up to 3.5 mm long; individual cymes 1–3-flowered; peduncle ca. 1 mm long; pedicel 1–2 mm long. Calyx ca. 2.5 mm across, deeply lobed, lobes ovate-deltoid, ca. 1 mm long, 1.3 mm broad; corolla green or pale yellow, rotate, ca. 5 mm across, petals obovate, ca. 2 mm long; stamens shorter than petals. Fascicles of female inflorescences 2–5-flowered; pedicel 2–4 mm long. Fruit red, globose, ca. 5 mm in diameter. Pyrenes 4, oblong, 3.5–4.5 mm long, 2–2.5 mm wide, palmately striate and sulcate.

Phenology: flowering from April to June; fruiting in September and October.

Habitats: dense forests, bamboo forests, open thickets, on granites; 2400–4000 m.

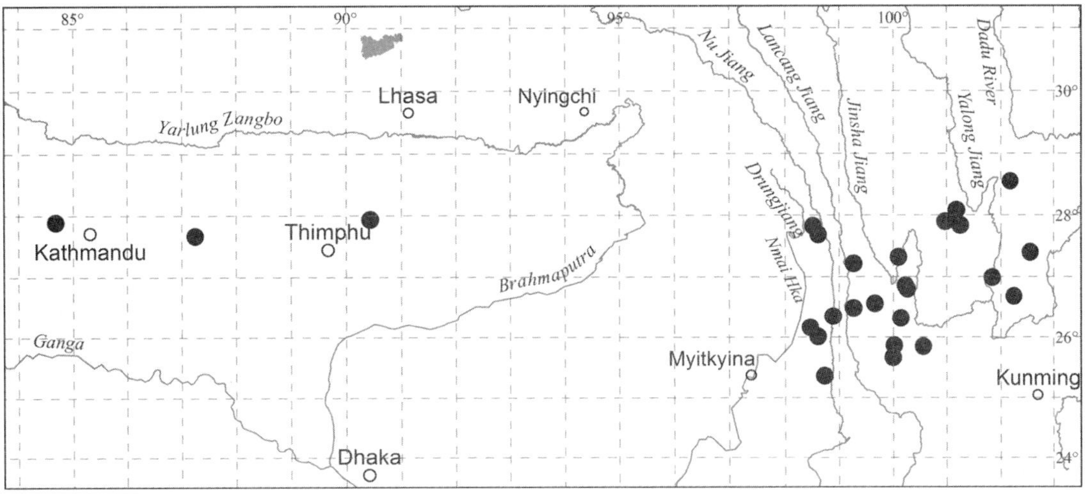

Map 47–24: **Ilex delavayi** Franch.

Distribution: disjunct between S Hengduan/U Irrawaddy and Bhutan/E Nepal (Map 47–24); also in other parts of Sichuan (Jinyang) and C Yunnan of China.

E NEPAL: **Kosi**, Sankhuwasabha, 1990, *M. Minaki et al. 9040296* (E); Sankhuwasabha, Saldim Khola Valley, 3000 m, 1991, *D. G. Long et al. 680* (E). **BHUTAN**: **Upper Mo Chu** Distr., S side of Pari La, between Gasa and Kohina, 3390 m, 1984, *I. W. Sinclair & D. G. Long 5060* (E, K). **S HENGDUAN**: **Binchuan**, Kuanshang, 1942, *H. C. Wang 2016* (PE); **Cawarong**, Salween-Drungjiang Divide, 3050–3350 m, 1921, *G. Forrest 23040* (K); **Dali**, Canshan, Huadianba, 3100 m, 1981, *S. B. E. C. 0917* (KUN); **Gongshan**, Champutung, 3500 m–3700 m, 1940, *K. M. Feng 7837* (KUN, PE); **Heqing**, Lianping, Ma'er shan, 3600 m, 1940, *R. C. Ching 23961* (KUN, PE); **Huili**, near the city, 2700–3600 m, 1914, *Handel-Mazzetti 888* (WU); **Jianchuan**, Laojunshan, 1958, *W. T. Wang 466* (KUN); **Lanping**, Laojun Shan, 3510 m, 1960, *Nanshui Beidiao Exped. 9810* (KUN, PE); **Lushui**, Pianma Pass, 3100 m, 1978, *Nujiang Prefecture Exped. 1825* (KUN); **Mianning**, Tuowu, 2400 m, 1959, *Sichuan Econ. Plant Surv. (Xi 59) 1755* (KUN, PE); **Miyi**, Baipo Shan, 2800 m, 1983, *Qinghai-Xizang Exped. 11909* (KUN, PE); **Muli**, Gumatian, 3400 m, 1937, *T. T. Yü 5722* (KUN, PE); Muli, Mt E of Yungning, 1922, *G. Forrest 21235* (PE); **Puge**, Tuomugou, 2500 m, 1959, *Sichuan Econ. Plant Surv. (Liang 59) 5501* (KUN, PE); **Shangri-la** (Zhongdian), Habaxue Shan, 2800 m, 1955, *K. M. Feng 21016* (PE); **Weixi**, Wudiping, 3200 m, 1940, *K. M. Feng 4602* (KUN, PE); Weixi, 2900 m, 1934, *H. T. Tsai 59761* (PE); **Yangbi**, Yangjiang, 3400 m, 2009, *Yunnan Exped. 1357* (PE); **Yulong** (Lijiang), *R. C. Ching 20479* (paratype of *I. delavayi* var. *comberiana*: PE); 3350 m, 1913, *G. Forrest 10237* (PE). **U IRRAWADDY**: **Chaungmaw Hka Valley**, 8,000 ft, 1919, *F. Kingdon-Ward 3176* (paratype of *I. delavayi* var. *exalata*: E); Western Spur of **Imaw Bum**, 9,000–10,000 ft, 1919, *F. Kingdon-Ward 3298* (paratype of *I. delavayi* var. *exalata*: E); **Tengchong**, Shweli-Salween Divide, 1924, *G. Forrest 25365* (PE).

25. Ilex nothofagifolia Kingdon-Ward, Gard. Chron. Ser. 3. 81: 195. 1927 & Pl. Hunting, 223. 1930; Hu, J. Arnold Arbor. 31: 48. 1950; Hong, Phytotaxa 230(2): 162. 2015. Type: Myanmar, Seinghku Wang (Advance Base), 2740 m, 8 Jun. 1926, *F. Kingdon-Ward 6878* (lectotype designated by Hong, 2015: K).

Ilex intricata Hook. f. var. *oblata* W. E. Evans, Notes Roy. Bot. Gard. Edinburgh 13: 163. 1921. ≡ *Ilex oblata* (W. E. Evans) H. F. Comber, Notes Roy. Bot. Gard. Edinburgh 18: 55. 1933. Type: Upper Burma, Lai Kam Valley, 2438 m, 19 Sept. 1919, *R. Farrer 1299* (holotype: E).

Evergreen shrubs or small trees 0.5–8 m tall. Branchlets densely covered with longitudinal rows of brown corky warts. Leaves persisting even on three year old growth; petiole slender, 4–5 mm long; leaf blade chartaceous, broadly elliptic, rarely ovate or obovate, 0.7–1.4 cm long, 0.6–1 cm broad, base obtuse, margin serrate, with 4–7 teeth on each side, apex obtuse but cuspidate, midrib impressed adaxially, slightly elevated abaxially, lateral veins 3 or 4 pairs. Inflorescences on previous year's growth, small, 1–3-flowered fascicles; individual cymes 1-flowered. Male flowers: pedicel ca. 4 mm long; calyx ca. 2 mm across, deeply 4-lobed, glabrous; corolla subrotate, white, ca. 4 mm across. Infructescence usually with one fruit, rarely 2–4 together. Fruiting pedicel 3–4 mm long; fruit globose, red when mature, 3–5 mm in diameter. Pyrenes 4, broadly ellipsoid, ca. 3 mm long, 1.5–2 mm wide, longitudinally 3–4-striate, almost smooth, endocarp thickly leathery. Figure 47–9: e-h.

Phenology: flowering in May and June; fruiting from September to December.

Habitats: common in evergreen broad-leaved forests or conifer forests; 1900–3000 m.

Distribution: Yarlung Zangbo- Brahmaputra, S Hengduan, and U Irrawaddy (Map 47–25); also in S Yunnan of China, other parts of Myanmar (Kachin State), and N Vietnam.

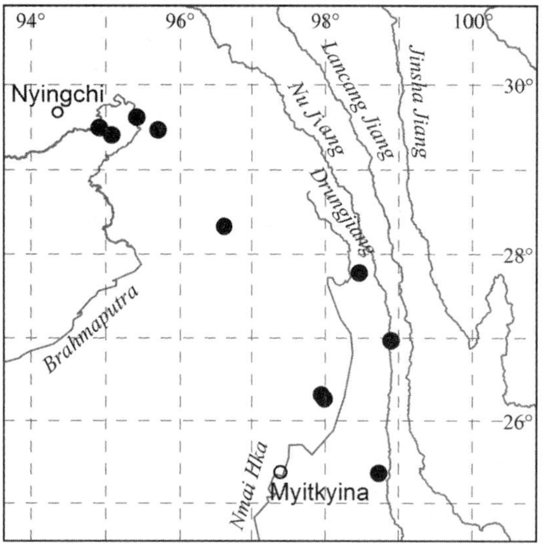

Map 47–25: **Ilex nothofagifolia** Kingdon-Ward

YARLUNG ZANGBO –BRAHMAPUTRA: **Bangxing**, 1900 m, 1982, *S. Z. Cheng & B. S. Li 3322* (PE) ; **Delei** Valley, 2440–2743 m, 1928, *F. Kingdon-Ward 8167* (K); **Gutang**, Xingkai to Jiangbietang, 2200–2500 m, 1982, *B. S. Li & S. Z. Cheng 1083* (PE); **Lage** to Hami, 2200 m, 1974, *Qinghai-Xizang Exped. 74–3987* (KUN, PE); Lage to Paiqu (Doxong), 2600 m, 1974, *Qinghai-Xizang Exped. 2686* (PE). **S HENGDUAN**: **Fugong**, Yaping, 2467m, 2004, *Gaoligong Shan Biodiversity Surv. 20262* (E); **Gongshan**, Gongshan to Drungjiang, 2200– 2400 m, 1982, *Qinghai-Xizang Exped. 8252* (PE); **Lushui**, N' Maikha-Salween Divide, 2134–2440 m, 1919, *G. Forrest 17847*(E, K). **U IRRAWADDY**: **Tengchong**, Shweli-Salween Divide, 2134–2440 m, 1924, *G. Forrest 24227* (K).

26. Ilex intricata Hook. f., Fl. Brit. India 1: 602. 1875; Hu, J. Arnold Arbor. 31: 49. 1950; Chen *et al.* in Wu, Raven & Hong, Fl. China 11: 401. 2008; Hong, Phytotaxa 230(2): 161. 2015. Type: Sikkim, Lachen, 3050–3350 m, 29 May 1849, *J. D. Hooker* s. n. [**lectotype here designated**: K (K000669511)]; loc. eodem, 3350 m, 29 May 1849, *J. D. Hooker* s. n. (paratype: K).

Ilex xizangensis Y. R. Li, Acta Bot. Yunnan. 3: 353. 1981. Type: China, Xizang (Tibet), Mêdog, Nage, 3200 m, 31 Jun. 1974, *Qinghai-Xizang Exped. 74–3852* (holotype: KUN; isotype: PE).

Evergreen shrubs, prostrate, glabrous throughout. Branchlets occasionally verrucose. Leaves small, crowded, persisting even on fourth year's growth. Petiole 1–3 mm long; leaf blade leathery, elliptic or obovate-elliptic, 1–3 cm long, 0.5–1.2 cm broad, base cuneate, margin serrate, with 3–6 teeth on each side, apex obtuse or rounded, midrib deeply impressed adaxially, prominent abaxially, lateral veins 2–4 pairs. Inflorescences on previous year's growth, small, 1–3-flowered fascicles, individual cymes 1-flowered; flowers 4-merous. Male flowers: pedicel ca. 2 mm long; calyx ca. 2 mm across, deeply 4-lobed; corolla rotate, ca. 5 mm across. Female pedicel ca. 1 mm long. Fruiting pedicel 2–3 mm long; fruit globose, red, ca. 5 mm in diameter. Pyrenes 4, oblong, ca. 4 mm long, ca. 2 mm wide, abaxially palmately striate and sulcate, striate and slightly rugose on sides, endocarp woody.

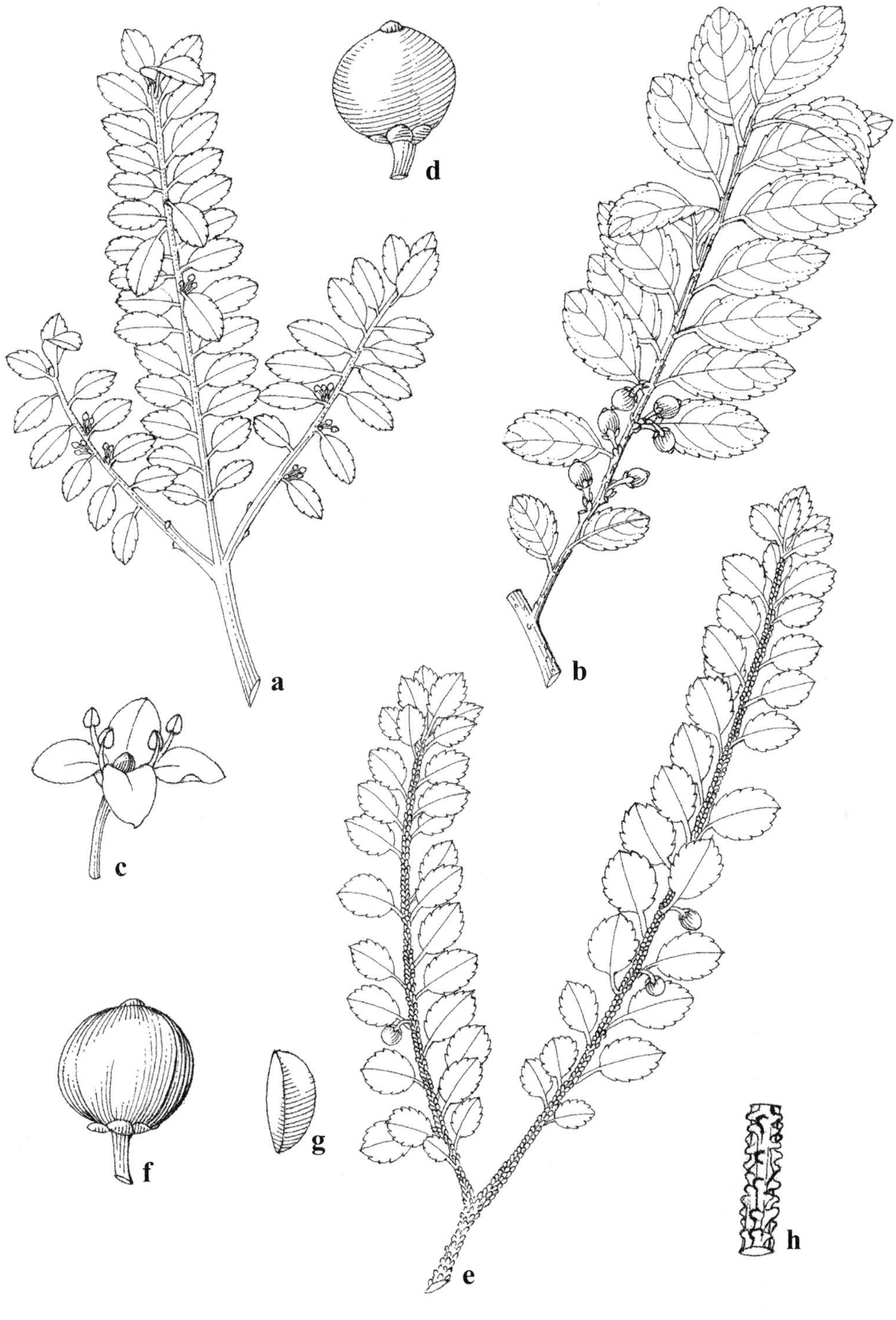

Figure 47–9: **a–d**, **Ilex intricat**a Hook. f.: **a**, male flowering branch; **b**, fruiting branch; **c**, male flower; **d**, fruit. **e–h**, **Ilex nothofagifolia** Kingdon-Ward: **e**, fruiting branch; **f**, fruit; **g**, pyrene; **h**, a section of branch, showing corky verruculose ridges. From *Fl. Reipubl. Popularis Sin*. 45(2), 1999.

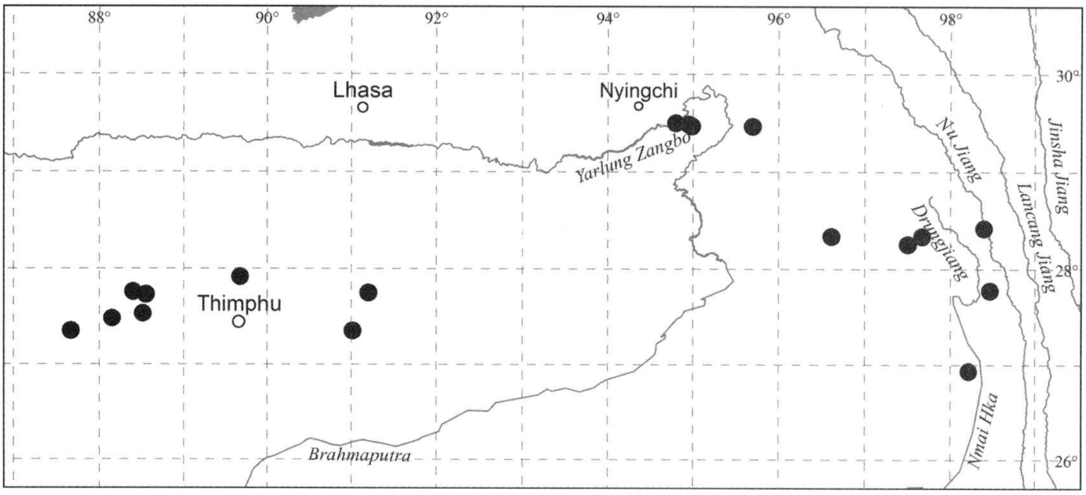

Map 47–26: **Ilex intricata** Hook. f.

Figure 47–9: a-d.

Phenology: flowering in May and June; fruiting from July to November.

Habitats: conifer (mainly *Abies*) forests, often on rocks; 2900–3960 m.

Distribution: **endemic**: E Nepal, Sikkim & Darjiling, Bhutan, L Yarlung Zangbo, Yarlung Zangbo- Brahmaputra, S Hengduan, and U Irrawaddy (Map 47–26).

E NEPAL: **Mechi**, Taplejung, Yalung Ridge, 3350–3660 m, 1981, *A. D. Schilling 2529* (K). **SIKKIM & DARJILING**: **Sikkim**, Jongri, 3350 m, 1875, *C. B. Clarke 25882 A* (K) & *25882B* (K); Bakhim bungalow to Phedang, 3000–3700 m, 1983, *B. N. Starling et al. AGSES 11* (K); W Distr., 3200 m, 1992, *D. G. Long et al. 157* (E); Zemu Valley, *G. H. Cave 181* (K). **BHUTAN**: **Bumphaang**, Thrumseng La, 3515 m, 1987, *HEB-APB 204* (K); **Gasa** Dzong, Mo Chu Valley, 3050 m, 1949, *F. Ludlow et al. 17382* (BM); NW of **Mongar**, above Sengor, 3500 m, 1979, *A. J. C. Grierson & D. G. Long 1909* (K); **Tashigang** Distr., Chorten Lora, Pang La, 3300 m, 1994, *S. Bowes Lyon 10099* (E). **L YARLUNG ZANGBO**: **Mainling**, Paiqu, Deyang, 3600 m, 1983, *B. S. Li et al. 5715* (PE); Mainling, Doshong (Doxong) La, 3050 m, 1924, *F. Kingdon-Ward 6249* (K). **YARLUNG ZANGBO –BRAHMAPUTRA**: **Delei** Valley, 1928, *F. Kingdon-Ward 8291* (K); **Gutang**, 3400 m, *B. S. Li & S. Z. Chen 1227* (PE); **Nage**, 3250 m, 1974, *Qinghai- Xizang Exped. 3779* (KUN, PE); **S HENGDUAN**: **Cawarong**, Sichito, 3960 m, 1922, *G. Forrest 22811* (type of *I. intricata* f. *macrophylla* H. F. Comber: E, K); **Gongshan**, Gongshan to Drungjiang, 2900 m, 1982, *Qinghai-Xizang Exped. 8805* (KUN, PE). **U IRRAWADDY**: **Adung Valley**, 3050–3350 m, 1931, *F. Kingdon-Ward 9624* (BM); **Nam Tamai Valley**, 3960m, 1937, *F. Kingdon-Ward 13400* (BM); **North Triangle** (Wung Bum, above Ahkail), 3200 m, 1953, *F. Kingdon-Ward 21558* (BM); **Seingku Valley**, 3350 m, 1926, *F. Kingdon-Ward 7044* (K).

27. Ilex sikkimensis Kurz, J. Asiat. Soc. Bengal. Nat. Hist. 44(3): 202. 1875; Li in Wu, Fl. Xizang. 2: 110. 1983; Chen *et al.* in Wu, Raven & Hong, Fl. China 11: 403. 2008. Type: India, Sikkim, Rungbee, 17 Jul. 1875, *G. King* s. n. (holotype: CAL, n. v., photo, K).

Ilex sikkimensis Kurz var. *coccinea* H. F. Comber, Notes Roy. Bot. Gard. Edinburgh 18: 58. 1933. Type: China, Yunnan, Shweli-Salween Divide, 25°40′N, 98°30′E, 3050 m, Dec. 1924, *G. Forrest 26058* (holotype: E).

Evergreen trees, 5–15 m tall. Bark covered with lenticels. Branchlets longitudinally striate-sulcate. Leaves persisting on previous year's growth. Petiole 2–5 cm long, adaxially sulcate, abaxially rugose; leaf blade thinly leathery, elliptic, obovate-elliptic, or oblanceolate, 8–20 cm long, 3–6 cm broad, base rounded or obtuse, mar-

Figure 47–10: **a–c**, **Ilex sikkimensis** Kurz: **a**, fruiting branch; **b**, fruit; **c**, pyrene. **d** & **e**, **Ilex intermedia** Loes.: **d**, flowering branch; **e**, flower. From *Fl. Reipubl. Popularis Sin.* 45(2), 1999.

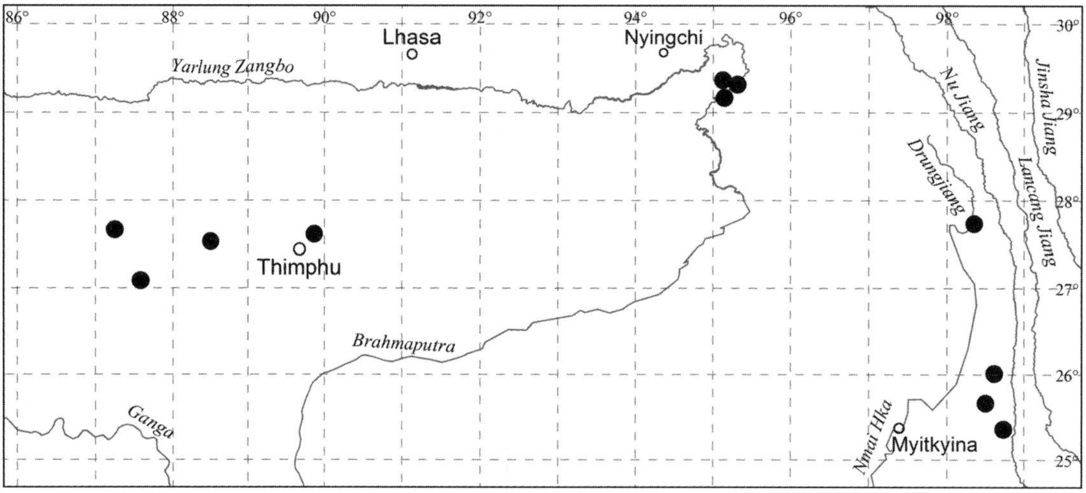

Map 47–27: **Ilex sikkimensis** Kurz

gin serrulate, apex acuminate or acute, glabrous, midrib deeply impressed adaxially, lateral veins 12–15 paired. Inflorescences on previous year's growth, fasciculate, pseudopaniculate or racemelike cymes. Male cymes 3–7-flowered; calyx 2–2.5 mm across, deeply 4-lobed; corolla rotate, ca. 5 mm across, petals ovate or elliptic, connate at base. Female cymes 3-flowered. Fruiting peduncle ca. 1.5 mm long; fruiting pedicel 6–10 mm long, glabrous; fruit red, globose, 4–5 mm in diameter. Pyrenes 4, ovoid- trigonous, 2.5–3 mm long, 1.5–1.8 mm wide, 3–5-striate and sulcate, endocarp stony. **Chromosomes**: 2n=36. Figure 47–10: a-c.

Phenology: flowering from April to July; fruiting from October to December.

Habitats: evergreen broad-leaved forests; 1000–3100 m.

Distribution: E Nepal, Sikkim & Darjiling, Bhutan, Yarlung Zangbo- Brahmaputra, S Hengduan, and U Irrawaddy (Map 47–27); also in Manipur of India, and NW Myanmar (Kachin State).

Notes: About the type of *Ilex sikkimensis*: Kew Herbarium (K) keeps a letter from Dr. U. C. Bhattacharyya (the then deputy director of Calcuta Herbarium, CAL) to Dr. Lucas in Kew. It runs: "Of the three specimens of *I. Sikkimensis*, the one bearing acc. no. 84257, is from Senchul, near Darjeeling, which appears to have been collected by T. Anderson and bears an annotation in Kurz's hand. The other two are George King's collections from Rungbee in 1875 and Tongloo in 1876, respectively. The Rungbee specimen bears King's annotation. We have not been able to trace the type specimen of *Ilex sikkimensis* in CAL." Among the three collections, the collecting date of Anderson's one is not clear, but S. Andrews' note is "14 / 9 / 88". Thus, only the specimen from Rungbee could be the type.

E NEPAL: **Kosi**, Sankhuwasabha, Arun Valley, 1967, *J. D. A. Stainton 5937* (BM); **Mechi**, Taplejung, Tamur Valley, 2286 m, 1967, *J. D. A. Stainton 5881* (BM). **SIKKIM & DARJILING**: **Sikkim**, Bakhim to Yoksum, ca. 2600 m, 1983, *B. N. Starling et al. (AGSES) 32* (K); Tongloo, 3050 m, 1875–1876, *G. King* s. n. (K). **BHUTAN**: **Punakha** Distr., 2620 m, 1982, *A. J. C. Grierson & D. G. Long 4600* (E, K). **YARLUNG ZANGBO -BRAHMAPUTRA**: **Baibung**, 2300–2400 m, 1983, *B. S. Li & S. Z. Chen 4411* (PE); **Hami**, 2080 m, 1974, *Qinghai-Xizang Exped. 74–4086* (KUN, PE); **Mêdog**, Madê, 850–1200 m, 1983, *B. S. Li & S. Z. Chen 2654* (PE). **S HENGDUAN**: **Lushui**, Pianma, 2790 m, 2005, *Gaoligong Shan Biodiversity Survey 23309* (E). **U IRRAWADDY:** W side of **Drungjiang**, 2000 m, 1959, *K. M. Feng 24304* (KUN); **Tengchong**, Shweli-Salween Divide, 2440–2743 m, 1918, *G. Forrest 17573* (E, K).

28. Ilex intermedia Loes., Nova Acta Acad. Caes. Leop. – Carol. Nat. Cur. 78: 273. 1901; Hu, J. Arnold Arbor. 31: 78. 1950; Hong, Phytotaxa 230 (2): 160. 2015. Type: China, Hupeh (Hubei), S Patung (Badong), *A. Henry 5549* [**lectotype here designated**: K; isolectotype: P, W].

Ilex dunniana H. Lév., Repert Spec. Nov. Regni Veg. 9: 458. 1911. Type: China, Kouy-Tcheou (Guizhou), Long-ly, May 1908, *J. Cavalerie 3000* (holotype: E; isotype: K).

Evergreen trees up to 25 m tall. Branchlets puberulent or glabrescent; previous year's growth glabrous, 3–5 mm in diameter, with persistent leaves. Petiole 1.1–1.6 cm long; leaf blade leathery, oblong or oblong-elliptic, 6–13 cm long, 2–4.5 cm broad, base cuneate, obtuse or rarely rounded, margin crenulate or coarsely serrate, apex obtuse, acute or shortly acuminate with acumen 3–5 mm long, glabrous or puberulent near base and along midrib adaxially, midrib flat or slightly impressed adaxially, elevated abaxially, lateral veins 5–8 pairs. Male inflorescences fasciculate or pseudopaniculate, individual cymes 1–3-flowered; flowers 4-merous; calyx ca. 1.5 mm across, glabrous; corolla rotate, ca. 6 mm across; stamens as long as petals. Female inflorescences racemelike, central axis 3–5 mm long, extending to 4–8 mm long in fruit, cymes 1-fruited. Fruiting pedicel 5–8 mm long, usually glabrous, rarely puberulent; fruit red, globose or oblate, 5–6 mm in diameter, tuberculate. Pyrenes 4, palmately striate, endocarp stony. Figure 47–10: d & e.

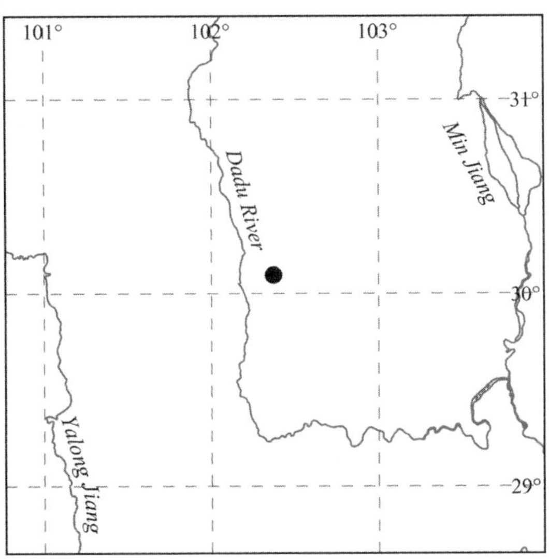

Map 47–28: **Ilex intermedia** Loes.

Phenology: flowering from April to June; fruiting from August to October.

Habitats: fo -rests or thickets; 600–2200 m.

Distribution: S Hengduan (Map 47–28); also in other parts of China (Chongqing, Guizhou, W Hubei, Hunan, N Jiangxi, NE Yunnan, and other parts of Sichuan).

S HENGDUAN: **Tianquan**, Fengzihe, below Xianghuagang, 1500 m, 1933, *T. H. Tu 4727* (PE).

29. Ilex cyrtura Merr., Brittonia 4: 101. 1941; Hu, J. Arnold Arbor. 31: 77. 1950, p. p., excl. specim. Guangxi & Guizhou; Chen *et al.* in Wu, Raven & Hong, Fl. China 11: 416. 2008, p. p., excl. specim. Guangdong, Guangxi and Guizhou; Hong, Phytotaxa 230(2): 155. 2015. Type: Upper Burma [Myanmar], Adung Valley, Apr. 1931, *F. Kingdon-Ward 9427* [holotype: A (photo, K); isotype: BM].

Ilex gracilis C. J. Tseng, Bull. Bot. Res. Harbin 1 (1–2): 28. 1981. Type: China, Yunnan, Gongshan, Taron-taru Divide, Valley of Bucahwang, 1600 m, 6 Sept. 1938, *T. T. Yü 20137* (holotype: PE).

Evergreen trees up to 12 m tall. Branchlets glabrous or very sparsely pubescent; leaves persistent on previous year's growth. Petiole 0.8–1.2 cm long, puberulent; leaf blade subleathery, oblong, elliptic or obovate-elliptic, 6–11 cm long, 2–4 cm broad, base obtuse or cuneate, margin serrate, apex long-caudate with acumen 1.5–2.5 cm long, midrib impressed and sparsely puberulent adaxially, elevated abaxially, lateral veins 7- or 8-paired. Inflorescences on previous year's growth, fasciculate, pubescent; flowers 4-merous. Male flowers: pedicel ca. 1 mm long; calyx deeply 4-lobed, sparsely ciliate; corolla yellow, rotate, ca. 6 mm across, petals oblong, sparsely ciliate, slightly connate at base. Infructescence usually a fascicle of 3 fruits; axis 1–5 mm long. Fruiting pedicel 2–3 mm long, hispidulous; fruit globose, 4–5 mm in diameter. Pyrenes 4, broadly ellipsoid, ca. 3.5 mm long, 2.5–3 mm wide, palmately striate, esulcate; endocarp woody. Figure 47-11, a-c.

Phenology: flowering from April to June;

fruiting from September to December.

Habitats: forests, particularly evergreen forests, or thi-ckets; 1200–3000 m.

Distribution: **endemic**; S Hengduan and U Irrawaddy (Map 47–29).

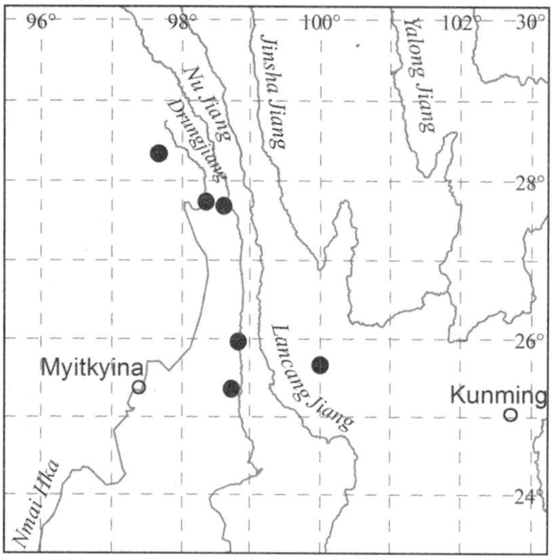

Map 47–29: **Ilex cyrtura** Merr.

S HENGDUAN: **Lushui**, 2743–3050 m, 1924, *G. Forrest 26057* (K); **Yangbi**, 1984, *1984 Sino-Amer. Bot. Exped. 526* (E); **U IRRAWADDY: Drungjiang** Township, 1700 m, 1982, *Qinghai- Xizang Exped. 9563* (PE); Drungjiang Valley, Pre to Lungen, 1700 m, 1938, *T. T. Yü 19486* (KUN, PE); **Tengchong**, Shweli-Salween Divide, 2440 m, 1924, *G. Forrest 24048* (K); Tengchong, Shweli-Salween Divide, *G. Forrest 17716* (K).

30. Ilex corallina Franch., Bull. Soc. Bot. France 33: 452. 1886; Hu, J. Arnold Arbor. 31: 64. 1950; Hong, Phytotaxa 230(2): 154. 2015. Type: China, Yunnan, Dali, Tamitang, 17 Apr. 1884, *Delavay 927* [holotype: P (P03272859); isotypes: E, K, KUN, P)].

Ilex corallina Franch. var. *pubescens* S. Y. Hu, J. Arnold Arbor. 31: 66. 1950. Type: China, Yunnan, Salween Valley, Latsa, 2000 m, 7 Dec. 1938, *T. T. Yü 22890* (holotype: A; isotypes: KUN, PE).

Ilex wangiana S. Y. Hu, J. Arnold Arbor. 31: 54. 1950. ≡ *Ilex corallina* Franch. var. *wangiana* (S Y. Hu) Y. R. Li, Bull. Bot. Res. Harbin 5(1): 16. 1985. Type: China, Yunnan, Weixi (Wei-se), Kang-pu, 1932 m, Jul. 1935, *C. W. Wang 64164* (holotype: A; isotype: KUN, PE); Weixi, Mekong Valley, 27°50′N, Nov. 1917, *G. Forrest 15487* (not 15437) (paratype: E).

Ilex macrostigma C. Y. Wu ex Y. R. Li, Bull. Bot. Res. Harbin 5(1): 20. pl. 6–2. 1985. **syn. nov**. Type: China, Yunnan, Heqing, Songqui (not Songsong), Luhe, 29 Apr. 1939, *K. M. Feng 858* (not 458) (holotype: KUN; isotypes: KUN).

Evergreen small trees or shrubs, 1–12 m tall, glabrous throughout. Branchlets slender; leaves persisting on third year's growth. Petiole 4–9 mm long; leaf blade leathery, ovate, ovate-elliptic or ovate-lanceolate, 5–13 cm long, 1.5–5 cm broad, base rounded or obtuse, margin crenate-serrate or spinosely toothed in juvenile form, teeth nigrescent, apiculate, apex acute or shortly acuminate with acumen up to 1 cm long, midrib impressed, glabrous or rarely puberulent adaxially, elevated abaxially, lateral veins 7–10-paired. Inflorescences on previous year's growth, fasciculate, subsessile; flowers 4-merous. Individual male cymes 1–3-flowered; peduncle ca. 1 mm long; pedicel ca. 2 mm long; calyx ca. 2 mm across, deeply 4-lobed, lobes ciliate; corolla pale green, rotate, 6–7 mm across, glabrous. Infructescences a fascicle of 4–8 fruits; peduncle absent. Fruiting pedicel ca. 2 mm long, glabrous or sparsely puberulent; fruit red, globose, 3(–4) mm in diameter. Pyrenes 4, orbicular-trigonous, 2–2.5 mm long, ca. 1.5 mm wide, obscurely striate-sulcate; endocarp woody. Figure 47–11: d-h.

Phenology: flowering in May and June; fruiting from September to December.

Habitats: thickets or forests; 1100–2500 m.

Distribution: N and S Hengduan (Map 47–30); also in other parts of China: Hubei, Guizhou, Chongqing, and other parts of Sichuan and Yunnan

N HENGDUAN: **Pingwu**, Douya to Wangjiawan, 1120 m, 1958, *H. L. Tsiang 10157* (PE); **Wenxian**, Fangjiaba, 750 m, 1958, *Y. Q. He 1342* (PE). **S HENGDUAN**: **Baoxing**, Robi,

Figure 47–11: **a–c**, **Ilex cyrtura** Merr.: **a**, fruiting branch; **b**, fruit; **c**, pyrene. **d–h**, **Ilex corallina** Franch.: **d**, male flowering branch; **e**, fruiting branch; **f**, fruit; **g**, dorsal view of pyrene; **h**, ventral view of pyrene. From *Fl. Reipubl. Popularis Sin.* 45(2), 1999.

1938, *C. Pei 8248* (PE); **Bijiang**, Pihe, 2000 m, 1978, *Z. G. He 78–001* (KUN); **Binchuan**, Niujing to Jizhu Shan, 1946, *T. N. Liou 21813* (PE); **Butuo**, Jiaojihe, 1500 m, 1959, *Sichuan Econ. Plant Surv. (Liang 59) 5785* (KUN, PE); **Heqing**, Jiangying, 1940, *R. C. Ching 24627* (KUN, PE); **Lanping**, 2800 m, 1933, *H. T. Tsai 56220* (PE); **Lushui**, Laowo, 1650 m, 1981, *Z. X. Tang 53* (PE); **Muli**, Bo'ao, 3900 m, 1959, *S. K. Wu 2689* (KUN); **Shangri-la** (Zhongdian), Paiti, 2200 m, 1937, *T. T. Yü 11443* (KUN, PE); **Shimian**, 1955, *C. C. Hsieh 39770* (PE); **Tianquan**, Lianghekou, 1600 m, 1953, *H. L. Tsiang 33793* (PE); **Weixi**, Gonghe, 1600 m, 1956, *P. I. Mao 359* (KUN, PE); **Xichang**, Fushan,1850 m, 1964, *L. N. Zhao 2084* (PE); **Yangbi**, SE of the town, 1940, *R. C. Ching 25241* (KUN, PE); **Yanyuan**, 54 km N of Guabi, 2500 m, 1960, *S. Jiang 5564* (PE); **Yongping**, Yuntai Shan, 1700 m, 1981, *Qinghai-Xizang Exped. 36* (KUN, PE); **Yulong** (Lijiang), Shigu, Xinrong, 1860–2000 m, 1981, *Qinghai-Xizang Exped. 381* (KUN, PE).

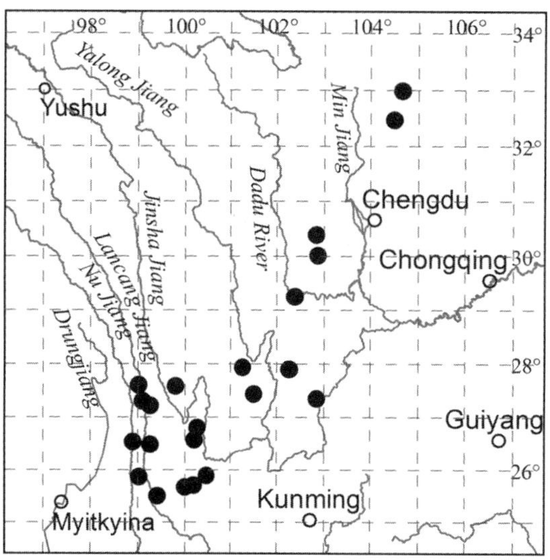

Map 47–30: **Ilex corallina** Franch.

31. Ilex wattii Loes., Nova Acta Acad. Caes. Leop. -Carol. Nat. Cur. 78: 322. (Monog. Aquif. 1: 322) 1901: Hu, J. Arnold Arbor. 31: 76. 1950; Chen *et al.* in Wu, Raven & Hong, Fl. China 11: 416. 2008. Type: India, Manipur, Mao, 2130 m, 22 Feb. 1882, *G. Watt 6165* (**lectotype here designated**: K; isolectotypes: E, P).

Evergreen trees up to 8 m tall, glabrous throughout. Branchlets longitudinally striate; leaves persisting for second year. Petiole 0.8–1 cm long; leaf blade papery or subleathery, elliptic or elliptic-lanceolate, 6–11 cm long, 2–3.5 cm broad, base obtuse, or rounded, margin serrate, apex acuminate, acumen 1–2 cm long, midrib deeply grooved and glabrous adaxially, elevated abaxially, lateral veins 10–11-paired. Inflorescences on previous year's growth, fasciculate; flowers 4-merous. Male cymes 1–3-flowered, nearly sessile; pedicel 1–2 mm long, glabrous; calyx glabrous, ca. 2.5 mm across; corolla rotate, ca. 5 mm across. Female cymes 1-flowered. Fruiting pedicel 1–3 mm long; fruit globose, 6–7 mm in diameter. Pyrenes 4, obovoid-oblong, ca. 4 mm long, 3 mm wide, palmately striate and sulcate, endocarp stony. Figure 47–12: a-c.

Phenology: flowering from February to April; fruiting in July and August.

Habitats: thickets; ca. 2000 m.

Distribution: U Irrawaddy (Map 47–31); also in Manipur of India.

U IRRAWADDY: **Tengchong**, 1924, *G. Forrest 24095* (E, PE); Tengchong, Shweli-Salween Divide, 1918, *G. Forrest 17604* (E).

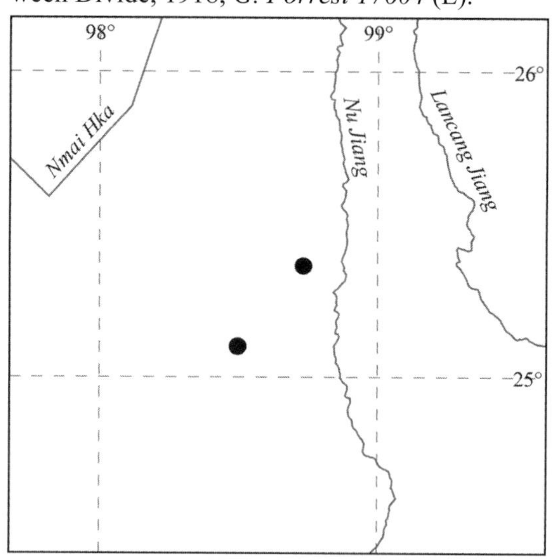

Map 47–31: **Ilex wattii** Loes.

32. Ilex pseudo-odorata Loes., Nova Acta Acad. Caes. Leop.-Carol. Nat. Cur. 78: 285. 1901; Basu in Singh *et al.*, Fl. India 5: 66. 2000.

Type: s. loc., *Wallich 9023* (holotype: K).

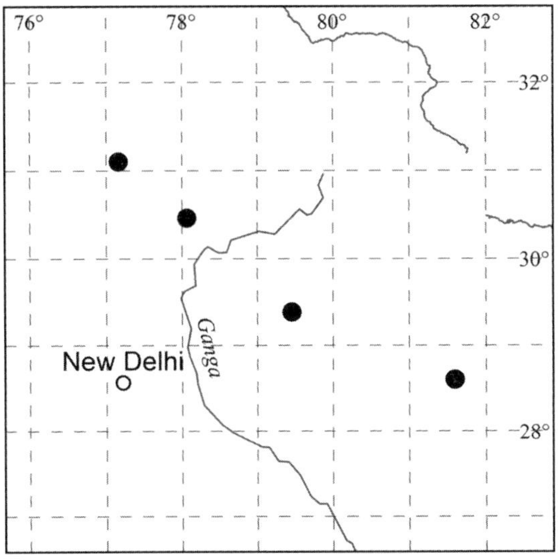

Map 47–32: **Ilex pseudo-odorata** Loes.

Evergreen trees up to 12 m tall, glabrous throughout. Petiole 1–2 cm long; leaf blade elliptic, obvate-elliptic, 8–21 cm long, 2.5–7.5 cm broad, base cuneate, rarely rounded, margin serrate, apex shortly acuminate, less often acute or cuspidate-acuminate, midrib impressed adaxially, lateral veins 8–13-paired. Inflorescences densely fasciculate; flowers 4-merous. Male cymes 7-flowered; peduncle 1–4 mm long; pedicel 3–4 mm long; corolla white. Female inflorescences racemelike; peduncle ca. 8 mm long; pedicel 4–5 mm long. Fruit globose, 3–3.5 mm in diameter.

Phenology: flowering and fruiting from March to July.

Habitats: 600–2000 m.

Distribution: **endemic**; U Ganga & Indus and W Nepal (Map 47–32).

U GNAGA & INDUS: **Garhwall**, *Falconer 358* (K); **Nainital**, 1220 m, *R. Strachey & J. E. Winterbottom 2* (K); **Shimla**, 1830 m, 1844, *M. P. Edgeworth* s. n. (K); **Mussoorie**, 1897, *P. W. Mackinon* s. n. (E, K); **Khunigadh**, 1370 m, 1896, *J. S. Gamble 25780* (K). **W NEPAL**: **Bheri,** Surkhet, Eri Valley, 610 m, 1968, *J. D. A. Stainton 6134* (BM, photo, K).

33. Ilex odorata Buch.-Ham. ex D. Don, Prodr. Fl. Nepal. 189. 1825; Hooker, Fl. Brit. India 1: 599. 1875; Basu in Singh *et al.*, Fl. India, 5: 64. 2000. Type: Nepal (or N India?), Chillong, 12 Apr. 1802, *Buchanan-Hamilton* s. n. (holotype: BM).

Ilex theifolia Wall. ex Hook. f., Fl. Brit. India I: 601. 1875. Type: *Wallich 6391* (*Myrsine*) (holotype: K).

Small evergreen trees, 5–7 m tall. Bark brownish gray, smooth. Branchlets puberulent. Petiole 0.5–1.5 cm long; leaf blade leathery, elliptic or oblong, 3–9.5 cm long, 1.5–3.2 cm broad, base rounded or broadly cuneate, margin obscurely serrate, apex abruptly acuminate

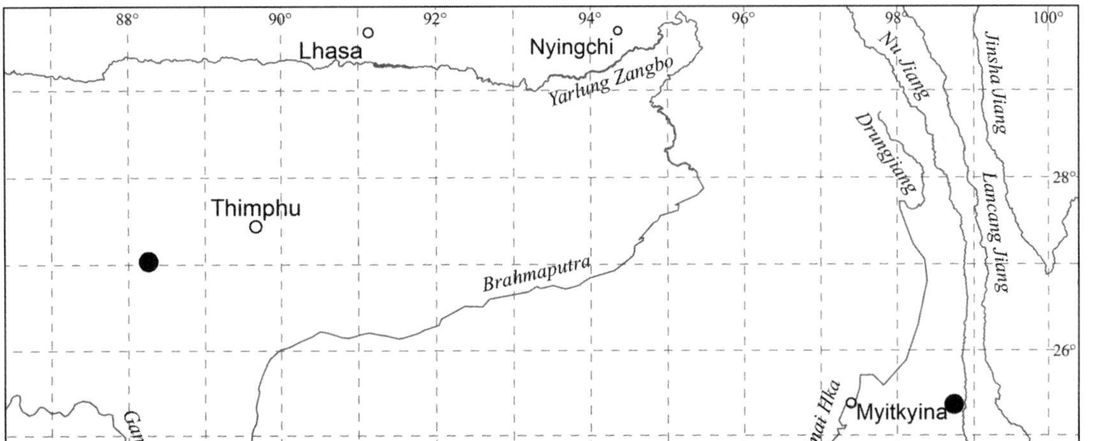

Map 47–33: **Ilex odorata** Buch.-Ham. ex D. Don

with acumen 0.5–1.5 cm long, lateral veins 7–12-paired. Inflorescences on previous year's growth, fasciculate. Male flowers 4- or 5-merous; calyx 1.7–2 mm across, glabrous, lobes ovate; corolla white, petals oblong-lanceolate, 2.5–3 mm long, 1–2 mm broad. Fruiting pedicel 3–5 mm long, puberulent. Fruit globose, 4–5 mm in diameter. Pyrenes 4 or 5, trigonous, 2.7–4 mm long, 1.8–2 mm wide, 3-striate.

Phenology: flowering from April to June; fruiting from October to December.

Habitats: forests, 1000–2400 m.

Distribution: Nepal (?), Sikkim & Darjiling, and U Irrawaddy (Map 47–33); also in NE India (Assam, Meghalaya, Manipur).

Notes: This species is still poorly known at least to the present author. Only a few specimens were available to me; the records from Nepal and those provided by Basu (2000) are not certain.

SIKKIM & DARJILING: **Darjiling**, 1877, *W. Johnston* s. n. (K). **U IRRAWADDY**: **Tengchong**, Shweli-Salween Divide, 2130–2438 m, 1917, *G. Forrest 17547* (W).

34. Ilex medogensis Y. R. Li, Acta Bot. Yunnan. 3(3): 351. fig. 1. 1981. Type: China, Tibet, Mêdog, Miri, 800 m, 4 Sept. 1974, *Qinghai-Xizang Exped. 74–5041* (holotype: KUN; isotype: PE).

Evergreen trees, ca. 8 m tall. Branchlets gray, angular, glabrous, without lenticels. Petiole slender, 1–1.5 cm long, adaxially sulcate, glabrous; leaf blade leathery, glabrous, broadly elliptic or oblong, 10–13 cm long, 4–6 cm broad, base nearly rounded or broadly cuneate, margin serrate, apex caudate-acuminate with acumen 1–1.5 cm long, lateral veins 8–10-paired. Flowers not seen. Infructescences on previous year's growth, fasciculate, nearly sessile, 1–4-fruited (?). Fruiting pedicel 1–2 mm long; persistent calyx 4-lobed; fruit globose, 4–5 mm in diameter. Pyrenes 4, abaxially sulcate.

Phenology: fruiting in September.

Habitats: broad-leaved forests; ca. 800 m.

Distribution: **endemic**; confined to Yarlung Zangbo-Brahmaputra (Mêdog) (Map 47–34).

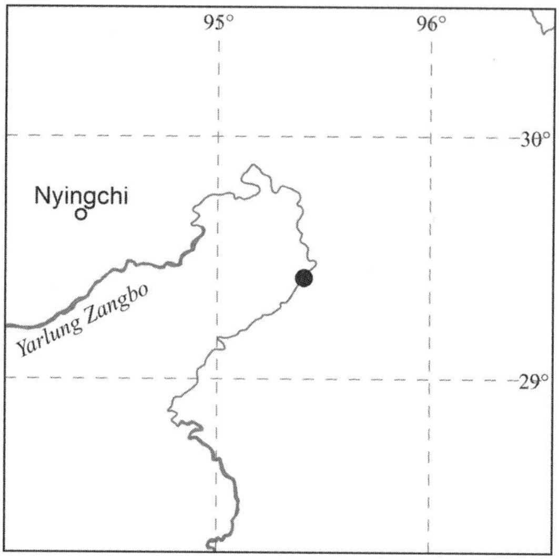

Map 47–34: **Ilex medogensis** Y. R. Li

Sect. 4. Lauroilex S. Y. Hu, J. Arnold Arbor. 31: 214. 1950.

Trees or shrubs. 5–20 m tall. Leaf blade entire. Inflorescences cymose or umbelliform. Flowers 5–7-merous. Pyrenes 5–7, smooth; endocarp leathery.

Two species; one in the Pan-Himalaya.

35. Ilex venulosa Hook. f., Fl. Brit. India 1: 602. 1875; Hu, J. Arnold Arbor. 31: 216. 1950; Chen *et al.* in Wu, Raven & Hong, Fl. China 11: 419. 2008. Type: India, Khasia Mts, *Wallich 4321* (holotype: K).

Ilex venulosa Hook. f. var. *simplicifrons* S. Y. Hu, J. Arnold Arbor. 31: 217. 1950, **syn. nov.** Type: China, Yunnan, hills S of Tengyueh (Tengchong), 25°N, Mar. 1913, *G. Forrest 9800* (holotype: A; isotypes: E, K).

Evergreen trees or shrubs, 5–20 m tall. Branchlets with conspicuous lenticels; leaves persisting on previous year's growth. Petiole robust, 1.3–2.2 cm long; leaf blade leathery, oblong-elliptic or ovate, 10–20 cm long, 3–7 cm broad, base rounded or obtuse, margin entire, apex caudate-acuminate with acumen 2–3 cm long, midrib deeply impressed adaxially, conspicuously elevated abaxially, lateral veins 15–22-paired. Inflorescences on previous year's growth, pseudo-paniculate or rarely fasciculate,

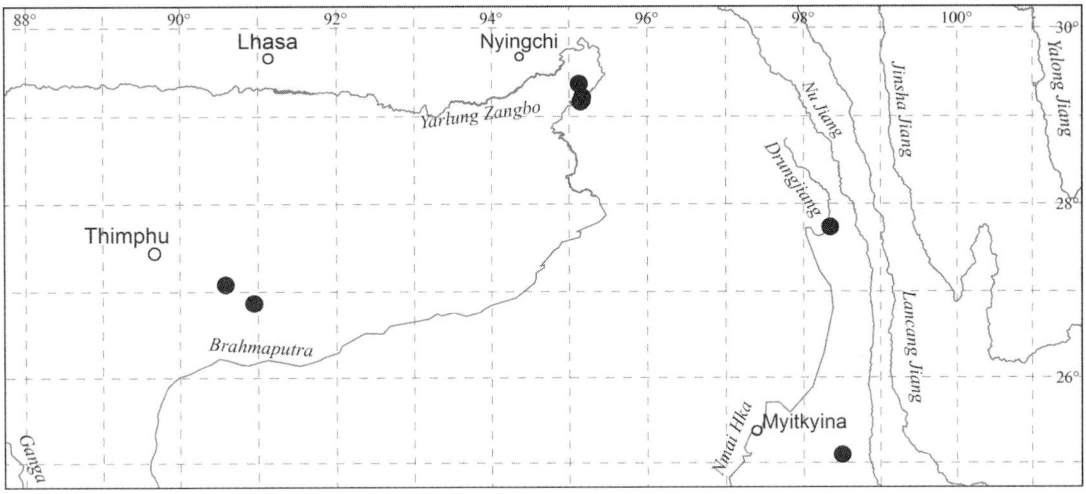

Map 47–35: **Ilex venulosa** Hook. f.

very rarely solitary at base of current year's shoots, central axis 0.4–3 cm long; flowers 5- or 6-merous. Individual branches of male pseudo-panicles cymose, 2–4 times trichotomous; peduncle 1–1.7 cm long; pedicel ca. 2 mm long; calyx ca. 3 mm across; corolla white, ca. 6 mm across. Individual branches of female pseudo-panicles cymose, trichotomous; fruiting peduncle 0.2–1.8 cm long; pedicel 2–3 mm long. Fruit red, globose, 3–4 mm in diameter. Pyrenes 5–7, oblong-ovoid, ca. 2 mm long, 1 mm wide, 3-striate, endocarp leathery. Figure 47–12: d-g.

Phenology: flowering from April to June; fruiting from September to December.

Habitats: evergreen forests, rarely in thickets; 1300–2400 m.

Distribution: Bhutan, Yarlung Zangbo-Brahmaputra, and U Irrawaddy (Map 47–35); also in other parts of N Myanmar (Kachin Hills and Bhamo), S Yunnan of China, and Haga Hills and Meghalaya of India.

Notes: The peduncle of femal inflorescences in this species was found to vary in length continuously between 2 and 18 mm. The isotype in K has peduncle 2–6 mm long, not just 2 mm long as the protologue of Hu's variety described.

BHUTAN: **Gaylegphug**, 1680 m, 1982, *A. J. C. Grierson & D. G. Long 4025* (E, K); **Tongsa**, 1660 m, 1982, *A. J. C. Grierson & D. G. Long 4252* (E, K). **Y YARLUNG ZANGBO-BRAHMAPUTRA**: **Baibung**, Didong, Xirang, 1800 m, 1983, *B. S. Li & S. Z. Cheng 4083* (PE): Baibung, Gelin, 1800 m, 1983, *B. S. Li et al. 3702* (PE); **Damu**, 1800 m, 1993, *H. Sun & Z. K. Zhou 4251* (KUN); **Dexing**, 1700 m, 1993, *H. Sun & Z. K. Zhou 6062* (KUN); **Hami**, 1800 m, 1992, *H. Sun & Z. K. Zhou 0916* (KUN). **U IRRAWADDY: Drungjiang** Township, 1300–1400 m, 1982, *Qinghai-Xizang Exped. 9214* (KUN, PE); **Tengchong**, 1830–2130 m, 1924, *G. Forrest 26168* (PE).

Sect. 5. Pseudoaquifolium S. Y. Hu, J. Arnold Arbor. 31: 218. 1950.

Trees or shrubs, 2–9 (–30) m tall. Leaf blade entire, rarely crenulate-serrate. Inflorescences fasciculate. Flowers (4–) 6–8-merous. Pyrenes (4–) 6–8, striate, esulcate, rarely sulcate.

More than 35 species in five series according to Hu's system; six species in the Pan-Himalaya (2 endemic).

36. Ilex longecaudata H. F. Comber, Notes Roy. Bot. Gard. Edinburgh 18: 54. 1933; Hu, J. Arnold Arbor. 31: 245. 1950; Chen *et al.* in Wu, Raven & Hong, Fl. China 11: 426. 2008. Type: China, Yunnan, Shweli-Salween Divide, 25°20′N, 3050 m, Jul. 1917, *G. Forrest 15657* (holotype: E; isotypes:A, K).

Ilex longecaudata H. F. Comber var. *glabra* S. Y. Hu, J. Arnold Arbor. 31: 246. 1950. **syn.**

Figure 47–12: **a–c**, **Ilex wattii** Loes.: **a**, fruiting branch; **b**, fruit; **c**, pyrene. **d–g**, **Ilex venulosa** Hook. f.: **d**, fruiting branch; **e**, male flower; **f**, fruit; **g**, pyrene. From *Fl. Reipubl. Popularis Sin.* 45(2), 1999.

nov. Type: China, Yunnan, Ping-pien-hsien (Pingbian CO), *H. T. Tsai 61730* (holotype: A; isotypes: KUN, PE).

Evergreen trees up to 30 m tall. Branchlets slender, previous year's growth with evident lenticels and leaves. Petiole slender, 6–12 mm long; leaf blade leathery, ovate-elliptic or elliptic, 4–9 cm long, 1–2.5 cm broad, base obtuse or rounded, margin entire, apex long-caudate with acumen 1–2 cm long, punctate abaxially, midrib impressed and puberulent adaxially, elevated abaxially, lateral veins 7–8-paired. Inflorescences on previous year's growth, puberulent; flowers 4- or 5-merous. Male inflorescences pseudo- paniculate, central axis 0.4–1.5 cm long, puberulent; individual branches cymose, 3-flowered, rarely 1-flowered; peduncle ca. 4 mm long; pedicel ca. 2 mm long; calyx ca. 2.5 mm across; corolla green-yellow. Female inflorescences pseudo-paniculate or fasciculate; fruiting peduncle 4–6 mm long; pedicel 3–5 mm long. Fruit red, globose, 3–4 mm in diameter. Pyrenes 5, ellipsoid, ca. 2.5 mm long, 1.3 mm wide, endocarp leathery. Figure 47–13: a-c.

Phenology: flowering in May and June; fruiting from September to December.

Habitats: evergreen forests; 1300–3100 m.

Distribution: Yarlung Zangbo-Brahmaputra, S Hengduan, and U Irrawaddy (Map 47–36); also in S Yunnan of China.

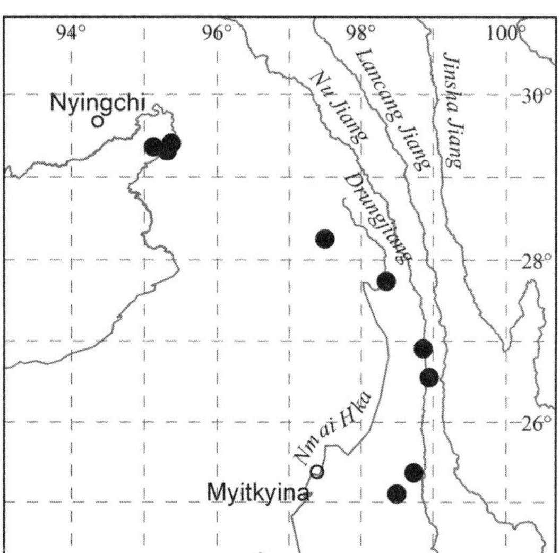

Map 47–36: **Ilex longecaudata** H. F. Comber

YARLUNG ZANGBO–BRAHMAPUTRA: **Dexing**, Dege, 2200–2400 m, 1983, *B. S. Li & S. Z. Cheng 2837* (PE); **Gedan**, Xinkai, 2200 m, 1993, *H. Sun & Z. K. Zhou 4807* (KUN); **Hami**, 2080 m, 1974, *Qinghai-Xizang Exped. 74–4032* (PE); **Mêdog**, 80 k, 2200 m, 2012, *D. Y. Hong et al. H12009* (PE). **S HENGDUAN**: **Bijiang**, Biluoxue Shan, 2300 m, 1978, *Bijiang Exped. 676* (KUN); **Fugong**, Shangbai, 2300 m, 1982, *Qinghai-Xizang Exped. 7090* (KUN, PE). **U IRRAWADDY**: **Drungjiang**, 1300 m, 1979, *s. coll.* s. n. (KUN); **Nam Tamai Valley**, 1830–2134m, 1937, *F. Kingdon-Ward 13505* (BM); **North Triangle** (Wang Bum), 2440 m, 1953, *F. Kingdon-Ward 20873* (BM); **Tengchong**, 2200 m, 1983, *L. Qin 770586* (KUN).

37. Ilex wilsonii Loes., Nova Acta Acad. Cass. Leop.-Carol. Nat. Cur. 89: 287. 1908; Hu, J. Arnold Arbor. 31: 252. 1950. Type: China, W Hupeh (Hubei), Fang (Fang CO?), Jun. 1901. *E. H. Wilson 2101 A* (holotype: A; isotype: K).

Evergreen trees up to 10 m tall, glabrous throughout. Branchlets nearly terete, but slightly longitudinally rugose, or angular; leaves persisting for three years. Petiole 5–9 mm long; leaf blade thickly leathery, ovate or obovate-oblong, 3–7 cm long, 1.5–3.5 cm broad, base obtuse or rounded, margin entire, apex abruptly caudate, acumen 6–10 mm long and tip obtuse, midrib flat adaxially, elevated abaxially, lateral veins 7–8-paired. Inflorescences on previous year's growth, fasciculate; flowers 4-merous. Individual branches of male fascicles cymose or umbelliform, 3–5-flowered; peduncle 3–8 mm long; pedicel 1–1.5 mm long; calyx ca. 1.5 mm across, lobes ciliate; corolla white, 4–5 mm across. Individual branches of female fascicles 1–3-flowered; fruiting pedicel 4–7 mm long. Fruit globose, ca. 4 mm in diameter. Pyrenes 4, ovoid-trigonous, ca. 3 mm long, 1.5 mm wide, 3-striate but not sulcate, endocarp leathery. Figure 47–13: d-g.

Phenology: flowering in May and June; fruiting in August.

Habitats: forests; ca. 1300 m.

Distribution: S Hengduan (Map 47–37); also in E and W China: Anhui, Guizhou, Hubei,

Figure 47–13: **a-c**, **Ilex longecaudata** H. F. Comber: **a**, fruiting branch; **b**, fruit; **c**, ventral and lateral view of pyrene. **d–g**, **Ilex wilsonii** Loes.: **d**, male branch; **e**, fruiting branch; **f**, fruit; **g**, ventral view of pyrene. From *Fl. Reipubl. Popularis Sin.* 45(2), 1999.

Hunan, Jiangxi, other parts of Sichuan, Taiwan, and Zhejiang.

S HENGDUAN: **Tianquan**: Xingou, 1300 m, 1959, *Sichuan Econ. Plant Surv. (Ya'an) (59) 710* (KUN, PE).

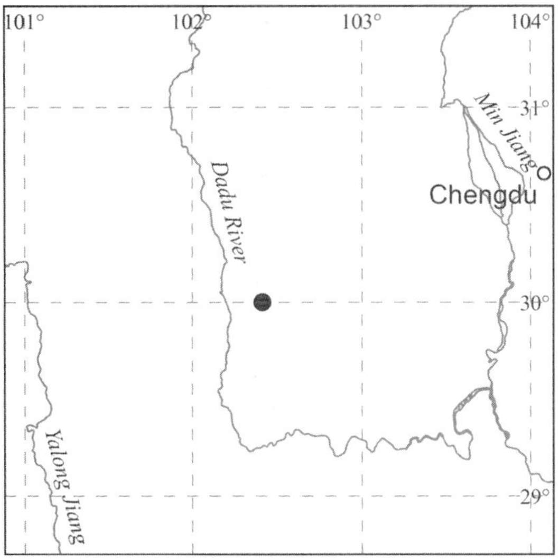

Map 47–37: **Ilex wilsonii** Loes.

38. Ilex sublongecaudata C. J. Tseng & S. Liu ex Y. R. Li, Bull. Bot. Res. Harbin 5(1): 23. pl. 7. fig. 2. 1985; Chen *et al.* in Wu, Raven & Hong, Fl. China 11: 429. 2008. Type: China, Yunnan, Longling, 2600 m, 28 Aug. 1941, *C. W. Wang 90264* (holotype: KUN; isotype: PE).

Small evergreen trees or shrubs ca. 2 m tall. Branchlets sulcate, puberulent. Petiole 4–6 mm long, canaliculate, puberulent adaxially, glabrous and rugose abaxially; leaf blade leathery, elliptic, 4–6 cm long, 1.5–2 cm broad, base obtuse, margin entire, apex caudate-acuminate with acumen 5–10 mm long, not punctate abaxially, midrib flat or slightly impressed and puberulent adaxially, elevated abaxially, lateral veins 7–9-paired. Flowers not seen. Infructescences on current year's branchlets, fasciculate, individual fascicles 1–3-fruited. Fruiting pedicel 2.5–4.5 mm long, glabrous; persistent calyx 4–5-lobed, lobes triangular, ciliate; fruit globose, 3.5–4 in diameter. Pyrenes 4 or 5, oblong- ellipsoid, ca. 3.5 mm long, 1.5 mm wide, abaxially 3–4-striate, sulcate or esulcate.

Phenology: flowering in June; fruiting in August.

Habitats: forests or edges of forests; 1700–2600 m.

Distribution: S Hengduan (Map 47–38); also in SW Yunnan of China (Longling CO).

S HENGDUAN: **Fugong**, Guqian to Qiaomigulu, 1700–1800 m, 1982, *Qinghai-Xizang Exped. 82–3137* (KUN, PE).

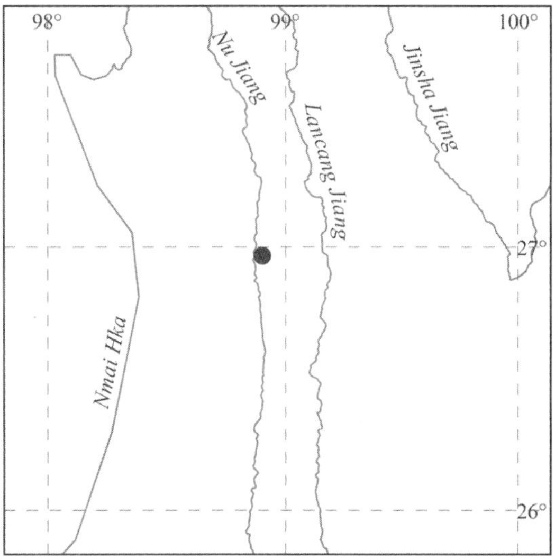

Map 47–38: **Ilex sublongecaudata** C. J. Tseng & S. Liu ex Y. R. Li

39. Ilex wuana T. R. Dudley, Holly Soc. J. 6(4): 14. 1988 ("wuiana"); Chen *et al.* in Wu, Raven & Hong, Fl. China 11: 421. 2008. ≡ *Ilex mamillata* C. Y. Wu ex Y. R. Li, Bull. Bot. Res. Harbin 5(1): 22. 1985, non C. Y. Wu ex C. J. Tseng (1984). Type: China, Yunnan, Yanshan, forests, 1200 m, 14 Oct. 1939, *C. W. Wang 84381* (holotype: KUN).

Evergreen trees, up to 12 m tall. Branchlets angular, sulcate, puberulent or glabrous. Petiole 5–8 mm long; leaf blade leathery, oblong-elliptic, or elliptic, 5–8 cm long, 1.5–3 cm broad, base obtuse or broadly cuneate, margin entire, apex acuminate, midrib adaxially impressed, puberulent, abaxially elevated, lateral veins 9–10-paired. Flowers not seen. Infructescences on previous year's growth, umbelliform, sessile, 1–4-fruited. Fruiting pedicel 6–8 mm long, puberulent; persistent calyx 6-lobed, lobes broadly triangular, ciliate; fruit oblate, ca. 4 mm long, 5 mm in di-

Figure 47–14: **a–d**, **Ilex forrestii** H. F. Comber: **a**, male flowering branch; **b**, fruiting branch; **c**, pyrene; **d**, cross section of pyrene. **e–g**, **Ilex wuana** T. R. Dudley: **e**, fruiting branch; **f**, fruit, top view and bottom view; **g**, pyrene. **h–j**, **Ilex tsiangiana** C. J. Tseng: **h**, fruiting branch; **i**, fruit; **j**, pyrene. From *Fl. Reipubl. Popularis Sin.* 45(2), 4–6. 1999.

ameter. Pyrenes 6, ellipsoid, 3–4 mm long, 1.5–2 mm wide, abaxially 3-striate and 2-sulcate. Figure 47–14: e-g.

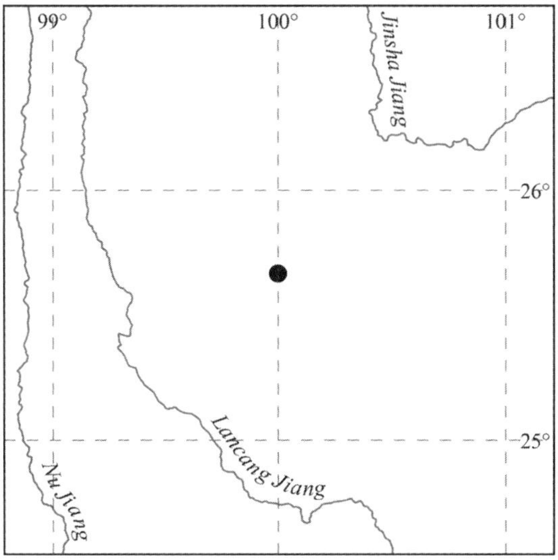

Map 47–39: **Ilex wuana** T. R. Dudley

Phenology: fruiting in October.

Habitats: forests; 1200–2200 m.

Distribution: S Hengduan (Map 47–39); also in C and S Yunnan (Fuming, Guangnan, Wenshan, and Yanshan) and W Guangxi of China.

S HENGDUAN: **Yangbi**, 1940, *R. C. Ching 25197* (paratype of *Ilex mamillata* C. Y. Wu ex Y. R. Li: KUN).

40. Ilex forrestii H. F. Comber, Notes Roy. Bot. Gard. Edinburgh 18: 46. 1933; Hu, J. Arnold Arbor. 31: 255. 1950; Hong, Phytotaxa 230(2): 159. 2015. Type: China, Yunnan, Mekong Salween Divide, Salween Valley, 28°12′N, 3350–3660 m, Nov. 1917, *G. Forrest 15047* (holotype: E; isotypes:A, K).

Ilex wardii Merr., Brittonia 4: 102. 1941. Type: Upper Burma, Adung Valley, 6 Nov. 1931, *F. Kingdon-Ward 9212* (holotype: A; isotype: BM).

Evergreen trees or shrubs, 2–9 m tall. Branchlets gray, striate and longitudinally rugose, puberulent. Leaves persisting on previous year's growth. Petiole 5–12 mm long; leaf blade leathery, oblong-oblanceolate or elliptic, 4.5–10 cm long, 1.2–3.5 cm broad, base rounded, obtuse or rarely cuneate, margin serrate, or lower part entire or nearly entire, apex acuminate with acumen 1–1.5 cm long, midrib adaxially impressed and puberulent, abaxially elevated, lateral veins 10–12-paired. Inflorescences on previous year's growth, nearly fasciculate, subsessile. Male inflorescences often pseudo- paniculate, individual branches 3-flowered; flowers 4- or 5-merous; peduncle ca. 3 mm long; pedicel ca. 2 mm long, puberulent; calyx ca. 2 mm across, deeply lobed, lobes ciliate; corolla white, ca. 6 mm across, petals oblong-obovate, ciliate. Infructescences raceme-like, central axis up to 8 mm long. Fruiting pedicel 3–5 mm long, puberulent; fruit red,

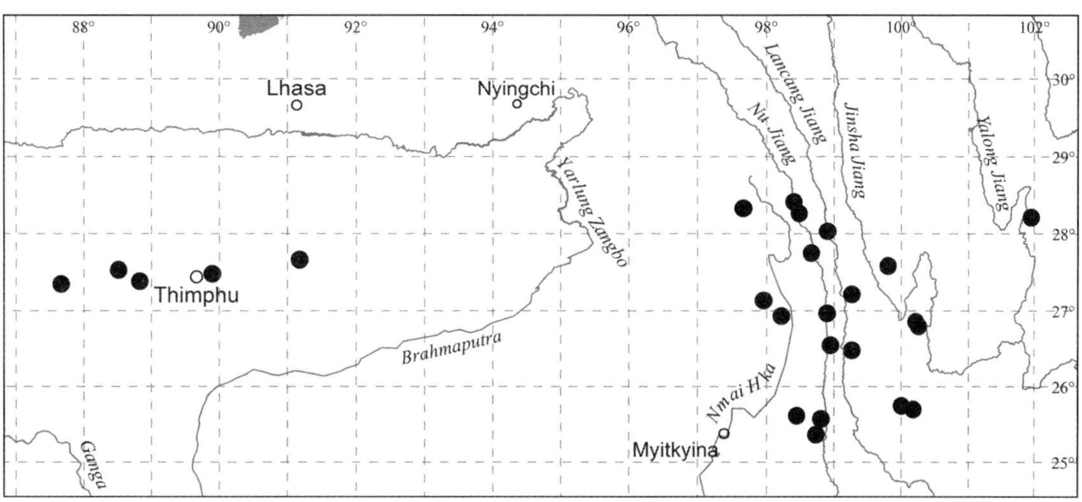

Map 47–40: **Ilex forrestii** H. F. Comber

globose, 4–5 mm in diameter. Pyrenes 5–7, ca. 2 mm long, 1 mm wide, smooth, endocarp leathery. Figure 47–14: a-d.

Phenology: flowering in May and June; fruiting from September to December.

Habitats: forests, rarely thickets; 1700–3400 m.

Distribution: **endemic**; E Nepal, Sikkim & Darjiling, Bhutan, M Yarlung Zangbo, S Hengduan, and U Irrawaddy (Map 47–40).

E NEPAL: **Mechi**, Taplejung, 2730 m, 1992, *M. Suzuki et al. 9263093* (BM); **SIKKIM & DARJILING**: **Sikkim,** 2130–3050 m, *J. D. Hooker* s. n. (BM). **BHUTAN**: **Pangkar**, Kuru Chu, Mt Lhuntse, 2740 m, 1949, *F. Ludlow et al. 18738* (BM); **Thimphu**, 2800 m, 1915, *R. E. Cooper 3914* (BM); **Trashi Yangtsi**, 2740 m, 1949, *F. Ludlow et al. 20634* (BM). **M YARLUNG ZANGBO**: **Yadong**, Chumbi, Tha Poo, 1884, *King's Collector 466* (BM). **S HENGDUAN**: **Bijiang**, Lumbering Ground, 2700 m, 1978. *Bijiang Exped. 0710* (KUN); **Cawarong**, Salween-kiu-chiang Divide, 3050–3350 m, 1921, *G. Forrest 20093* (paratypes of *Ilex forrestii*: A, E, K); **Dali**, 25°40'N, 3050 m, 1917, *G. Forrest 15624* (paratype of *Ilex forrestii*: K); **Deqên**, Cizhong, 2600–2700 m, 1940, *K. M. Feng 5642* (KUN, PE); **Fugong**, 2700 m, 1933, *H. T. Tsai 54449* (KUN, PE); **Gongshan**, Cikai, Heipu Shan, 1940, *K. M. Feng 8196* (K, PE); **Lanping**, Biluoxue Shan, 3070 m, 1964, *S. K. Wu 8943* (KUN); **Lushui**, Shweli-Salween Divide, 2440 m, 1919, *G. Forrest 18133* (paratype of *Ilex forrestii*: K); **Mianning**, Maoniu Shan, 2450 m, 1960, *S. Jiang 5822* (PE); **Shangri-la** (Zhongdian), Chutso, on Yangtze bank, 2800 m, 1939, *K. M. Feng 3252* (KUN, PE); **Weixi**, Yezi, 2500 m, 1935, *C. W. Wang 70416* (KUN, PE); **Yangbi**, Malutang, 1940, *R. C. Ching 22486* (PE); **Yulong** (Lijiang), Kaitze, 2500 m, 1939, *K. M. Feng 2634* (KUN, PE); Lijiang Range, 1933, *Mc/L. C. 244* (K). **U IRRAWADDY**: **Mungku Hkyet**, 2440–2740 m, 1937, *F. Kingdon-Ward 13007* (BM); **North Triangle** (Urung Bum, above Ahkail), 2440–2740 m, 1953, *F. Kingdon-Ward 20865* (BM); **Tengchong**, Qipanshi, 2400 m, 1964, *S. K. Wu 7176* (KUN); Tengchong, Shweli-Salween Divide, 1924, *G. Forrest 24449* (paratypes of *Ilex forrestii*: E, K).

41. Ilex tsiangiana C. J. Tseng, Bull. Bot. Res. Harbin 1(1): 36. 1981; Chen *et al.* in Wu, Raven & Hong, Fl.China 11: 432. 2008. Type: China, Yunnan, Dali, Diancang Shan, *Y. Tsiang 11680* (holotype: PE).

Evergreen trees (?). Branchlets glabrous; terminal buds pubescent or glabrescent; previous year's growth with sparse lenticels. Petiole 7–10 mm long; leaf blade leathery, elliptic, 3.5–5.5 cm long, 1.2–2 cm broad, base obtuse, margin crenulate, apex shortly acuminate, nearly without acumen, glabrous, midrib adaxially impressed, abaxially elevated, lateral veins 8–10-paired. Flowers not seen. Infructescences fasciculate, with 1–3 fruit. Fruiting pedicel 4–5 mm long, puberulent; persistent bracteoles triangular; persistent calyx 5-lobed, lobes triangular; fruit red or red-purple, globose, 3–4 mm in diameter. Pyrenes 5, ellipsoid-oblong, ca. 2 mm long, 1 mm wide, 1-striate, esulcate; endocarp leathery. Figure 47–14: h-j.

Distribution: **endemic**; confined to S Hengduan (Yunnan: Dali) (Map 47–41).

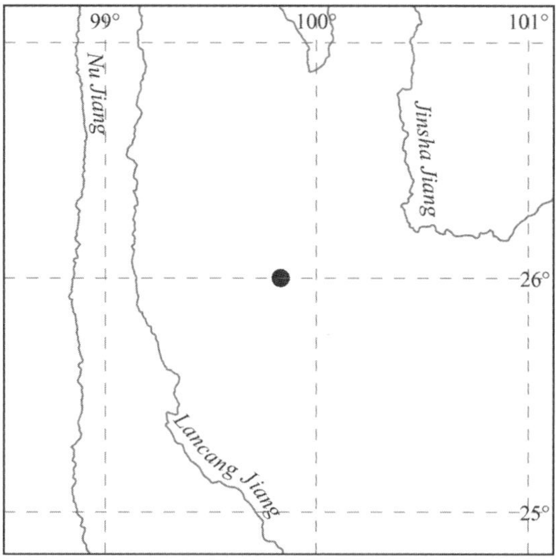

Map 47–41: **Ilex tsiangiana** C. J. Tseng

HELWINGIACEAE

Deciduous or evergreen shrubs or small trees, dioecious. Winter buds ovoid; scales 4. Leaves simple, alternate, petiolate; margin glandular-serrate or crenate-serrate; stipules 2, visible only when young, entire or filiform-divided. Inflorescences inserted on adaxial surface of leaf blade, very rarely on upper part of young branches, umbellate. Flowers 3- or 4(5)-merous; calyx small; petals deltoid-ovate, valvate; floral disc flat, fleshy. Male flowers 3–20 per umbel; stamens 3 or 4 (5), anthers 2-locular. Female flowers 1–4 per umbel; style very short; stigma lobes 3 or 4, rarely 5, reflexed; ovary inferior, 3- or 4 (5)-locular; ovules 1 per locule, pendulous, apotropous, with dorsal raphe. Fruit a berry, drupe-like, crowned with persistent calyx and style, with grooves and ridges when dry. Seeds (stones) 1–4 (5); embryo straight.

One genus, *Helwingia* Willd.

HELWINGIA Willd.

Willdenow, Sp. Pl. 4: 634,716, 1806, nom. cons., non Adanson, 1763; Soong in Fang & Hu, Fl. Reipubl. Popularis Sin. 56: 20. 1990; Xiang & Boufford in Wu, Raven & Hong, Fl. China 14: 227. 2005. Type: *Helwingia japonica* (Thunb.) D. Dietr. (*Osyris japonica* Thunb.).

The description is the same as that of the family; distributed from Nepal eastwards to Japan: Bhutan, S and C China, N India, Japan, S Korea Peninsula, N Myanmar, Nepal, Thailand, and N Vietnam.

Notes: 1) Soong's (1990) treatment recognizes five species, while Xiang and Boufford (2005 in Fl. China, Vol. 14) recognize four species. The present author tends to consider that the genus may well consist of only one extremely variable species, *Helwingia japonica* (Thunb.) D. Dietr. Yes, the typical *H. japonica* is distinctly different from the typical *H. chinensis*. The former is a deciduous shrub with papery, ovate, and densely glandular-serrate leaves, whereas the latter is an evergreen shrub with leathery, linear-lanceolate to ovate-lanceolate, and crenulate-serrate leaves. However, *H. himalaica* comprises intermediate forms with subleathery, obovate-lanceolate leaves, with some forms that seem deciduous, while others seem evergreen. In the three species mentioned above 18 infraspecific taxa have been described, which indicates that the group is very variable. Furthermore, *H. himalaica* was treated as a variety in *H. japonica*, and *H. chinensis* var. *crenata* (Lingelsh. ex Limpr.) Fang was transferred by Li (1944: J. Arnold Arbor. 25: 310) into *H. himalaica*, also as a variety, which shows that "boundary between the species" is problematic. Li (1944) realized this problem, saying "Wangerin [Pflanzenr. 41 (IV. 229): 37. 1910] treats the genus *Helwingia* as having three species, but I fail to notice clear-cut specific differences between them." However, I have no convincing evidence so far to treat this genus as monospecific. To answer this fascinating question population observations and sampling, statistic analysis and molecular phylogenetic analysis are necessary.

2) As I stated above, this taxonomic account of the genus is tentative and temperate. For this reason I left infraspecific taxa untouched. This genus is waiting for further research and a solid taxonomy.

1a. Evergreen shrubs or small trees; leaf blade leathery or nearly leathery, lanceolate or linear-lanceolate, rarely ovate-lanceolate, 0.8–2(–3) cm broad, with length/width ratio 4.5–9; marginal teeth 5–17 on each side; petiole 0.5–2 cm long, rarely longer .... 3. *H. chinensis*

1b. Deciduous shrubs or small trees, sometimes evergreen; leaf blade papery, rarely thickly papery, broadly ovate to oblong-lanceolate, or obovate-lanceolate, 2–7 cm broad, with length/width ratio 1.4–5; marginal teeth usually more than 17 on each side, rarely less; petiole usually 2–4(–7) cm long.

2a. Leaf blade papery, ovate or wide-ovate, 5–9 (–13) cm long, 2.5–5 (–7) cm broad, with length / width ratio 1.4–2.7, apex caudate, acumen 1–2.5 cm long; always deciduous .... 1. *H. japonica*

2b. Leaf blade thickly papery, oblong-lanceolate or obovate-lanceolate, 7–12 (–18) cm long, 2–4 (–6) cm broad, with length / width ratio 2.5–5, apex acuminate; usually deciduous but sometimes evergreen .... 2. *H. himalaica*

1. Helwingia japonica (Thunb.) D. Dietr., Nachur Vollst. Lex. Gärtn. 3: 660. 1817; Xiang & Boufford in Wu, Raven & Hong, Fl. China 14: 227. 2005. ≡ *Osyris japonica* Thunb., Fl., Jap. 31. 1784. Type: Japan, not seen.

Deciduous shrubs or small trees, 0.8–4 m tall. Bark dark brown or grayish black, smooth; young branches glabrous, smooth with conspicuous leaf scars. Stipules thread-like, divided apically or undivided. Petiole 2–7 cm long; leaf blade mostly ovate or broadly ovate, less often obovate-elliptic or ovate-lanceolate, 5–9 (–13) cm long, 2.5–5 (–7) cm broad, papery, veins 5–8, midrib and veins conspicuous, slightly raised abaxially, sometimes papillose, base broadly cuneate or rounded, margin densely glandular-serrate, with 17–40 serrations on each side, apex caudate-acuminate or acuminate. Umbels in between middle and lower 1/3 of leaf midrib, very rarely on upper part of young branches. Male umbels 3–12 (–18)-flowered; flowers pedicellate; petals light green, 1–5 mm long. Female umbels 1–4-flowered; ovary globose or ovoid; stigma 3–5-fid. Fruit nearly sessile, green to red, black when mature, globose or ellipsoid, 4–8 mm in diameter. Seeds 3–5. Figure 47–15: a–f.

Phenology: flowering and fruiting from June to August.

Habitats: forests or thickets; 1400–4100 m.

Distribution: L Yarlung Zangbo, Yarlung Zangbo-Brahmaputra, N and S Hengduan (Map 47–42); also in E Asia: E China, Japan, S Korea Peninsula, Thailand, and N Vietnam.

L YARLUNG ZANGBO: **Guxiang**, 2680 m, 1965, *J. S. Ying & D. Y. Hong 0133* (PE); **Yi'ong**, 2400 m, 1976, *s. coll. 76–743* (PE). **YARLUNG ZANGBO-BRAHMAPUTRA**: **Shangzayü**, Quzi, 2100 m, 1973, *Qinghai-Xizang Exped. 73–969* (PE); **Zayü** Town, 2600 m, 1973, *Qinghai-Xizang Exped. 73–216* (PE). **N HENGDUAN**: **Heishui**, Shimai Gou, 2750 m, 1957, *X. Li 73340* (PE); **Jinchuan**, Anning, 2950 m, 1958, *X. Li 77200* (PE); **Jiuzhaigou** (Nanping), Jiuzhaigou, 3000 m, 1975, *Sichuan Plant Surv. 8823* (PE); Changhai, 2600 m, 1984, *W. L. Chen et al. 8676* (PE); **Lixian**, Jiabi Gou, 2400 m, 1989, *Q. S. Cao 1431* (PE); Miyaluo, 3000 m, 1958, *P. X. Li 10559* (PE); **Maoxian**, Guanyin Shan, 2650 m, 1930, *F. T. Wang 22033* (PE); **Pingwu**, Wangba, 2000 m, 1958, *H. L. Tsiang 10620* (PE); **Wenchuan**, Wolong, 2500 m, 1975, *Sichuan Plant Surv. 8153* (PE); **Wenxian**, Baishui Jiang Nature Reserve, 2000–2230 m, 2007, *D. E. Boufford & Y. Jia 37668* (PE); **Xiaojin**, Shitianjia Gou, 2800 m, 1958, *S. S. Chang & Y. X. Ren 6814* (PE). **S HENGDUAN**: **Baoxing**, Ganyang, 3000 m, 1936, *K. L. Chu 3534* (K); **Butuo**, Jiaojihe, 1400 m, 1959, *Sichuan Econ. Plant Surv. (Liang 59) 5877* (PE); **Dali**, Dali Range, 1910, *G. Forrest 7180* (PE); Cangshan, Fengyandong, 1946, *T. N. Liou 23045* (PE); **Dêqên**, Atuntze, 3200 m, 1937, *T. T. Yü 8292* (PE); **Eryuan**, Fengxiang, 1940, *R. C. Ching 23242* (PE); **Gongshan**, Sijitong, 2500 m, 1935, *C. W. Wang 67113* (PE); **Heqing**, Lianping, 2650 m, 1940, *R. C. Ching 23941* (PE); **Jianchuan**, 3050–3352 m, 1922, *G. Forrest 21942* (K); **Kangding**, 2600 m, 1982, *D. Y. Hong & Z. H. Zhong PB82129* (PE); Yuling Township, 3050 m, 1953, *H. L. Tsiang 35687* (PE); **Lanping**, 3000 m, 1933, *H. T. Tsai 54019* (PE); **Luding**, E slope of Gongga Shan, 2200 m, 1982, *K. Y. Lang et al. 551* (PE); Hailuogou, 2001, *H. N. Qing et al. 312* (PE); **Lushui**, Yaojiaping to Pianma, 2700 m, 1960, *NW Yunnan Subgroup 10314* (PE); **Mianning**, Tuowu,

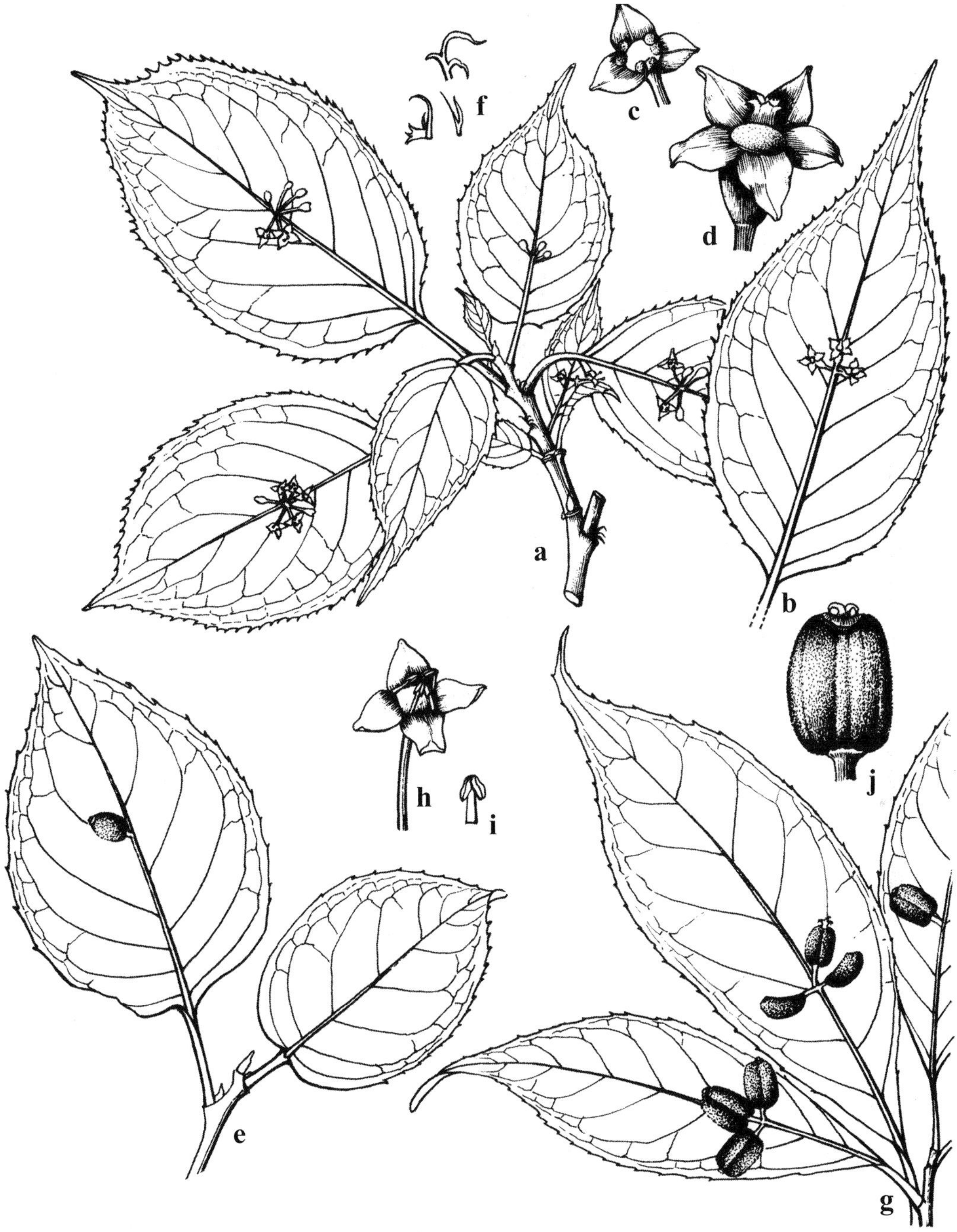

Figure 47–15: **a–f**, **Helwingia japonica** (Thunb.) D. Dietr.: **a**, male flowering branch; **b**, female inflorescence on leaf; **c**, male flower; **d**, female flower; **e**, fruiting branch; **f**, stipules. **g–j**, **Helwingia himalaica** Hook. f. & Thomson: **g**, fruiting branch; **h**, male flower; **i**, a stamen; **j**, fruit. From *Fl. Reipubl. Popularis Sin.* 56, 1990.

2400–2600 m, 1959, *Sichuan Econ. Plant Surv. (Xi 59) 1731* (PE); Yejin, 2900 m, 1959, *S. F. Zhu 20437* (PE); **Meigu**, Houbulaibo, 1959, *Sichuan Econ. Plant Surv. (Xi 59) 1741* (PE) & *1884* (PE); **Muli**, near the city, 2600 m, 1936, *T. T. Yü 7355* (PE); **Shangri-la** (Zhongdian), Haba Shan, 3200 m, 1962, *Zhongdian Exped. 2060* (PE); Shangjiangkong, 2800 m, 2008, *Q. E. Yang & Q. Yuan Yangqe 2314* (PE); **Shimian**, 1955, *H. S. Kung 39889* (PE); **Tianquan**, Longtanxi, 2200 m, 1953, *H. L. Tsiang 35312* (PE); Daqiaopo, 1250 m, 1933, *T. T. Yü 1823* (PE); **Weixi**, Yezhi, 3400 m, 1935, *C. W. Wang 68259* (PE); Xuelong Shan, 3000–3200 m, 1940, *K. M. Feng 4722* (PE); **Yuexi**, Yuexi, Bao'an, 2000 m, 1959, *Sichuan Econ. Plant Surv. (Liang 59) 3849* (PE); **Yulong** (Lijiang), Yulong Snow Range, Lengziluo, 3000–3200 m, 1942, *K. M. Feng 8998* (PE).

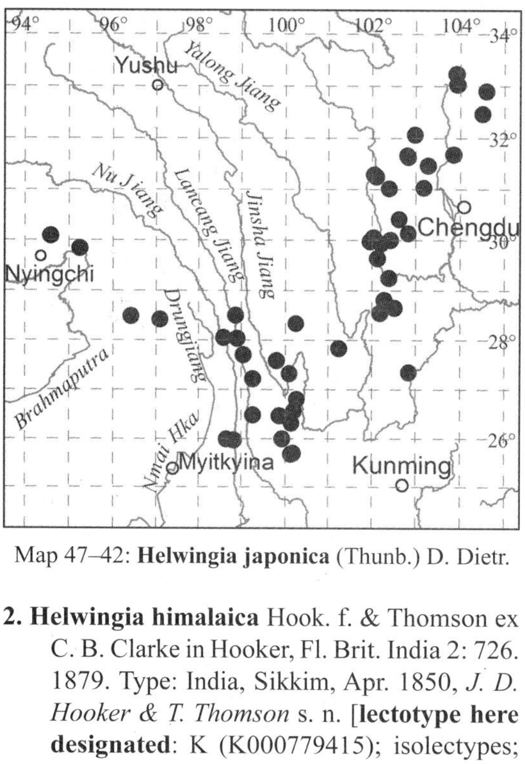

Map 47–42: **Helwingia japonica** (Thunb.) D. Dietr.

2. Helwingia himalaica Hook. f. & Thomson ex C. B. Clarke in Hooker, Fl. Brit. India 2: 726. 1879. Type: India, Sikkim, Apr. 1850, *J. D. Hooker & T. Thomson* s. n. [**lectotype here designated**: K (K000779415); isolectypes; K)].

Deciduous or evergreen shrubs or small trees, 1–5 m tall; young branches yellow-brown. Leaves all on current year's shoots or also some on previous year's growth; petiole 2–4 (–7) cm long; leaf blade thickly papery or thinly leathery, obovate-lanceolate, elliptic, or oblong-lanceolate, 7–12 (–18) cm long, 2–4 (–6) cm broad, both surfaces glabrous, base broadly cuneate or rounded, margin glandular-serrate with 18–37 serrations on each side, apex acuminate or caudate-acuminate. Umbels inserted between middle and lower 1/3 of midrib, rarely on upper part of young branches. Male umbels with 8–20 flowers; pedicel 5–8 mm long, flowers 4-merous, light green, purple at top; female umbels usually with (1–) 3 or 4 flowers; flowers 4- or 3-merous; stigma 3- or 4-fid, segments recurved, fruit usually 1–3 on midrib; fruit pedicel 1–2 (–6) mm long. Drupe red-purple, nearly globose, 6–9 mm long, 6–8 mm in diameter. Seeds 3 or 4, 3- or 4- ridged. Figure 47–15: g–j.

Phenology: flowering from April to June; fruiting from August to October.

Habitats: fo-rests, on rocks; 1500–2800 m.

Distribution: E Nepal, Sikkim & Darjiling, Bhutan, Yarlung Zangbo-Brahmaputra, S Hengduan, and U Irrawaddy (Map 47–43); also in C and S China, and N Vietnam.

E NEPAL: **Kosi**, Sankhuwasabha, Bakang, ca. 2600 m, 1981, *C. Grey-Wilson et al. 4007* (K); **Mechi**, near Taplejung, 2700 m, 1963, *H. Hara et al. 6300518* (K). **SIKKIM & DARJILING**: **Darjiling**, Goompakar, 2134 m, 1874, *J. S. Gamble 3045 A* (K) **BHUTAN**: SE of **Sengor**, above Namning, 2730 m, *A. J. C. Grierson & D. G. Long 2556* (K) ;**Tongsa**, 2250 m, 1979, *A. J. C. Grierson & D. G. Long 1113* (K). **YARLUNG ZANGBO-BRAHMAPUTRA**: **Baibung**, Gelin, 1800 m, 1983, *S. Z. Cheng et al. 3737* (PE); Gelin to De'ergong, 1700 m, 1983, *B. S. Li & S. Z. Cheng 3631* (PE); **Delei Valley**, 1520–1830 m, 1928, *F. Kingdon-Ward 8104* (K); **Mêdog**, Maniweng, 960 m, 1974, *Qinghai-Xizang Exped. 74–1861* (PE); between Maniweng and Hami, 1300 m, 1983, *B. S. Li & S. Z. Cheng 5246* (PE). **S HENGDUAN**: **Dali**, Wuweisi, 1943, *H. C. Wang 3833* (PE); **Dêqên**, Yongzhong, 2400–2800 m, 1981, *Acad. Sin. Inst. Bot. Exped. to Hengduan Mts 3443* (PE); **Fugong**, Guquan to Qiaomiqilu, 1700–1800 m, 1982, *Qinghai-Xizang Exped. 7153* (PE); **Gongshan**, Bingzhongluo, 2400–2600 m, 1982, *Qinghai-Xizang Exped. 7469* (PE); **Lushui**, Pianmagan, 2500–2700 m, 1981, *Acad. Sin. Inst. Bot. Exped. to Hengduan Mts 071* (PE) & *147* (PE); **Weixi**, Xuelong Shan, 2500 m,

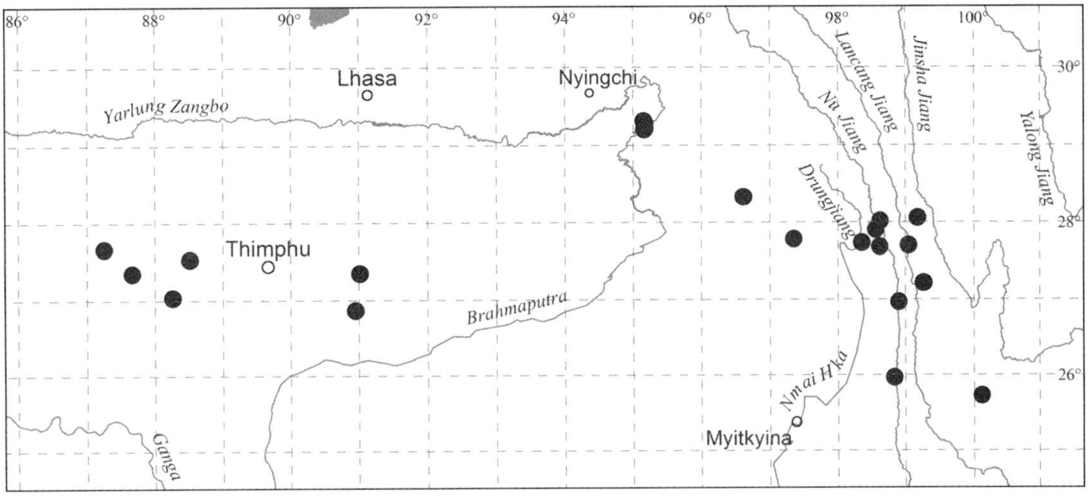

Map 47–43: **Helwingia himalaica** Hook. f. & Thomson ex C. B. Clarke

1955, *K. M. Feng 20907* (PE); Yezhi, 2500 m, 1940, *K. M. Feng 3887* (PE). **U IRRAWADDY**: **Drungjiang** Township, 1500–1700 m, 1982, *Qinghai-Xizang Exped. 9099* (PE); W flank of Gaoligong Shan, 1700 m, 1982, *Qinghai-Xizang Exped. 8873* (PE); Taron-Taru divide, 2300 m, 1938, *T. T. Yü 20017* (PE); **Seinghku Valley**, 1520 m, 1926, *F. Kingdom-Ward 6714* (K).

3. Helwingia chinensis Batal., Trudy Imp. S.- Petersburgsk. Bot. Sada 13: 97, 1893; Soong in Fang & Hu, Fl. Reipubl. Popularis Sin. 56: 31, 1990; Xiang & Boufford in Wu, Raven & Hong, Fl. China 14. 228. 2005. Type: China, Hupeh (Hubei), S Patung (Badong), March 1889, *A. Henry 6719B* (holotype: K).

Evergreen shrubs or small trees, up to 4 m tall. Bark dark-gray or pale gray-brown. Petiole 0.5–2 cm long, occasionally longer; leaf blade leathery or thinly leathery, linear-lanceolate to ovate-lanceolate, 5–12 cm long, 0.8–2(–3) cm broad, with length/width ratio 4.5–9, both surfaces glabrous, base cuneate or more or less rounded, margin sparsely glandular crenulate-serrate, rarely crenate, with no more than 17 serrations or crenulations on each side, apex acuminate. Umbels occurring near middle point of midrib on adaxial leaf surface or on upper part of young branches. Flowers of male umbels 4 or 5, 3–5-merous; pedicel 2–10 (–14) mm long. Flowers of female umbels 1–3, nearly sessile or with pedicel up to 2 mm long; stigma 3–5-fid. Fruit short-cylindrical, 5–7 (–9) mm in diameter. Seeds 3–5, 3–5-ridged. Figure 47–16.

Phenology: flowering from April to June; fruiting from August to November.

Habitats: forests or thickets; 1200–3200 m.

Distribution: N and S Hengduan and U Irrawaddy (Map 47–44); also in other parts of China (Chongqing, Guizhou, S Shaanxi, and other parts of Sichuan and Yunnan).

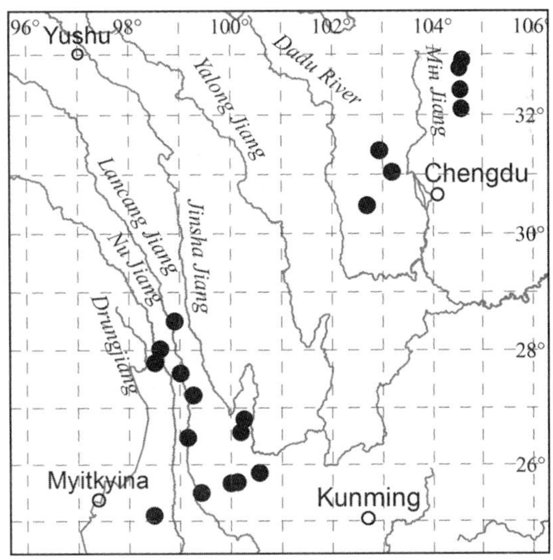

Map 47–44: **Helwingia chinensis** Batal.

N HENGDUAN: **Lixian**, Suoluogou, 2300 m, 1956, *X. Li 46811* (PE); **Pingwu**, Lijia, 1820 m, 1958, *H. L. Tsiang 10489* (PE); Wangba, 1900

Figure 47–16: **Helwingia chinensis** Batal.: **a** & **b**, male flowering branches; **c**, fruiting branch; **d**, male flower; **e**, fruit; **f**, leaf with a fruiting umbel. From *Fl. Reipubl. Popularis Sin.* 56, 1990.

m, 1958, *H. L. Tsiang 10545* (PE); **Wenchuan**, Wolong, 2000 m, 1975, *Sichuan Plant Surv. 8237 (PE) & 8240* (PE); **Wenxian**, Motianling, 1300–1340 m, 2007, *D. E. Boufford & Y. Jia 37636* (P, PE); Baishui Jiang Nature Reserve, 1500 m, 2006, *Baishui Jiang Exped. 0724* (PE) & *0680* (PE). **S HENGDUAN**: **Baoxing**, Longdong, 1500 m, 1959, *Sichuan Econ. Plant Surv. (Ya 59) 99* (PE); **Binchuan**, Jizhu Shan, 1940, *T. N. Liou 17807* (PE); **Dali**, Cangshan, Zhonghesi, 1940, *T. N. Liou 17605* (PE); **Dêqên**, by Mekong, 1900 m, 1959, *Kunming Inst. Bot. 23825* (PE); **Gongshan**, Bingzhongluo, 1800–3200 m, 1982, *Qinghai-Xizang Exped. 7600* (PE); Gongshan Town to Drungjiang, 1700 m, 1982, *Qinghai-Xizang Exped. 8080* (PE); **Heqing**, Ma'er Shan, 2546 m, 2010, *H. Sun et al. Sun H–072X–3181* (PE); **Lanping**, Yingpan, Luomoping, 2300 m, 1981, *Acad. Sin. Inst. Bot. Exped. to Hengduan Mts 685* (PE); **Weixi**, Dedeng, 1780 m, 1982, *Qinghai-Xizang Exped. 6518* (PE); Kanpu, 1928 m, 1935, *C. W. Wang 64128* (PE); **Yangbi**, near the city, 1940, *R. C. Ching 25267* (PE); Maiweng, 2400 m, 1964, *Z. H. Yang 101876* (PE); **Yongping**, Yuntai Shan, 2400 m, 1965, *Acad. Sin. Sichuan-Yunnan-Guizhou Subgroup 129* (PE); **Yulong** (Lijiang), Yangjian, 2500 m, 1955, *K. M. Feng 21609* (PE). **U IRRAWADDY**: **Tengchong**, 1520 m, 1912, *G. Forrest 7554* (K).

CAMPANULACEAE

Herbs lactiferous, perennial, often with rhizomes or caudices, less often annual. Leaves estipulate, simple, alternate, less often opposite or whorled. Flowers perfect, tetracyclic, protandrous, with a specialized secondary pollen presentation, solitary in an axillary or less often terminal position, or aggregated into terminal or axillary inflorescences appearing racemose or paniculate, rarely capitate, pedicellate or rarely sessile; bracts leaf-like or reduced, rarely absent; pedicel often bracteolate. Calyx and corolla jointly or corolla alone adnate to ovary, forming hypanthium. Calyx synsepalous, adnate to hypanthium, free in *Cyananthus,* and some species of *Codonopsis* and *Cyclocodon*; calyx lobes (3–) 5 (–10), valvate, sometimes with a reflexed appendage in each sinus (some groups in *Campanula*). Corolla sympetalous, always epigynous, actinomorphic; corolla lobes (4 or) 5 (–10), valvate. Stamens equaling number of corolla lobes, antisepalous, distinct; anthers free or connivent, tetrasporangiate, dithecal, introrsely dehiscent by longitudinal slits, basifixed. Gynoecium syncarpous, (2)3–5(–10)-locular with axile placentation; ovary at least partly inferior, but inferior to corolla while superior to calyx in some species of *Codonopsis* and *Cyclocodon*, and superior to both calyx and corolla in *Cyananthus*, crowned by annular or tubular nectary disc in *Adenophora* and *Codonopsis*; style solitary, pubescent with pollen-collecting hairs below apex; stigma typically with as many lobes as ovary locules. Fruit a capsule, commonly apically loculicidal or laterally poricidal, or a berry. Seeds usually small, numerous; embryo straight; endosperm copious, oily or rarely starchy.

The family Campanulaceae has two concepts, broad and narrow. In the broad sense it includes five monophyletic groups, one of which is the Campanuloideae (= Campanulaceae in the narrow sense). Thus, it is a matter of taste to use either. The Campanulaceae s. s. differs clearly from the other four in having flowers actinomorphic, locules and stigma segments usually 3–5 (vs. usually 2), and pollen spinuliferous. Therefore, we use here the Campanulaceae in the narrow sense.

Recent references: Hong, 2015, A Monograph of *Codonopsis* and Allied Genera (Campanulaceae), Beijing &: xx Science Press and Academic Press; Hong et al., 2011, Campanulaceae. In Wu, Raven & Hong, Fl. China 19: 505–563; Hong & Pan, 2012, Pollen morphology of the platycodonoid group (Campanulaceae) and its systematic implications. *J. Integr. Pl. Biol.* 54: 773–789; Lammers 2007a, Campanulaceae. Pp. 26–56. In: Kadereit & Jeffrey (eds.), The families and genera of vascular plants, Vol.8. Berlin: Springer; Lammers 2007b, World Checklist and Bibliography of Campanulaceae. London: Royal Botanic Garden, Kew. Pp. 675; Wang et al., 2013, Molecular phylogeny of the platycodonoid group (Campanulaceae s. s.) with special reference to the circumscription of *Codonopsis. Taxon* 62(3): 498–504; Wang et al., 2014, Evolution of the platycodonoid group with particular references to biogeography and character evolution. *J. Integr. Pl. Biol.*56(10): 995–1008; Wang et al., 2014, Phylogenetic analyses reveal three new genera of the Campanulaceae. *J. Syst. Evol.* 52(5): 541–550; Hong & Wang, 2015, A new taxonomic system of Campanulaceae s. s. *J. Syst. Evol.* 53(1):.

The family consists of three tribes, 51 genera and about 1000 species, cosmopolitan but mostly in the Old World. All the three tribes, 13 genera (one endemic) and 125 species (59 endemic) in the Pan-Himalaya.

1a. Fruit dehiscent by lateral pores (below insertion of calyx lobes) (tribe 3. Campanuleae).
  2a. Corolla rotate, lobed to near base; plants annual .... 10. *Asyneuma*
  2b. Corolla campanulate, mostly lobed less than 3/5 its length, very rarely over 3/5, and thus nearly rotate; plants mostly perennial.

3a. Corolla mostly 1.5–7.5 cm long; fruit dehiscent via 2 or 3 definite pores.

4a. Flowers without nectary disc .... 9. *Campanula*

4b. Flowers with a tubular (rarely annular) nectary disc .... 11. *Adenophora*

3b. Corolla 0.5–1.2 cm long; fruit dehiscent irregularly via ruptures or tears of membranous or papery pericap.

5a. Perennials with tuberiferous stolons; stems terete; pedicel 0.2–7 cm long .... 12. *Peracarpa*

5b. Annuals without stolons; stems 3-winged; flowers sessile or pedicel up to 8 mm long .... 13. *Homocodon*

1b. Fruit dehiscent by 3–5 apical valves (above insertion of calyx lobes) or a berry.

6a. Annuals or perennials; tap root fibrous or scarcely thickened; stigma lobes filiform or linear, much longer than broad (tribe 2. **Wahlenbergieae** Endl.) .... 8. *Wahlenbergia*

6b. Perennials, very rarely annuals; tap root usually thickened, carrot-shaped; stigma lobes ovoid or round, about as long as broad, less often clavate, slightly longer than broad (tribe 1. **Cyanantheae** Mesin.)

7a. Ovary superior to both calyx and corolla; calyx tube tubular, sometimes dilated into urceolate ....3. *Cyananthus*

7b. Ovary inferior, rarely superior or half-superior only to calyx; calyx tube various in shape.

8a. Fruit baccate; leaves opposite .... *2. Cyclocodon*

8b. Fruit capsular, rarely baccate (some *Codonopsis*); leaves alternate, less often opposite (*Pankycodon*).

9a. Fruit 5-loculicidal, valves opposite to calyx lobes; leaves verticillate or opposite, rarely alternate .... 1. *Platycodon*

9b. Fruit usually 3–4 (5)- loculicidal, valves alternate to calyx lobes; leaves usually aternate, less often opposite.

10a. Corolla rotate, lobed to near base; roots mostly globose, rarely elongate; filament base lamellar, orbicular, lanate .... 7. *Pseudocodon*

10b. Corolla campanulate, lobed for up to half way; roots attenuate; filament base narrow-triangular, glabrous or ciliate at dilated part.

11a. Disc or glands present; plants fetid except 2 spp. in sect. Campanumoea .... 4. *Codonopsis*

11b. Disc or glands absent; plants not fetid.

12a. Leaves alternate on main stems; hypanthium entirely adnate to ovary; capsule base obconical; seeds ovoid, wingless .... 5. *Himalacodon*

12b. Leaves all opposite; hypanthium free from ovary at top; capsule base semiglobose; seeds lunate, strongly compressed and wide-winged .... 6. *Pankycodon*

Trib.1. Cyanantheae Mesin. in Pl. Vasc. Gen. : Tab. Diagn. 273. Comm., 180. 1840. Type : *Cyananthus* Wall. ex Benth.

Canarineae Webb. ex Benth., Hist. Nat. Iles Canaries 2(3): 1, 1844. Type: *Canarina* L.

Ostrowskieae Fed. in Komarov, Fl. URSS 24: 471. 1957. Type: *Ostrowskia* Regel.

Platycodoneae Yeo, Taxon 42: 109. 1993. Type: *Platycodon* A. DC.

Campanumoeae Kolak., Bot. Zhurn. 79: 111. 1994. Type: *Campanumoea* Bl.

Fruit a loculicidal capsule dehiscent by apical valves (above the insertion of calyx lobes) (with an exception in *Ostrowskia*), or a berry. Pollen colpate or colporate, with elongate apertures. Chromosomes x=5, 6, 7, 8, 9, 17.

Ten genera, in Asia, Africa, and Oceania (New Guinea); seven in the Pan-Himalaya (one endemic).

1. PLATYCODON A. DC.

de Candolle, Monogr. Camp. 125. 1830. Type: *Platycodon grandiflorus* (Jacq.) A. DC. (*Campanula grandiflora* Jacq.)

Herbs perennial. Tap root thickened, carrot-like. Stems erect. Leaves cauline, 3- or 4-verticillate, opposite, or alternate, sessile or shortly petiolate. Flowers terminal, solitary, pedicellate. Calyx tube entirely adnate to hypanthium; calyx lobes 5, shorter than calyx tube. Corolla bowl-shaped; corolla lobes 5, shorter than or as long as tube. Stamens 5, free; filaments dilated into a triangular base, dilated part ciliate; anthers longer than filaments. Disc absent. Ovary semi-inferior, 5-locular; stigma 5-fid. Capsule apically 5-loculicidal, valves with septa, opposite to calyx lobes. Seeds numerous, black or dark brown, ellipsoid or short-cylindrical, longitudinally keeled.

Recent reference: Wang et al., 2013, Molecular phylogeny of the platycodonoid group (Campanulaceae s. str.) with special reference to the circumscription of *Codonopsis*. *Taxon* 62: 498–504; Hong & Wang, 2015, A new taxonomic system of the Campanulaceae s. s. *J. Syst. Evol.* 53(3): 203–209.

One species in East Asia; widely cultivated.

1. Platycodon grandiflorus (Jacq.) A. DC., Monogr. Camp. 125. 1830; Tsoong, Contr. Inst. Bot. Nat. Acad. Peiping 3(3):88. 1935; Hong & Ma in Wang & Wu, Vasc. Pl. Hengduan Mountains 2: 1962. 1994. ≡ *Campanula grandiflora* Jacq., Hort. Bot. Vindob. 3: 4. tab. 2. 1776. Type: Jacquin's tab. cited (**lectotype here designated**).

Stems erect, 0.2–1.2 m tall, usually glabrous, rarely densely puberulent, simple, rarely branched above. Leaf blade ovate, elliptic, or lanceolate, 2–7 cm long, 0.5–3.5 cm broad, adaxially glabrous, green, abaxially glabrous, rarely puberulent along veins, glaucous, base broadly cuneate or rounded, margin serrate, apex acute or acuminate. Calyx tube entirely adnate to hypanthium, hemispherical, obovoid, or obconic, rarely campanulate, glaucous; calyx lobes triangular or narrowly triangular, sometimes tooth-like. Corolla blue or purple, rarely pink or white, 1.5–4.5 cm long. Capsule globose, obconic, or obovoid, 0.7–2.5 cm long, 0.7–1.5 cm wide. **Chromosomes**: 2n=18. Figure 47–17.

Phenology: flowering and fruiting from July to October.

Habitats: sunny grassy places, thickets, rarely forests; below 2000 m.

Distribution: N and S Hengduan (Map 47–45); widely distributed in E Asia: E China, Japan, Korea Peninsula, and Russia (the Far East and SE Siberia).

N HENGDUAN: **Pingwu**, 1928, *W. P. Fang 4219* (PE). **S HENGDUAN: Ganluo**, 1160 m, 1959. *s. coll. 4450* (KUN); **Luding**, 1320–1380 m, 1959, *S. Jiang 1976* (PE); **Meigu**, 1959, *Sichuan Econ. Plant Surv. (Liang) 6016* (KUN).

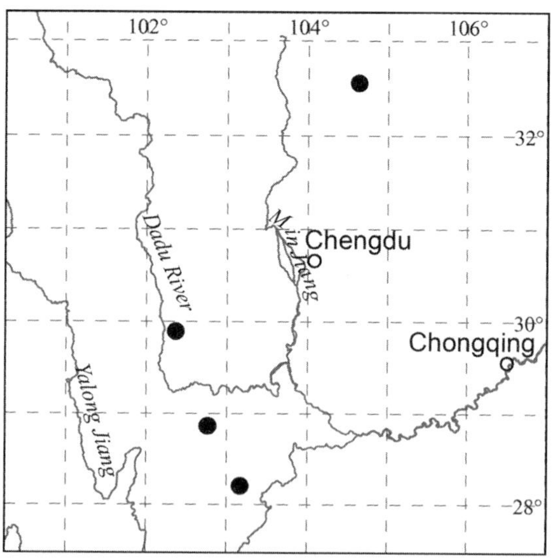

Map 47–45: **Platycodon grandiflorus** (Jacq.) A. DC.

Figure 47–17: **Platycodon grandiflorus** (Jacq.) A. DC. **a**, root; **b**, upper part of stem; **c**, a flower with calyx removed and corolla sectioned; **d**, ventral side of a stamen; **e**, dorsal side of a stamen; **f**, cross section of ovary; **g**, mature fruit, showing fruit valves opposite to calyx lobes. Redrawn by Ai-Li LI from *Icon. Cormophyt. Sin.* 4, 1975

2. CYCLOCODON Griff. ex Hook. f. & Thomson

Hooker & Thomson, J. Proc. Linn. Soc. Bot. 2: 17. 1857; Hong & Pan, Acta Phytotax. Sin. 36: 109. 1998; Hong, Monogr. *Codonopsis* Allied Gen. Camp. 206.. 2015. ≡ *Campanumoea* sect. *Cyclocodon* (Griff. ex Hook. f. & Thomson) C. B. Clarke in Hooker, Fl. Brit. India 3:436. 1881. Type: *Cyclocodon parviflorus* (Wall. ex A. DC.) Hook. f. & Thomson (*Codonopsis parviflora* Wall. ex A. DC.).

Herbs perennial, without caudex, nor rhizome. Roots attenuate. Stems erect, sometimes woody at base, mostly branched. Leaves opposite, occasionally in whorl of 3. Flowers often in dichasium or solitary, terminal or axillary; bracteoles often present, filiform or leaf-like. Calyx adnate to hypanthium, or completely free from it, and thus epigynous or hypogynous to ovary; calyx lobes 4–6 (–7), from nearly entire to ramiform. Corolla 4–6-merous, epigynous, infundibular or rotate, lobed to 1/3–1/2. Stamens 4–6; filaments dilated below, dilated part glabrous or ciliate. Ovary 4–6-locular, inferior to corolla, from half-inferior to superior to calyx; stigma (4–) 5–6 (–8)-fid, segments clavate; ovules extremely numerous. Fruit a berry. Seeds ovoid or nearly globose, compressed in *C. parviflorus*; seed-coat irregularly reticulate, secondary ornamentation on radial walls spiral/rope-like. Pollen 3-colporate; sexine spinulose, spinules longer than 2 μm. Chromosomes 2n=18.

Recent references: Hong & Pan, 1998, The restoration of the genus *Cyclocodon* (Campanulaceae) and its evidence from pollen and seed-coat. *Acta Phytotax. Sin.* 36(2): 106–110; Wang et al., 2013, Molecular phylogeny of the platycodonoid group (Campanulaceae s. str.) with special reference to the circumscription of *Codonopsis*. *Taxon* 62(3): 498–504; Hong, 2015, A Monograph of *Codonopsis* and Allied Genera (Campanulaceae). Science Press and Academic Press; Hong & Wang, 2015, A new taxonomic system of Campanulaceae s. s. *J. Syst. Evol.* 53(3): 203–209.

Cyclocodon has been proved to be a distinct genus (Hong & Pan, 1998; Wang et al., 2013); it differs from *Codonopsis* in having flowers often in dichasium of 3, stigma usually (4-)5- or 6-fid with clavate lobes, calyx lobes ramiform to nearly entire and often inserted at a different level, distant from the other floral parts, seed-coat irregularly reticulate with rope-like ornamentation on radial walls, and pollen 3-colporate, with long sexine spinules (>2 μm high).

Three species, distributed from the Himalaya extending eastwards to Japan (Kyushu and Ryukyu Islands) and southeastwards to Philippines, Indonesia, and Papua New Guinea.

1a. Flowers often 4-merous; calyx inserted on pedicel, 1–4 mm distant from hypanthium, resembling involucres .... 1. *C. parviflorus*

1b. Flowers usually 5- or 6-merous; calyx at least partly adnate to hypanthium.

2a. Flowers all terminal; bracteoles often absent, rarely present and then stalked, more or less leaf-like; calyx lobes shortly denticulate to nearly entire; filaments glabrous; berry often white, less often yellow or pale red .... 2. *C. lancifolius*

2b. Flowers both axillary and terminal; bracteoles present and filiform; calyx lobes ramiform; filaments hairy below; berry dark purple .... 3. *C. axillaris*

1. Cyclocodon parviflorus (Wall. ex A. DC.) Hook. f. & Thomson, J. Proc. Linn. Soc., Bot. 2: 18. 1857; Hong & Pan, Acta Phytotax. Sin. 36(2): 110. 1998; Hong & Lammers in Wu, Raven & Hong, Fl. China 19: 528. 2011. ≡ *Codonopsis parviflora* Wall. ex A. DC., Monogr. Camp. 123. 1830. ≡ *Campanumoea parviflora* (Wall. ex A. DC.) Benth. & Hook. f. ex C. B. Clarke in Hooker, Fl. Brit. India 3: 436. 1881. ≡ *Campanula punduana* D. Dietr., Syn. Pl. 1: 758.

1839. Type: Bangladesh, Pundua, *Wallich 1300* [**lectotype here designated**: K (K000814184); isolectotypes: K (K000814182, K000814183, K000814185)].

Cyclocodon distans Griff., Not. Pl. As. 4: 277. pl. 481. fig. 1. 1854. Type: Griffith's pl. cited.

Cyclocodon adnatus Griff., Not. Pl. As. 4: 278. 1854. Type: not designated.

Plants 1–2 m tall. Stems sometimes woody below, branched, glabrous. Petiole up to 6 mm long; leaf blade lanceolate, 5–11 cm long, 1.5–3 cm broad, those on branches smaller, both surfaces glabrous, base rounded or broad-cuneate, margin serrate, teeth apiculate at apex, apex acuminate. Flowers terminal, often in dichasium of 3; pedicel 1–4 cm long, glabrous, with or without a pair of leaf-like or linear bracteoles. Calyx entirely free from hypanthium and ovary, and thus hypogynous, just below ovary or 1–4 mm distant from ovary, forming an anthophore; calyx lobes usually 4, free from each other, linear, 3.5–8 mm long, with 1–4 pairs of teeth. Corolla white, infundibular, 5–8 mm long; corolla lobes as long as tube. Stamens 4; filaments subulate, very slightly dilated below. Ovary globose or obovoid, 4-locular; stigma 4-fid. Berry globose, 4–6 mm in diameter. Seeds light brown, rounded, compressed. **Chromosomes**: 2n=18. Figure 47–18: a–c.

Phenology: flowering and fruiting from August to January.

Habitats: thickets or grassy slopes; 200–3000 m.

Distribution: Sikkim & Darjiling, Bhutan, Yarlung Zangbo-Brahmaputra, and U Irrawaddy (Map 47–46); also in E India (Manipur, Meghalaya, and Nagaland) and other part of China (S Yunnan).

Notes: Griffith did indicate a collection from Khasia (Meghalaya) for his new species, *Cyclocodon distans*, but I have not been able to find it, though I have visited many major herbaria. Therefore, I designate here his original plate as type.

SIKKIM & DARJILING: **Darjiling**, Rungbee, 1370 m, 1869. *C. B. Clarke 9015C* (BM, K); **Sikkim**, Pachum, 1520–1830 m, *J. D. Hooker* s. n. (K). **BHUTAN**: **Deothang** (from Grierson & Long, *Fl. Bhutan*, 2(3): 1384. 2001). **YARLUNG ZANGBO-BRAHMAPUTRA**: **N Cachar Hills**, Hafbry, 760 m, 1908, *G. Craib 329* (K); **Mêdog**, Xiran, 600 m, 1992, *H. Sun et al. ETM. 1627* (KUN); **Kameng**, Toppi, 250 m, 1977, *S. S. R. Bennet & H. B. Naithani 2926* (PE). **U IRRAWADDY**: **Mahtum**, 1370 m, 1939, *R. Kaulback 378* (BM); **Nam Tamai Valley**, 1220 m, 1938, *R. Kaulback* s. n. (BM). **Tengchong**: 1180–1350 m, 1960, *W. Q. Yin 60–1473* (KUN).

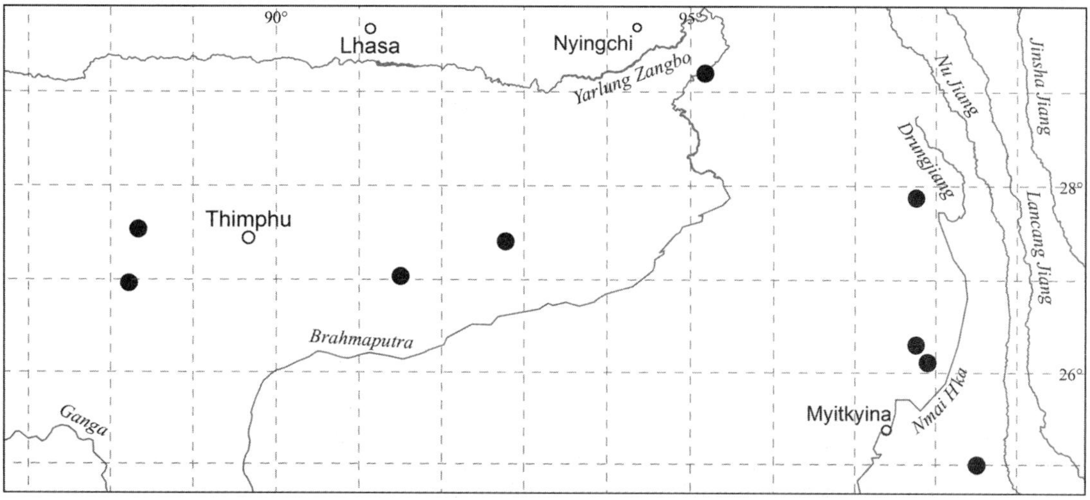

Map 47–46: **Cyclocodon parviflorus** (Wall. ex A. DC.) Hook. f. & Thomson

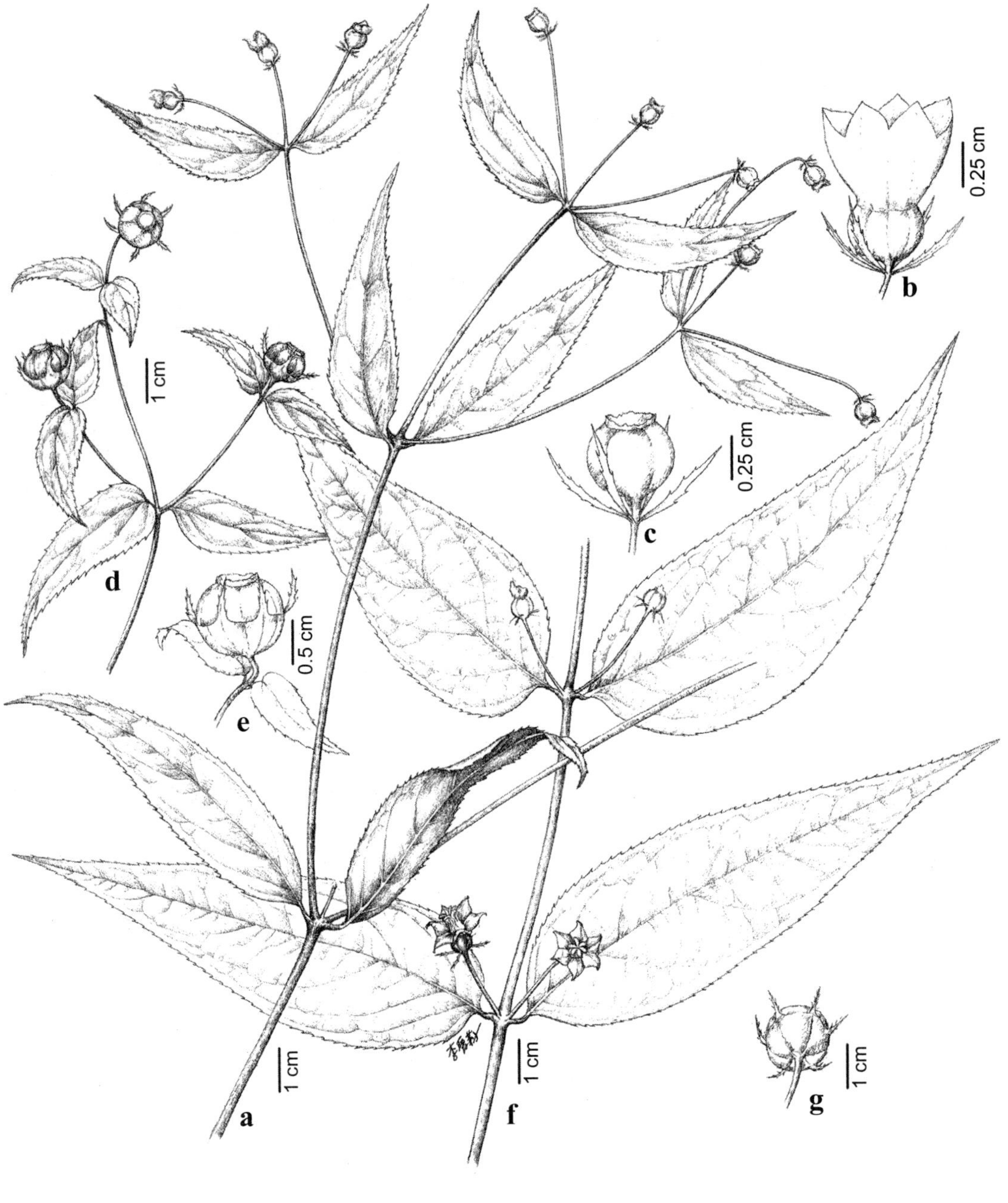

Figure 47–18: **a–c**. **Cyclocodon parviflorus** (Wall. ex A. DC.) Hook. f. & Thomson: **a**, upper part of stem; **b**, a flower with four calyx lobes on pedicel; **c**, a berry. **d** & **e**. **Cyclocodon lancifolius** (Roxb.) Kurz: **d**, upper part of stem; **e**, a berry at the top of branch. **f** & **g**. **Cyclocodon axillaris** (Oliv.) W. J. de Wilde & Duyfjes: **f**, a section of stem, showing axillary flowers; **g**, berry. Drawn by Ai-Li LI.

2. Cyclocodon lancifolius (Roxb.) Kurz, Flora 55: 303. 1872; Hong, Monogr. *Codonopsis* Allied Gen. Camp. 210. 2015. ≡ *Campanula lancifolia* Roxb., Fl. India 2: 96. 1824. ≡ *Campanumoea lancifolia* (Roxb.) Merr., Enum. Philip. Pl. 3: 587. 1923. ≡ *Codonopsis lancifolia* (Roxb.) Moeliono in van Steenis, Fl. Males. Ser. 1. 6: 120. 1960, p. p. Type: Bangladesh, Chittagong, Oct. 1810, *Roxburgh 217* (holotype: BM).

Cyclocodon celebicus (Bl.) D. Y. Hong, Acta Phytotax. Sin. 36: 109. 1998. ≡ *Campanumoea celebica* Bl., Bijdr. 727. 1826. ≡ *Campanula celebica* (Bl.) D. Dietr., Syn. Pl. 1: 758. 1839. ≡ *Codonopsis celebica* (Bl.) Miq., Fl. India Bat. 2: 565. 1835. ≡ *Codonopsis lancifolia* subsp. *celebica* (Bl.) Moeliono in van Steenis, Fl. Males. Ser. 1, 6: 121. 1960. ≡ *Cyclocodon lancifolius* subsp. *celebicus* (Bl.) K. E. Morris & Lammers, Novon 9: 387. 1999. Type: Indonesia, Sulawesi, *Reinwardt* s. n. (LINN, n. v.).

Codonopsis truncata Wall. ex. A. DC., Monogr. Camp. 120. 1830; Wallich, Cat. 1301. 1820, nom. nud. ≡ *Companumoea truncata* (Wall. ex A. DC.) Diels, Bot. Jahrb. Syst. 29: 606. 1901. ≡ *Cyclocodon truncatus* (Wall. ex A. DC.) Hook. f. & Thomson, J. Proc. Linn. Soc. Bot. 2: 18. 1857. Type: India, Darjiling, *Wallich 1301* (**lectotype here designated**: K; isolectotype: BM).

Codonopsis leucocarpa Miq., Fl. India Bat. 2: 565. 1857. Type: Indonesia, Sumatra, in de kloof van den Singalang, *Teijsmann* **s.** n. (holotype, n. v.).

Plants erect or scandent, 0.3–4 m long. Stems branched. Petiole up to 5 mm long; leaf blade ovate-lanceolate to lanceolate, 5–10 cm long, 2–4 cm broad, those on branches smaller, glabrous or sparsely puberulent along veins on both surfaces, base usually rounded, margin serrate, teeth pointed towards apex. Flowers terminal, usually in dichasium of 3; pedicel 0.8–2 cm long, hispidulous, usually without bracteoles, rarely with stalked and more or less leaf-like bracteoles. Calyx adnate to basal ca. 1/3 of hypanthium; calyx tube glabrous; calyx lobes mostly 5 or 6, linear to linear-elliptic, 3–10 mm long, 1–3 mm broad, nearly entire or with 2–4 pairs of serrate teeth. Corolla pale blue, pale lavender, or white, infundibular, 7–10 mm long, 5- or 6-lobed to middle. Stamens 5 or 6, glabrous; filaments slightly dilated below. Ovary globose; stigma 5- or 6-fid. Berry white, oblate, 5- or 6-locular, up to 8 mm long, ca. 12 mm wide. Figure 47–18: d & e.

Phenology: flowering and fruiting from July to February.

Habitats: forests or thickets, often by stream or river, mostly on granites, less often on marbles or serpentines; 600–3100 m.

Distribution: E Nepal, Sikkim & Darjiling,

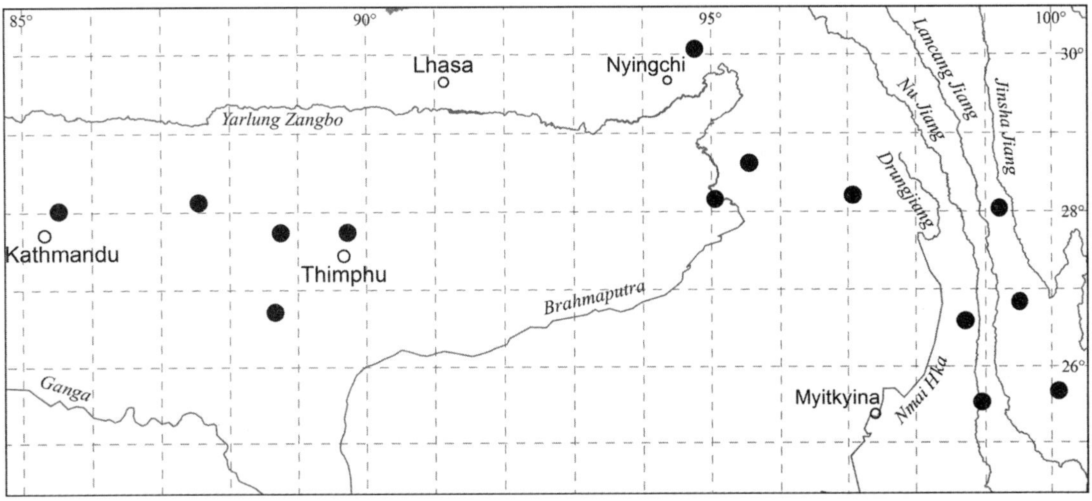

Map 47–47: **Cyclocodon lancifolius** (Roxb.) Kurz

Bhutan, L Yarlung Zangbo, Yarlung Zangbo-Brahmaputra, S Hengduan, and U Irrawaddy (Map 47–47); also widely distributed in S and SE Asia: Bangladesh, Indonesia, Laos, Malaysia, Philippines, Thailand, Vietnam, and S Yunnan of China; also in Oceania: Papua New Guinea.

C NEPAL: **Bagmati**, Nuwakot, 1800 – 1900 m, 1972, *H. Kanai et al. 721668* (TI). **E NEPAL**: **Kosi**, Arun Valley, 1370 m, 1956, *J. D. A. Stainton 1366* (BM, PE). **SIKKIM & DARJILING**: **Darjiling**, Lebong, 1520 m, 1881, *J. S. Gamble 9784* (K); **Sikkim**, Luycham, 1830 m, 1875, *C. B. Clarke 25457* (K). **BHUTAN**: **Punakha** (from Grierson & Long, *Fl. Bhutan* 2(3): 1384, 2001). **L YARLUNG ZANGBO**: **Nyingchi**, Dongjiu to Layue, 1953, *P. C. Tsoong 6259* (PE). **YARLUNG ZANGBO-BRAHMAPUTRA**: **Dexing**, 950 m, 2012, *D. Y. Hong et al. H12010* (PE); **Digdong**, 800–1000 m, 1973, *Biol. Inst. Tibet Exped. 1611* (HNWP); Jialasa to Suila, 1800 m, 1982, *S. Z. Cheng & B. S. Li 3033* (PE); **Mêdog** to Deergong, 1700 m, 1974, *Qinghai-Xizang Exped. 74–4430* (KUN, PE); **Xiazayü**, 1700 m, 1973, *Qinghai-Xizang Exped. 73–873* (KUN, PE). **S HENGDUAN**: **Fugong**, Lishadi 1610 m, 2005, *Gaoligong Shan Biodiversity Survey 27544* (CAS); **Gongshan**, Binzhongluo, 1740 m, 2004, *Gaoligong Shan Biodiversity Survey 22204* (CAS); **Yangbi**, Huiminqiao to Dacun, 1946, *T. N. Liou 22728* (PE). **U IRRAWADDY**: s. loc., 3050 m, 1924, *G. Forrest 24903* (K); **Tengchong**, Mingkwong Valley, 1830 m, 1912, *G. Forrest 8806* (K).

3. Cyclocodon axillaris (Oliv.) W. J. de Wilde & Duyfjes, Thai Forest Bull. Bot. 40: 22. 2012; Hong, Monogr. *Codonopsis* Allied Gen. Camp. 218. 2015. ≡ *Campanumoea axillaris* Oliv., Hook. Icon. Pl. 3: 8, t. 1775. 1888. Type: China, Sichuan, Omei (Emei), 1550 m, Dec. 1887, *R. E. Faber 253* (holotype: K; isotype: K).

Codonopsis albiflora Griff., Not. Pl. As. 4: 279. 1854, ambiguous name. Type: India, Mergue, about Banlaw and Yengboo, Feb. 1835, *Griffith 1013* (n.v., not preserved?).

Plants usually glabrous throughout. Stems erect or ascending, up to 3 m tall; branches many, horizontal or pendulous. Leaves opposite, rarely in whorl of 3; petiole 3–10 mm long; leaf blade ovate to lanceolate, 6–16 cm long, 1–7 cm broad, base rounded, margin serrate, dentate, or crenate. Flowers usually solitary, sometimes in dichasium of 3, usually both terminal and axillary; pedicel or peduncle 1–10 cm long, with a pair of filiform bracteoles at upper part. Calyx adnate to hypanthium up to lower part; calyx lobes (4) 5 (–7), filiform or linear, margin with several ramiform teeth. Corolla white or pale red, rotate, 7–12 mm long, 5- or 6 (7)-cleft to middle; corolla lobes ovate to ovate-triangular. Stamens 5 or 6; filaments as long as anthers, dilated base villous-ciliate. Stigma (4) 5–6 (–8)-fid. Berry dark purple when mature, globose, (4)5- or 6-locular, 5–10 mm in diameter Seeds ovoid. **Chromosomes**: 2n=18. Figure 47–18 f & g.

Phenology: flowering and fruiting from July to November.

Habitats: forests, thickets, or grasslands; below 1700 m.

Distribution: N Hengduan (Map 47–48); also widely distributed in E and SE Asia: S China, Indonesia, Japan (Ryukyu and Kyushu), Laos, Philippines, Thailand, and Vietnam.

N HENGDUAN: **Wenchuan**, Sanjiang, 1050 m, 2007, *X. J. He et al. 129841* (SZ).

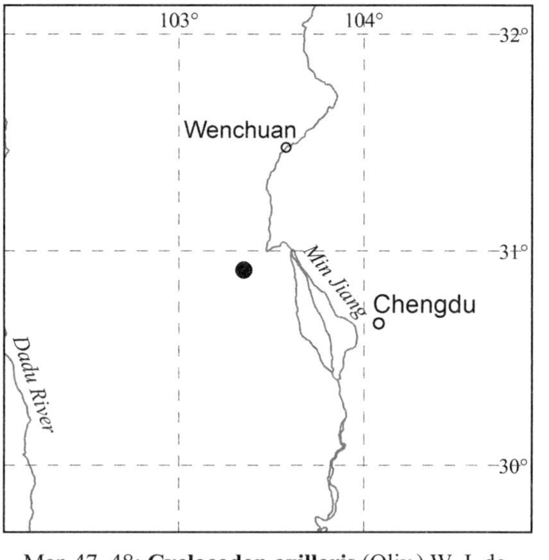

Map 47–48: **Cyclocodon axillaris** (Oliv.) W. J. de Wilde & Duyfjes

3. **CYANANTHUS** Wall. ex Benth.

Bentham in Royle, Ill. Bot. Himal. Mts. 309. 1836, nom. cons., non Rafinesque (1815); Hong & Ma, Acta Phytotax.Sin. 29(1): 25. 1991; Shrestha, Acta Phytotax. Sin. 35(5): 396. 1997. Type: *Cyananthus lobatus* Wall. ex Benth. (lectotype designated by Reveal & Hoogland, Taxon 40:650. 1991)

Herbs perennial or annual. Caudex thick if present, lignified, covered with numerous scales, usually branched. Stems caespitose. Leaves alternate, but sometimes 4 or 5 leaves in a whorl-like group below pedicel. Flowers solitary, terminal only on main stem or also on branches, rarely in dichasium at apex of stem and branches, pedicellate. Calyx tubular or tubular-campanulate, (4- or) 5-lobed. Corolla tubular-campanulate, (3–) 5-lobed. Stamens 5, less often 4; anthers often connivent and surrounding upper part of gynecium. Ovary superior, conical, (3–) 5-locular. Fruit a capsule, loculicidal. Seeds numerous, brown-red or brown-black. Pollen long-colpate. Chromosomes 2n=10, 12, 14, and 28.

Recent references: Zhou et al., 2013, Phylogenetic and biogeographic analyses of the Sino-Himalayan endemic genus *Cyananthus* (Campanulaceae) and implications for the evolution of its sexual system. *Mol. Phylogenet. Evol.* 68: 482–497; Chen et al., 2014, Systematic significance of cytology in *Cyananthus* (Campanulaceae) endemic to Sino-Himalayan region. *J. Syst. Evol.* 52(3): 260–270; Hong & Wang, 2015, A new taxonomic system of Campanulaceae s. s. *J. Syst. Evol.* 53(3): 203–209.

The genus *Cyananthus* comprises three monophyletic groups (Zhou et al., 2013; Hong & Wang, 2015). We treat these three groups here as three sections which are distinct from each other. Nineteen species in total in the Pan-Himalaya and adjacent regions; all in the Pan-Himalaya (11 endemic).

1a. Plants annual; caudex lacking (sect. 3. ***Annui***).

2a. Plants usually diffuse, less than 20 cm tall; flowers almost sessile, typically 4-merous; corolla usually less than 1 (–1.5) cm long .... 19. *C. hookeri*

2b. Plants erect or ascending, taller; flowers pedicellate, typically 5-merous; corolla usually over 1.5 cm long.

3a. Corolla pale yellow or white; calyx covered with red-brown setae, base of setae swollen, black verrucose, calyx lobes obovate-oblong, broadest near middle .... 16. *C. lichiangensis*

3b. Corolla blue or blue-purple; calyx villous, calyx lobes nearly linear or triangular, broadest at base.

4a. Leaves glabrous or sparsely puberulent; flowers in dichasia; calyx lobes mostly linear-triangular, 1/3–2/5 as long as calyx tube .... 18. *C. fasciculatus*

4b. Leaves hairy; flowers solitary, terminal on main stem and branches, rarely 3–5fascicled; calyx lobes triangular, 1/4–1/3 as long as calyx tube .... 16. *C. inflatus*

1b. Plants perennial, with a robust caudex densely covered with persistent and straw-coloured scales.

5a. Calyx with brown-black setae, sometimes mixed with other hairs; corolla lobes nearly orbicular, broadly ovate, or oblong, as long as or slightly longer than broad (sect. 2. ***Cyananthus***).

6a. Calyx with mixed white and brown-black setae; stems sparsely arachnoid or glabrous.

7a. Pedicel 2–9 cm long; stems glabrous; corolla lobed to middle .... 14. *C. ligulosus*

7b. Pedicel absent or nearly absent; stems sparsely arachnoid; corolla lobed to 1/3–2/5.

8a. Leaf blade cordate to triangular-ovate, base cordate or rounded, margin obviously crenulate .... 12. *C. cordifolius*

8b. Leaf blade linear-elliptic, base cuneate, margin entire or sinuous .... 13. *C. sherriffii*

6b. Calyx with only brown-black setae; stems villous, scaberulose or sericeous.

9a. Leaf blade usually shorter than 10 mm; corolla throat densely villous .... 11. *C. microphyllus*

9b. Leaf blade usually longer than 10 mm; corolla throat glabrous or sparsely villous.

10a. Leaf blade usually oblanceolate, spatulate, or obovate, margin with 3–9 obtuse lobes above middle .... 8. *C. lobatus*

10b. Leaf blade elliptic or ovate-lanceolate, margin entire or sinuous-crenulate above middle.

11a. Leaves several in a whorl-like group below pedicel; leaf blade adaxially glabrous, abaxially densely sericeous; pedicel 5–12 mm long .... 15. *C. sericeus*

11b. Leaves all alternate, not grouped below pedicel; leaf blade hispidulous on both surfaces; pedicel 1.2–4 cm long or absent, rarely up to 1.5 cm long.

12a. Pedicel usually absent, less often up to1.5 cm long; leaves usually ovate to elliptic, less often linear-lanceolate, base cuneate to decurrent into short petiole .... 10. *C. integer*

12b. Pedicel 1.2–4 cm long; leaves usually linear-lanceolate, base often amplexicaul .... 9. *C. pedunculatus*

5b. Calyx glabrous or covered with hairs other than brown-black setae; corolla lobes usually oblong, distinctly longer than broad (sect. 1. ***Stenolobi***).

13a. Flowers several in a raceme-like inflorescence; leaves ovate-lanceolate, abaxially densely sericeous; margin strongly revolute .... 7. *C. longiflorus*

13b. Flowers solitary and terminal; leaves rhombic, flabellate, spatulate, or ovate, rarely lanceolate, abaxially rarely sericeous, margin spreading or revolute.

14a. Leaves usually rhombic, flabellate, or rarely spatulate, (2–)2.5–4.8(–5.5) mm long, obviously petiolate, base usually truncate or broadly cuneate, margin strongly revolute.

15a. Flowers 1.6–2.4 cm long; leaves rhombic, base broadly cuneate, margin usually entire .... 1. *C. delavayi*

15b. Flowers 2.5–5.5 cm long; leaves flabellate to rhombic, base mostly truncate, margin serrate to sinuous-crenulate, rarely entire .... 2. *C. formosus*

14b. Leaves usually spatulate, elliptic, ovate, or obovate-lanceolate, (3–)4.5–16 mm long, attenuate into short petiole, or leaves sessile, base cuneate or rounded, margin spreading or slightly revolute.

16a. Plants except corolla densely covered with long yellow-brown appressed hairs; scales on caudex 10–15 mm long, ca. 3 mm broad; corolla 3–4 cm long, lobed to 1/3 .... 4. *C. wardii*

16b. Plants glabrous or covered with other types of hairs; scales on caudex much smaller; corolla 2–2.5(–3.8) cm long, lobed to middle.

17a. Corolla yellow or flavo-purple; leaf blade spatulate, nearly as long as wide; petiole narrowly winged, (0.5–)1.3–5 mm long .... 3. *C. macrocalyx*

17b. Corolla blue, blue-purple, yellow or white; leaf blade elliptic, ovate, narrowly ovate, or narrowly obovate, obviously longer than wide; petiole lacking or less than 1(–1.7) mm long.

18a. Corolla blue or blue-purple; leaves shortly petiolate or sessile, leaf blade elliptic, narrowly elliptic, or oblanceolate, base cuneate .... 5. *C. incanus*

18b. Corolla yellow or white; leaves sessile, leaf blade broadly ovate or obovate-lanceolate, base rounded .... 6. *C. flavus*

Sect. 1. Stenolobi (Franch.) Y. S. Lian in Hong, Fl. Reipubl. Popularis Sin. 73(2): 13. 1983. ≡ *Cyananthus* (unranked) *Stenolobi* Franch., J. Bot. (Morot) 1: 258. 1887. Type : *Cyananthus incanus* Hook. f. & Thomson (lectotype here designated).

Plants perennial. Caudex densely clothed with persistent and straw-coloured scales. Leaves alternate. Flowers solitary, terminal on main stem and branches, subtended by a whorl-like group of 3–5 leaves. Calyx glabrous or covered with hairs other than brown-black setae, often more or less swollen and sac-like after anthesis. Corolla lobes usually oblong, longer than broad.

Seven species; all in the Pan-Himalaya (3 endemic).

1. Cyananthus delavayi Franch., J. Bot. (Morot) 1: 280. 1887; Hong & Ma, Acta Phytotax. Sin. 29(1): 42. 1991; Shrestha, Acta Phytotax. Sin. 35(5): 403. 1997. Type: China, Yunnan, above Lankong, 09. Oct. 1884, *M. Delavay 137* [lectotype designated by Shrestha, 1997: P (P00235637); isolectotypes: K, P (P00235636)]; loc. eodem, 03. Sept. 1884, *M. Delavay 96* [paratypes: P (P00235631, P00235630, P00235632 & P00235633)].

Cyananthus microrhombeus C. Y. Wu in Wu & Li, Yunnan Trop. Subtrop. Fl. Res. Rep. 1: 85. 1965. Type: China, Yunnan, Shangri-la, Tangwu, 3700 m, 04 Aug. 1937, *T. T. Yü 12621* (holotype: KUN; isotype: PE).

Stems prostrate or ascending, 7–20 cm long, branched, hispid. Leaves alternate; petiole 1–3 mm long, white hispid; leaf blade nearly orbicular, ovate, or deltoid, 2–5 mm long, 1–7 mm broad, adaxially hispidulous, abaxially silvery sericeous, base cuneate to truncate, margin revolute, entire, sinuous, or crenulate. Flowers solitary, terminal on main stem and some branches, each subtended by a whorl-like group of 3–5 leaves; pedicel 3–9 mm. Calyx brown hirsute or glabrescent; calyx tube cylindrical at anthesis, 4.5–9 mm long, 3.5–7 mm broad, enlarged in fruit; calyx lobes narrowly triangular, 1.5–4 mm long, 1–3 mm broad, both surfaces hirsute. Corolla blue, tubular-campanulate, 1.6–2.4 cm long, inside densely villous at throat; corolla lobes linear-oblong, 6–12 mm long. Style extending to corolla throat; stigma 5-fid. Capsule conical, longer than calyx. Seeds brown, ellipsoid or short-cylindrical, ca. 1.8 mm long, 0.3 mm wide. **Chromosomes**: 2n=12.

Phenology: flowering in September.

Habitats: grassy calcareous slopes, meadows with shrubs, forests, forest margins, or thickets; 1900–4000 m.

Distribution: S Hengduan (Map 47–49); also in C Yunnan (Kunming) of China.

Notes: "*M. Delavay 96*" in P includes three collections: one was collected on September 03, 1884 from Lankong, and forms the syntypes of *Cyananthus delavai* Franch., the second was collected on August 26, 1884 from near Hokin, and is the type of *C. macrocalyx* Franch., and the third was collected on August 04, 1884 from Dali and forms the syntypes of *Cyananthus incanus* var. *leiocalyx* Franch. For avoiding confusion I attach here a sheet number to each specimen cited.

S HENGDUAN: **Jianchuan**, Shizhong Shan, 1975, *B. Y. Qiu 61135* (KUN); **Lanping**, Biluoxue Shan, 3600 m, 1964, *S. K. Wu 9036* (KUN); **Muli**, Yazui Forest Farm, 3600–3800 m, 1983, *Qinghai-Xizang Exped. 13238* (PE); **Ninglang**, Ningli, 3800 m, 1981, *Z. X. Tang et al. 1054* (PE); **Shangri-la** (Zhongdian), Tianci, 3200 m, 1998, *Q. E. Yang & H. Z. Kong 98–121*

(PE); **Xiangcheng**, Dalongwu, 3700 m, 1937, *T. T. Yü 13286* (KUN, PE); **Xichang, 1932,** *T. T. Yü 1285* (PE); **Xide,** 2800 m, 1976, *Sichuan Plant Surv. 14273* **(PE); Yanyuan,** 2800 m, **1932,** *T. T. Yü 1741* **(PE).**

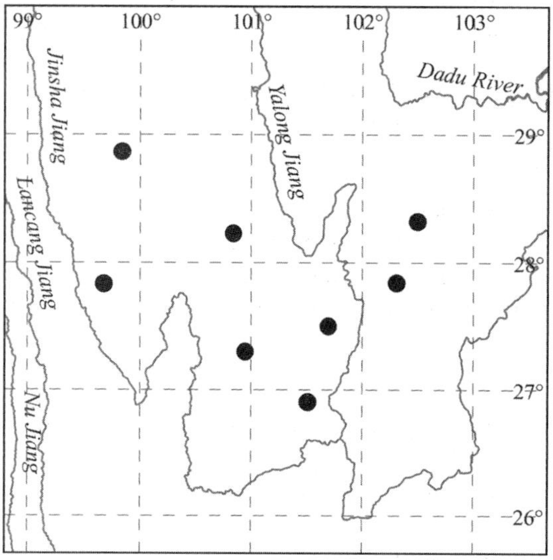

Map 47–49: **Cyananthus delavayi** Franch.

2. Cyananthus formosus Diels, Notes Roy. Bot. Gard. Edinburgh 5: 172. 1912; Hong & Ma, Acta Phytotax. Sin. 29(1): 43. 1991; Shrestha, Acta Phytotax. Sin. 35(5): 402. 1997. Type: China, Yunnan, E flank of Lichiang Range (Lijiang), 27°20′N, 3200–3600 m, Aug. 1906, *G. Forrest 2726* (lectotype designated by Shrestha, 1997: E; isolectotype: P).

Cyananthus chungdienensis C. Y. Wu in Wu & Li, Yunnan Trop. Subtrop. Fl. Res. Rep. 1: 83. pl. 30–3. 1965. Type: China, Yunnan, Chungdian (Shangri-la), Haba, 3500 m, 08 Sept. 1937, *T. T. Yü 13524* (holotype: KUN; isotype: PE).

Roots carrot-shaped, up to 1 cm wide. Caudex often branched. Stems prostrate to ascending, 5–25 cm long, simple or with short branches. Leaves alternate; petiole 2–7 mm long; leaf blade ovate or rhombic, 3–9 mm long, 2–6 mm broad, abaxially densely white hirsute, base obtuse or nearly truncate, margin revolute, crenulately 3–5-lobed or -parted, rarely entire, apex truncate. Flowers solitary, terminal on main stem and branches, subtended by a whorl-like group of 4 or 5 leaves; pedicel 3–5 mm long. Calyx densely hirsute with long pale brown stiff hairs; calyx tube cylindrical, 7–12 mm long; calyx lobes triangular, 4–6 mm long, 2–3.5 mm broad, both sides hirsute, margin ciliate. Corolla dark blue or purple-blue, 2.5–5.5 cm long, inside densely villous at throat; corolla lobes narrowly obovate or oblong, 1/3–1/2 as long as tube, apex with a tuft of hairs outside. Style extended up to corolla throat. **Chromosomes**: 2n=12. Figure 47–19: a.

Phenology: flowering in August and September.

Habitats: grassy slopes, edges of forests, glades, or screes; in limestone areas; 2800–4600 m.

Distribution: confined to S Hengduan (Map 47–50); also in NE Yunnan (Dongchuan) of China.

S HENGDUAN: Heqing, Lianping, 3600 m, 1940, *R. C. Ching 23998* (KUN, PE); **Jianchuan**, 3660–4270 m, 1922, *G. Forrest 23175* (PE); **Muli**, Shaoxiangliangzi, 3600 m, 1937, *T. T. Yü 7789* (PE); **Shangri-la** (Zhongdian), Haba, 3900–4200 m, 1962, *Zhongdian Exped. 1655* (KUN, PE); **Yanyuan**, Bailing, 3600 m, 1983, *Qinghai-Xizang Exped. 12689* (PE); **Yulong** (Lijiang), Yulong Snow Range, 1910, *G. Forrest 6245* (PE).

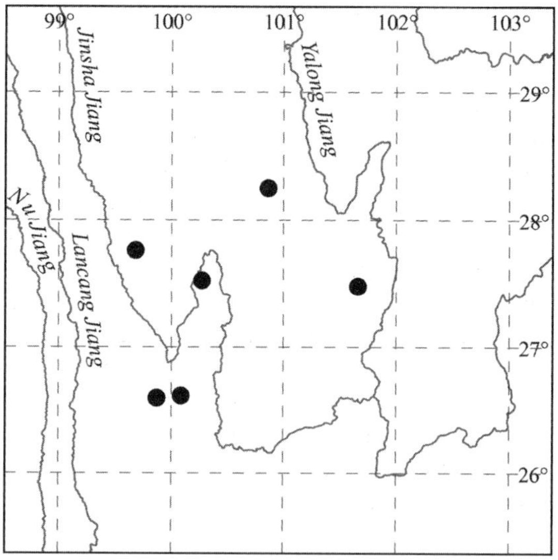

Map 47–50: **Cyananthus formosus** Diels

Figure 47–19: **a**. **Cyananthus formosus** Diels. **b–d**. **Cyananthus macrocalyx** Franch.: **b**, habit, showing root, caudex, and some of stems; **c**, flower; **d**, corolla opened, showing connivent stamens and style. Arrows indicate scales on caudex. **a** from *Icon. Cormophyt. Sin.* 4, 1975; **b–d** from *Fl. Reipubl. Popularis Sin.* 73(2), 1983.

3. Cyananthus macrocalyx Franch., J. Bot. (Morot) 1: 279. 1887; Ballard, Bot. Mag. 164: t. 9562. 1939; Hong & Ma, Acta Phytotax. Sin. 29(1): 43. 1991, p. p., excl. syn. "*C. wardii*". Type: China, Yunnan, near Hokin (Heqing), Mt Konalapo, 26 Aug. 1884, *M. Delavay 96* [(lectotype designated by Shrestha, 1997: P (P00237895); isolectotypes: K, P (P00237894)].

Cyananthus macrocalyx Franch. subsp. *spathulifolius* (Nannf.) K. K. Shrestha, Acta Phytotax. Sin. 35(5): 412. 1997. **syn**. **nov**. ≡ *Cyananthus spathulifolius* Nannf., Acta Horti Gothob. 5: 30. 1929. Type: China, Sichuan, "Sze-ch'uan, reg. bor.: Tsipula, in prato alpino", ca. 4000 m, 05. Aug. 1922, *H. Smith 3042* (holotype: E; isotype: PE).

Cyananthus leiocalyx (Franch.) Cowan subsp. *lucidus* K. K. Shrestha, Acta Phytotax. Sin. 35(5): 409. fig. 10. 1997. Type: China, Qinghai (not Xizang), Nangqên, 32°15′N, 96°40′E, 06. Sept. 1900, *V. Ladygin 530* (holotype LE; isotype: LE).

Cyananthus neurocalyx C. Y. Wu in Wu & Li, Yunnan Trop. Subtrop. Fl. Res. Rep. 1: 86. pl. 31–3. 1965. Type: China, Yunnan, Dali (?), *H. T. Tsai 53761*, p. p.(holotype: KUN).

Cyananthus pilosus (C. Marquand) K. K. Shrestha, Acta Phytotax. Sin. 35(5): 405. fig. 4. 1997. ≡ *Cyananthus macrocalyx* Franch. var. *pilosus* C. Marquand, Kew Bull. 1924: 251. 1924. Type: China, Yunnan, Tsekon, Jun. 1905, *P. Monbeig* s. n. (holotype: K; isotype: K).

Stems ascending, 5–20 cm long. Leaves alternate; petiole 1–4 mm long; leaf blade nearly orbicular, rhombic, or spatulate, 5–10 mm long, 1–6 mm broad, adaxially sparsely hispidulous, abaxially densely white hirsute, base attenuated into petiole, margin slightly revolute, entire or sinuous-serrate. Flowers solitary, terminal, subtended by a whorl-like group of 4 or 5 leaves; pedicel 4–10 mm long. Calyx brown hirsute; calyx tube cylindrical at anthesis, 7–13 mm, basally subglobose in fruit; calyx lobes triangular, both surfaces brown hirsute, or sometimes glabrous outside. Corolla yellow, sometimes purple- or red-veined, less frequently purple basally and yellow apically, tubular-campanulate, 2–3 cm long, inside densely villous at throat; corolla lobes oblanceolate, ca. 2/3 as long as tube. Style extended up to throat. Capsule exserted above calyx. Seeds oblong, ca. 1.3 mm long. **Chromosomes**: 2n=10. Figure 47–19: b-d.

Phenology: flowering in July and August; fruiting in August and September.

Habitats: alpine meadows, grassy slopes, rarely in forests; 2500–5000 m.

Distribution: E Nepal, Sikkim & Darjiling, Bhutan, L Yarlung Zangbo, Yarlung Zangbo-Brahmaputra, Tangut, N and S Hengduan, and U Irrawaddy (Map 47–51); also in Assam of India

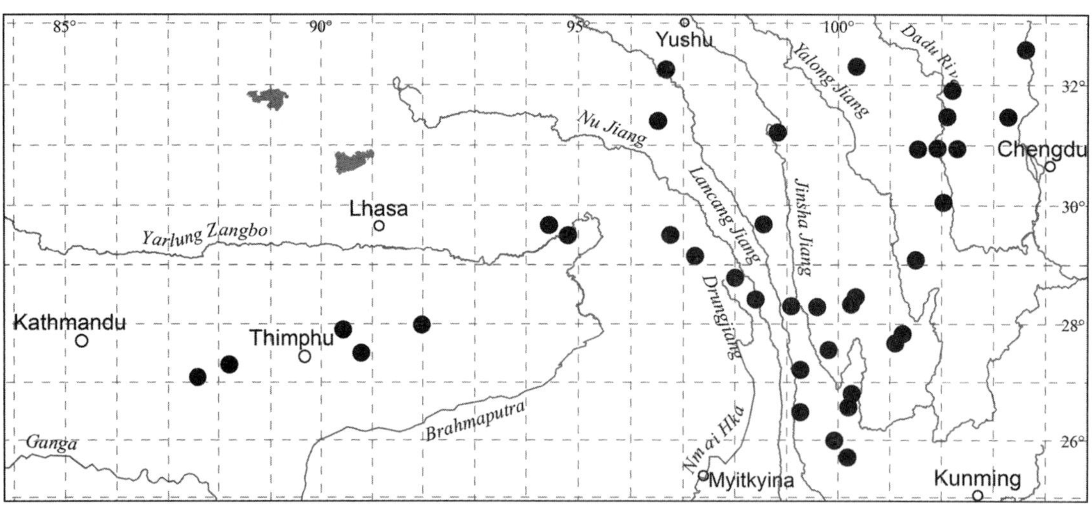

Map 47–51: **Cyananthus macrocalyx** Franch.

and NE Yunnan of China.

E NEPAL: Mechi, Tamur Valley, 3960 m, 1956, *J. D. A. Stainton 945* (E). **SIKKIM & DARJILING: Sikkim**, West Distr., 4420 m, 1992, *D. G. Long et al. 666* (E). **BHUTAN: Kiki-la**, Pummang, 3960 m, 1915, *R. E. Cooper 4118* (E); **Upper Mangde Chu,** 4270 m, 1949, *L. S. Hicks 16808* (E). **L YARLUNG ZANGBO**: **Mainling**, Paiqu, 4300 m, 1983, *B. S. Li et al. 5973* (PE): **Nyingchi**, Sheji-la, 4360 m, 1975, *Qinghai-Xizang Suppl. Exped. 751096* (PE). **YARLUNG ZANGBO-BRAHMAPUTRA**: **Cona**, 4600–4650 m, 1975, *Qinghai-Xizang Exped. 751530* (KUN); **Goyü** Pass, 4160 m, 2012, *D. Y. Hong et al. H12042* (PE); **TANGUT**: **Baxoi**, Rawu, 4100 m, 1973, *s. coll. 950* (PE); **Riwoqê**, Butuoka, 4450 m, 1991, *J. S. Yang 91–727* (PE); **Sêrtar**, Sêrqu, 1974, *G. C. Zhang 6703* (PE). **N HENGUAN**: **Baiyü**, Anzi, 4500 m, 1982, *Z. X. Tang & al. 731* (PE); **Barkam**, 1958, *C. L. Wu 32786* (PE); **Danba**, 1930, *Z. P. Huang et al. 1491* (PE); **Dawu**, Qianning, 3800 m, 1961, *J. S. Ying 9819* (PE); **Jinchuan**, Wanli, 4300 m, 1958, *X. Li 78196* (PE); **Lixian**, 1952, *Z. He & Z. L. Zhou 13690* (PE); **Markam**, 4100 m, 1992, *J. R. Chen 92355 A* (PE); **Songpan**, Yancong, 3615 m, 1983, *Z. X. Tang 1579* (PE); **Xiaojin**, Shepiliangzi, 4200 m, 1975, *Sichuan Plant Surv. 9748* (PE). **S HENGDUAN: Cawarong**, 3500 m, 1935, *C. W. Wang 65947* (PE); **Dali**, 1910, *G. Forrest 6986* (PE); **Daocheng**, Gonggaling, 3800 m, 1937, *T. T. Yü 13136* (KUN, PE); **Dêqên**, Baima Shan, 4500 m, 1981, *Qinghai-Xizang Exped. 2701* (PE); **Eryuan**, Fengxiang, 1940, *R. C. Ching 23196* (KUN, PE); **Heqing**, Xiushuihe, 3800 m, 1940, *R. C. Ching 24298* (PE); **Jiulong**, Metigonga, 4000 m, 1937, *T. T. Yü 6833* (PE); **Kangding**, Dapao Shan, 1951, *W. K. Hu & Z. He 10712* (PE); **Lanping**, 3500 m, 1932, *H. T. Tsai 53761* (PE); **Meigu**, Huangmaochang, 3800 m, 1976, *Sichuan Plant Surv. 13236* (PE); **Muli**, Mt Mitzuga, 4420 m, 1932, *J. F. Rock 23975* (PE); **Shangri-la** (Zhongdian), Xianrendong, 3450 m, 1937, *T. T. Yü 12223* (PE); Xiaozhongdian, 3200 m, 2010, *D. Y. Hong et al. H10021* (PE); **Weixi**, 2500 m, 1935, *C. W. Wang 67829* (KUN, PE); **Xiangcheng**, Rawu, 3900 m, 1981, *Qinghai-Xizang Exped. 3984* (PE); **Yanyuan**, Yuanbao, 3900 m, 1983, *Qinghai-Xizang Exped. 12243* (PE); **Yulong** (Lijiang), Yulong Snow Range, 4100 m, 1981, *Inst. Bot. Exped. To Hengduan Mts 2256* (PE). **U IRRAWADDY: Ridong**, 4000–4200 m, 1982, *Qinghai-Xizang Exped. 10749* (PE).

4. Cyananthus wardii C. Marquand, J. Linn. Soc., Bot. 48: 196. 1929; Lian in Hong, Fl. Reipubl. Popularis Sin. 73(2): 18. 1983, pro. syn.; Shrestha, Acta Phytotax. Sin. 35(5): 415. 1997. Type: China, SE Tibet, Nambu La, 3350–3650 m, 13. Aug. 1924, *F. Kingdon-Ward 6082* (holotype: K).

Roots ca. 5 mm in diameter Caudex secondary- or tertiary-branched, densely scaly; scales elliptic or spatulate, 10–15 mm long, 2–4 mm broad. Stems several from each branch of caudex, erect, 10–18 cm long, sparsely yellow-arachnoid, lower half with small, straw-coloured, elliptic scales. Leaves alternate, sessile or nearly so; leaf blade elliptic or spatulate, 5–11 mm long, 3–5 mm broad, both surfaces densely brown-yellow arachnoid, base cuneate, margin entire. Flowers solitary, terminal, subtended by a whorl-like group of 4 or 5 leaves; pedicel 2–4 mm long, brown-yellow arachnoid. Calyx densely brown-yellow arachnoid; calyx tube cylindrical, 10–12 mm long, 6–8 mm broad; calyx lobes ovate or triangular, 1/3–1/2 as long as tube. Corolla tubular, 3–4 cm, outside glabrous but inside densely villous; corolla lobes obovate, ca. 1/2 as long as tube, outside with a tuft of yellow villous hairs at apex. Stamens ca. 1/2 as long as corolla tube. Style slightly exceeding corolla tube.

Phenology: flowering in August.

Habitats: 2700–5000 m.

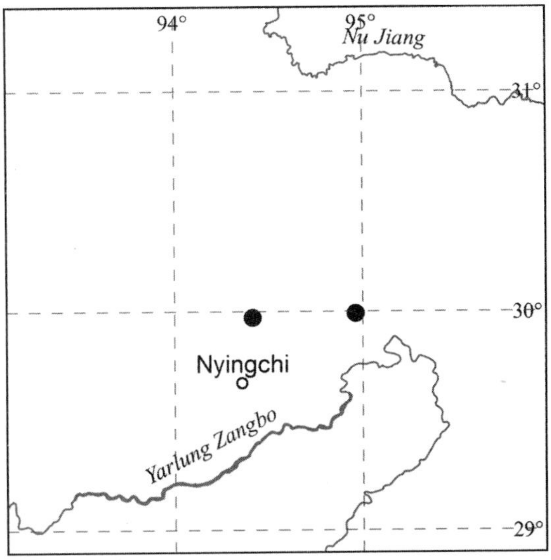

Map 47–52: **Cyananthus wardii** C. Marquand

Distribution: **endemic;** confined to L Yarlung Zangbo (Map 47–52).

L YARLUNG ZANGBO: **Nyingchi**, Tongyuk Dzong (Dongjug), 2730 m, 1947, *F. Ludlow et al. 12303* (BM, E).

5. Cyananthus incanus Hook. f. & Thomson, J. Proc. Linn. Soc., Bot. 2: 20. 1857; Lian in Hong, Fl. Reipubl. Papularis Sin. 73(2): 18. pl. 3: 4–6, 1983; Shrestha, Acta Phytotax. Sin. 35(5): 405. 1997. Type: India, Sikkim, 3660–3960 m, 23. Jul. 1849, *J. D. Hooker* s. n. (lectotype designated by Shrestha, 1997: K; isolectotypes: P, W).

Cyananthus dolichosceles C. Marquand, Bull. Misc. Inform. Kew 1924: 250. 1924. Type: China, Sze-Chuan (Sichuan), 2700–3200 m, Jul. 1903, *E. H. Wilson 3983* (lectotype designated by Shrestha, 1997: K; isolectotypes: BM, P).

Cyananthus leiocalyx (Franch.) Cowan, New Fl. Silva 10: 187. 1938. ≡ *Cyananthus incanus* var. *leiocalyx* Franch., J. Bot. (Morot) 1: 279. 1887. Type: China, Yunnan, Dali, 4000 m, 04 Aug. 1884, *M. Delavay 96* (lectotype designated by Shrestha, 1997: P; isolectotype: K).

Cyananthus neglectus C. Marquand, Bot. Mag. 147: tab. 8909. 1938. Type: Marquand's tab. cited.

Cyananthus petiolatus Franch., Bull. Soc. Phil. Paris 8: 147. 1891. ≡ *Cyananthus incanus* subsp. *petiolatus* (Franch.) D. Y. Hong & L. M. Ma, Acta Phytotax. Sin. 29(1): 44. 1991. Type: China, Sichuan, Tachienlu (Kangding). *M. Soulie* s. n. (lectotype designated by Shrestha, 1997: P; isolectotype: P).

Cyananthus pilifolius C. Y. Wu in Wu & Li, Yunnan Trop. Subtrop. Fl. Res. Rep. 1: 87, pl. 31–4. 1965. Type: China, Yunnan, Dêqên, Baima Shan, 3900 m, 10. Aug. 1937, *T. T. Yü 9586* (holotype: KUN; isotype: PE).

Stems erect or ascending, simple, less often branched below, white villous. Leaves alternate; petiole up to 1.7 mm long; leaf blade elliptic, narrowly elliptic, or oblanceolate, 4–12 mm long, 1.5–6 mm broad, both surfaces white hirsute, base cuneate, margin slightly revolute, nearly entire or sinuous. Flowers solitary, terminal on main stem, rarely also on branches, subtended by a whorl-like group of 4 or 5 leaves; pedicel 4–13 mm long, hirsute. Calyx densely or sparsely hirsute mixed with yellow-brown stiff hairs, rarely glabrous; calyx tube narrowly ellipsoid, 5–10 mm long; calyx lobes triangular, 2–4 mm long, inside densely white hirsute, margin long ciliate. Corolla dark blue or blue-purple, tubular, 2.5–3.8 cm long, inside densely villous at throat; corolla lobes narrowly obovate or oblong, ca. 2/3 as long as tube. Style extended up to corolla throat. Capsule exceeding calyx, 10–13 mm long. Seeds

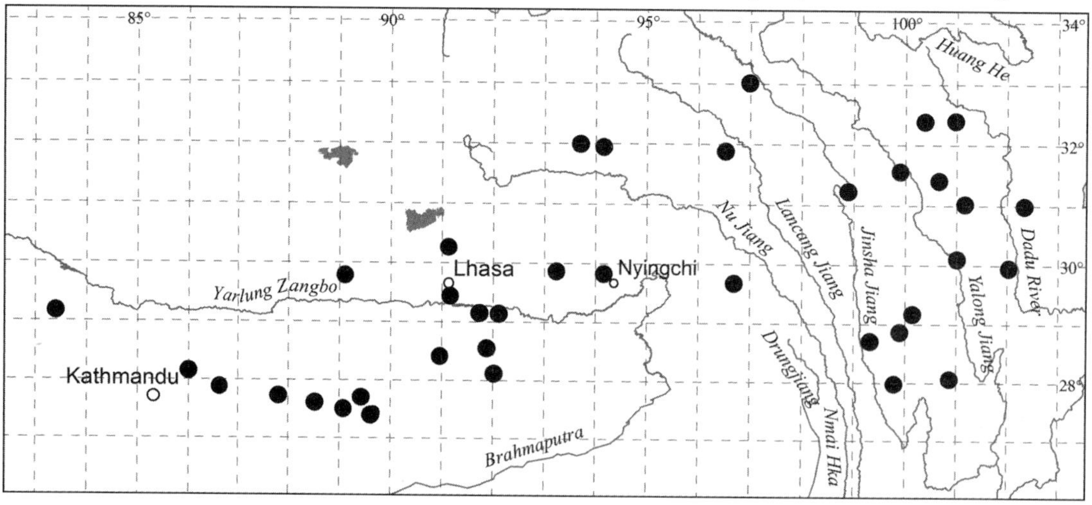

Map 47–53: **Cyananthus incanus** Hook. f. & Thomson

Figure 47–20: **a–c**, **Cyananthus incanus** Hook. f. & Thomson: **a**, stem; **b**, flower; **c**, calyx. **d** & **e**, **Cyananthus flavus** C. Macquand subsp. **montanus** (C. Y. Wu) D. Y. Hong & L. M. Ma: **d**, upper part of stem; **e**, flower. **f–h**, **Cyananthus longiflorus** Franch.: **f**, habit with arrow indicating caudex; **g**, two leaves, showing adaxial (lower) and abaxial (upper) surfaces; **h**, flower. From *Fl. Reipubl. Popularis Sin.* 73(2), 1983.

pale brown, oblong. **Chromosomes**: 2n=10. Figure 47–20: a–c.

Phenology: flowering from July to September; fruiting in September and October.

Habitats: grassy slopes, glades with grasses, thicket margins, or alpine meadows; 2700–5300 m.

Distribution: **endemic**; U Yarlung Zangbo, W and E Nepal, Sikkim & Darjiling, Bhutan, M and L Yarlung Zangbo, Yarlung Zangbo-Brahmaputra, Tangut, and N and S Hengduan (Map 47–53).

U YARLUNG ZANGBO: Nyalam, near city, 4100 m, 1981, *C. C. Ni et al. 2046* (PE). **W NEPAL: Karnali**, Dolpa, ca. 4500 m, 1973, *Grey-Wilson & Phillips 457* (BM). **E NEPAL: Mechi**, Taplejung Distr., 3700 m, 1985, *Curzan 117* (BM); **Sagarmatha**, Solukhumbu, 4950 m, 2005, *DNEP BX 196* (E). **SIKKIM & DARJILING**: **Sikkim**, Bikbari Valley, 4270 m, 1983, *AGSES 516* (K); Yalung, 4173 m, 1978, *D. Binns et al. 92* (E). **BHUTAN: Pomu-la**, 3720 m, 03. Aug. 1971, *Ramesh Bedi 608* (BM); **Somana**, 4371 m, 1971, *Ramesh Bedi 978* (BM); **Thimphu**, above Paro Pass, 3960 m, 1987, *Bigger 3029* (E); **Yala**, Pyala, 1938, *B. J. Gould 1348* (BM). **M YARLUNG ZANGBO**: **Lhasa**, Naiguoma Shan, 5000 m, 1965, *Y. T. Chang & K. Y. Lang 1845* (PE); **Lhozhag**, Lakang-Sêngera, 3500 m, 1975, *s. coll. 75–1107* (PE); **Lhünzhub**, Poindo, 4050 m, 1972, *Xizang Chinese Mater. Medic. Herbs Exped. 1923* (PE); **Namling**, Rendü, 4400 m, 1975, *Qinghai-Xizang Exped. 7406* (PE); **Nedong**, Hongmu-la, 4280 m, 1974, *Qinghai-Xizang Exped. 2900* (PE); **Yadong**, in vicinity of Pagri, 1953, *P. C. Tsoong 5801* (PE). **L YARLUNG ZANGBO**: **Gongbo'gyamda**, Sumdo-Mila Pass, 4700 m, 1965, *Y. T. Chang & K. Y. Lang 1417* (PE); **Mainling**, Zêtanggan, 4400 m, 1972, *Xizang Chinese Mater. Medic. Herbs Exped. 3874* (PE); **Nyingchi**, Naixi, 3100 m, 1965, *Y. T. Chang & K. Y. Lang 1270* (PE); **Qusum**, Budanla Shan, 4400 m, 1982, *C. C. Ni et al. 2715* (PE). **YARLUNG ZANGBO-BRAHMAPUTRA: Cona**, Boshan, 3300–3400 m, 1975, Q*inghai-Xizang Suppl. Exped. 751926* (PE); **Gujin**, 4100 m, 1980, *C. C. Ni & al. 1122* (PE); **Lhünzê**, Zêtang to Ritang, 4200 m, 1975, *Qinghai-Xizang Syppl. Exped. 751475* (PE); Gêxi, 4300 m, 1960, *J. W. Zhang 1200* (PE). **TANGUT**: **Baqên**, Ya'ngamdo, 1976, *D. D. Tao 11067* (PE); **Baxoi**, Rawu to Bêma, 4200 m, 1973, *Qinghai-Xizang Exped. 73–1269* (PE); **Garzê**, 1951, *W. K. Hu 13143* (PE); **Nangqên**, Bêca, 3900 m, 7. Sept. 1996, *T. N. Ho et al. 2949* (PE); **Sêrtar**, Wungda, 4040 m, 1961, *Q. L. Zhang 310164* (PE); **Sogxian**, Rongbo-Rabung, 3700–4500 m, 1976, *D. D. Tao 11000* (PE); **Yushu**, Xiao Surmang, 4300–4500 m, 1964, *s. coll. 545* (PE); **Zamtang**, Shangzhai, 3700 m, *s. coll. 1807*. **N HENGDUAN**: **Baiyü**, Changtai, 3950 m, 1982, *Z. X. Tang et al. 748* (PE); **Dawu**, Taining, 3700 m, 1934, *H. Smith 11984* (PE); **Luhuo**, Songlinkou, 3870 m, 1982, *Z. X. Tang 892* (PE); **Xiaojin**, Dawaliangzi, 4300 m, 1958, *S. S. Chang & Y. X. Ren 6917* (PE); **Yajiang**, 4000 m, 1974, *Q. Q. Wang 5017* (PE). **S HENGDUAN**: **Daocheng**, Sumdo, 4700 m, 1973. *Sichuan Veget. Surv. 3865* (PE); **Dêrong**, Zhege Shan, 4450 m, 1981, *Qinghai-Xizang Exped. 3494* (PE); **Kangding**, Dapao Shan, 3540 m, 1934. *C. S. Liu 1171* (PE); **Muli**, Shaoxiangliangzi, 1937, *T. T. Yü 7779* (PE); **Shangri-la** (Zhongdian), Gaza, 4030 m, 1981, *Z. X. Tang 712* (PE); **Xiangcheng**, Rawu, 3900–4000 m, 1981, *Qinghai-Xizang Exped. 3921* (PE).

6. Cyananthus flavus C. Marquand, Bull. Misc. Inform. Kew 1924: 247. 1924; Tsoong, Contr. Inst. Bot. Nat. Acad. Peiping 3(3): 106. 1935; Shrestha, Acta Phytotax. Sin. 35(5): 414. 1997. Type: China, Yunnan, "Lichiang（Lijiang） Range", 27°20'N, 3300–3600 m, Jul. 1906, *G. Forrest 2530* (lectotype designated by Shrestha, 1997: K; isolectotypes: BM, E).

Roots carrot-shaped, up to 5 mm in diameter. Caudex multibranched. Stems erect or ascending, 5.5–16 cm tall, simple or with short branches, densely white villous to nearly glabrous. Leaves alternate, sessile or nearly so; leaf blade elliptic, obovate-lanceolate, or broadly ovate, 5–16 mm long, 3–8 mm broad, both surfaces hirsute or glabrous, base cuneate, margin entire or slightly sinuate, slightly revolute. Flowers solitary, terminal, subtended by a whorl-like group of 4 or 5 leaves; pedicel 1–2 cm long, usually glabrous. Calyx glabrous except yellow-hirsute inside lobes; calyx tube broadly cylindrical, 8–12 mm long, base rounded or truncate, enlarged in fruit; calyx lobes triangular, 3–4 mm long, ca. 3 mm broad. Corolla white or yellow with various shades, tubular, 2.5–3 cm long, outside glabrous, inside densely white or yellow villous at throat; corolla lobes obovate-elliptic, 14–16 mm long, 5–7 mm broad, apex often very sparsely brown villous. Style exceeding corolla tube. Capsule nearly equal to calyx in length. **Chromosomes**: 2n=12.

Distribution: S Hengduan (Map 47–54); also in NE Yunnan of China.

1a. Flowers yellow, pale yellow, or greenish yellow; leaves broadly ovate, apex typically obtuse or rounded .... 6a. subsp. *flavus*

1b. Flowers white; leaves elliptic or obovate-lanceolate, apex typically acute .... 6b. subsp. *montanus*

6a. Cyananthus flavus subsp. **flavus**

Leaves broadly ovate, apex obtuse. Flowers yellow, pale yellow, or greenish yellow.

Phenology: flowering in July and August.

Habitats: grassy slopes; 3100–3600 m.

Distribution: **endemic;** confined to S Hengduan.

S HENGDUAN: Shangri-la (Zhongdian), Annan Chang, 1939, *K. M. Feng 2041* (KUN); **Yulong** (Lijiang), Xuesongcun, 3200 m, 1937, *T. T. Yü 15344* (PE).

6b. Cyananthus flavus subsp. **montanus** (C. Y. Wu) D. Y. Hong & L. M. Ma, Acta Phytotax. Sin. 29: 46. 1991; Shrestha, Acta Phytotax. Sin. 35(5): 414. fig. 14. 1997. ≡ *Cyananthus montanus* C. Y. Wu in Wu & Li, Yunnan Trop. Subtrop. Fl. Res. Rep. 1: 89. 1965. ≡ *Wahlenbergia mairei* H. Lév., Repert. Spec. Nov. Regni Veg. 12: 285. 1913. ≡ *Atropanthe mairei* (H. Lév.) H. Lév., Bull. Acad. Géog. Bot. 25 : 37. 1915. ≡ *Cyananthus mairei* (H. Lév.) Cowan, New Flora & Silva 10 : 188. 1938, non H. Lév., 1915 (=*Codonopsis bulleyana*). ≡ *Cyananthus albiflorus* Chamb., Notes Roy. Bot. Gard. Edinburgh 35: 252. 1977, based on *Wahlenbergia mairei*. Type: China, Yunnan, Paturages des h. plateaux à Ta-hai, 3200 m, Jul. 1912, *E. E. Maire* s. n. (holotype : E).

Leaves elliptic or obovate-lanceolate, apex acute. Flowers white. Figure 47–20: d&e.

Phenology: flowering from June to September; fruiting in September and October.

Habitats: grassy slopes, forests, or meadows with shrubs; 2700–3400 m. **Distribution**: S Hengduan; also in NE Yunnan of China.

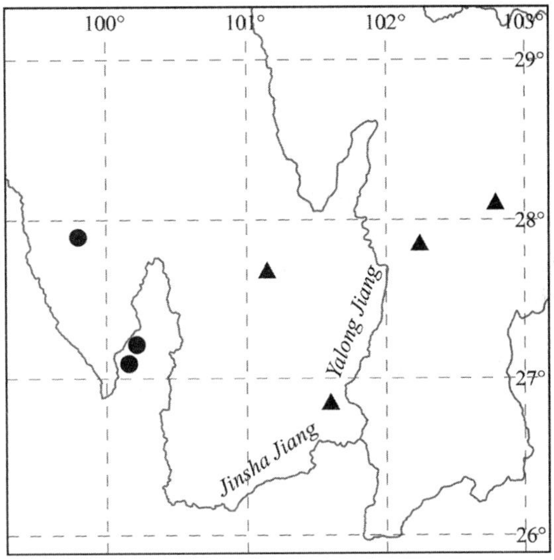

Map 47–54: **Cyananthus flavus** C. Marquand
● subsp. **flavus**, ▲ subsp. **montanus**

S HENGDUAN: Muli, Mianya, 3247 m, 2010, *E. D. Liu et al. 2560* (PE); **Xichang**, 1932, *T. T. Yü 1301* (PE); **Yanbian**, Yankou, 3150 m, 2002, *X. F. Gao et al. 3747* (PE); **Zhaojue**, Nidi, 3400 m, 1982, *Z. X. Tang et al. 163* (PE).

7. Cyananthus longiflorus Franch., J. Bot. (Morot) 1: 280. 1887; Marquand, Bot. Mag. 158: tab. 9387, 1935; Lian in Hong, Fl. Reipubl. Popularis Sin. 73(2): 23. pl. 4: 5–6. 1983. Type: China, Yunnan, Tali (Dali), Tsang-chan, 26 Sept. 1884, *M. Delavay 138* (loctotype designated by Shrestha, 1997: P; isolectotypes: E, K).

Cyananthus argenteus C. Marquand, Bull. Misc. Inform. Kew 1924: 253. 1924. Type: China,

Yunnan, Dali (Tali), Cangshan (Mt Tsang), 16 Oct. 1914, *C. Schneider 2774* (lectotype designated by Shrestha, 1997: K).

Cyananthus obtusilobus C. Marquand, Bull. Misc. Inform. Kew 1924: 254. 1924. Type: China, Yunnan, "Mountain of Tsang, near Tali" (Dali, Cangshan), 2800 m, 16 Oct. 1914, *C. Schneider 2753* (holotype: K).

Stems erect, 4–22 cm tall, woody toward base, densely gray-white lanate. Leaves alternate; petiole 2–4 mm long; leaf blade elliptic or ovate, 5–15 mm long, 2–8 mm broad, adaxially sparsely puberulent or glabrescent, abaxially densely silvery sericeous, base cuneate, margin revolute, entire. Flowers solitary, terminal on main stem and branches, nearly sessile, subtended by a whorl-like group of 3–5 leaves. Calyx villous; calyx tube cylindrical, 10–15 mm long; calyx lobes triangular or lanceolate, 3–7 mm long, pubescent throughout. Corolla blue-purple, tubular-campanulate, 3.5–5 cm long, inside densely villous at throat; corolla lobes obovate-oblong, from as long as to 1/2 as long as tube, with a tuft of setae at apex. Style extending almost up to corolla throat. Capsule slightly longer than calyx at maturity. Seeds oblong, ca. 1.3 mm long. **Chromosomes**: 2n= 12. Figure 47–20: f–h.

Phenology: flowering from July to September.

Habitats: pine forests, dry grassy slopes, or sand dunes; 2800–4300 m.

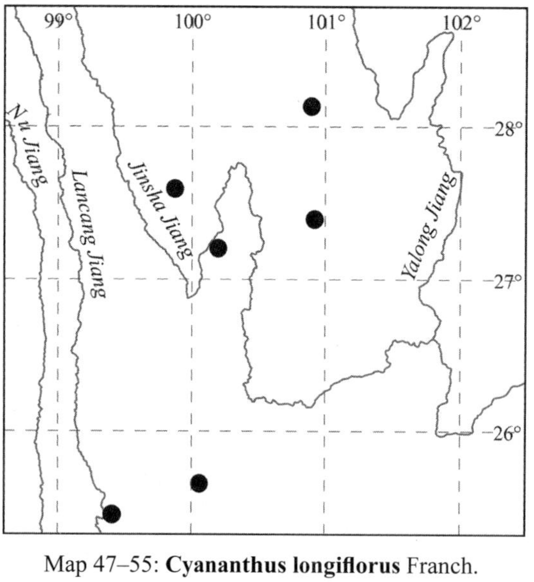

Map 47–55: **Cyananthus longiflorus** Franch.

Distribution: **endemic**; confined to S Hengduan (Map 47–55).

S HENGDUAN: Dali, 3960 m, 1929, *G. Forrest 27995* (PE); **Longyang**, Baoshan to Shuizai, 1941, *H. C. Wang 766* (PE); **Muli**, 3000 m, 1993, *Q. E. Yang 9305* (PE); **Ninglang**, Paomaping, 1981, *Tian & Zhang (Sample) 108* (PE); **Shangri-la** (Zhongdian), Haba, 2900 m, 1962, *Zhongdian Exped. 2024* (KUN, PE); **Yulong** (Lijiang), Longshan, 1983, *Qinghai-Xizang Exped. 14978* (PE).

Sect. 2. Cyananthus

Plants perennial. Caudex branched, densely clothed with persistent and straw-coloured scales. Calyx densely covered with brown-black setae, or sometimes mixed with other types of hairs. Corolla lobes nearly orbicular, broadly ovate or oblong, as long as or slightly longer than broad.

Eight species, all in the Pan-Himalaya (7 endemic).

8. Cyananthus lobatus Wall. ex Benth. in Royle, Ill. Bot. Himal. Mts. tab. 69. 1835; Hooker, Bot. Mag. 106: tab.6485, 1880; Shrestha, Acta Phytotax. Sin. 35(5): 421. fig. 21. 1997. Type: Nepal, Gosainkunda, *Wallich's collector 1473* (lectotype designated by Shrestha, 1997: K; isolectotypes: P, W).

Cyananthus hayanus C. Marquand, New Fl. Silva 8: 207. 1936 ('*hayana*'), **syn. nov.** Type: Nepal, Panjen, 4500 m, 18. August 1932, *K. Sharma E 427* (holotype: K; isotype: E).

Roots carrot-shaped, up to 1.3 cm in diameter. Stems decumbent or ascending, 10–50 cm long, sparsely villous toward apex. Leaves alternate, sessile or with petiole less than 5 mm long; leaf blade obovate, oblanceolate, spatulate, or rhombic, 7–35 mm long, 5–18 mm broad, abaxially villous, base cuneate, margin slightly revolute, 3–9-parted or -lobed toward apex. Flowers solitary, terminal on main stem, less often also on branches; pedicel 1–3 cm long, brown hirsute. Calyx tube cylindrical, 3–15 mm long, densely brown-red to brown-black hispid; calyx lobes triangular to lanceolate, 1/4–1/2 as long as tube, both sides brown hispid. Corolla pale blue to blue-purple, tubular-campanulate, 3–5.5

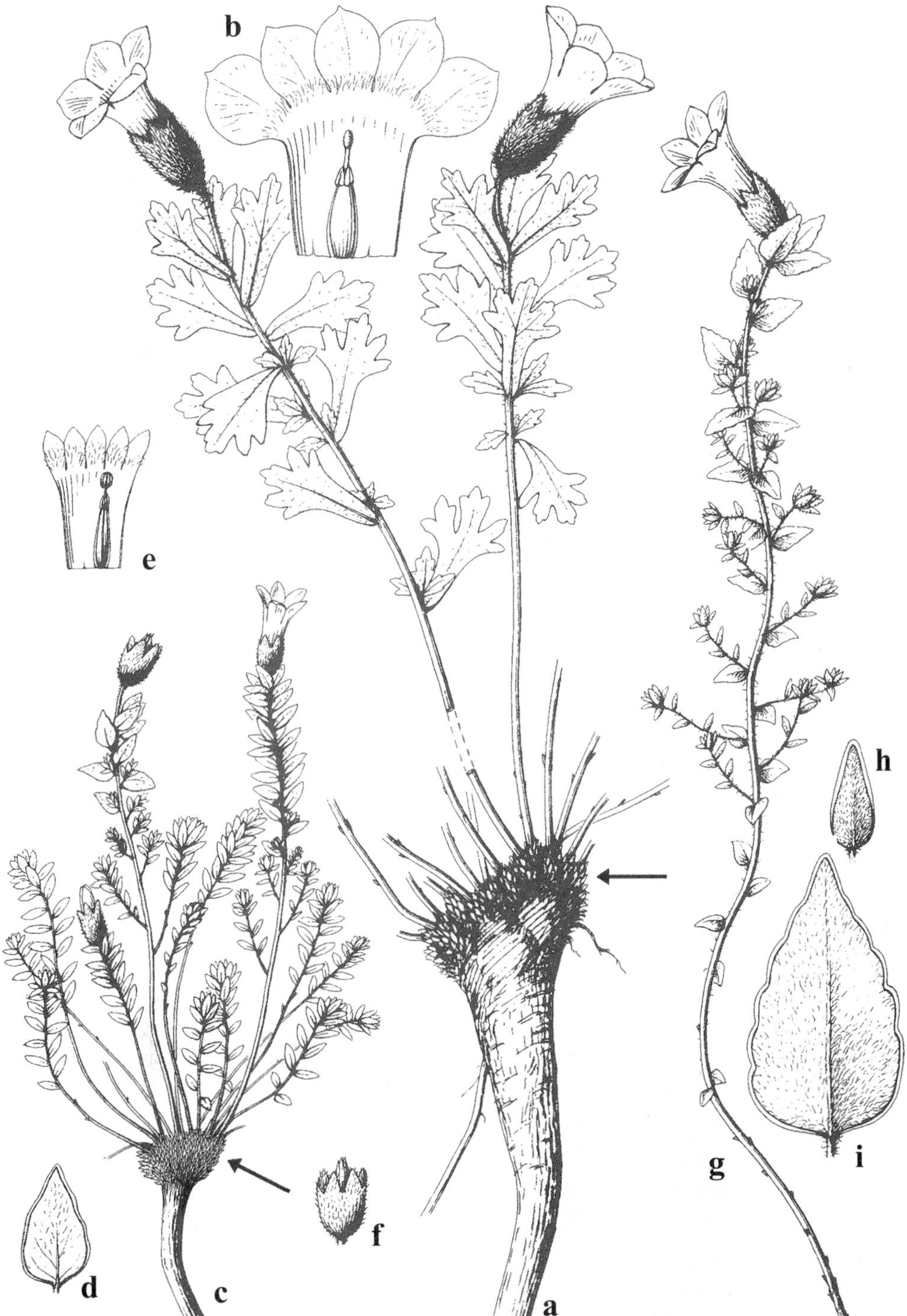

Figure 47–21: **a** & **b**, **Cyananthus lobatus** Wall. ex Benth.: **a**, habit; **b**, corolla opened, showing connivent anthers; **c–f**, **Cyananthus microphyllus** Edgew.: **c**, habit; **d**, adaxial surface of leaf, showing revolute margin; **e**, corolla opened, showing connivent anthers; **f**, calyx. **g–i**, **Cyananthus cordifolius** Duthie: **g**, stem with terminal flower; **h** & **i**, abaxial side of leaf, showing revolute margin. Arrows indicate caudex with numerous scales. From *Fl. Reipubl. Popularis Sin.* 73(2), 1983.

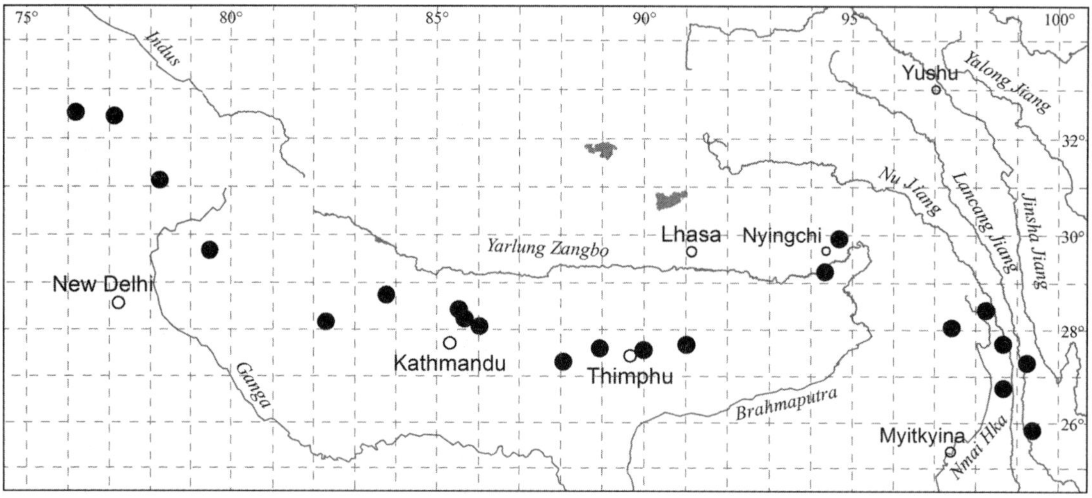

Map 47–56: **Cyananthus lobatus** Wall. ex Benth.

cm long, inside densely villous at throat; corolla lobes suborbicular, almost 1/2 as long as tube, apex with a tuft of short brown hairs in center. Style extending up to middle part of corolla. **Chromosomes**: 2n= 14. Figure 47–21: a & b.

Phenology: flowering in August and September.

Habitats: grassy slopes, meadows with shrubs, or forests; 2800–4500 m.

Distribution: U Ganges & Indus, U Yarlung Zangbo, W, C and E Nepal, Sikkim & Darjiling, Bhutan, M and L Yarlung Zangbo, Yarlung Zangbo-Brahmaputra, S Hengduan, and U Irrawaddy (Map 47–56); also in Punjab of India.

9. Cyananthus pedunculatus C. B. Clarke in Hooker, Fl. Brit. India 3: 434. 1881; Lian in Hong, Fl. Reipubl. Popularis Sin. 73(2):10. 1983; Hong, Lammers & Klein in Wu, Raven & Hong, Fl. China 19: 510. 2011. Type: India, Sikkim, Lachong, 3900 m, 05 Sept. 1849, *J. D. Hooker* s. n. (lectotype designated by Shrestha, 1997: K; isolectotype: P).

Cyananthus himalaicus K. K. Shrestha, Brittonia 44: 253. 1992, **syn.nov**. Type: Nepal, S of Gurjakhan; (Gajurkhani), 28°36′N, 83°13′E, 3200 m, 18 Aug. 1954, *Stainton, Sykes & Williams 3918* (holotype: BM; isotype: E).

Roots carrot-shaped, up to 1.2 cm in diameter. Stems ascending or sprawling, 10–20 cm long, usually simple, hispidulous. Leaves alternate, subsessile; leaf blade elliptic to linear-lanceolate, 7–16 mm long, 2–6 mm broad, both surfaces hispidulous, base rounded or cuneate, margin entire or shallowly sinuous above middle. Flowers solitary, terminal; pedicel 1.2–4 cm long, brown or black villous. Calyx densely brown-black hirsute; calyx tube cylindrical, 0.9–1.5 cm long, 0.5–1 cm wide; calyx lobes lanceolate-triangular, 2/5–1/2 as long as tube. Corolla purple-blue, funnelform-campanu- late, 2.5–4.5 cm long, throat glabrous or sparsely villous; corolla lobes nearly orbicular, ca. 1/2 as long as tube, apex with brown hirsute tuft. Style extending up to corolla throat. **Chromosomes**: 2n=14.

Phenology: flowering in August and September.

Habitats: alpine scrubs or meadows with shrubs; 3600–4900 m.

Distribution : **endemic**; U Yarlung Zangbo, E Nepal, Sikkim & Darjiling, Bhutan, and M Yarlung Zangbo (Map 47–57).

U YARLUNG ZANGBO: **Dinggyê**, Riwu, 4400–4700 m, 1990, *B. S. Li et al. 13035* (PE). **E NEPAL: Kosi,** Sankhuwasabha, 4100 m, 1991, *D. G. Long et al. 625* (E); **Mechi**, Taplejung, 1989, *KEKE 665* (K); **Sagarmatha,** Solukhumbu, 3810 m, 2005, *DNEP Bx 263* (E). **SIKKIM & DARJILING: Sikkim**, Uanqua Pass, 4570 m, 1909, *Smith & Cave 2266* (K) ; Sangla, 4890 m, 1911, *Ribu & Rohmoo 5342* (K). **BHUTAN: Chira**, 4270 m, 1938, *B. J. Gould 1312* (K). **M YARLUNG ZANGBO: Yadong**, Xiayadong, 1974, *Qinghai-Xizang Exped. 74–2369* (KUN, PE).

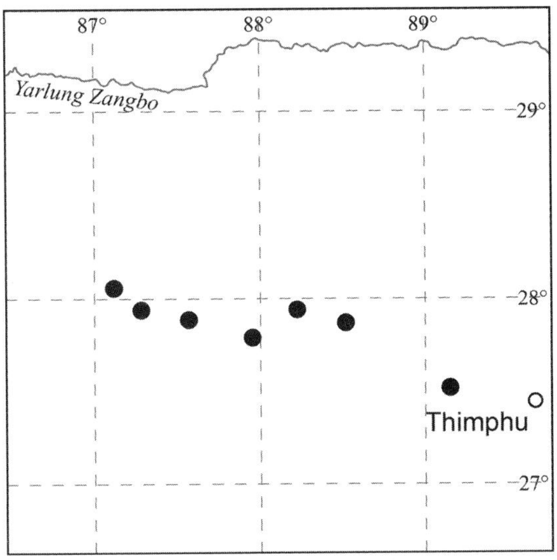

Map 47–57: **Cyananthus pedunculatus** C. B. Clarke

10. Cyananthus integer Wall. ex Benth. in Royle, Ill. Bot. Himal.: tab. 69. 1835; Haridasan & Mukherjee in Hajra & Sanjappa, Fasc. Fl. India 22: 90. 1996; Shrestha, Acta Phytotax. Sin. 35: 420. fig. 19. 1997. Type: India, Kumaon, *Wallich 1472* [lectotype designated by Shrestha, 1997: K; isolectotypes: BM, E, P (P00237846 & P 00237845), W].

Stems several dozens caespitose, prostrate, simple, 10–50 cm long, glabrous or villous. Leaves nearly sessile with petiole less than 2 mm long; leaf blade elliptic, ovate or lanceolate, 10–20 mm long, 3–10 mm broad, densely hispidulous on both surfaces, base cuneate or decurrent into petiole, margin serrate, apex acute. Flowers sessile or shortly pedicelled, pedicel up to 1.5 cm long, yellow-brown hirsute. Calyx black hirsute, cylindrical, 12–15 mm long, 7–9 mm wide; calyx lobes ovate-lanceolate, 1/3 or more as long as whole calyx, apex acute. Corolla pale blue, tubular-campanulate, 2.5–3.5 cm long; corolla lobes oblong, 6–11 mm long, 4–8 mm broad, apex obtuse or rounded. Stamens as long as corolla tube.

Phenology: flowering and fruiting from August to October.

Habitats: moist rocks in thi-ckets; 2800–3600 m.

Distribution: **endemic**; confined to U Ganga & Indus (Map 47–58).

U GANGES & INDUS: Tehri Garhwal, Jungnath, 3050 m, *R. Strachey & J. E. Winterbottom 3* (P); **Kidarkenta**, 3050–3350 m, *J. S. Gamble 1194* (K); **Uttarkashi**, above Mana, 3050–3660 m, 1844, *M. Edgeworth 455* (K).

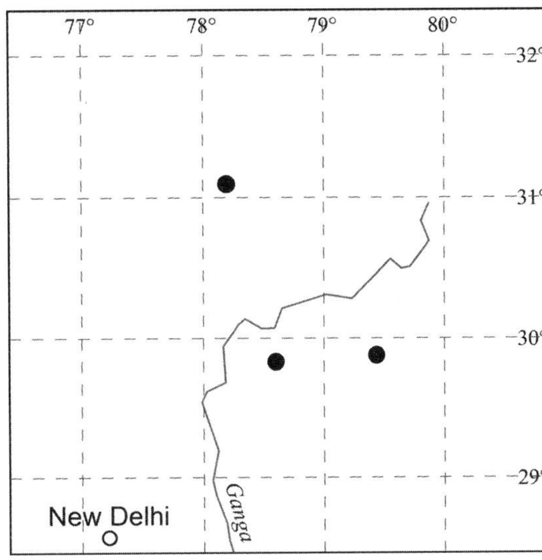

Map 47–58: **Cyananthus integer** Wall. ex Benth.

11. Cyananthus microphyllus Edgew., Trans. Linn. Soc. London 20: 81. 1846; Hong & Ma, Acta Phytotax. Sin. 29(1): 48. 1991; Shrestha, Acta Phytotax. Sin. 35(5): 417. 1997. Type: W Himalaya, above Mana, 3050–3660 m, Sept.–Oct. 1844, *M. Edgeworth 456* (holotype: K)

Cyananthus microphyllus Edgew. subsp. *williamsonii* K. K. Shrestha, Acta Phytotax. Sin.35(5): 419. fig.18. 1997. **syn. nov**. Type: W Nepal, Phoksumdo Tal, 5000 m, 21 Sept. 1952. *O. Polunin, W. Sykes & L. Williams* 3537 (holotype: BM; isotype: P).

Cyananthus nepalensis Kitam., Acta Phytotax. Geobot. (Kyoto) 15: 109. 1954. Type: Nepal, circa Shiar Khola, Tsumje, 2500 m, inter frutices Juniperini, 24 Jul. 1953, *S. Nakao* s. n. (holotype: KYO, n. v.)

Stems decumbent or ascending, 5–25 cm long, branched toward base, glabrous or sparsely

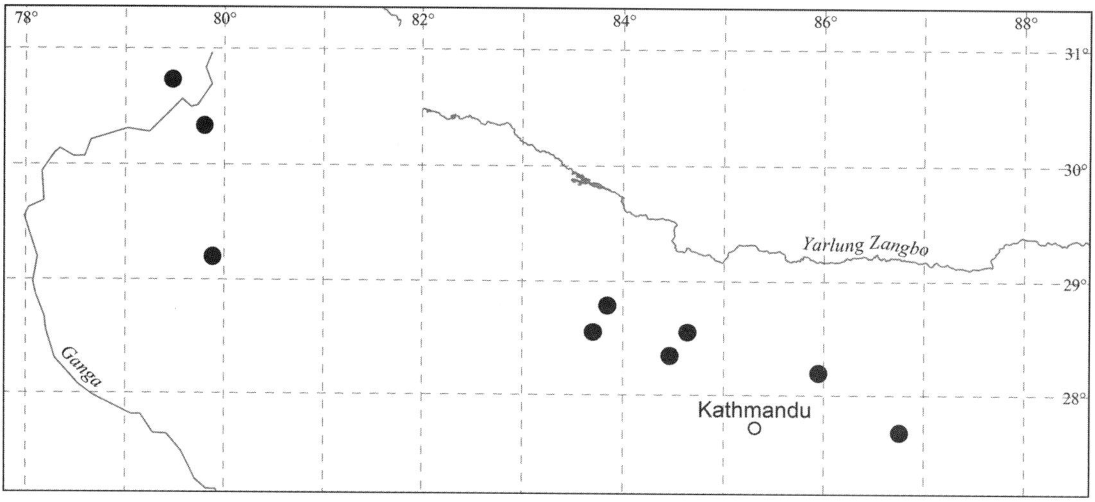

Map 47–59: **Cyananthus microphyllus** Edgew.

puberulent. Leaves alternate, almost sessile; leaf blade ovate, lanceolate, or linear-elliptic, 5–13 mm long, 1.5–5 mm broad, adaxially glabrous, abaxially sericeous, base rounded to subcordate, margin revolute, entire or sinuous. Flowers solitary and terminal; pedicel 3–10 mm long, brown- or black-setose. Calyx brown- or black-setose; calyx tube cylindrical, 7–10 mm long, base truncate; calyx lobes triangular, 1/2–2/3 as long as tube, both sides hairy. Corolla blue-purple or blue, cylindrical-campanulate, 1.8–2.8 cm long, outside glabrous or brown- or black-setose, inside densely fimbriate white- villous; corolla lobes obovate-oblong, nearly as long as tube. Style extending up to corolla throat. Seeds shiny brown, oblong, ca. 2 mm long, 0.5–0.7 mm wide. **Chromosomes**: 2n=14. Figure 47–21: c–f.

Phenology: flowering in July and August; fruiting in September.

Habitats: meadows or grassy slopes; 3000–5500 m.

Distribution: **endemic;** U Ganga & Indus, U Yarlung Zangbo, C and E Nepal (Map 47-59).

U GANGA & INDUS: **Bidang**, Dad Valley, 1901, *J. F. Duthie 24622 (a)* (K); **Kumaon**, Ralum River, 3350 m, *R. Strachey & Winterbottom 2* (K); **Tehri Garhwal**, Mt Badrinath, 3250 m, 1975, *F. Billiet & J. Leonard 6647* (K). **U YARLUNG ZANGBO: Nyalam**, 5500 m, 1992, *J. R. Chen 92212* (PE). **C NEPAL: Dhawalagiri**, Mustang, 3450–3960 m, 1995, *M. Mikage & K. Yonekura 9552411* (E); **Gandaki**, Marsyandi Valley, 3660 m, 1983, *R. J. D. McBeath 1552* (E); Gorkha, 3380 m, 1994, *M. Suzuki et al. 8485356* (PE). **E NEPAL**: **Sagarmatha,** Solukhumbu, 1985, *H. Ohba et al. 8581107* (E).

12. Cyananthus cordifolius Duthie, Bull. Misc. Inform. Kew, 1912: 37. 1912; Lian in Hong, Fl. Reipubl. Popularis Sin. 73(2): 13. pl. 1: 7–9.1983; Shrestha, Acta Phytotax. Sin. 35(5): 421. fig. 20. 1997. Type: W Nepal, Nampa gadh, 3350 m, 28 Jul. 1886, *J. F. Duthie 5730* (lectotype designated by Shrestha, 1997: K).

Stems slender, decumbent, 14–27 cm long, sparsely arachnoid, sterile-branched at middle. Leaves alternate, sessile or subsessile; leaf blade cordate to triangular-ovate, 5–15 mm long, 3–9 mm broad, adaxially glabrous or sparsely puberulent, abaxially sparsely villous, base cordate to or rounded, margin serrate, apex acute or obtuse; leaves on sterile branches much smaller. Flowers solitary and terminal; pedicel 5–18 mm long, brown or black setose. Calyx densely brown- or black-setose; calyx tube cylindrical, 11–16 mm long, 5–8 mm broad, base truncate; calyx lobes lanceolate-triangular, slightly shorter than tube, inside black- and white-villous. Corolla blue or pale blue, cylindrical-campanulate, 2.5–3.5 cm long, densely villous at throat; corolla lobes oblong, 12–15 mm long, 5–6 mm broad. Style extending up to corolla throat. **Chromo-**

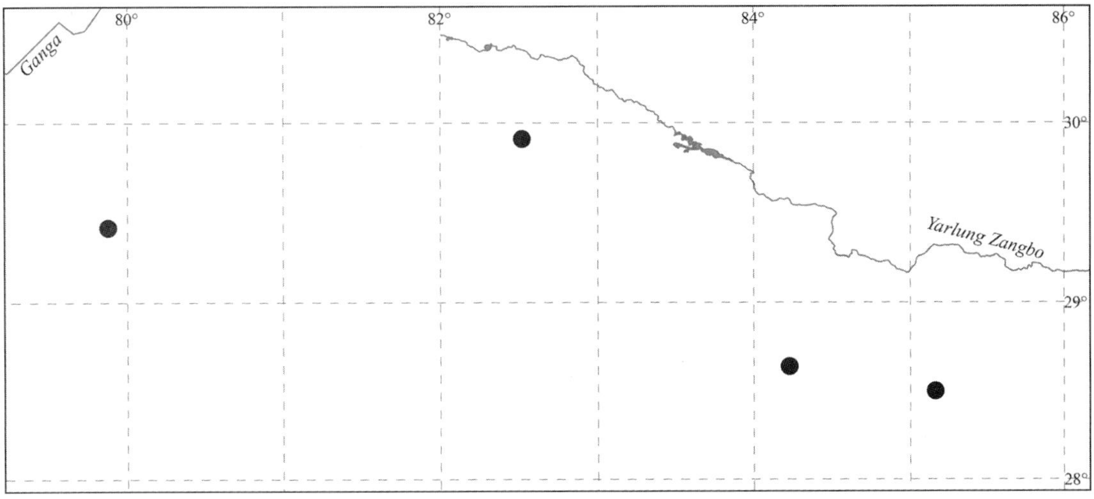

Map 47–60: **Cyananthus cordifolius** Duthie

somes: 2n=14. Figure 47–21: g–i.

Phenology: flowering in August.

Habitats: thickets; 3300–5000 m.

Distribution: **endemic**; U Ganga & Indus, U Yarlung Zangbo, and C Nepal (Map 47–60).

U GANGA & INDUS: Kumaon, Kutti Yangti Valley, 3660–3960 m, 1886, *J. F. Duthie 3114* (BM, K). **U YARLUNG ZANGBO: Gyirong**, Gongdan, 4400 m, 1975, *Qinghai-Xizang Exped. 75–7185* (KUN, PE); Gyirong Town, 3700 m, 2013, *PE-Tibet Exped 3827* (PE). **C NEPAL: Gandaki**, Manang, 4270 m, 1983, *R. J. D. McBeath 1461* (E).

13. Cyananthus sherriffii Cowan, New Fl. & Silva 10: 181. 1938; Hong & Ma, Acta Phytotax. Sin. 29(1): 48. 1991; Shrestha, Acta Phytotax. Sin. 35(5): 416. fig. 16. 1997. Type: China, SE Tibet, Gyandro, Chagul-Charma, 28°20′N, 92°54′E, 4900 m, 01 Nov. 1936, *F. Ludlow & G. Sherriff 2819* (holotype: BM).

Caudex somewhat woody. Stems 5–15 cm tall, sparsely arachnoid. Leaves alternate, sessile or nearly sessile; leaf blade linear or narrowly elliptic, 4–8 mm long, 1–3 mm broad, adaxially sparsely arachnoid-villous along midvein, abaxially silvery-sericeous, base cuneate, margin entire or sinuous. Flowers solitary, terminal on longer stems, subtended by a whorl-like group of 4 or 5 leaves 8–12 mm long, 2–3.5 mm broad; pedicel 5–8 mm long. Calyx glaucous and densely brown- or black-setose; calyx tube cylindrical, 10–15 mm long, 5–7 mm wide, base truncate; calyx lobes triangular or lanceolate, ca. 1/3 as long as tube, both sides pubescent. Corolla blue or gray-blue, cylindrical-campanulate, ca. 2.5 cm long, inside villous at throat; corolla lobes oblong, apex with a tuft of brown villous hairs. Ovary nearly as long as calyx; style extending up to corolla throat

Phenology: flowering in July and August; fruiting from September to early November.

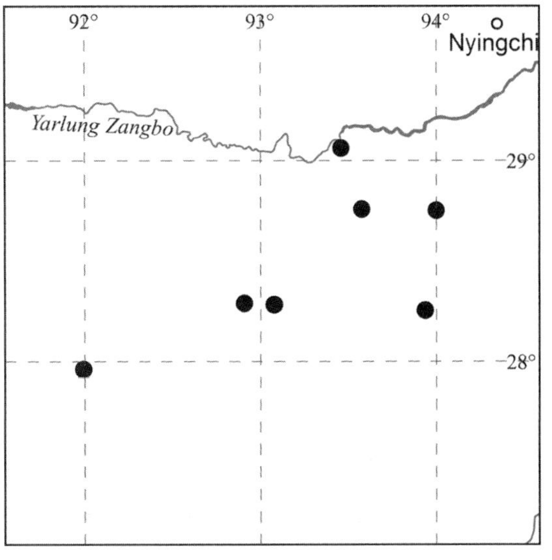

Map 47–61: **Cyananthus sherriffii** Cowan

Habitats: alpine meadows or thickets; 3200– 5300 m.

Distribution: **endemic**; confined to L Yarlung Zangbo and Yarlung Zangbo-Brahmaputra (Map 47–61).

L YARLUNG ZANGBO: Nangxian, Bimbi-la, 4270–4720 m, 1936, *F. Ludlow & G. Sherriff 2557* (E); Gyêmdong, 4619 m, 2012, *FLPH Tibet Exped. 12–1094* (PE); **Mainling**, Molo, 3200 m, 1938, *F. Ludlow et al. 5655* (BM). **YARLUNG ZANGBO–BRAHMAPUTRA**: **Cona**, Langpo, 2013, *Y. S. Chen et al. 13–0921* (PE); **Kashong-la**, Chagul Chu, 4270 m, 1936, *F. Ludlow & G. Sherriff 2402* (BM); **Le-la**, Chagul-Charma, 4570 m, 1936, *F. Ludlow & G. Sherriff 2458* (BM).

14. Cyananthus ligulosus D. Y. Hong, **sp. nov.** Type: China: Tibet, Gyirong, Gyirong Gou, above Zhuotang Village, 28°32′N, 85°14′E, 3520 m, 9 Oct. 2011, *Y. S. Chen, S. R. Zhang, S. X. Yu & R. Wei 181* (holotype: PE);

Figure 47–22: **a & b**, **Cyananthus ligulosus** D. Y. Hong: **a**, stem; **b**, lower part of plant, showing root, caudex, and lower part of stems. Drawn by Ai-Li LI.

Roots over 10 cm long, ca. 3 mm in diameter. Caudex short, ca. 6 mm long, covered with numerous pale gray scales. Stems caespitose, ascending or procumbent, 6–30 cm long, glabrous, simple. Leaves alternate; petiole 1–2 mm long; leaf blade ovate or elliptic, 4–14 mm long, 3–8 mm broad, adaxially glabrous, abaxially hirsute, base nearly truncate, rounded or broad-cuneate, margin entire, slightly revolute, apex acute or obtuse; (3–) 5 leaves at top of stem in a whorl, much narrower than those below. Flowers solitary, terminal: pedicel 2–9 cm long, sparsely villous. Calyx tube short-tubular, covered densely with brown-black setae, mixed with a few white setae; calyx lobe ca. 1/3 as long as calyx tube,narrow-triangular.

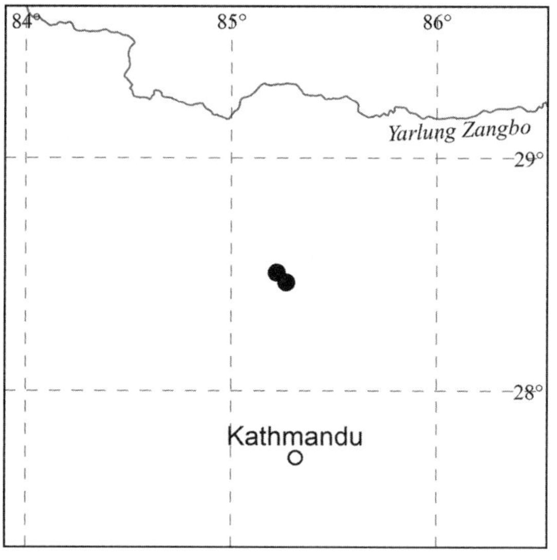

Map 47–62: **Cyananthus ligulosus** D. Y. Hong

Corolla blue, tubular, 2.5–4 cm long, lobed to ca. 1/2 of corolla total length; corolla lobes ligulate, densely white villous near base. Style extending up to corolla throat. Stamens glabrous, 1/2 as long as style; anthers connivent. Figure 47–22.

Phenology: flowering in October and November.

Habitats: grassy slopes by forests, or thickets; 3500–3800 m.

Distribution: **endemic**; confined to U Yarlung Zangbo (Gyirong) (Map 47–62).

Notes: *C. ligulosus* certainly belongs to the section *Cyananthus* with brown-black setae on calyx, but is distinct from all the other species in this group. It has three diagnostic characters: 1) a whorl of (3–) 5 leaves at the top of stem, which are much narrower than the others; 2) corolla lobes ligulate, ca. 4 times longer than broad (vs. oblong to nearly orbicular, nearly as long as broad); 3) very long pedicel.

U YARLUNG ZANGBO: **Gyirong**, Gyirong Gou, 3809 m, 2011, *Y. S. Chen et al. 491* (PE).

15. Cyananthus sericeus Y. S. Lian, Acta Phytotax. Sin. 17(3): 122. fig. 1. 1979 & in Hong, Fl. Reipubl. Popularis Sin. 73(2): 10. pl. 2: 1–4, 1983. Type: China, S Tibet, Cona, 3500–3600 m, 10 Sept. 1975, *Qinghai-Xizang Suppl. Exped. 751932* (holotype: PE).

Stems erect or ascending, 20–30 cm tall, simple, densely sericeous when young. Leaves alternate, sessile; leaf blade of middle and upper leaves narrowly elliptic, 8–12 mm long, 2–4 mm broad, adaxially glabrous or sparsely hairy, abaxially densely retrorsely and appressed sericeous. Flowers solitary, terminal, subtended by a whorl-like group of 3 or 4 leaves; pedicel 5–12 mm long, brown- or black-villous. Calyx densely brown or black hirsute; calyx tube cylindrical, 1.5–2 cm long, 0.9–1.1 cm wide; calyx lobes lanceolate-triangular, ca. 1/2 as long as tube, margin white villous. Corolla blue, with dark-blue veins, tubular-campanulate, 3.5–4 cm long, glabrous; corolla lobes broadly obovate, ca. 1/2 as long as tube. Ovary slightly shorter than calyx. Style almost extending up to corolla throat. Figure 47–23.

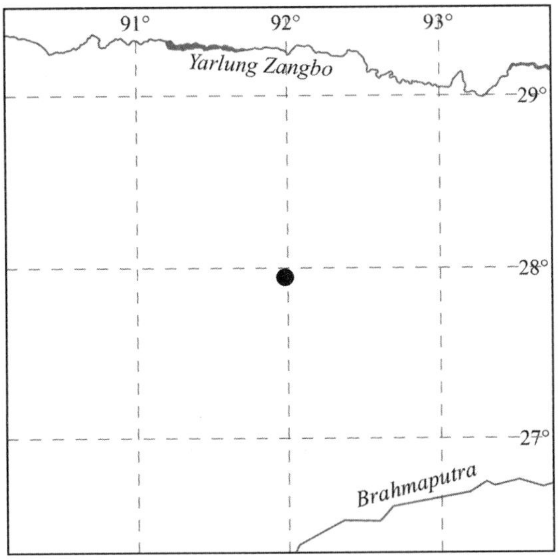

Map 47–63: **Cyananthus sericeus** Y. S. Lian

Phenology: flowering in September

Habitats: rocky crevices; 3500–3600 m.

Distribution : **endemic**; confined to Yarlung Zangbo–Brahmaputra (Cona) (Map 47–63).

Sect. 3. Annui (Y.S. Lian) K. K. Shrestha, Acta Phytotax. Sin. 35(5): 425. 1997. ≡ *Cyananthus* sect. *Stenolobi* subsect. *Annui* Y.S. Lian in Hong, Fl. Reipubl. Popularis Sin. 73(2): 183. 1983. Type: *Cyananthus inflatus* Hook. f. & Thomson.

Cyananthus subg. *Micranthus* K. K. Shrestha, Acta Phytotax. Sin. 35: 430. 1997. **syn. nov.** Type: *Cyananthus hookeri* C. B. Clarke.

Plants annual, lacking caudex. Flowers 5- but sometimes 4- or even 3- merous.

Four species in the Pan-Himalaya and adjacent regions; all in the Pan-Himalaya (one endemic).

16. Cyananthus lichiangensis W. W. Smith, Notes Roy. Bot. Gard. Edinburgh 8: 109. 1913; Tsoong, Contr. Inst. Bot. Nat. Acad. Peiping 3(3): 108. 1935; Shrestha, Acta Phytotax. Sin. 35(5): 426. fig. 25. 1997. Type: China, Yunnan, W flank of Lichiang (Lijiang), 3900 m, Aug. 1910, *G. Forrest 6369* (lectotype designated by Shrestha, 1997: E; isolectotype: K).

Stems caespitose, erect, 10–30 cm tall, glabrous. Leaves alternate; petiole 2–5 mm long, villous; leaf blade ovate or triangular, 5–13 mm long, 5–11 mm broad, both surfaces hispidulous, base truncate or rounded, margin revolute, entire, sinuous, or crenulate. Flowers solitary, terminal on both main stem and branches, subtended by a whorl-like group of 4 or 5 leaves; pedicel 2–5 mm long. Calyx red-brown setose outside, base of setae swollen, often black verrucose; calyx tube cylindrical, 8–10 mm long, 6–9 mm wide; calyx lobes obovate-oblong, ca. 1/3 as long as tube, broadest near middle, outside sparsely red-brown setose, inside appressed red-

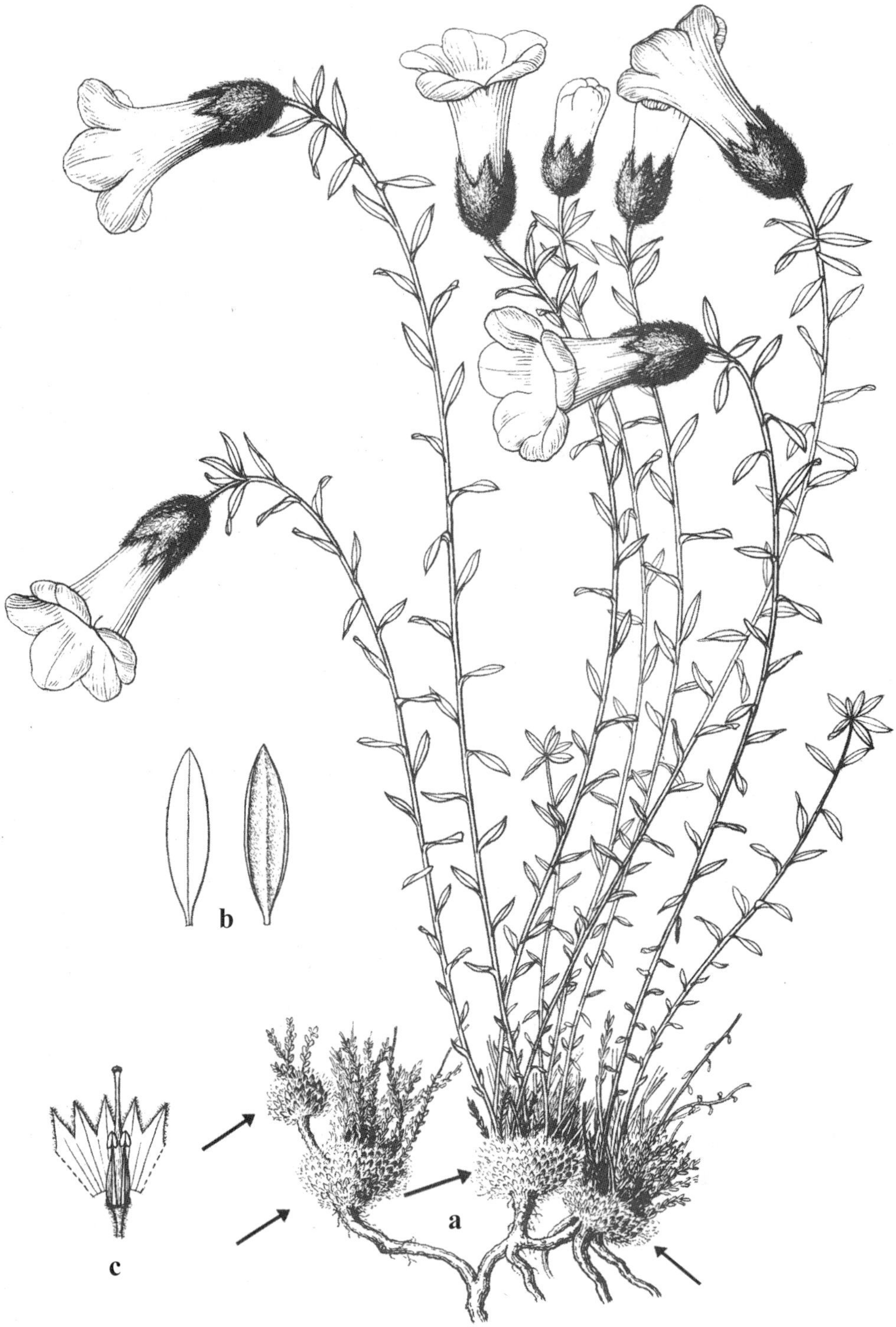

Figure 47–23: **Cyananthus sericeus** Y. S. Lian: **a**, habit ; **b**, adaxial (left) and abaxial (right) surfaces of leaves, showing revolute margin; **c**, flower with calyx removed and corolla opened. Arrows indicate caudex with numerous scales. From *Fl. Reipubl. Popularis Sin.* 73(2), 1983.

brown villous. Corolla pale or greenish yellow, cylindrical-campanulate, 1.6–2 cm long, inside densely villous at throat; corolla lobes oblong, 1/3–1/2 as long as tube. Style extending up to corolla throat. Capsule exceeding calyx at maturity. Seeds short-cylindrical, ca. 1 mm long, 0.5 mm wide. **Chromosomes**: 2n=14.

Phenology: flowering in August and September; fruiting in September and October.

Habitats: grassy slopes, meadows with shrubs, rarely in *Larix* forests; 3000–4400 m.

Distribution: **endemic**; L Yarlung Zangbo, N and S Hengduan (Map 47–64).

L YARLUNG ZANGBO: Nyingchi, Sheji-la, 4300 m, 2011, *D. Y. Hong et al. H11015* (PE). **N HENGDUAN: Danba,** Haiziping, 3800 m, 1958, *No.8 Forest Manag. Team 2538* (PE); **Dawu**, Songlinkou, 1951, *W. K. Hu 10957* (PE); **Jinchuan**, Kasa, 4400 m, 1958, *X. Li 78689* (PE). **S HENGDUAN: Cawarong**, Reesu-la, 3600 m, 1935, *C. W. Wang 65940* (PE); **Daocheng**, Chonji-Lhorong, 4200 m, 1981, *Qinghai-Xizang Exped. 5479* (PE); **Dêqên**, Baima Shan, 4050 m, 1937, *T. T. Yü 9351* (PE); **Kangding**, Dapao Shan, 2900–3280 m, 1934, *C. S. Liu 1118* (PE); **Muli**, Shaoxiangliangzi, 3400 m, 1937, *T. T. Yü 7796* (KUN, PE); **Shangri-la** (Zhongdian), Wongshui, 3900 m, 1937, *T. T. Yü 12780* (KUN, PE); **Yulong** (Lijiang), Yulong Snow Range, 3300 m, 1959, *K. M. Feng 22701* (KUN).

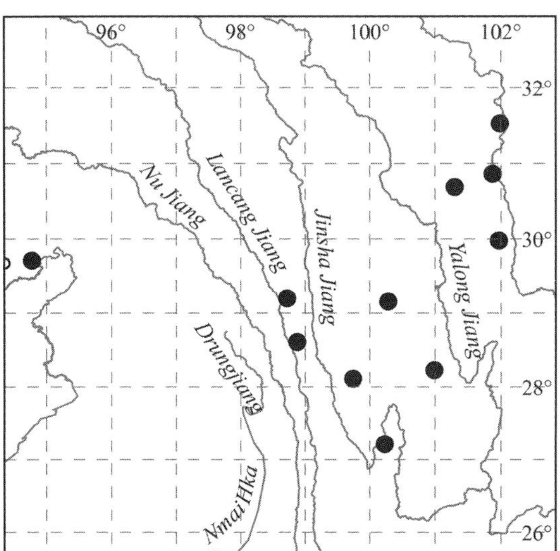

Map 47–64: **Cyananthus lichiangensis** W. W. Smith

17. Cyananthus inflatus Hook. f. & Thomson, J. Proc. Linn. Soc., Bot. 2: 21. 1857; Lian in Hong, Fl. Reipubl. Popularis Sin. 73(2): 26. pl. 5: 1–4. 1983; Shrestha, Acta Phytotax. Sin. 35(5): 427. fig. 26. 1997. Type: India, Sikkim, Naman Choon, 3350–4880 m, *J. D. Hooker* s. n. (lectotype designated by Shrestha, 1997: P; isolecto- types: BM, K, W).

Cyananthus forrestii Diels, Notes Roy. Bot. Gard. Edinburgh 5: 175. 1912. Type: China, Yunnan, "Ming Kwang valley, north of Ma chang", 25°15′N, Oct. 1903, *G. Forrest 955* (holotype: E).

Cyananthus pseudo-inflatus P. C. Tsoong, Contr. Inst. Bot. Nat. Acad. Peiping 3(3): 109. 1935. Type: China, Sichuan (Hsikang), Kangding, 27 Aug. 1935, *C. S. Liou 1304* (holotype: PE).

Stems erect or ascending, up to 80 cm tall, branched, sparsely villous or glabrous. Leaves alternate; petiole 2–7 mm long, villous; leaf blade orbicular, ovate, or triangular, 5–15 mm long, 3–10 mm broad, adaxially hispidulous, abaxially hirsute, base rounded or truncate, margin entire or obscurely crenulate. Flowers usually solitary, terminal on main stem and branches, subtended by a whorl-like group of 3–5 leaves; pedicel 2–5 mm long, brown hirsute. Calyx densely brown hirsute outside; calyx tube urceolate at anthesis, 7–12 mm long, much inflated after flowering; calyx lobes deltoid, 1/4–2/5 as long as tube, both sides brown hirsute. Corolla pale blue, cylindrical-campanulate, 1.5–2.5 cm long, inside densely villous at throat; corolla lobes narrowly obovate, nearly 1/2 as long as tube. Style extending up to corolla throat. Capsule ovoid, exceeding calyx in height at maturity. Seeds ellipsoid, ca. 0.5 mm long, 0.2 mm wide. **Chromosomes**: 2n=14, 28. Figure 47–24: a–c.

Phenology: flowering from August to October; fruiting in September and October.

Habitats: alpine meadows, grassy or shrubby slopes, or forests; 1800–4900 m.

Distribution: U Yarlung Zangbo, C and E Nepal, Sikkim & Darjiling, Bhutan, M and L

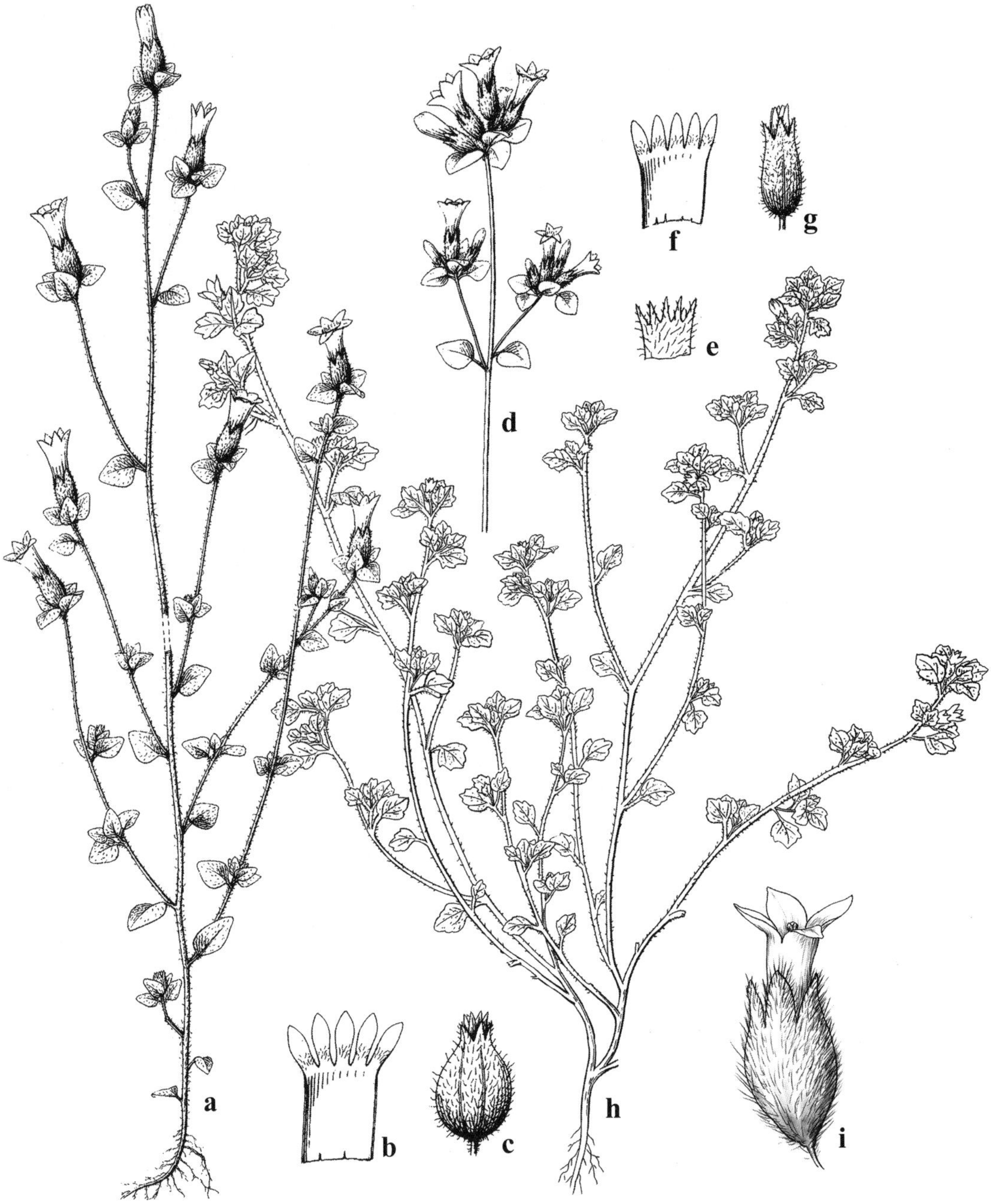

Figure 47–24: **a–c**. **Cyananthus inflatus** Hook. f. & Thomson: **a**, habit; **b**, corolla opened; **c**, dilated calyx. **d–g**, **Cyananthus fasciculatus** C. Macquand; **d**, upper part of stem, showing three dichasia; **e**, calyx opened; **f**, corolla opened; **g**, fruit. **h–i**, **Cyananthus hookeri** C. B. Clarke: **h**, habit; **i**, a 4-merous flower. **a–g** from *Fl. Reipubl. Popularis Sin.* 73(2), 1983; **h** & **i** redrawn by Ai-Li LI from *Icon. Cormophyt. Sin.* 4, 1975.

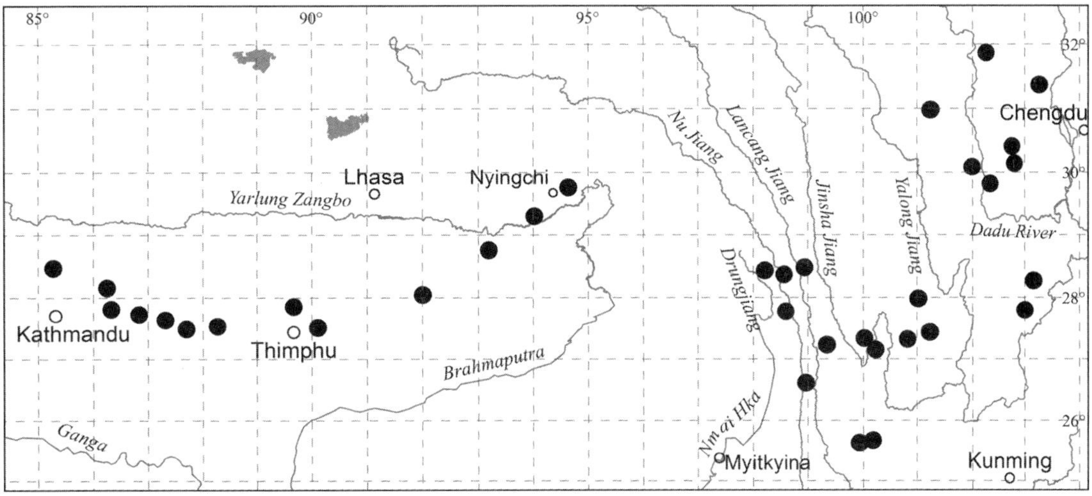

Map 47–65: **Cyananthus inflatus** Hook. f. & Thomson

Yarlung Zangbo, Yarlung Zangbo-Brahmaputra, N and S Hengduan, and U Irrawaddy (Map 47–65); also in other parts of Yunnan and S Sichuan (Leibo) of China.

U YARLUNG ZANGBO: **Gyirong**, Sale -Kagabu, 3100 m, 1990, *B. S. Li et al. 13442* (PE); **Nyalam**, Quxam, 3700–3800 m, 1972, *Xizang Chinese Mater. Medic. & Herbs. Exped. 1577* (PE). **C NEPAL**: **Janakpur**, Dolakha, 2000–2700 m, 1983, *H. Ohba et al. 8351171* (KUN). **E NEPAL: Kosi**, Sankhuwasabha, ca. 3500 m, 1981, *Grey-Wilson et al. 4205* (K); **Mechi**, Taplejung, 4500 m, 22. 1985, *Gray 32* (K); **Sagarmatha**, Solukhumbu, 3350 m, 1926, *F. Kingdon-Ward 7576* (E, K). **SIKKIM & DARJILING**: **Sikkim**, Phedang, 3760 m, 1983, *AGSES 398* (K). **BHUTAN**: **Upper Mo Chu**, 2850 m, 1984, *I. W. J. SinClair & D. G. Long 5037* (K); **Tongsa**, 3300 m, 1984, *I .W. J. SinClair & D. G. Long 5677* (K). **M YARLUNG ZANGBO** : **Yadong**, Xiayadong, 4100 m, 1974, *Qinghai-Xizang Exped. 74–2384* (KUN, PE). **L YARLUNG ZANGBO**: **Mainling**, Hongwei Lumbering Ground, 1974, *Qinghai-Xizang Exped. 74–1981* (PE); **Nyingchi**, Lunang, 3100 m, 1980, *C. C. Ni et al. 1674* (PE). **YARLUNG ZANGBO-BRAHMAPUTRA**: **Cona**, Boshan, 3800 m, 1975, *Qinghai-Xizang Suppl. Exped. 751942* (KUN, PE); **Lhünze**, Tsari, 3960 m, 1938, *F. Ludlow et al. 6371* (E). **N HENGDUAN: Barkam**, Ermu Shan, 3600 m, 1957, *X. Li 71791* (PE); **Dawu**, Taining, 3700 m, 1934, *H. Smith 11740* (PE); **Lixian**, 1952, *Z. He & Z. L. Zhou 14136* (PE); **Wenchuan**, Balang Shan, 3200 m, 1986, *W. T. Wang* s. n. (PE). **S HENGDUAN**: **Baoxing**, Qiaoqi, 2800 m, 1963, *K. C. Kuan et al. 2937* (PE); **Bijiang**, 4100 m, 1964, *S. K. Wu 8851* (KUN) ; **Cawarong**, Gulsaigul, 3200 m, 1935, *C. W. Wang 65820* (PE); **Dali**, Cangshan, 1940, *T. N. Liou 20929* (PE); **Dêqên**, Baima Shan, 3000 m, 1935, *C. W. Wang 69795* (KUN, PE); **Gongshan**, Junzi-la, 3900 m, 1938, *T. T. Yü 23278* (KUN, PE); **Kangding**, Zheduo Shan, 3650 m, 1995, *D. Y. Hong et al. 95078* (PE); **Luding**, Ganhaizi, 2980 m, 1963, *K. C. Kuan et al. 1815* (PE); **Meigu**, 2650 m, 1993, *Q. E. Yang 93037* (PE); **Muli**, Zhabulang, 3900–4000 m, 1983, *Qinghai–Xizang Exped. 13698* (PE); **Ninglang**, Xiaochangping, 3300 m, 1981, *Tian & Zhang (sample) 107(6)* (PE); **Shangri-la** (Zhongdian), Kongxinshu, 3400 m, 1937, *T. T. Yü 13987* (PE); **Tianquan**, Erlang Shan, 2500 m, 1953, *W. K. Hu 35416* (PE); **Weixi**, 2800 m, 1935, *C. W. Wang 67639* (PE); **Yanbian**, Yankou, 2700 m, 2002, *X. F. Gao et al. 3667* (PE); **Yangbi,** Meiwong, 2400 m, 1964, *Z. H. Yang 101847* (PE); **Yulong** (Lijiang), Xuesongcun, 4000 m, 1955, *K. M. Feng 21482* (PE) ; **Zhaojue**, Zhaojue to Jinyang, 2650 m, 1985, *H. Li 0136* (KUN). **U IRRAWADDY**: **Ridong**, 3800–3900 m, 1982, *Qinghai-Xizang Exped. 10180* (PE).

18. Cyananthus fasciculatus C. Marquand, Bull. Misc. Inform. Kew 1924: 247. 1924; Lian in Hong, Fl. Reipubl. Popularis Sin. 73(2): 26. pl. 5:5–8. 1983; Shrestha, Acta Phytotax. Sin. 35(5): 428. fig. 27. 1997. Type: China, Yunnan, NW flank of Lichiang (Lijiang) Range, 27°20′N, 100°10′E, 3600 m, Oct. 1922, *G. Forrest 22485* (lectotype designated by Shrestha, 1997: K; isolectotypes: BM, E, W).

Stems erect, 30–100 cm tall, branched, glabrous or sparsely villous. Leaves alternate; petiole 5–10 mm long, glabrous or sparsely villous; leaf blade ovate or suborbicular, 4–16 mm long, 4–15 mm broad, adaxially sparsely hispidulous, abaxially usually glabrous, base rounded, truncate, or subcordate. Flowers in dichasia, aggregated at apex of main stem and branches, subtended by a whorl-like group of 4 or 5 leaves; pedicel 2–4 mm long, glabrous. Calyx brown setose; calyx tube subcylindrical, broader toward round base, 5–8 mm long, 3–5 mm wide; calyx lobes usually 5 (4 on some lower branches), linear-triangular, ciliate. Corolla pale blue, cylindrical-campanulate, 14–17 mm long, inside villous near throat; corolla lobes (4 or) 5, obovate-oblong, ca. 5 mm long, 1.5–3 mm broad. Style exceeding corolla tube. Capsule exceeding calyx at maturity. Seeds ellipsoid, ca. 0.5 mm long, 0.2 mm wide. Figure 47–24: d–g.

Phenology: flowering from August to October; fruiting from October to December.

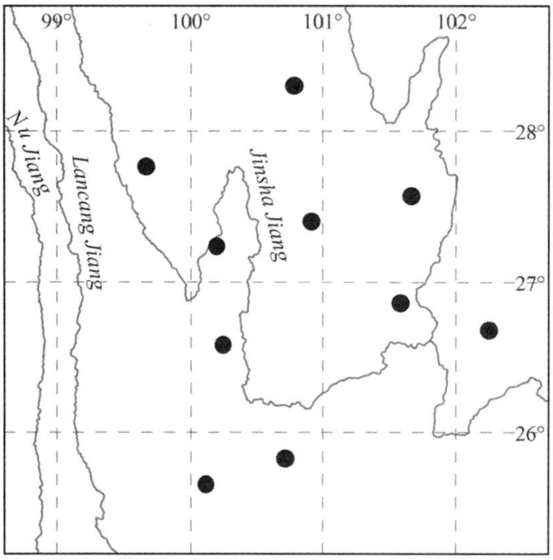

Map 47–66: **Cyananthus fasciculatus** C. Marquand

Habitats: forests, thickets, or grassy slopes; 2400–3700 m.

Distribution: confined to S Hengduan (Map 47–66); also in central Yunnan of China

S HENGDUAN: Binchuan, Jizushan, 1946, *T. N. Liou 21980* (PE); **Dali**, Huadianba, 2800 m, 1993, *Q. E. Yang 93058* (PE); **Heqing**, Ma'er Shan, 2546 m, 2010, *H. Sun et al. Sun H-072X-3245* (KUN); **Huili**, near the city, 3100–3400 m, 1914, *Handel-Mazzetti 5171* (W); **Muli**, Wachin, 3000 m, 1937, *T. T. Yü 14472* (KUN, PE); **Ninglang**, Xiayanglantang, 3320 m, 1981, *s. coll. Sample 96 (9)* (PE); **Shangri-la** (Zhongdian), Haba, 1939, *K. M. Feng 2376* (KUN, PE); **Yanbian**, 2600 m, 1932, *T. T. Yü 1795* (PE); **Yanyuan**, Niuwei Shan, 2850 m, 1993, *Q. E. Yang 9308* (PE); **Yulong** (Lijiang), Xuesongcun, 3000 m, 1937, *T. T. Yü 15607* (KUN, PE).

19. Cyananthus hookeri C. B. Clarke in Hooker, Fl. Brit. India 3: 435. 1881; Tsoong, Contr. Inst. Bot. Nat. Acad. Peiping 3(3): 107. 1937; Hong, Lammers & Klein in Wu, Raven & Hong, Fl. China 19: 512. 2011. Type: E Nepal, Wallanchoon, 3950 m, *J. D. Hooker 249* (lectotype designated by Shrestha, 1997: K; isolectotype: P).

Cyananthus hookeri var. *densus* C. Marquand, Bull. Misc. Inform. Kew 1924: 249. 1924. Type: China Yunnan, Dêqên, Doker-la, 28°20′N, 3900 m, Aug. 1917, *G. Forrest 14847* (holotype: K).

Cyananthus hookeri var. *grandiflorus* C. Marquand, Bull. Misc. Inform. Kew 1924: 249. 1924. Type: China, Yunnan, Lijiang Range, 27°20′N, 100°10′E, Oct. 1922, *G. Forrest 22477* (holotype: K; isotype: P).

Cyananthus hookeri var. *hispidus* Franch., J. Bot. (Morot) 1: 281. 1887. Type: China, Yunnan, *Delavay 208* (holotype: P).

Cyananthus cronquistii K. K. Shrestha, Kew Bull. 49(1): 143. 1994. Type: China, Yunnan, Dêqên (Atuntze), 28°35′N, 99°10′E, 4750 m, Sept. 1921, *G. Forrest 20165* (holotype: E; isotypes: K, W).

Stems usually caespitose, ascending or diffuse, 4–20 cm long, glabrous to densely villous. Leaves alternate; petiole 1–4 mm long; leaf blade rhombic or ovate, 3–7 mm long, 2–6 mm broad, adaxially hispidulous, abaxially hirsute, base rounded or truncate, margin crenate or sometimes entire. Flowers solitary, terminal on main stem and branches, subtended by a whorl-like group of 3–5 leaves, almost sessile. Calyx densely hirsute outside, rarely nearly glabrous; calyx tube ovoid, 3–8 mm long; calyx lobes (3) 4 (5), triangular, 1/3–1/2 as long as tube, both sides hirsute. Corolla purple-blue, cylindrical, 7–10 (–15) mm long, outside glabrous, inside densely villous at throat; corolla lobes (3) 4 (5), narrowly obovate or oblong, apex often with 3 or 4 brown-yellow villous hairs. Stamens 4. Style extending above corolla throat; stigma 4-fid. Capsule ovoid, exceeding calyx at maturity. Seeds narrowly ovoid, ca. 1.2 mm long, 0.4 mm wide. **Chromosomes**: 2n=12, 14. Figure 47–24: h–i.

Phenology: flowering from July to September; fruiting from September to November.

Habitats: grassy slopes, thickets, or forests; 2700–5350 m.

Distribution: U Yarlung Zangbo, E Nepal, Sikkim & Darji-ling, Bhutan, M and L Yarlung Zangbo,Yarlung Zangbo-Brahmaputra, Tangut, N and S Hengduan (Map 47–67); also in E Qinghai (Tongren) and SE Gansu (Taohe Basin) of China.

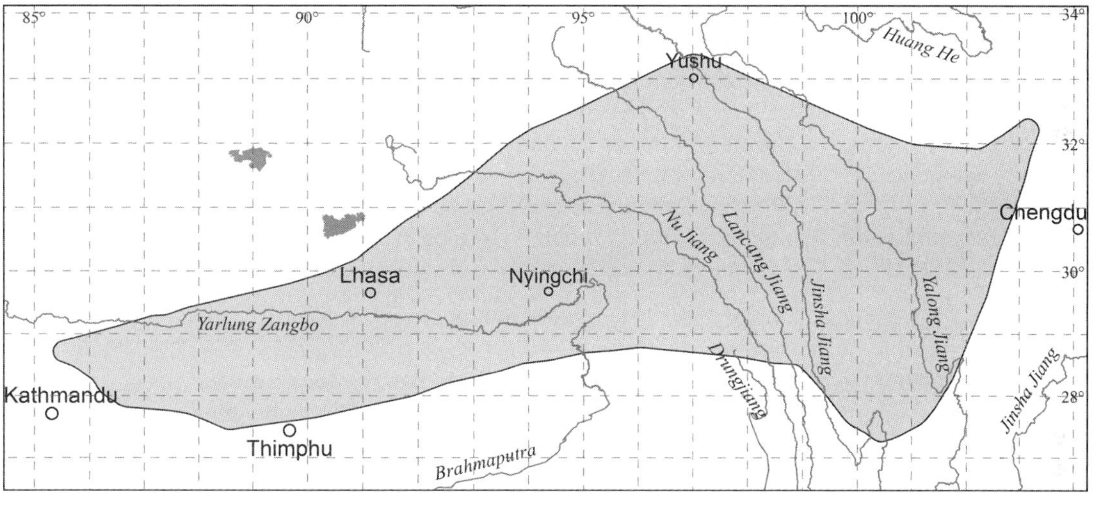

Map 47–67: **Cyananthus hookeri** C. B. Clarke

4. CODONOPSIS Wall.

Wallich in Roxburgh, Fl. India 2: 103. 1824, p. p.; de Candolle, Monogr. Camp. 120. 1830, p.p., quoad sect.1; Hooker & Thomson, J. Proc. Linn. Soc. London 2: 11. 1857, p. p. excl. *Codonopsis purpurea*; Clarke in Hooker, Fl. Brit. India 3: 430. 1881, p. p.; Komarov, Trudy Imp. S.-Peterburgsk. Bot. Sada 29(1): 102. 1908, p. p.; Tsoong, Contr. Inst. Bot. Nat. Acad. Peiping 3(3): 90. 1935, p. p.; Shen & Hong in Hong, Fl. Reipubl. Popularis Sin. 73(2): 32. 1983, p. p.; Hong, Lammers & Klein in Wu, Raven & Hong, Fl. China 19: 513. 2011, p. p. Hong, Monogr. *Codonopsis* Allied Gen. Camp. 61. 2015. ≡ *Wahlenbergia* sect. *Megasanthes* (G. Don) A. DC. in de Candolle, Prodr. 7: 424. 1839. ≡ *Codonopsis* sect. *Megasanthes* G. Don, Gen. Syst. Gard. 3: 736. 1834. Type: *Codonopsis viridis* Wall.

Campanumoea Bl., Bijdr. Fl. Ned. India 13: 726. 1826. Type: *Campanumoea javanica* Bl. [= *Codonopsis javanica* (Bl.) Hook. f.]

Glossocomia D. Don, Prodr. Fl. Nepal. 158. 1825. Type: *Glossocomia tenera* D. Don (=*Codonopsis thalictrifolia* Wall., 1824)

Leptocodon (Hook. f. & Thomson) Lem., Ill. Hort. 3. Misc. 49. 1856. ≡ *Codonopsis* subg. *Leptocodon* Hook. f. & Thomson, Ill. Himal. Pl. t. 16, A. 1855. Type: *Leptocodon gracilis* (Hook. f. & Thomson) Lem. (=*Codonopsis gracilis* Hook. f. & Thomson).

Perennial herbs. Roots usually thickened, various in shape, mostly fleshy and carrot-shaped, rarely fusiform or tuberous, rarely woody. Caudex often short. Stems erect, ascending, scandent, or twining. Leaves alternate, opposite, or (3-) 4-verticil-like in sect. *Lanceolatae*. Flowers solitary, less frequently in cyme of 2–4, terminal only on main stem or on both main stem and branches, sometimes leaf-opposed, rarely axillary. Calyx tube usually adnate to hypanthium and thus calyx epigynous, often 10-ribbed, calyx lobes 5, in a few cases inserted at base of hypanthium and thus appearing as 5 sepals and calyx hypogynous or nearly hypogynous. Corolla always epigynous, broad-campanulate, campanulate, funnel-shaped, tubular-campanulate, tubular, or urceolate, 5-lobed for 1/8–1/2 of length, red-purple, blue-purple, blue-white, yellow-green or green, often with dark purple veins, sometimes with dark rings or spots. Disc present, various in colour, annular or pentagonal, or interrupted and appearing as 5 glands alternate to stamens. Stamens 5; filaments often dilated at base, glabrous or hairy; anthers basifixed, connective usuall glabrous, rarely setose. Ovary at least half-inferior to corolla, usually 3-locular (5-locular in *C. javanica*); ovules numerous; style hairy; stigma (2- or) 3-fid (4- or 5-fid in *C. javanica*), lobes broad, oblong or ovate-deltoid. Fruit usually a capsule, occasionally a berry, with persistent calyx, lower part semi-globose or obconical, often 10-ribbed, upper part rostrate, capsule apically 3-loculicidal. Seeds numerous, usually ellipsoid, oblong or ovoid, wingless, rarely winged, smooth, those in species with berries irregular or ovoid-globose, reticulate; embryo straight, embedded in copious endosperm. Plants mostly fetid. Pollen 5–7-colpate; colpi long, with colpus length/polar axis length ≥ 0.5; sexine with short spinules (≤ 1 μm). Chromosomes 2n=16.

Recent references: Hong & Pan, 2012, Pollen morphology of the platycodonoid group (Campanulaceae s. str.) and its systematic implications. *J. Integr. Plant Biol.* 54: 773–789; Wang et al., 2013, Molecular phylogeny of the platycodonoid group (Campanulaceae s. str.) with special reference to circumscription of *Codonopsis*. *Taxon* 62: 498–504; Hong, 2015, A Monograph of *Codonopsis* and Allied Genera (Campanulaceae). Science Press & Academic Press

Codonopsis as here circumscribed is monophyletic perhaps except *Codonopsis chimiliensis*, which will be discussed under the treatment of that species. However its subdivision is still problematic, and thus the subdivision into four sections as here used is tentative.

Fourty-six species in East, Southeast and Central Asia: Afghanistan, Bhutan, China, N and NE India, Indonesia, Japan, Kazakstan, Korean Peninsula, Kyrgyzstan, Laos, Malesia, W Mongolia, Myanmar, Nepal, Pakistan, Philippines, Russia (the Far East), Tajikistan, Thailand, SE Uzbekistan, and Vietnam; 40 species in the Pan-Himalaya (22 endemic).

1a. Disc consisting of 5 lamellar glands, glands alternate with stamens; calyx almost free from hypanthium; corolla long-tubular; stems twining; plants very fetid (sect. 3. ***Leptocodon***: spp., 37–38).
- 2a. Plants glabrous throughout except for sparsely villous young leaves; calyx lobes linear-oblong, often reflexed, distant from each other, glabrous, base attenuated into stipe .... 37. *C. gracilis*
- 2b. Plants more or less hairy, rarely villous only on leaf margin; calyx lobes ovate to ovate-lanceolate, upright, overlapping each other, villous at least on margin, base obtuse to cordate .... 38. *C. hongii*

1b. Disc continuous, circular or pentagonal; calyx usually adnate to hypanthium; corolla usually campanulate, rarely tubular; stems erect or twining; plants fetid or not.

3a. Fruit a fleshy berry; leaf blade cordate (sect. 4. ***Campanumoea***: spp., 39–40).

4a. Leaves opposite; calyx hypogynous or nearly hypogynous .... 39. *C. javanica*

4b. Leaves alternate; calyx epigynous .... 40. *C. inflata*

3b. Fruit a capsule; leaf blade various in shape, less frequently cordate.

5a. Stems erect or ascending, rarely scandent, very occasionally also twining (sect. 1. ***Erectae***: spp., 1–19).

6a. Stems simple, without many usually sterile pinnate-like branches from lower part.

7a. Stems simple and erect; flowers solitary and terminal; leaves evenly distributed; corolla purple or dark purple, without dark markings.

8a. Corolla nearly funnel-shaped, tubular below, nearly rotate above; stems sparsely to densely villous .... 2. *C. elliptica*

8b. Corolla campanulate; stems glabrous .... 1. *C. lixianica*

7b. Stems often branched; leaves mostly basal and thus stems somewhat scapiform, or evenly distributed; corolla white, yellowish, blue, pale purple, often with dark markings on veins or yellow spots.

9a. Corolla tubular or tubular-campanulate; stems mostly ascending or scandent, rarely also twining, main stem not scapiform; leaves evenly distributed on main stem.

10a. Leaf blade 0.6–1.7 cm wide; calyx lobes 5–7 mm long, 1.4–2.6 mm broad; corolla gray or pale purple, lobed to ca. 1/6, with a dark purple ring just below throat... 5. *C. gongshanica*

10b. Leaf blade 2–7 cm wide; calyx lobes 13–16 mm long, 6–8.5 mm broad; corolla whitish, yellow or pale blue,lobed to ca.1/4, without a coloured ring.

11a. Corolla 1.8–4 cm long; fruiting calyx lobes enlarged, up to 2.5 cm long, 1.5 cm broad; stamens villous .... 34. *C. benthamii*

11b. Corolla 1.2–2 cm long, fruiting calyx lobes not enlarged, up to 1.6 cm long, 0.9 cm broad; stamens glabrous .... 10. *C. subsimplex*

9b. Corolla campanulate; stems mostly erect, less frequently ascending, main stems scapiform; leaves located at lower part of stem.

12a. Calyx entirely free from hypanthium; calyx lobes 2–3 cm long, 0.6–1 cm broad .... 3. *C. chimiliensis*

12b. Calyx adnate to whole hypanthium; calyx lobes up to 1.2 cm long, 0.6 cm broad.

13a. Stems often branched at lower and middle parts; petiole 1–10 cm long; calyx lobes 5–7 mm long, 2–3 mm broad, with sinus beween them truncate .... 6. *C. subscaposa*

13b. Stems usually without branches at lower and middle parts; petiole less than 1 cm long; calyx lobes 8–12 mm long, 4–6 mm broad, with sinus between them narrow and pointed .... 4. *C. meleagris*

6b. Stems with many usually sterile branches, such branches with many nearly opposite leaves and thus resembling a pinnate leaf.

14a. Stems, lower surface of leaves and calyx densely white hairy; stems mostly with several flowers .... 15. *C. canescens*

14b. Stems, leaves and calyx all green, glabrous or sparsely hairy, not white; stems single or several-flowered.

15a. Stems and leaves glabrous or nearly glabrous; calyx entirely glabrous or calyx lobes rarely hispidulous. .... 12. *C. alpina*

15b. Stems and leaves always hairy; calyx hispidulous at least in upper part of lobes, very rarely entirely glabrous.

16a. Calyx lobes usually denticulate above, 4–6 mm long, 2–3 mm broad, sinus between calyx lobes broad and truncate .... 13. *C. chlorocodon*

16b. Calyx lobes entire, rarely denticulate, much larger, sinus between calyx lobes narrow, acute or obtuse.

17a. Leaf blade usually less than 1.5, rarely up to 3 cm long; sterile branches aggregated at lower part of main stem; main stem scapiform, with usually one, less frequently 2–3 flowers.

18a. Corolla tubular-campanulate, 2.5–4.8 cm long, shallowly lobed to 1/6–1/8; anthers villous at connective .... 18. *C. thalictrifolia*

18b. Corolla campanulate or funnel-shaped, 1.8–3 cm long, lobed to ca. 1/3; anthers glabrous.

19a. Sterile branches and leaves nearly evenly distributed at lower and middle parts of stem; leaf blade 1.5–3 cm long; corolla with a red-purple ring below throa .... 11. *C. campanulata*

19b. Sterile branches and leaves mostly aggregated towards lower part of stem; leaf blade 0.5–1.8 cm long; corolla without a ring below throat.

20a. Corolla broadly campanulate, pale purple or pink; calyx lobes 3.5–6.5 mm broad .... 16. *C. foetens*

20b. Corolla tubular below, suddenly dilated from middle and thus infundibular, pale blue above, purplish below; calyx lobes 5–7 mm broad .... 17. *C. bulleyana*

17b. Leaf blade usually bigger; sterile branches mostly evenly distributed along main stem, main stem not scapiform, usually with several flowers.

21a. Corolla tubular, 1.2–1.5 cm long; capsule 0.8–1 cm in diameter .... 14. *C. bhutanica*

21b. Corolla campanulate, 1.5–3 cm long; capsule usually more than 1 cm in diameter.

22a. Calyx lobes subulate, widest at base, 2–4 mm wide, always reflexed; leaf blade ovate-lanceolate to lanceolate, with length/width ratio mostly over 2 .... 9. *C. obtusa*

22b. Calyx lobes ovate to ovate-lanceolate, widest above base, 4–8.5 mm broad, reflexed or not; leaf blade ovate to ovate-lanceolate, with length/width ratio mostly less than 2.

23a. Calyx lobes always hispidulous entirely or at upper part; corolla yellow-green, green-purple, or bluish .... 19. *C. viridiflora*

23b. Calyx lobes glabrous or ciliate, or rarely hispidulous only at apex; orolla pale blue, purple inside.

24a. Calyx lobes revolute, 0.5–1(–1.2) cm long, more or less ciliate; corolla infundibular .... 7. *C. ovata*

24b. Calyx lobes spreading but reflexed, 1–1.8 cm long, totally glabrous; corolla campanulate .... 8. *C. clematidea*

5b. Stems twining, sometimes also scandent (sect. 2. ***Codonopsis***: spp., 20–36).

25a. Corolla tubular; leaf blade dentate or lobed, very rarely nearly entire; calyx lobes toothed or crenulate, less often entire.

26a. Corolla 1–1.3 cm long, 0.3–0.4 cm wide; leaf blade cordate-ovate, cordate at base, 2–3.5 cm long; plants pilose except corolla 32. *C. microtubulosa*

26b. Corolla much larger; leaf blade rounded to cuneate, or cordate at base, 3–10 cm long; plants puberulent, less often hirsute.

27a. Leaves on main stem mostly opposite, less frequeutly alternate; petiole less than 1 cm long; filaments villous .... 31. *C. tubulosa*

27b. Leaves on main stem mostly alternate; petiole longer than 1.5 cm; filaments glabrous or villous.

28a. Leaf blade mostly cordate at base; corolla broadly tubular, evenly broad, lobed to 1/5–1/4; filaments glabrous .... 23. *C. deltoidea*

28b. Leaf blade mostly rounded or cuneate, less frequently cordate at base; corolla narrowly tubular, slightly dilated above, lobed to 1/7–1/6; filaments villous .... 30. *C. benthamii*

25b. Corolla campanulate or tubular-campanulate; leaf blade entire or shallowly incised; calyx lobes entire.

29a. Calyx lobes linear to linear-lanceolate, 10–20 mm long, 2–3 mm broad, with length/width ratio over 5, distant from each other .... 28. *C. viridis*

29b. Calyx lobes wider, with length/width ratio less than 3, close to each other.

30a. Calyx more or less hairy; leaf blade cordate at base.

31a. Leaf blade (2.5–)3–16 cm long, (1–)3–9 cm broad; calyx tube scaberulose; calyx lobes triangular-lanceolate, densely puberulent; corolla broad-campanulate .... 27. *C. affinis*

31b. Leaf blade 0.5–3 cm long, 0.5–2.5 cm broad; calyx tube hirsute entirely or along ribs; calyx lobes ovate-orbicular or rhomboid-ovate, hirsute; corolla globose-campanulate.

32a. Petiole 5–20 mm long; corolla glabrous outside; stigma segments 2-lobed .... 33. *C. subglobosa*

32b. Petiote less than 2 mm long; corolla hirsute outside; stigma segments entire .... 34. *C. bragaensis*

30b. Calyx glabrous; leaf blade cuneate, rounded or cordate at base.

33a. Leaf blade of lower cauline leaves cordate, very rarely truncate or rounded at base.

34a. Leaf blade densely white-scaberulose on lower surface; calyx lobes strongly reflexed.. 21. *C. hemisphaerica*

34b. Leaf blade glabrous, sparsely hirsute or hispidulous on lower surface; calyx lobes upright or spreading.

35a. Calyx perigynous or hypogynous, calyx lobes clearly inserted below corolla, with space between them; stems and leaves always hairy .... 20. *C. pilosula*

35b. Calyx epigynous, calyx lobes and corolla inserted at same level; stems and leaves glabrous or sparsely hirsute.

36a. Leaf blade shallowly cordate to rounded at base, 1.5–3.5(–5) cm long, 1.3–3 cm broad; corolla 1.7–2.5 cm long .... 19. *C. viridiflora*

36b. Leaf blade deeply cordate at base, 2–10 cm long, 2.4–7 cm broad; corolla 0.7–1.8 cm long.

37a. Corolla 7–9 mm long, white; fruit 8–10 mm in diameter, superior part 2.5–3.5 mm long .... 36. *C. micrantha*

37b. Corolla 17–18 mm long, purple, dark blue, or white but purple-red inside; fruit 13–18 mm in diameter, superior part much longer .... 35. *C. cordifolioidea*

33b. Leaf blade of lower cauline leaves cuneate or rounded, occasionally cordate at base.

38a. Stems and leaves more or less hairy; leaf blade crenulate or serrate.
  39a. Calyx lobes ciliate, reflexed .... 26. *C. henryi*
  39b. Calyx lobes glabrous, upright .... 22. *C. bomiensis*
38b. Stems and leaves glabrous; leaf blade entire
  40a. Leaf blade 0.5–4 cm long, 0.5–1.5 cm broad, apex acute or obtuse .... 29. *C. farreri*
  40b. Leaf blade 6–7 cm long, 2–2.5 cm broad, apex shortly acuminate.
    41a. Calyx lobes upright; corolla lobed to 1/5; filaments glabrous .... 24. *C. rotundifolia*
    41b. Calyx lobes strongly reflexed; corolla lobed to 3/5; filaments villose-ciliate at dilated part .... 25. *C. reflexa*

Sect. 1. Erectae (Kom.) D. Y. Hong, Fl. Reipubl. Popularis Sin. 73(2): 48. 1983. ≡ *Codonopsis* ser. *Erectae* Kom., Trudy Imp. S. -Petersburgsk. Bot. Sada 29(1): 113. 1908. Type: *Codonopsis tha-lictrifolia* Wall.

Roots carrot-shaped. Stems erect or ascending, rarely scandent, very occasionally also twining. Leaves alternate, those on lower branches nearly opposite, thus branches resembling a pinnate leaf. Calyx always epigynous except for *C. chimiliensis*. Seeds oblong, wingless.

Twenty-two species, distributed from Central Asia and NE Afghanistan eastwards to central China (W Hubei Province); 19 in the Pan-Himalaya (13 endemic). It reaches the highest altitude among the four sections, up to 5490 m (*C. bhutanica*), and even to 5560 m (*C. thalictrifolia* subsp. *thalictrifolia*).

1. Codonopsis lixianica D. Y. Hong, Plant Divers. Resour. 36(3): 292. fig. 5. 2014; Hong, Monogr. *Codonopsis* Allied Gen. Camp. 65. 2015. Type: China, Sichuan (NW), Lixian, Mt Huangtuliang, 3400 m, 24. Jul. 1957, *W. K. Hu & D. Z. Deng 1988* (holotype: SZ).

Roots carrot-shaped, ca. 5 mm in diameter. Stems erect, simple, glabrous, ca. 35 cm tall. Lower leaves nearly opposite, middle and upper ones alternate; petiole 4–10 mm long, sparsely villous; leaf blade of lower leaves ovate, those of middle and upper ones narrowly ovate to ovate-lanceolate, 1.6–6.3 cm long, 1.8–3.2 cm broad, adaxially glabrous, abaxially very sparsely white-hirsute, base truncate to rounded, margin crenate, apex obtuse to acute. Flowers solitary, terminal only on main stem, or also on small branches; small branches with a pair of very small bracteole-like leaves. Calyx tube adnate to whole hypanthium, glabrous; calyx lobes inserted at same level as corolla, ovate-lanceolate, broadest at middle-lower part, ca. 11 mm long, 7 mm broad, adaxially sparsely white-hispidulous, apex acute. Corolla dark purple, campanulate, glabrous, ca. 1.8 cm long, 1.5 cm wide; corolla lobes deltoid-ovate, ca. 8 mm long, 10 mm broad. Disc annular, yellow. Filaments glabrous, 7 mm long, slightly dilated at lower part; anthers 4 mm long. Fruit not seen. Figure 47–25: a–c.

Phenology: flowering in July.

Habitats: ca. 3400 m.

Distribution: **endemic**; confined to N Hengduan (Map 47–68).

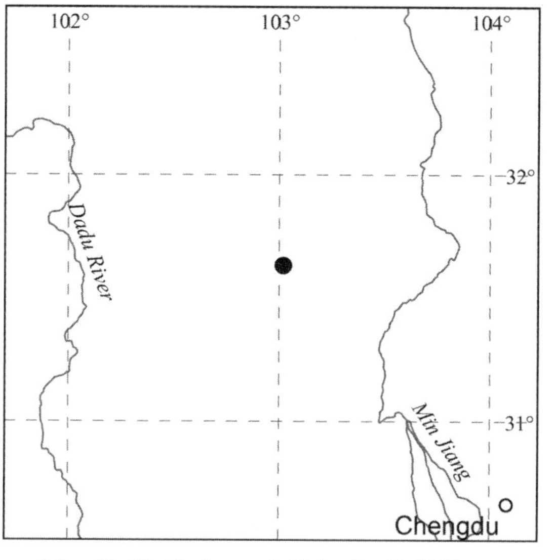

Map 47–68: **Codonopsis lixianica** D. Y. Hong

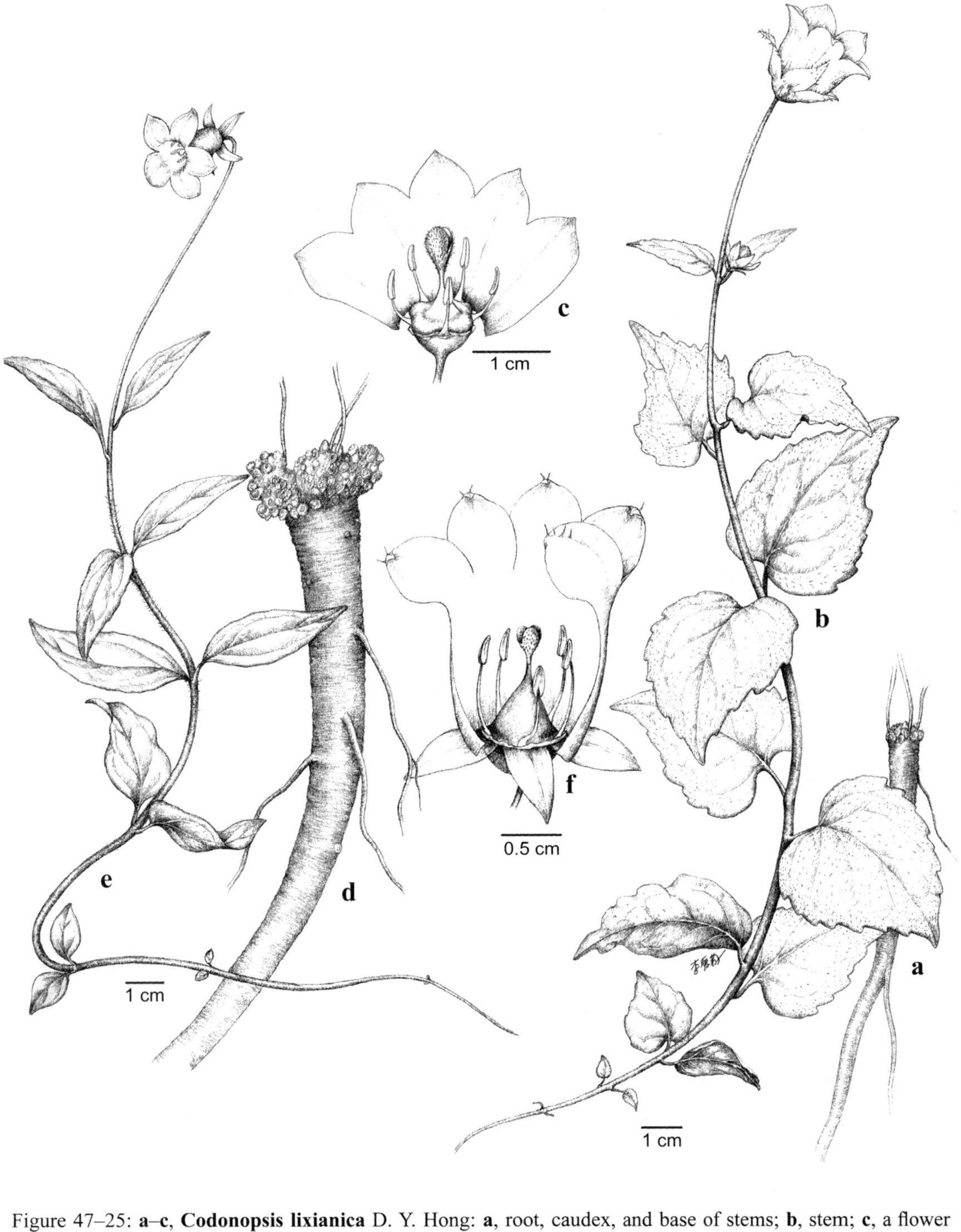

Figure 47–25: **a–c**, **Codonopsis lixianica** D. Y. Hong: **a**, root, caudex, and base of stems; **b**, stem; **c**, a flower with calyx removed and corolla opened, showing stamens, disc, style, and stigma. **d–f**, **Codonopsis elliptica** D. Y. Hong: **d**, root, caudex and base of tems; **e**, stem; **f**, a flower with corolla opened, showing stamens and superior part of pistil. From Hong in *Plant Diversity and Resources* 36(3), 2014, with permission.

2. Codonopsis elliptica D. Y. Hong, Plant Divers. Resour. 36(3): 292. fig. 6. 2014; Hong, Monogr. *Codonopsis* Allied Gen. Camp. 66. 2015. Type: China, Sichuan, Beichuan, Jianshe Commune, Heshangtou, 3800 m, sunny slopes, fl. purple, 9 Aug. 1978, *Beichuan Exped. 0383* (holotype: SM; isotype: PE).

Root carrot-shaped, over 15 cm long, ca. 1.3 cm in diameter. Stems several from one caudex, erect, simple, without any branch, 20–35 cm tall, sparsely to rather densely villous. Basal leaves much smaller than those at other parts, sometimes scale-like, those on lower and middle stem opposite or nearly opposite, upper ones alternate, sessile or with short petiole up to 5 mm long; petiole glabrous or sparsely villous; leaf blade elliptic, less frequently ovate-lanceolate, uppermost ones sometimes linear-elliptic, 3–6 cm long, 0.8–2 cm broad, adaxially glabrous, abaxially sparsely hirsute along veins or totally glabrous, base cuneate, margin entire or slightly crenulate, apex obtuse or acute. Flowers solitary and terminal. Calyx tube adnate to hypanthium up to middle part, hemispheric, glabrous; calyx lobes inserted at same level as corolla, narrowly triangular, or lanceolate, 5–7 mm long, 3–4 mm broad, glabrous or very sparsely hispidulous at apex. Corolla blue or purple, infundibular , tubular below, nearly rotate above, glabrous, 2–3.4 cm long, lobed to 1/3–2/5; corolla lobes orbicular, 1.1 cm long, 1.2 cm broad. Stamens glabrous, filaments slightly dilated at base. Stigma 3-fid; stigma lobes orbicular. Fruit not seen. Figure 47–25: d–f.

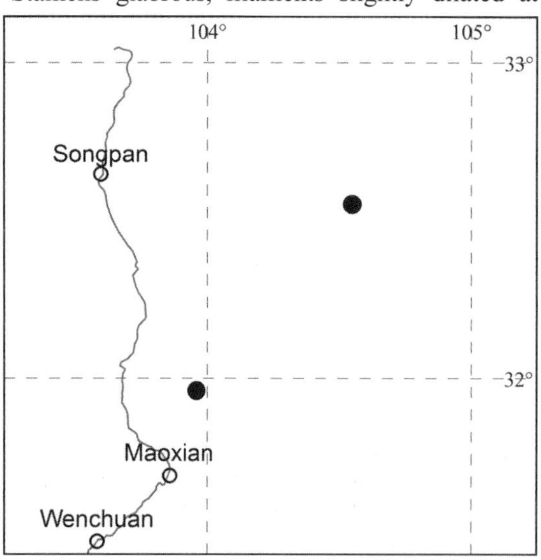

Map 47–69: **Codonopsis elliptica** D. Y. Hong

Phenology*:* flowering in July and August.

Habitats*:* meadows or forests; ca. 3800 m.

Distribution: **endemic**: confined to N Hengduan (Map 47–69).

N HENGDUAN*:* **Pingwu**, Huya, 1978, *Miandi Exped. 401* (SM).

3. Codonopsis chimiliensis J. Anthony, Notes Roy. Bot. Gard. Edinburgh 15: 184. pl. 217. 1926; Hong, Lammers & Klein in Wu, Raven & Hong, Fl. China 19: 520. 2011; Hong, Monogr. *Codonopsis* Allied Gen. Camp. 67. 2015. Type: "N. E. Burma, western flank of the Chimi-li, N'Maikha-Salween Divide", 26°23'N, 98°48'E, 3350–3660 m, Aug. 1924, *G. Forrest 24887* (lectotype designated by Hong, 2015: E; isolectotypes: K, P, W); loc. eodem, 26°16'N, 98°48'E, 4270

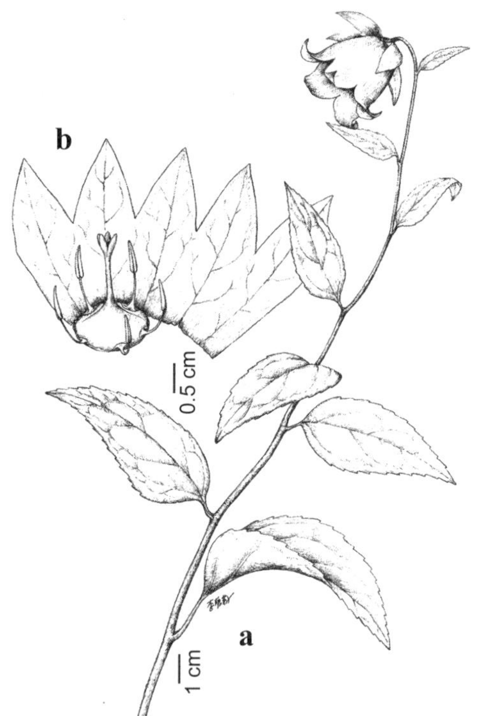

Figure 47–26: **Codonopsis chimiliensis** J. Anthony: **a**, middle and upper part of stem; **b**, a flower with calyx removed and corolla opened, showing stamens, style, and stigma. From Hong, *Monogr. Codonopsis Allied Gen. Camp.* 2015.

m, Sept. 1925, *G. Forrest 27223* (paratypes: E, K, P, W); loc. eodem, 26°24'N, 98°48'E, 3350 m, Jul. 1925, *G. Forrest 26998* (paratypes: E, K).

Plants glabrous throughout. Roots carrot-shaped. Stems erect, multibranched, 0.6–1 m tall. Leaves alternate; petiole 0.5–3 cm long; leaf blade ovate to ovate-lanceolate, 5–14 cm long, 3–7 cm broad, base rounded or decurrent, apex acute to short-acuminate. Flowers solitary, terminal on main stem and branches; pedicel 3–10 cm long. Calyx entirely free from hypanthium; calyx lobes oblong-lanceolate, 2–3 cm long, 0.6–1 cm broad, entire, apex acute. Corolla white, greenish-white or pale creamy-yellow, with pale purple veins, epigynous, campanulate, lobed to middle; corolla tube1.2–1.7 cm long, 1–1.2 cm wide; corolla lobes triangular, 1.2–1.5 cm long, 0.8–1.2 cm broad, apex acute. Ovary superior to calyx, but inferior to corolla. Stamens glabrous; filaments filiform, ca. 8 mm long. Fruit not seen. Figure 47–26.

Phenology: flowering from July to September.

Habitats: grassy slopes or meadows by scrubs; 2800–4270 m.

Distribution: **endemic**; confined to the border area between China (Yunnan) and Myanmar (Map 47–70).

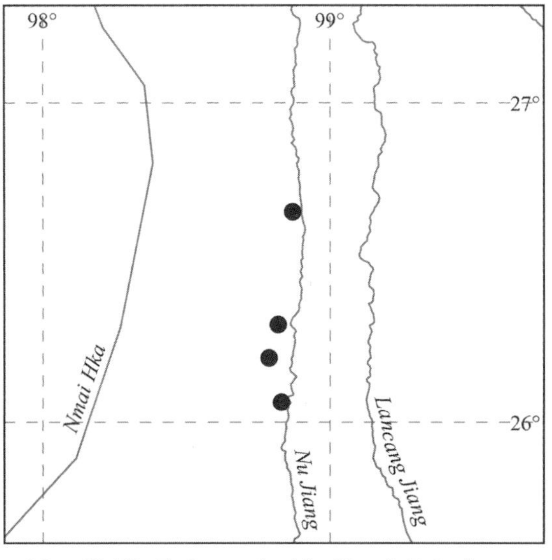

Map 47–70: **Codonopsis chimiliensis** J. Anthony

S HENGDUAN: Bijiang, W slope of Gaoligong Shan, 2900 m, 1978, *Bijiang Exped. 1584* (KUN); **Lushui,** Pianma, 2800 m, 1964, *S. K. Wu 8361* (KUN). **U IRRAWADDY**: **Hpimaw** Pass, 3200 m, 1919, *RA 1279* (E); Jungle Camp above Hpimaw, 1938, *Nam Mu Pa 17443* (K).

4. **Codonopsis meleagris** Diels, Notes Roy. Bot. Gard. Edinburgh 5: 172. 1912; Tsoong, Contr. Inst. Bot. Nat. Acad. Peiping 3(3): 97. fig. 9. 1935; Stapf, Bot. Mag. 154: t. 9237. 1931; Hong, Monogr. *Codonopsis* Allied Gen. Camp. 68. 2015. Type: China, NW Yunnan, Lichiang (Lijiang), 27°20′ N, 3350–3660 m, Aug. 1906, *G. Forrest 2674* (lectotype designated by Hong, 2015: E; isolectotypes: BM, CAL).

Roots carrot-shaped, 15–25 cm long, 0.5–1.8 cm in diameter. Stems erect, 20–90 cm tall, simple, very rarely with long branches from lower part, sparsely hairy, later glabrescent. Leaves on main stem alternate, those on branches opposite, sessile or with petiole less than 1 cm long; leaf blade ovate, oblong, elliptic, or lanceolate, 3–11 cm long, 1.5–3.8 cm broad, adaxially green, glabrous or sparsely hispidulous, abaxially gray-green, densely hirsute, base cuneate, margin crenate or nearly entire, apex acute or obtuse. Flowers solitary, terminal or axillary, long-pedicellate. Calyx tube adnate to whole hypanthium, semiglobose, glabrous or sparsely scaberulose; calyx lobes ovate or ovate-triangular, 8–12 mm long, 4–6 mm broad, glabrous, margin nearly entire or serrate, apex acute. Corolla yellow with red-purple veins, or red-purple with yellow spots, or pale purple with dark purple veins, broadly campanulate, 3–4 cm long, 3–4 cm wide, lobed to middle, both surfaces glabrous; corolla lobes often sparsely villous at apex. Stamens glabrous; filaments slightly dilated at base. Capsule semiglobose below, conical above, 1–2.2 cm long, 0.8–1 cm wide. Seeds ovoid, brown-yellow. Fi-gure 47–27: d.

Phenology: flowering and fruiting from July to October.

Habitats: grassy slopes or open woods; 2600–4000 m.

Distribution: **endemic**; confined to S Hengduan (Map 47–71).

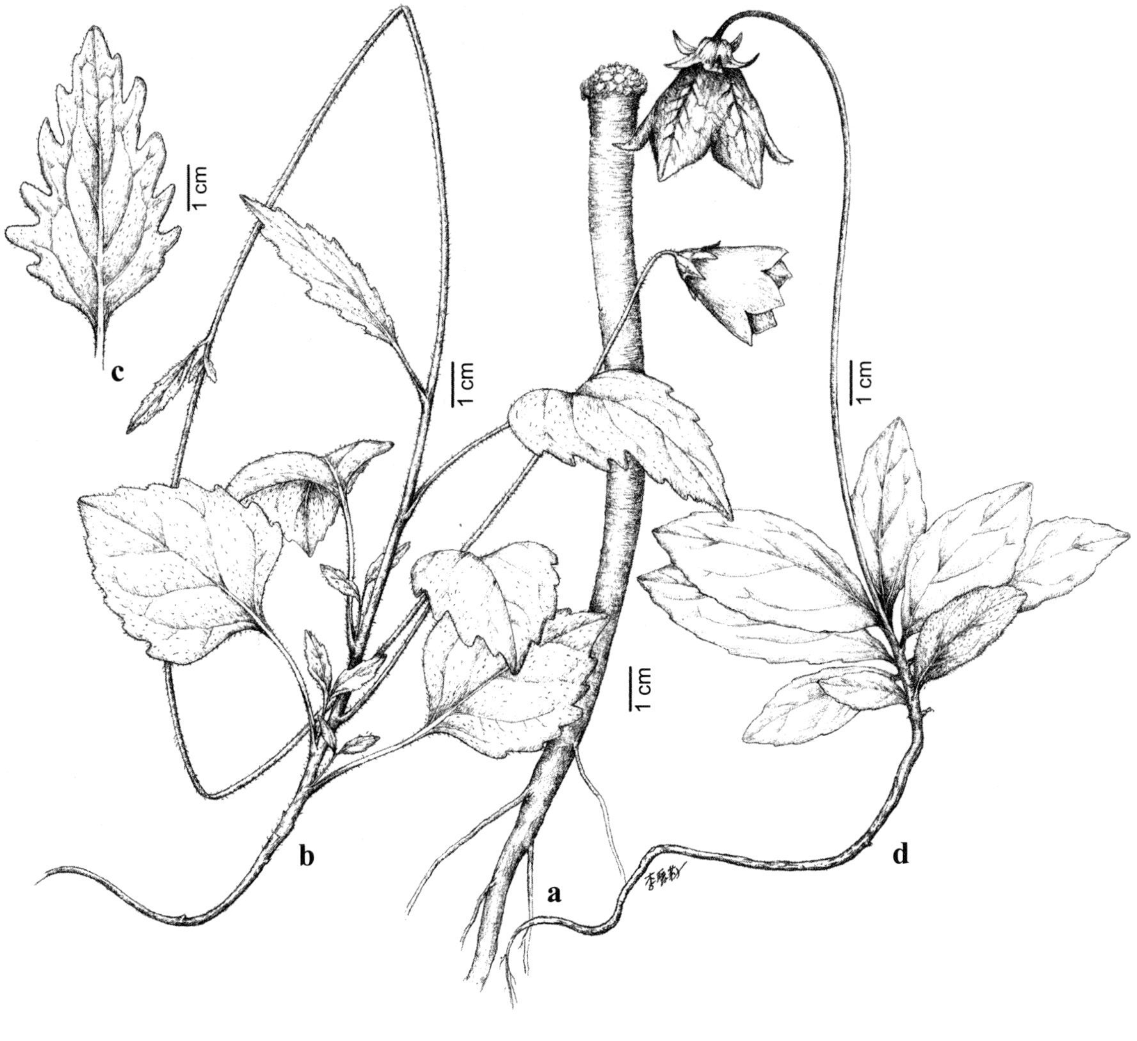

Figure 47–27: **a–c**, **Codonopsis subscaposa** Kom.: **a**, root and caudex; **b**, stem; **c**, a different form of leaf. **d**, **Codonopsis meleagris** Diels: stem. From Hong, *Monogr. Codonopsis Allied Gen. Camp.*, 2015.

S HENGDUAN: Jianchuan, Jianchuan-Mekong Divide, 1922, *G. Forrest 23198* (E); **Shangri-la** (Zhongdian): s. loc., 3400 m, 1914, *C. Schneider 3252* (G, K, PE); **Yulong** (Lijiang), Ganhaizi, 1939, *R. C. Ching 21124* (KUN, PE); Yulong Snow Range, 1959, *K. M. Feng 22567* (HNWP, KUN, PE).

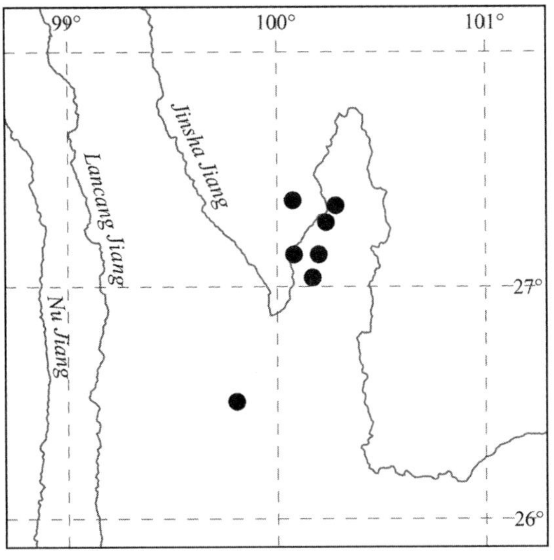

Map 47–71: **Codonopsis meleagris** Diels

5. Codonopsis gongshanica Q. Wang & D. Y. Hong, Phytotaxa 188(3): 145. 2015. Type: China, Yunnan, Gongshan, Bingzhongluo, Gawagapu Shan, alpine meadows, 3400–3800 m, 4 Aug. 2013, *X. H. Jin et al. ST1992* (holotype: PE; isotype: PE).

Roots carrot-shaped, 9–25 cm long, 0.4–1.3 cm in diameter. Stems erect, simple, very rarely with branches in lower and middle parts, 20–45 cm tall, glabrous or sparsely villous at base. Leaves alternate, subrosulate; petiole 0.2–0.5 cm long, glabrous; leaf blade broad-ovate, ovate, or lanceolate, 1.7–2.9 cm long, 0.6–1.7 cm broad, glabrous, base cuneate, margin entire to slightly sinuate, apex slightly obtuse. Flowers solitary, terminal on main stem. Calyx tube entirely adnate to hypanthium; calyx lobes lanceolate, 5–7 mm long, 1.4–2.6 mm broad, margin entire or slightly sinuate, apex obtuse. Corolla gray or pale purple, with a dark purple ring just below throat, short-tubular, 0.9–1.3 cm long, 0.6–0.9 cm wide, shallowly lobed to ca. 1/6, glabrous; corolla lobes ovate-deltoid. Stamens glabrous; filaments 4–6 mm long, slightly dilated at base; anthers 1–2 mm long. Fruit not seen. Figure 47–31: a–c.

Phenology: flowering in July and August.

Habitats: alpine scrubs or meadows 3400–4600 m.

Distribution: **endemic**: confined to S Hengduan (Map 47–72).

Notes: This species is related to *C. subsimplex* and *C. bhutanica* according to both gross morphology and molecular phylogenetic analysis (Wang & Hong, 2015). Howerer, it differs from them both in having a single flower on each stem, smaller calyx lobes, and gray or pale purple corolla, which has a dark purple ring just below the throat. Furthermore, it differs from the former in having much smaller leaf blade and shorter petiole, while from the latter in having usually simple stems.

S HENGDUAN: **Fugong**, Lumadeng, 3700 m, 2005, *Gaoligong Shan Biodiversity Survey 28529* (A, CAS).

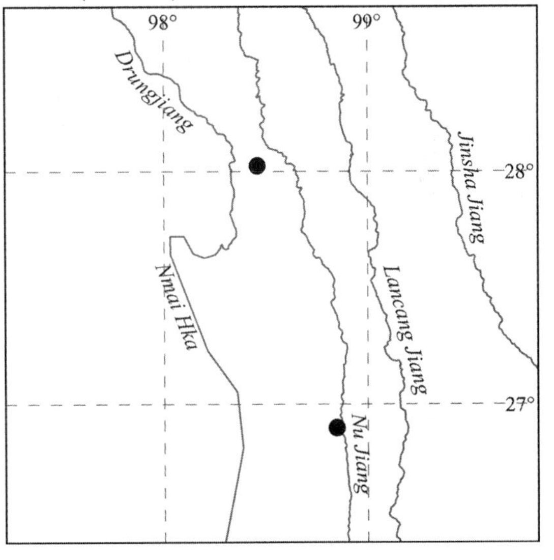

Map 47–72: **Codonopsis gongshanica** Q. Wang & D. Y. Hong

6. Codonopsis subscaposa Kom., Trudy Imp. S.-Peterburgsk. Bot. Sada 29(1): 114. 1908; Tsoong, Contr. Inst. Bot. Nat. Acad. Peiping 3(3): 101. fig. 11B. 1935; Anonymous, Icon. Cormophyt. Sin. 4: 380. fig. 6173. 1975; Hong, Monogr. *Codonopsis* Allied Gen.

Camp. 70. 2015. Type: China, W Sichuan, near Ta-tsien-lu (Kangding), 2740–3960 m, *A. E. Pratt 474* (lectotype designated by Hong, 2015: K; isolectotypes: BM, E, G, LE, P, PE).

Roots carrot-shaped, 15–20 cm long, 0.5–1.5 cm in diameter. Stems erect, 40–100 cm tall, simple or branched below, sparsely pubescent, later glabrescent. Leaves on main stem alternate, those on branches opposite; petiole 1–10 cm long, sparsely villous; leaf blade ovate to lanceolate, 2–13 cm long, 1.5–7.5 cm broad, adaxially green, sparsely hispidulous, abaxially gray-green, glabrous or sparsely villous along veins, base cuneate, margin sparsely dentate or crenate, or nearly entire, apex acute or obtuse. Flowers often solitary and terminal on main stem and branches, or in terminal cyme of 2–4, long-pedicellate. Calyx tube adnate to whole hypanthium, semiglobose, sparsely villous; calyx lobes narrow-deltoid, 5–7 mm long, 2–3 mm broad, glabrous, margin serrate, rarely entire, apex acute; sinus between lobes broad and truncate. Corolla variously coloured: yellow or green but with red-purple veins or red-purple with yellow spots, mostly pale blue with dark purple veins, or white above and brown-purple below, broad-campanulate, 1.5–3 cm long, 2–4 cm wide, lobed to near middle; corolla lobes glabrous on both sides or sparsely villous at apex. Stamens glabrous; filaments slightly dilated at base. Capsule semiglobose below, conical above. Seeds ovoid, brown-yellow, smooth. Figure 47–27: a–c.

Phenology: flowering from July to September; fruiting in September and October.

Habitats: grassy slopes, thickets, or open woods; 2000–5000m.

Distribution: **endemic**; Tangut, N and S Hengduan (Map 47–73).

TANGUT: **Zamtang**, Hongxin, 3600 m, 1975, *Sichuan Plant Surv. 9363* (CDBI, PE). **N HENGDUAN: Barkam**, Ermu Shan, 3600 m, 1957, *X. Li 71796* (PE, SZ); **Danba**, Shachong, 3300 m, 1958, *No. 8 Forest Managing Team 5622* (SZ); **Dawu**: Fangdashen Shan, 3400 m, 1958, *s. coll. 146* (SZ); Gaka, 3200 m, 20. 1960, *Medic. Plants Surv. 15322* (CDBI, SM); **Hongyuan**, Shuajingsi, 3200 m, 1957, *Z. Y. Zhang & H. F. Zhou 23561* (SZ); Zhegu Shan 3300 m, 2007, *X. J. He et al. 146764* (SZ); **Jinchuan**, Chuosijia (Guanyinqiao), 3600 m, 1958, *P. X. Li 10491* (PE); Wanli, 3400 m, 1958, *X. Li 78178* (PE, SZ); **Litang**, Labo, 4100 m, 1973, *Sichuan Veget. Surv. 3718* (CDBI, KUN, PE); **Lixian**, Dabanzhao, 3200 m, 1957, *D. Z. Deng 1945* (SZ); Miyaluo, 3700 m, 1956, *D. P. He 45862* (SZ); **Xiaojin**, Lianghekou, 3500 m, 1958, *S. S. Chang & Y. X. Re 6158* (CDBI, PE, SZ); **Yajiang**, Jianziwan Shan, 3200–3500 m, 1983, *K. Y. Lang et al. 2905* (KUN); Tiemenggou, 3700 m, 1959, *S. Jiang & C. L. Jin 2640* (KUN, PE, SZ). **S HENGDUAN**: **Daocheng**, Daocheng to Julong road, 3609 m, 2010, *Kunming & Edinburgh Exped. to Sichuan 281* (E); Julong, 3600 m, 1973, *Sichuan Veget. Surv. 2143* (CDBI, KUN); **Huili**, Shandi, 2200 m, 1960, *S. D. Li 60–1004* (SZ); **Jiulong**, Sanyanlong, 3500 m, 1979, *Jiulong Exped. 380* (SM); Tanggu, 3800 m, 1980, *Q. Q. Wang 22989* (CDBI); **Kangding,** Cheto La, 3500 m, 1934, *H. Smith 10924* (PE); Zhedotan, 3400 m, 2012. *D. Y. Hong & Q. Wang H12072* (PE); **Muli**, Bar Pasture, 5000 m, 1978, *C. S. Cao 6566* (CDBI, SZ); Gibboh Shan, 3960 m, 1932, *J. F. Rock 24168* (E,K, PE); Yungning-Muli, 3900–4000 m, 1915, *Handel-Mazzetti 7234* (E, K, WU). **Xiangcheng**:

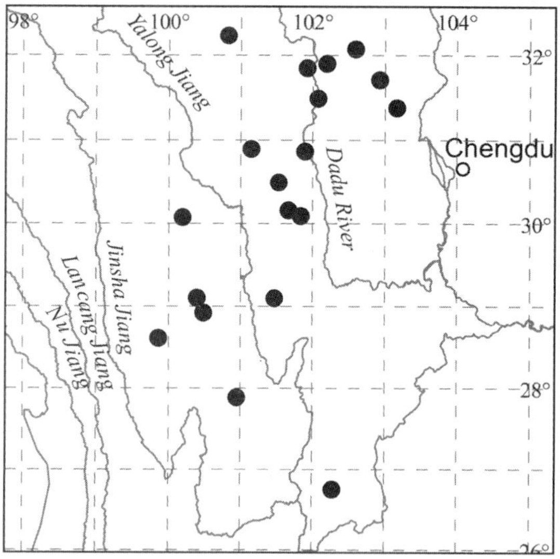

Map 47–73: **Codonopsis subscaposa** Kom.

Maxionggou, 4000–4300 m, 1998, *D. E. Boufford et al. 28482*(A, E); Ra'o, 4500 m, 1973, *Sichuan Plant Exped. (Xiang) 2894* (CDBI, KUN, PE); **Shangri-la** (Zhongdian), Dongwang, 3700–3800 m, 1998, *D. E. Boufford et al.29265* (E); Rengtsa, 3700 m, 1937, *T. T. Yü. 12772* (KUN, PE).

7. Codonopsis ovata Benth. in Royle, Ill. Bot. Himal. 253. t. 69. fig. 3. 1834; Hooker & Thomson, J. Proc. Linn. Soc. Bot. 2: 15. 1857, p. p., excl. pl. Soongar.; Komarov, Trudy Imp. S. -Peterburgsk. Bot. Sada 29(1): 115. 1908, p. p., excl. specim. Chinen.; Hong, Monogr. *Codonopsis* Allied Gen. Camp. 74. 2015. Type: India, Kashmir, *Royle* s. n.(holotype: LIV, n. v.)

Roots carrot-shaped, 15 cm long, ca. 1.5 cm in diameter. Stems erect, or sometimes ascending, sparsely villous, 30–80 cm tall, branches mostly or only aggregated at lower part, lower branches leafy, mostly sterile, upper ones more sparsely leafy, fertile. Leaves on main stem alternate; petiole absent to 8 mm long; leaf blade orbicular, ovate to ovate-lanceolate, 0.6–2.8 cm long, 0.6–2.2 cm broad, both adaxially and abaxially white-puberulent, base shallow-cordate, rounded to broad-cuneate, margin entire, apex rounded to acute; those on branches nearly opposite, similar to those on main stem. Flowers solitary, terminal only on main stem or also on upper branches, rarely also on lower branches; pedicel 1–15 cm long, mostly sparsely villous, rarely glabrous. Calyx tube adnate to whole hypanthium, rounded at base, glabrous, rarely sparsely villous along ribs; calyx lobes ovate to ovate-lanceolate, 5–10(–12) mm long, 4–8 mm broad, glabrous or hispidulous at apex, margin mostly entire, rarely serrate, more or less ciliate, revolute. Corolla light blue or purple-blue, purple inside, funnel-shaped, glabrous, lobed to 1/3, 1.6–3 cm long, 2–3 cm wide. Stamens totally glabrous; filaments slightly dilated at base. Capsule with inferior part semiglobose, superior part conical, ca. 1.7 cm long, 1.2 cm in diameter. Seeds not seen. **Chromosomes: 2n=16.** Figure 47–28.

Phenology: flowering in July and August; fruiting in September.

Habitats: bushes among rocks or alpine mea-dows; 2130–4370 m.

Distribution: N Pakistan, Jammu & Kashmir, and U Ganga & Indus (Map 47–74); also in Kyrgyzstan, Tajikistan and Uzbekistan.

Notes: *Codonopsis ovata* has been often confused in herbaria and literature with *C. clematidea* and *C. obtusa*. The most diagnostic characters of *C. ovata* are in calyx lobes which are revolute, ciliate, 5–10(–12) mm long, 5–8 mm broad. With reflexed and glabrous calyx lobes and ovate leaf blade, *C. clematidea* is easily distinguished from the other two. *C. obtusa* is characterized with lanceolate leaf blade (length/width over 2) and reflexed, subulate, and nearly always hispidulous calyx lobes. Nasir (1984: Fl. Pakist. 155: 2) failed to record *C. ovata* in Pakistan because of his misidentifying specimens of this species as *C. clematidea*.

N PAKISTAN: **Baltistan**, Deosai, Mir Panzil Pass, 1940, *R. R. Stewart 19955* (PE); Gieuwal, 4570 m, *Slolivzka* s. n.(CAL); Kashmir, Panchtami, 1938. *M. Nath* s. n. (KUH); Kishenganga, Barai Pass, 3510 m, 1935, *F. Ludlow & G. Sherriff 1454* (E); Kishenganga Valley, Folowai, 2130 m, 1935, *F. Ludlow & G. Sherriff 1431* (E); Kishenganga Valley, Kel-Taubat, 1980 m, 1939, *R. R. Stewart 17859* (PE); Nanga Parbat, 3300 m, 1993, *Nüsser 357* (B); **Chitral**, Barir Kafristan, 1976, *M. Shah et al. 2063* (ISL);

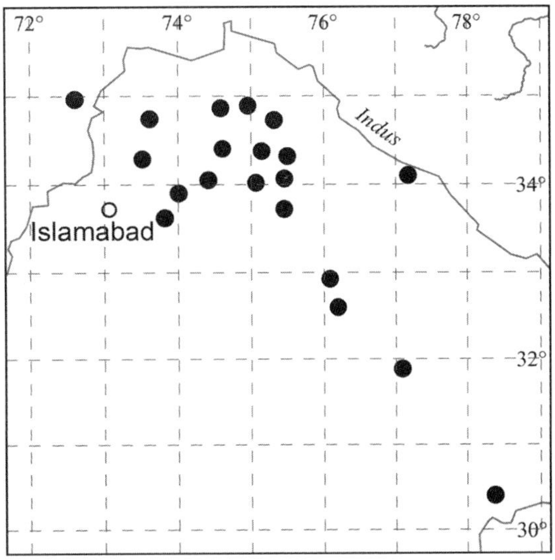

Map 47–74: **Codonopsis ovata** Benth.

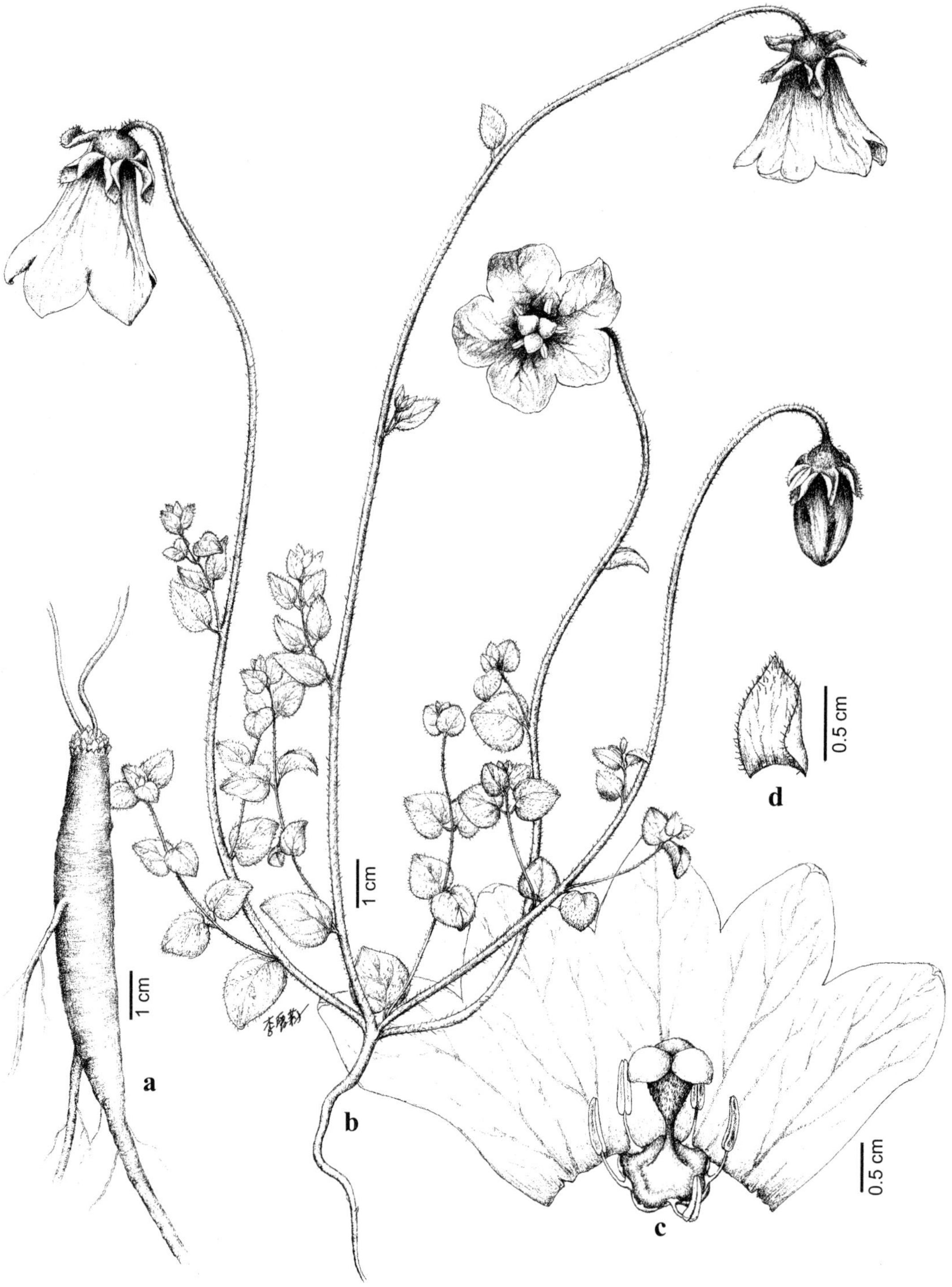

Figure 47–28: **Codonopsis ovata** Benth.: **a**, root, caudex, and base of stems; **b**, stem; **c**, a flower with calyx removed and corolla opened, showing stamens, disc, style, and stigma; **d**, a revolute calyx lobe. From Hong, *Monogr. Codonopsis Allied Gen. Camp.*, 2015.

Molikhoo, Joormoli /Sahat top, 2845 m, 2007, *H. Ali 5723* (KUH); **Gilgit**, Nattar Sokes, 3960 m, 1954, *R. R. Stewart 26406* (BM); **Hazára**, Bantara Gali, Poonch, 2740–3050 m, 1952, *R. R.Stewart & E. Nasir 24112* (KUH, W) ; Kagan, Silán, 1899, *Inayat* s. n.(CAL); Kagan Valley, Paya, ca. 3050 m, 1984, *Muqarrab Shah & Jamohed 41* (E); Kagan Valley, Sarnl, 1897, *Inayat 21946* (CAL); Siam Valley, 1896, *Inayat 19896* (CAL); **Swat**, Bishigram, 3350 m, 1953, *R. R. Stewart & A. Rahman 24928* (BM). **JAMMU & KASHMIR: Aphawab,** 3350 m, 1933, *Jvenning K45* (K); **Baltal,** 2900 m, 1848, *T. Thomson* s. n. (K); Butin Pantsar, Wangat Valley, 3350 m, 1940. *P. M. Pinfold 270* (BM, E); **Gagongiv,** Sim Valley, 3350 m, 1940, *F. Ludlow & G. Sherriff 7977* (BM); above **Gulmarg,** 3540 m, *G. L. B. Fuller 128* (K); **Khilanmarg,** ca. 3050 m, 1956, *O. Polunin 56/138* (B, BM, E); **Kolahoi Valley,** Earthy, ca. 3660 m, 1956, *O. Polunin 56/570* (B, BM, E); **Ladakh,** 3420 m, 1976, *Hans Hartmann 2061* (G); **Liddar** Valley, 3960–4270 m, 1893, *J. F. Duthie 13355* (BM, E); **Margan Pass,** 3960–4270 m, 1936, *M. K. Timins 189* (BM); **Tilpatra** Forest, Nr. Rampur, Jhelum Valley, 2930 m, 1940, *F. Ludlow & G. Sherriff 7737* (BM, E); **Tragbal,** 3350 m, 1876, *C. B. Clarke 29251A* (BM) & *C. B. Clarke 29251 B* (K); **Upper Chenab Valley,** 1879, *R. Ellis 265* (K); **Zojpal,** Kolahoi Glacier, 1913, *M. A. Evershed* s. n. (BM). **U GANGA & INDUS: Chamba**, 1880, *R. Ellis* s. n.(K); Rahla Kulu, 3660 m, 1916 *R. E. Cooper 5242* (E); Sural Valley, 2440–3660 m, 1899, *Harsuku* s.n. (K); **Pangs**, Dunai to Bindrabani, 2900 m, 1897, *J. H. Lace 1620* (E); **Tehri-Garhwal**, 3660–3960 m, 1883, *J. F. Duthie 609* (BM).

8. Codonopsis clematidea (Schrenk) C. B. Clarke in Hooker, Fl. Brit. India 3: 433. 1881; Komarov, Trudy Imp. S. - Peterburgsk. Bot. Sada 29(1): 115, 1908; Tsoong, Contr. Inst. Bot. Nat. Acad. Peiping 3(3): 93. 1935; Hong, Monogr. *Codonopsis* Allied Gen. Camp. 78. 2015. ≡ *Wahlenbergia clematidea* Schrenk in Fischer & Meyer, Enum. Pl. Nov. Songar. 1: 38. 1841. ≡ *Glossocomia clematidea* (Schrenk) Fisch. in Regel, Gartenflora 5: 226. t. 167. fig. 2. 1856. Type: "Songarei, Alatau, Dschillkaragai", 21 Jun. 1840, *Schrenk* s. n. (holotype: LE); loc. eodem, 11. Jul. 1840, *Schrenk* s. n. (paratype: LE).

Codonopsis ovata Benth. var. *cuspidata* Chipp, J. Linn. Soc. Bot. 38: 385. 1908, p.p., excl. specim. "China: Tachielu (Kangding), *Soulie, 593" (= Codonopsis pilosula* subsp. *handeliana).* Type: " Tibet Occ." (India: Kashmir), Yulchung, Zanskar River, *2750–3350 m,* 02. July 1848, *T. Thomson* s. n. (lectotype designated by Hong, 2015: K).

Roots carrot-shaped, 25–45 cm long, 1–3 cm in diameter. Stems single or caespitose, erect or ascending, rarely trailing, 50–100 cm long, branches mostly at lower part and sterile but those at upper part usually fertile. Leaves on main stem alternate, those on branches opposite; petiole up to 2.5 cm long, sparsely hispidulous; leaf blade ovate to lanceolate, 1–2.8(–5.2) cm long, 0.8–1.5(–3.2) cm broad, adaxially sparsely or densely hispidulous, abaxially densely hispidulous, base slightly cordate or rounded, margin entire, apex acute. Flowers solitary, terminal on main stem and branches, long-pedicellate; pedicel sparsely white hispidulous. Calyx tube adnate to whole hypanthium, semiglobose, glabrous or rarely sparsely white-scaberulose; calyx lobes ovate, elliptic or ovate-lanceolate, 10–18 mm long, 5–8.5 mm broad, glabrous or apex sparsely hispidulous, strongly reflexed from anthesis, margin entire, apex acute; sinus between calyx lobes narrow and pointed. Corolla purplish or bluish with dark blue veins and purple spots inside, or purple, campanulate, glabrous, 1.8–2.8 cm long; corolla lobes 1/3–1/4 as long as corolla length. Stamens glabrous; filaments, slightly dilated at base. Capsule semiglobose below, conical above, 1.5–2 cm long, 1.2–1.5 cm wide, apex acute; persistent calyx lobes much enlarged. Seeds narrowly ellipsoid, slightly compressed, pale brown-yellow, smooth. **Chromosomes**: 2n=16. Figure 47–29.

Phenology: flowering and fruiting from July to October.

Habitats: forests, edges of forests, or meadows; 1500–4600 m.

Distribution: N Pakistan, Jammu & Kashmir, and U Ganga & Indus (Map 47–75); also widely distributed in NE Afghanistan, China (Xinjiang), Kazakhstan, Kyrgyzstan, W Mongo-

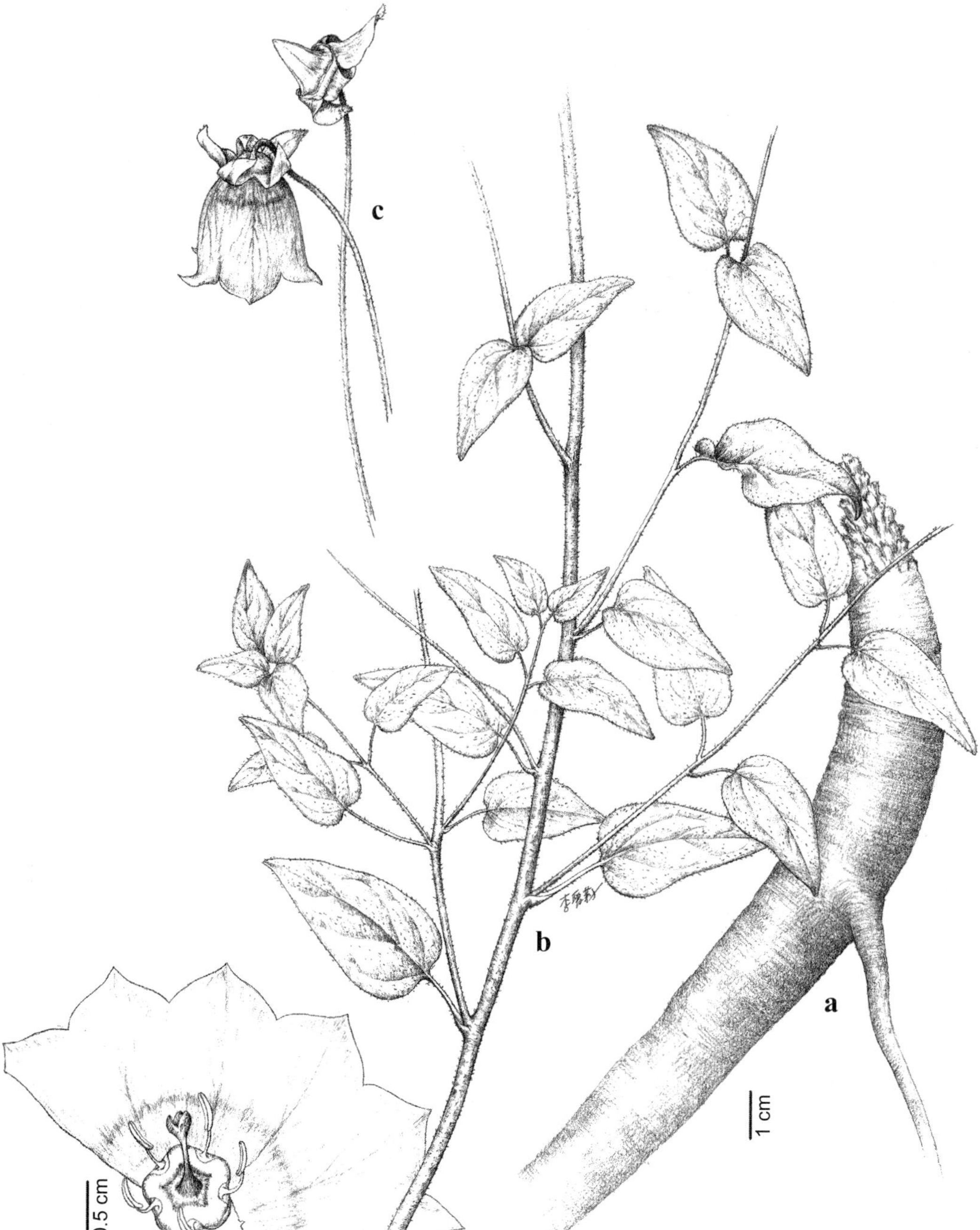

Figure 47–29: **Codonopsis clematidea** (Schrenk) C. B. Clarke: **a**, root and caudex; **b**, middle part of stem; **c**, two terminal flowers (flowering and fruiting); **d**, a flower with calyx removed and corolla opened, showing stamens, disc, style, and stigma. From Hong, *Monogr. Codonopsis Allied Gen. Camp.,* 2015.

lia, and Tajikistan.

N PAKISTAN: Baltistan, Astor, 3050 m, 1946, *R. R. Stewart & E. Nasir* s. n.(K); Deosai, ca. 3660 m, 1987, *Muqarrab Sha & Jamohed 273* (BM, E); Deosai Plains, 4270 m, 1966, *M. A. Gid-deji* s. n.(BM); Doghani, Shyoh River, 2440 m, 1928, *F. Ludlow 345* (BM,E); Haramosh, Kha-harew Valley, 2001, *S. W. Uhaw 21* (KUH); near Hushe Village, ca. 3050 m, 1955, *G. L. Webster & E. Nasir 5955* (PE); Karakoram, between Tur-mun and Makerum, 3350–3400 m, 1939, *R. S. Russell 1224* (BM); Karakoram, Barpu Glacier, 3050–3350 m, 1939, *R. S. Russell 1093* (BM); Karakoram, upper end of Hushe Valley, 3810–3960 m, 1955, *G. L. Webster & E. Nasir 6137* (PE); Kishenganga Valley, road to Nanga Parbat, 3050–3350 m, 1939, *R. R. Stewart 18486* (KUH, PE); Kishenganga Valley, Badwan, 2440 m, 1940, *R. R. Stewart 19534* (PE); Paskyum, 2830 m, 1928, *B. B. Osmaston 130* (K); Rama Valley, 3170 m, 1967, *T. E. Lankester & T. A. S. Pearson 1222* (BM); Samsam, 3050 m, 1943, *F. Ludlow & G. Sherriff 9188* (BM); Satpur, above Skardu, 2740–3050 m, 1892, *J. F. Duthie 11989* (BM, E); Shuko Gah-Nala Glacier, , 3510 m, 1950, *J. W. Thornley 52* (BM); Skardu, Satpara Nullah, 3400 m, 2002, *J. Alam et al. JA-11* (KUH); Solu Glacier, 3410 m, 1939, *R. S. Russell 1752* (BM); **Chitral**, Chumarkhan Pass, 3660 m, 1958, *J. D. A. Stainton 2907* (BM, E, W); Mastooj, Shandoor top, 4223 m, 2007, *H. Ali 6751* (KUH); Mastooj, Sor Laspur, 2007, *H. Ali 6682* (KUH); Yarkhun, 2440 m, 1958, *S. A. Bowes Lyon 915* (BM, E, W); **Gilgit**, Derachil, Karimabad, ca. 3600 m, 2005, *J. & N. Alam 3296* (KUH); Hunza, above Karimabad, 3500 m, 2001, *J. Alam 50* (KUH); Imit, 2900 m, 1954, *F. Schmid 2089* (G); Lower Naltar, ca. 2800 m, 2003, *J. Alam & Q. Ali 1867* (KUH); Minapir Glacier, 3050 m, 1961, *P. S. Lloyd & S. Megaw 26* (BM, E, PE); Nagar Vil-lage, 2380 m, 1960, *O. Polunin 6429* (B. BM, E, G); Naltar, 3050 m, 1986, *S. Malik et al. 378* (PMNH); ca. 9 km from Yasin, along Nazber nul-lah, 1982, *Saood Omer 230* (BM, K, KUH); **Ha-zara**, Besal, 3800 m, 1953, *F. Schmid 361* (BM, E, G); Kagan Valley, between Balakot and Babusar Pass, 1954, *J. Abei 113* (BM); Kagan Valley, be-tween Battakundi and Burawai, 1958, *B. L. Burtt & Roshan Din 925* (E); Naran, 2393 m, 1975, *I. D. A. Amin et al. 896* (PE); Saif-ul-Maluk-Sar, 3570 m, 1953, *F. Schmid 279* (G); **Swat**, Sho Nala, 3350 m, 1955, *A. Rahman 70* (BM); Ushu to Batani, 2440–2740 m, 1953, *R. R. Stewart & A. Rahman 25295* (BM). **JAMMU-KASHMIR**: **Chamba** (Pauri), 4270 m, 1899, *J. F. Duthie* s. n.(K); **Ladakh**, between Zoji Pass and Sonamarg, 3250 m, 1976, *F. Billiet & J. Leonard 6901* (K); Ladakh, 1 km from Kanji, 3960 m, 1984, *A. R. Brown & M. A. Rothera 79* (K); Ladakh, Pan-ichar, 2900 m, 1980, *Southampton Univ. 20* (K); **Zanskar** Distr., Karsha, 3510 m, 1981, *Grace, Bailey, Bull & Carr 81* (BM, K). **U GANGA & INDUS**: **Lahul**, Kyelang, 1888, *J. R. Drummond 25267*(E, K); Lahul, 3050 m, 1865, *H. Jaeschke* s. n.(K, WU); Lahul, Darcha, 3350 m, 1976, *J. D. A. Stainton 7664* (E); Lahul, Kyelang, 3200 m, 1936, *W. Koelz 8533* (PE); Lahul, Zingzingbar, 4570 m, 1941, *N. L. Bor 16373* (E, K) ; Randaug, 1888, *J. R. Drummond 25266* (K).

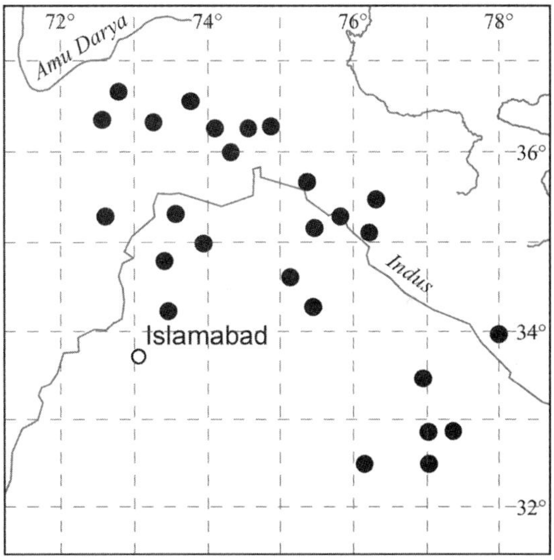

Map 47–75: **Codonopsis clematidea** (Schrenk) C. B. Clarke

9. Codonopsis obtusa (Chipp) Nannf., Acta Horti Gothob. 5: 28. 1930; Nasir in Nasir & Ali, Fl. Pakistan 155: 3. 1984; Hong, Monogr. *Co-donopsis* Allied Gen. Camp. 83. 2015. ≡ *Co-donopsis ovata* Benth. var. *obtusa* Chipp, J. Linn. Soc. Bot. 38: 385. 1908. ≡ *Codonopsis clematidea* (Schrenk) C. B. Clarke var. *obtusa* (Chipp) Kitam., Flowering Afgh. 383. 1960. Type: Afghanistan, Kurram Valley, 1879, 3050 m, *J. E. T. Aitchison 748* (lectotype designated by Hong, 2015: BM; isolectotypes: CAL, G, K , PE); Afghanistan, s. loc., *Griffith 3435* (paratypes: K, PE).

Roots carrot-shaped, up to 2 cm in diameter. Stems erect, 20–100 cm tall, glabrous or villous, multi-branched, lower branches sterile, middle and upper ones often fertile. Leaves on main stem mostly evenly distributed, alternate; petiole 2–8 mm long, hispidulous; leaf blade ovate to lanceolate, 1–5.2 cm long, 1–2.3 cm broad, hispidulous on both surfaces, base rounded to broad-cuneate, margin entire, apex acute to acuminate; leaves on branches nearly opposite, similar to those on main stem but smaller. Flowers terminal on main stem and upper branches; pedicel 1.5–30 cm long, hispidulous or villous. Calyx tube adnate to whole hypanthium, rounded to obconical-rounded at base, glabrous or hispidulous; calyx lobes subulate, broadest at base, very rarely slightly broader above, 6–15 mm long, 2–4 mm broad, entire, always ciliate, reflexed at or after anthesis, both surfaces hispidulous, rarely nearly glabrous. Corolla pale blue outside, deep blue-purple inside or just with 2 purple rings inside, campanulate, 1.5–3 cm long, lobed to 1/3 or 2/5; corolla lobes deltoid-orbicular. Stamens totally glabrous; filaments slightly dilated at base. Capsule obco-nical to globose-obconical in inferior part, long-conical in superior part, 1.8–2.5 cm long, 1–1.5 cm wide. Seeds unknown. **Chromosomes**: 2n=16. Figure 47–30.

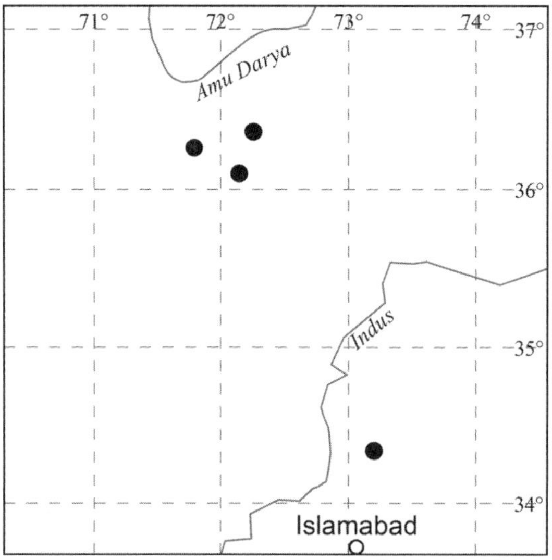

Map 47–76: **Codonopsis obtusa** (Chipp) Nannf.

Phenology: flowering from June to August; fruiting in August and September.

Habitats: damp meadows, open calcareous slopes or by stream; 1550–3400 m.

Distribution: N Pakistan (Map 47–76); also in NE Afghanistan, Tajikistan, and SE Uzbekistan.

N PAKISTAN: **Chitral,** Barum, SE of Tirich Mir, 2440 m, 1958, *J. D. A. Stainton 2772* (BM); Bashgalian, 2440 m, 1958, *J. D. A. Stainton 2702* (BM); Chitral, Kasaweer, 2206 m, 2007, *H. Ali 6175* (KUH); Tirich Gol, 2740 m, 1888, *S. A. Bowes Lyon 1064* (BM,E); **Hazara**: Mansehra, Saif-ul-Maluk, 1990, *S. Malik & J. Saqib 117* (PMNH).

10. Codonopsis subsimplex Hook. f. & Thomson, J. Proc. Linn. Soc. Bot. 2: 16. 1857; Clarke in Hooker, Fl. Brit. India 3: 432. 1881; Hong, Lammers & Klein in Wu, Raven & Hong, Fl. China 19: 519. 2011; Hong, Monogr. *Codonopsis* Allied Gen. Camp. 85. 2015. Type: India, Sikkim, Lachen, 4270 m, 15. Jul. 1849, *J. D. Hooker* s. n.(lectotype designated by Hong, 2015: K); Sikkim, Lachen, 3660–3960 m, 30. Jul. 1849, *J. D. Hooker* s. n. (paratypes: CAL, G, K, W).

Roots carrot-shaped, 15–35 cm long, 0.5–1.5 cm in diameter. Stems erect, ascending or trailing, few-branched, 35–70 cm long, nearly glabrous or sparsely white-villous. Leaves on main stem alternate, those on branches nearly opposite and much smaller; petiole 0.5–4.5 cm long, sparsely white-villous; leaf blade broadly ovate, ovate or elliptic, 2–12 cm long, 2–6 cm broad, adaxially hispidulous, abaxially hispidulous but hirsute along veins, base cuneate or rounded, margin crenate or dentate, apex acute or slightly obtuse. Flowers solitary or in cyme of 2–4, terminal on main stem and upper branches. Calyx tube adnate to whole hypanthium; calyx lobes ovate-lanceolate or ovate, 13–16 mm long, 6–8.5 mm broad, abaxially hirsute, margin usually denticulate, sparsely ciliate, apex acute or obtuse; sinus between calyx lobes narrow and pointed. Corolla whitish, pale yellow, yellow-green or pale blue, shortly tubular, glabrous, 1.2–2 cm long, 1.1–2 cm wide, shallowly lobed

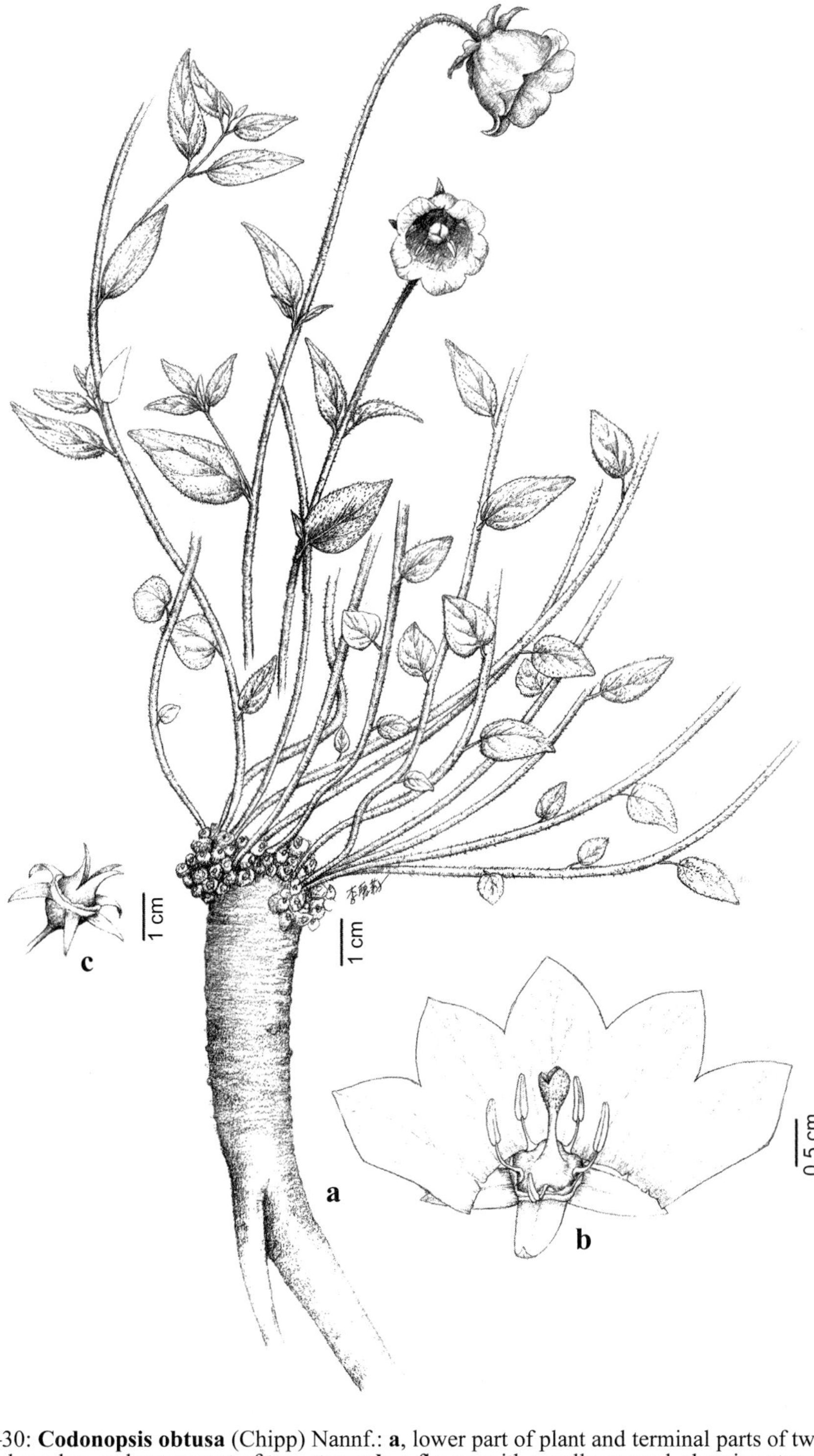

Figure 47–30: **Codonopsis obtusa** (Chipp) Nannf.: **a**, lower part of plant and terminal parts of two stems, showing root and caudex, and upper part of two stems; **b**, a flower with corolla opened, showing stamens, disc, style, and stigma; **c**, a 3-loculicidal capsule with persistent calyx. From Hong, *Monogr. Codonopsis Allied Gen. Camp.*, 2015.

to ca. 1/4; corolla lobes ovate-deltoid. Stamens glabrous; filaments slightly dilated at base. Capsule semiglobose or obovoid-conical below, short-conical above, 8–10 mm in diameter, rostrum 4–6 mm long. Seeds numerous, ellipsoid, ca. 1.5 mm long, yellow. **Chromosomes**: 2n=16. Figure 47–31: d–f.

Phenology: flowering in July and August; fruiting in September and October.

Habitats: mountain forests or scrubs; 1220–4270 m.

Distribution: **endemic**; U Yarlung Zangbo, W, C and E Nepal, Sikkim & Darjiling, Bhutan, and M Yarlung Zangbo (Map 47–77).

U YARLUNG ZANGBO: Gyirong, Gyirong-Gyirong Town, 3700 m, 2013, *PE-Tibet Exped. 3840* (PE). **W NEPAL**: **Seti**, Bajhang, 3274 m, 2009, *H. Ikeda et al. 20915092* (E). **C NEPAL**: **Bagmati**, Nuwakot, N of Gulu Bhanjyang, 3300 m, 1974, *J. H. de Haas 2156* (BM); Nuwakot, Kalanki, 3660–3960 m, *L. Dhevoj 0467* (BM, E); Rasuwa, Chilime Kharka-Thangjet, 4410 m, 1949, *O. Polunin 1331* (BM); Sindhupalchok, Gopte-Thale Patil, 3400 m, 1972, *H. Kanai et al. 723169* (BM, TI); **Gandaki**, Kaski, Annapurna Himal., Seti Khola, 3800 m, 1954. *Stainton, Sykes & William 6653* (BM, E); Lamjung, Lamjung Himal., 3960 m, 1954, *Stainton, Sykes & Williams 6352* (BM, E). **E NEPAL**: **Kosi**, Chhurchathange (3600 m) – Piling La (4350 m) – Thudam (3400 m), 3800 m, 1977, *H. Ohashi et al. 772283* (BM, E, PE, SZ, TI) & *773705* (TI); Sankhuwasabha, Jaljale Himal., 1991, *H. Ohba et al. 9110464* (BM, TI) & *9120394* (BM, TI); **Sagarmatha**, Solukhumbu, 3900 m, 1997, *M. Wakabayashi et al. 9715261* (E). **SIKKIM & DARJILING: Darjiling**, Singalila, Phalut, 3800 m, 1969, *H. Hara et al. 69713* (TI); 3350 m, 1870, *C. B. Clarke 12542* (CAL, K); **Sikkim**, Bijan, 1888, *King's Collector* s. n.(CAL); Gnatong, 3660–4270 m, 1913, *R. E. Cooper 750* (E); S of Imgri, 1887, *King's Collector* s. n.(CAL); Quatung, 1891, *Cummins* s. n.(CAL); Rathangoong, 1908, *Ribu 111* (CAL); Shaitanpokrikapûp, 3810 m, 1913, *R. E. Cooper 768* (E); Sherabthang, 3960 m, 1910, *W. W. Smith 4298* (B, CAL, E); Zemu Valley, 3960 m, 1909, *W. W. Smith & G. H. Cave 1653* (B, CAL). **BHUTAN**: **Ha,** Chelai La, 3660 m, 1949, *F. Ludlow et al. 19612* (BM, E); Ha to Chile La, 2750–3650 m, 1938, *B. J. Gould 1247* (K); **Mangde Chu**, Dur Chuteen, 1949, *F. Ludlow et al. 19456* (BM); Mangde Chu, Goktang La, 3660 m, *F. Ludlow et al. 16891* (BM, E); **Shingbe**, Me La (Cho La Valley), 3050 m, 1949, *F. Ludlow et al. 21052* (BM, PE); **Rinchen Chu**, 3350-3660 m, 1937, *F. Ludlow* & G. *Sherriff 3530* (BM); **Thimphu,** Parshang, 3660 m, 1914, *R. E. Cooper 1948* (E); **Tongsa**, **Yuto-la**, 3510 m, 1937, *F. Ludlow & G. Sherriff 3469* (BM). **M YARLUNG**

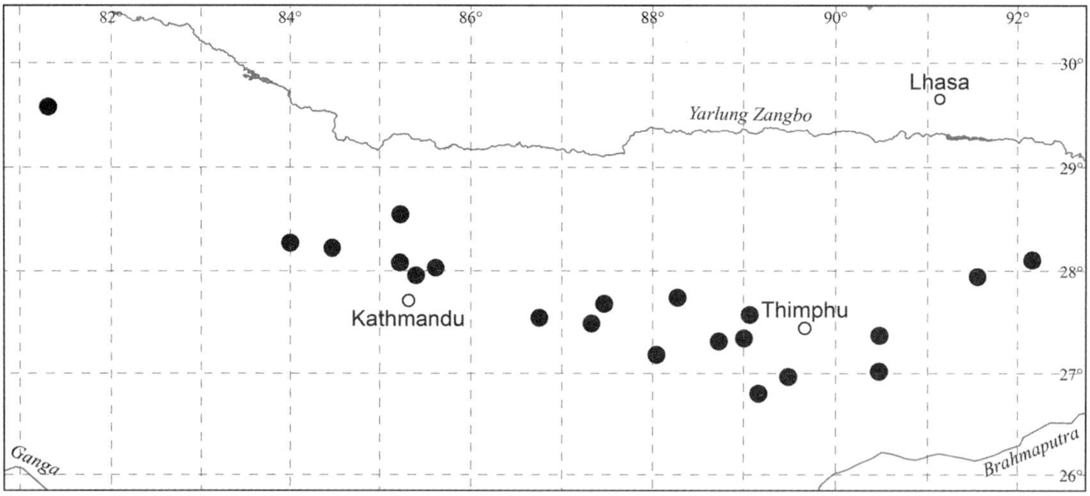

Map 47–77: **Codonopsis subsimplex** Hook. f. & Thomson

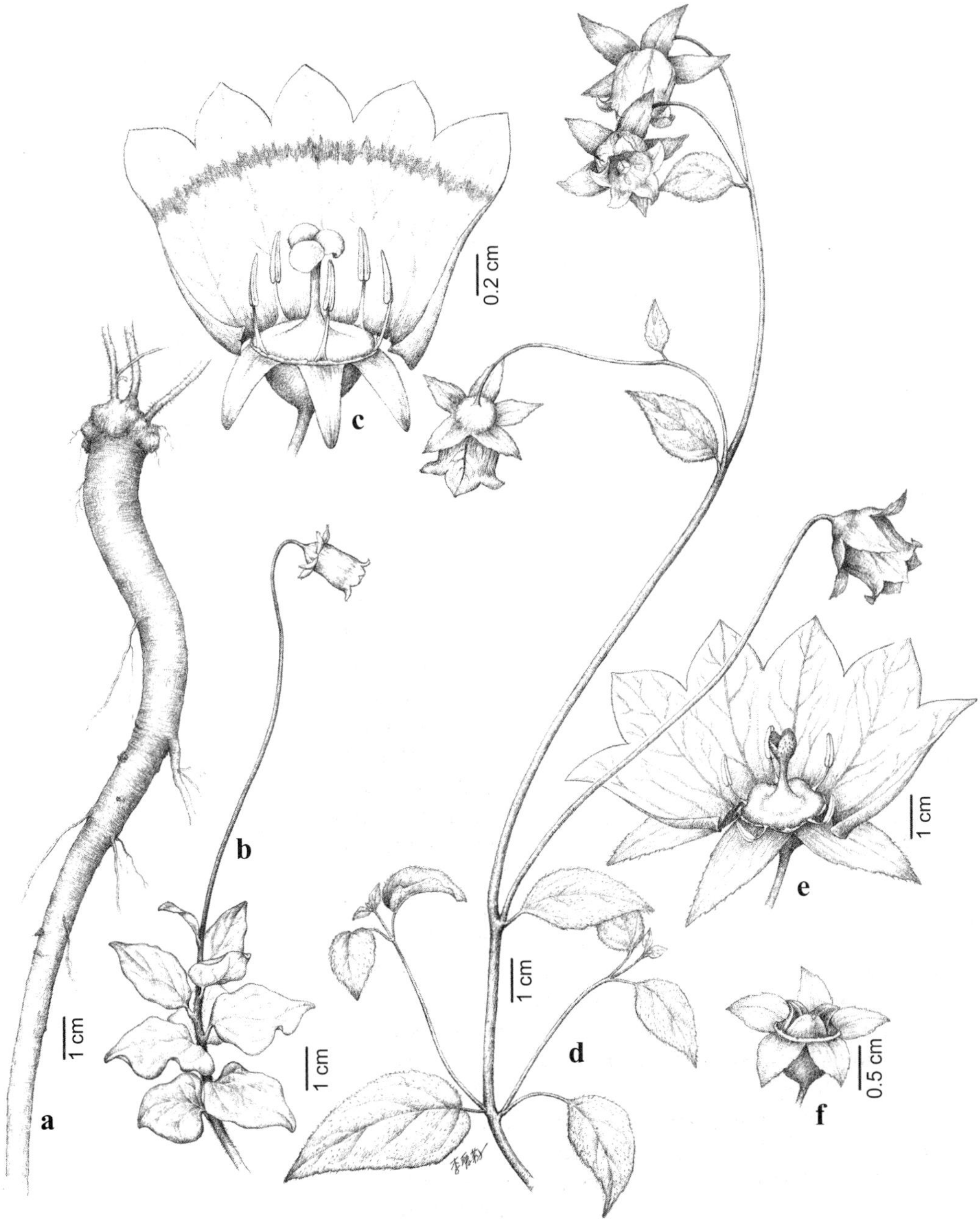

Figure 47–31: **a–c**, **Codonopsis gongshanica** Q. Wang & D.Y. Hong: **a**, root, caudex, and base of stems; **b**, middle and upper part of stem; **c**, a flower with corolla opened, showing stamens, disc, style, and stigma. **d–f**, **Codonopsis subsimplex** Hook. f. & Thomson: **d**, upper part of stem; **e**, a flower with corolla opened, showing stamens, disc, style,and stigma; **f**, a 3-loculicidal capsule with persistent calyx. **a–c** from Wang & Hong, Phytotaxa 188(3), 2015, with permission; **d–f** from Hong, *Monogr. Codonopsis Allied Gen. Camp.*, 2015.

ZANGBO: **Yadong**, Chumbi, 1882, *Dungboo* s. n.(K); Chumbi & Phari, 1879, *Dungboo* s. n.(K); Chumbitang, 1953, *P. C. Tsoong 5931* (PE); Tanggapu Gou , 3800 m, 1975, *s. coll. 75–914* (PE).

11. Codonopsis campanulata D. Y. Hong, Plant Divers. Resour. 36(3): 290. fig. 4. 2014; Hong, Monogr. *Codonopsis* Allied Gen. Camp. 87. 2015. Type: Nepal, Gandaki Zone, Marsyandi Valley, Manang Distr., Annapurna, slopes opposite to SW Braga, Jul. 1989, *C. Grey-Wilson* **s.** n.(holotype: K).

Roots unknown. Stems ascending or procumbent, not twining, 50–80 cm long, sparsely villous, with very slender, sterile branches from lower and middle parts. Leaves evenly distributed along most parts of stems, those on main stem alternate, those on branches nearly opposite; petiole 2–5 mm long; leaf blade hispidulous on both surfaces, base more or less cordate, margin entire or shallowly crenulate, apex obtuse or acute, basal leaves very small, scale-like. Flowers solitary, terminal on main stem; pedicel over 10 cm long, villous. Calyx tube adnate to whole hypanthium, and thus calyx lobes and corolla inserted at same level; calyx tube densely villous; calyx lobes lanceolate, 1–1.4 cm long, 0.4–0.5 cm broad, villous on abaxial side, entire, revolute. Corolla pale pink with a red-purple ring below throat, campanulate, 1.8–2.2 cm long, lobed to ca. 1/3; corolla lobes deltoid, 6–8 mm long, sparsely hispidulous outside. Stamens glabrous; slightly dilated at base. Stigma 3-fid, stigma segments broad-oblong. Fruit not seen. Figure 47–32: a–d.

Phenology: flowering in July.

Habitats: meadows with shrubs; ca. 3700 m.

Distribution: **endemic**; confined to U Yarlung Zangbo (Gyirong)and C Nepal (Gandaki Zone) (Map 47–78).

Notes: This newly described species resembles *Codonopsis foetens* Hook. f. & Thomson mostly in having campanulate corolla and sterile branches, but it differs from that species in having more evenly distributed and larger leaves (1.5–3 cm vs. 0.5–1.5 cm long), much longer stems (50–80 vs. 20–40 cm long), densely villous calyx tube (vs. sparsely hispidulous), and by a red-purple ring below corolla throat (vs. lacking such a ring).

U YARLUNG ZANGBO: **Gyirong**, Gyirong to Gyirong Town, 3700 m, 2013, *PE-Tibet Exped. 3840* (PE).

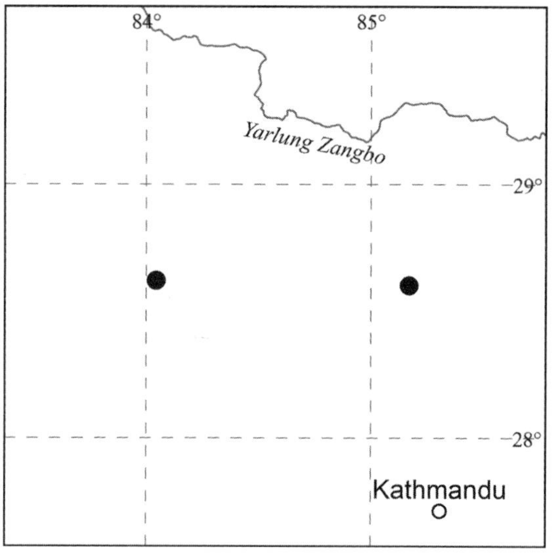

Map 47–78: **Codonopsis campanulata** D. Y. Hong

12. Codonopsis alpina Nannf., Notes Roy. Bot. Gard. Edinburgh 16: 154. pl. 227. & fig. 1g. 1931; Hong, Lammers & Klein in Wu, Raven & Hong, Fl. China 19: 522. 201 1; Hong, Monogr. *Codonopsis* Allied Gen. Camp. 87. 2015. Type: China, "Tibet austr.-orient., Sarong, on Doker-la, Mekong-Salween divide" (Dêqên, Yunnan), 28°20′ N, 3660 m, Aug. 1917, *G. Forrest 14659* (holotype: E; isotypes: K, W)

Codonopsis foetens Hook. f. & Thomson var. *major* Hand.-Mazz., Akad. Wiss. Wien Math. -Naturwiss. Kl. Anz. 61(20): 169. 1924. Type: China, NW Yunnan, Mekong-Salween, Mt Maya, 28°4′ N, 4050–4300 m, 03. Aug. 1916, *Handel-Mazzetti 9638* (lectotype designated by Hong, 2015: WU; isolectotypes: K, W).

Roots carrot-shaped, up to 30 cm long, 1.5 cm in diameter. Stems several from one caudex, simple or branched, main stem erect or ascending, 15–45 cm tall, or longer, glabrous or sparsely white-hirsute; branches aggregated at lower part,

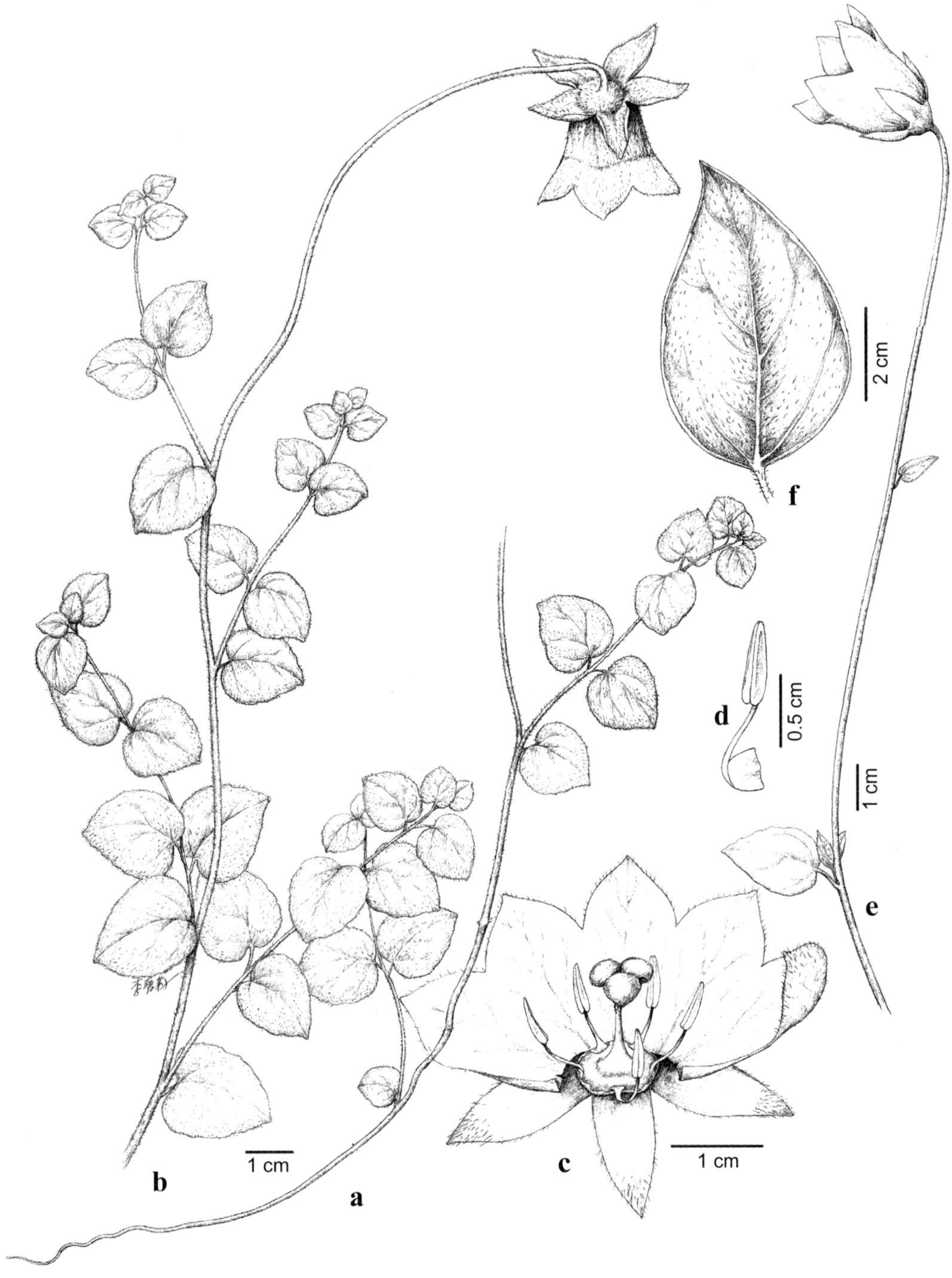

Figure 47–32: **a–d**: **Codonopsis campanulata** D. Y. Hong: **a**, lower part of stem; **b**, middle and upper part of stem; **c**, a flower with corolla opened, showing stamens, disc, style, and stigma; **d**, stamen. **e** & **f**, **Codonopsis alpina** Nannf.: **e**, upper part of stem; **f**, abaxial surface of leaf blade, showing revolute margin. **a–d** from Hong in *Plant Diversity and Resources* 36(3), 2014, with permission; **e** & **f** from Hong, *Monogr. Codonopsis Allied Gen. Camp.*, 2015.

leafy, sterile, sparsely white-hirsute. Leaves on main stem sparse and alternate, those on branches nearly opposite, rarely alternate; petiole 4–6 mm long, glabrous or sparsely hirsute; leaf blade ovate-cordate, 0.8–2.8 cm long, 0.8–2 cm broad, adaxially glabrous or sparsely white-hispidulous, abaxially sparsely white-hispidulous, base shallowly cordate or rounded, margin sinuous-crenate or entire, revolute, apex obtuse. Flowers solitary, terminal, usually only on main stem, less frequently also on upper branches; pedicel sparsely hirsute. Calyx tube adnate to whole hypanthum, obconical or semiglobose, glabrous; calyx lobes triangular-oblong, 7–11 mm long, 3.5–6.5 mm broad, clearly 3–5-nerved, both surfaces glabrous or abaxially very sparsely hispidulous at apex, margin entire, dentate or sparsely crenate, sometimes revolute, apex acute or obtuse; sinus between calyx lobes broad and obtuse. Corolla variously coloured: purple-green with purple veins, red, violet-blue, but mostly purple, campanulate, 2.5–3 cm long, ca. 2.5 cm wide, shallowly lobed, glabrous; corolla lobes deltoid-orbicular. Stamens glabrous; slightly dilated at base. Capsule 1.3–1.6 cm long, ca. 1.2 cm in diameter, obovoid-conical in inferior part. Figure 47–32: e & f.

Phenology: flowering from July to October; fruiting in October.

Habitats: alpine meadows or rocky grassy slopes; 3600–4420 m.

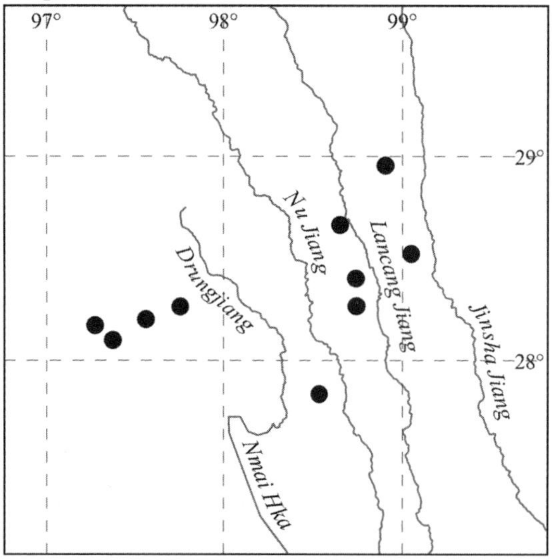

Map 47–79: **Codonopsis alpina** Nannf.

Distribution: endemic; N and S Hengduan, and U Irrawaddy (Map 47–79).

N HENGDUAN: Markam, s. loc., 3660 m, 1932, *J. F. Rock 22266* (E). **S HENGDUAN: Dêqên**, Baima Shan, 3660 m, 1912, *F. Kingdon-Ward 5330* (E); Mekong-Salween Divide, Sewalongba, 4000 m, 1938, *T. T. Yü 22576* (KUN, PE); **Gongshan**, Salween–Kiukiang (Drungjiang) Divide, Netahtsu, 3600 m, 1938, *T. T. Yü 20792* (KUN, PE). **U IRRAWADDY**, **Adung Valley**, 1931. *F. Kingdon-Ward 9884* (BM); **Nam Tamai Valley** (Adung Wang-Gamlang Wang), 3660 m, 1937. *F. Kingdon-Ward 13357* (BM); **Senighku Valley**, 3660–3960 m, 1926, *F. Kingdon-Ward 7563* (K).

13. Codonopsis chlorocodon C. Y. Wu in Wu & Li, Yunnan Trop. Subtrop. Fl. Res. Rep. 1: 82. pl. 30. fig. 2.1965; Hong, Lammers & Klein in Wu, Raven & Hong, Fl. China 19: 523. 2011; Hong, Monogr. *Codonopsis* Allied Gen. Camp. 92. 2015. ≡ *Codonopsis viridiflora* Maxim. var. *chlorocodon* (C. Y. Wu) S. H. Huang, in Wu & Chen, Fl. Yunnan. 5: 490. 1991. Type: China, Yunnan, Dêqên, Baima Shan, Gujianong, 3400 m, 9 Aug. 1937, *T. T. Yü 9461* (holotype: KUN; isotypes: PE); Dêqên, 2700 m, Sept. 1935, *C. W. Wang 69913* (paratypes: KUN, PE)

Roots carrot-shaped, over 20 cm long, 2.5 cm in diameter. Stems often single from caudex, erect, 0.6–1 m tall, lower branches many, sterile, rarely fertile, upper ones, if present, fertile but leafless, or with leaves linear and much smaller. Leaves on main stem alternate, those on branches nearly opposite, sessile or with petiole less than 1 cm long; leaf blade triangular-ovate to lanceolate, 1–2.5 cm long, 0.5–2 cm broad, both surfaces sparsely hispidulous, base shallowly cordate to rounded, margin entire or crenate, revolute. Flowers solitary, terminal on main stem, upper branches often also with 1 or 2 flowers. Calyx tube adnate to whole hypanthium, semiglobose, calyx lobes distant from each other, narrowly triangular-lanceolate, 4–6 mm long, 2–3 mm broad, glabrous or rarely very sparsely hispidulous apically, margin denticulate above, less frequently entire; sinus between calyx lobes truncate. Corolla green, pale yellow-green, or pale yellow,

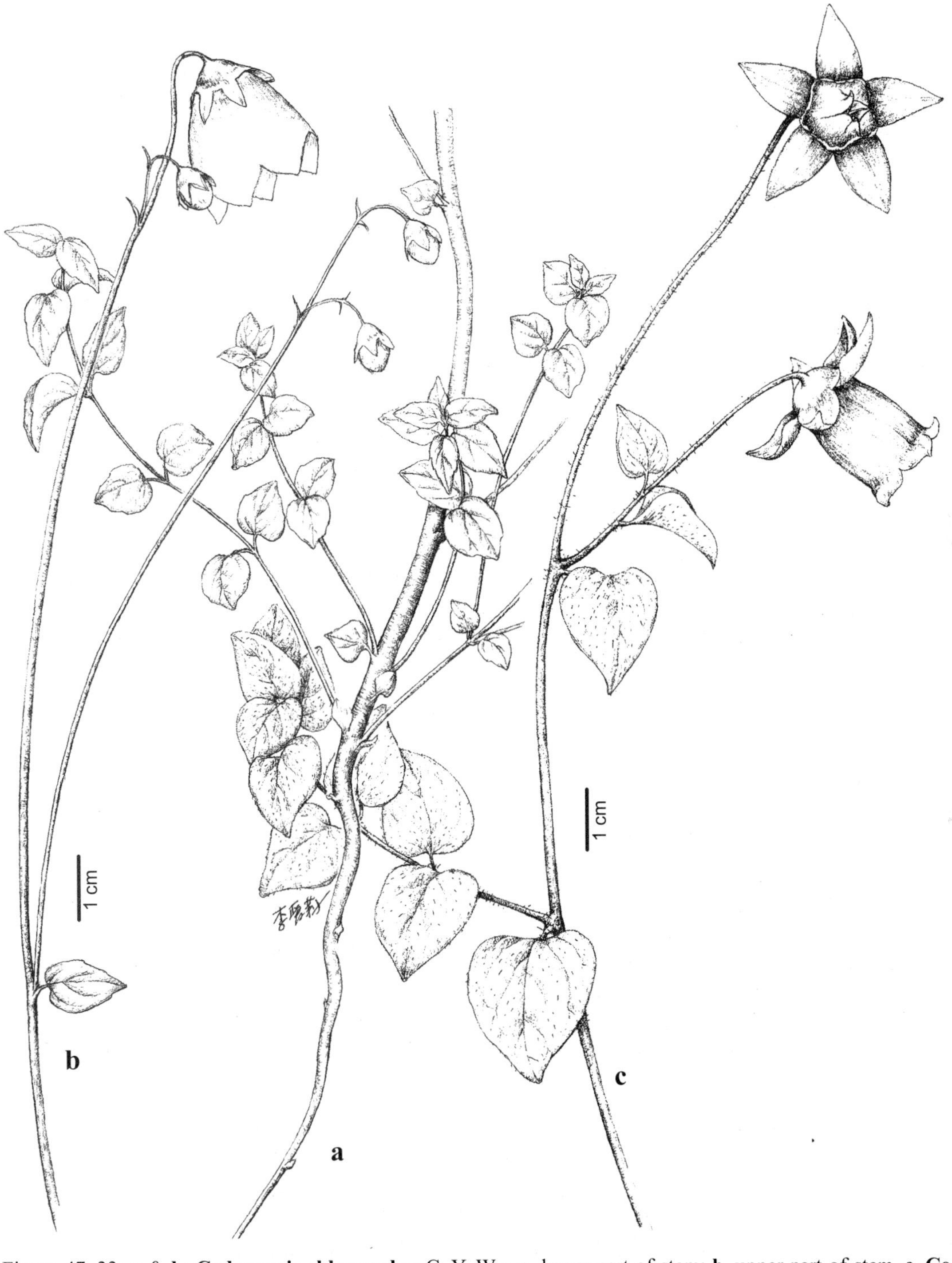

Figure 47–33: **a** & **b**: **Codonopsis chlorocodon** C. Y. Wu: **a**, lower part of stem; **b**, upper part of stem. **c**, **Codonopsis bhutanica** Ludlow: upper part of stem. From Hong, *Monogr. Codonopsis Allied Gen. Camp.*, 2015.

tubular-campanulate, 1.5–1.8 cm long, lobed to 1/3. Capsule ca. 1 cm in diameter, inferior part hemispheric, slightly obtuse below. Seeds ellipsoid, ca. 1.5 mm long, brown-yellow. Figure 47–33: a & b.

Phenology: flowering from June to August; fruiting in August and September.

Habitats: sunny grassy slopes or open thickets; 2700–4000 m.

Distribution: **endemic**; confined to N and S Hengduan (Map 47–80).

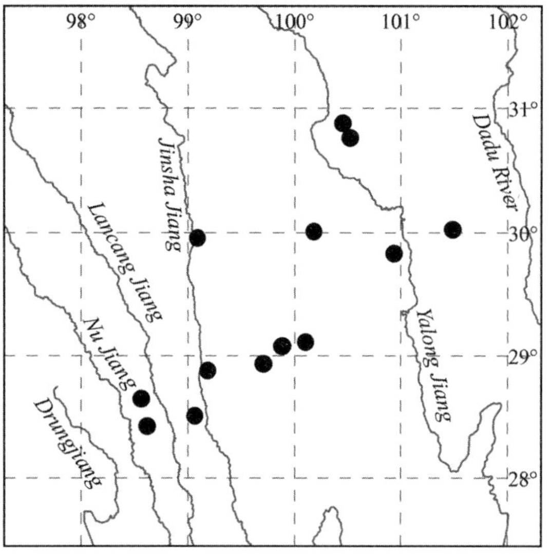

Map 47–80: **Codonopsis chlorocodon** C. Y. Wu

N HENGDUAN: Batang, s. loc. 1903, *J. A. Soulie 3703* (P) **; Dawu**, near Baimai (Qianning), 3300 m, 2012, *D. Y. Hong & Q. Wang H 12071* (PE); **Litang**, between Maoya Qusangzhi and Sumdo of Daocheng, 3680 m, 1964, *J. D. Li et al. 64–54* (PE, SM); **Yajiang**, No 4 Distr., 1986, *Z. T. Wang 868-362* (PE). **S HENGDUAN: Dêqên**, Baima Shan, 3600 m, 1981, *Qinghai-Xizang Exped. 3650* (CDBI, KUN, PE); Doker-la, Mekong-Salween Divide, 3960 m, 1913, *F. Kingdon-Ward 762* (E); **Daocheng,** Sumdo, 1964, *J. D. Li et al. 64–54* (PE, SM); **Dêrong**, Way to Riyu, 3400 m, 1981, *Qinghai-Xizang Exped. 3299* (CDBI, KUN, PE); **Kangding**, Xinduqiao, 3500 m, 1963, *K. C. Kuan et al. 824* (PE); **Xiangcheng**, Ma'an Shan, 3873 m, 2010, *Kham Exped. 10–0227* (PE); Reda, 3450–3650 m, *D. E. Boufford & al. 28739* (A, E, P, W).

14. Codonopsis bhutanica Ludlow, J. Roy. Hort. Soc. 97:127. 1972; Grey-Wilson, Plantsman 12(2): 71. 1990; Hong, Monogr. *Codonopsis* Allied Gen. Camp. 92. 2015. Type: Bhutan, Shingbe, Me La, 27°58′N, 91°37′E, 3810 m, 2 Jul. 1949, *F. Ludlow, G. Sherriff & J. H. Hicks 20786* (holotype: BM; isotypes: E, K, TI).

Codonopsis xizangensis D.Y. Hong, Acta Phytotax. Sin. 18: 246. fig. 1, 1–3. Type: China, Tibet, Cona, 1975, *C. Y. Wu & S. K. Chen 75-907* (holotype: PE; isotype: KUN).

Roots carrot-shaped, up to 20 cm long, 1.8 cm in diameter. Stems erect, ascending or procumbent, 20–50 cm long, branches several to many from lower and middle parts, usually sterile. Leaves on main stem alternate, those on branches opposite or nearly so; all leaves sessile or with petiole up to 5 mm long; leaf blade shallowly cordate, 1–2.8 cm long, 0.8–2.4 cm broad, glabrous or villous, or sparsely hispidulous, margin revolute. Flowers solitary, terminal on main stem, sometimes also on upper branches, rarely also on some slender, lower branches. Calyx tube adnate to whole hypanthum, glabrous or sparsely villous; calyx lobes long-ovate, 5–15 mm long, 3–5 mm broad, hispidulous entirely or only at upper part, or entirely glabrous, revolute; sinus between calyx lobes narrow and pointed. Corolla pale blue with dark blue at base, tubular-campanulate, 1.2–1.8 cm long, lobed to 1/3. Stamens glabrous. Capsule globose in inferior part, conical in superior part, 1.4–1.6 cm long, 0.8–1 cm in diameter. Seeds ellipsoid, ca. 1 mm long. Figure 47–33: c.

Phenology: flowering from July to September; fruiting in September and October.

Habitats: grassy slopes or thickets; 3100–5490m.

Distribution: **endemic**; C and E Nepal, Bhutan, L Yarlung Zangbo, and Yarlung Zangbo-Brahmaputra (Map 47–81).

C NEPAL: **Dhawalagiri**, Tegar (N of Mustang), 5490 m, 1954, *Stainton, Sykes & Williams 2237* (BM). **E NEPAL**: **Kosi**, Arun Valley (Phungchu), 3350 m, 1922, *s. coll. 296* (K).

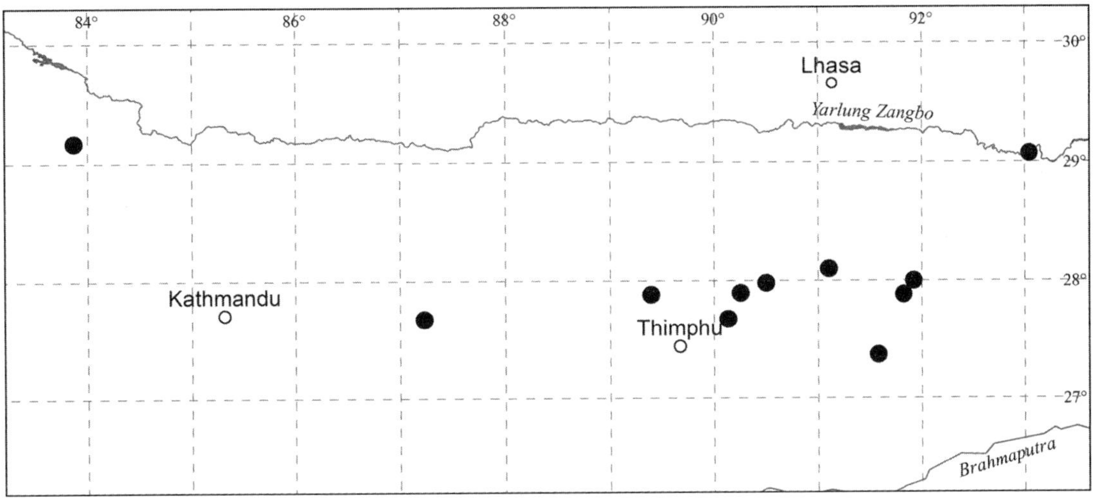

Map 47–81: **Codonopsis bhutanica** Ludlow

BHUTAN: **Mangde Chu**, Khoa La, 4420 m, 1949. *F. Ludlow et al. 17202* (paratypes of *C. bhutanica*: BM, TI); **Shingbe**, Me La, 4270 m, 1933. *F. Ludlow & G. Sherriff 417* (BM); **Tang Chu**, 4270–4570 m, 1937, *F.Ludlow & G. Sherriff 3539* (paratype of *C. bhutanica*: BM); **Tashigang**, Shingphel, Trashi Yangtsi, 3900 m, 2000, *S. A. Bowes Lyon 13198* (E) ; **Tsampa**, Marlung, 4420 m, 1949, *F. Ludlow et al. 19372* (paratypes of *C. bhutanica*: BM, K); **Thampela,** 4570 m, 1949, *F. Ludlow et al. 17132* (paratype of C. bhutanica: BM); **Upper Mo Chu**, below Chhew La, above Chebesa, 4130 m, 1984. *I. W. J. Sinclair & D. G. Long 5439* (BM, K); near Chhew La, 1973, *S. Tshering & R. Nawang* s. n.(BM); **Yale-la**: E of Chamolhari, 4570 m, 1938, *B. J. Gould 1134* (BM). **M YARLUNG ZANGBO**: **Lhozhag,** Lakang, 4450 m, 2013, *Y. S. Chen et al. 13–1591* (PE); **L YARLUNG ZANGBO**: **Nangxian**, by Yarlung Zangbo, 3100 m, 1977, *P. C. Kuo 22966* (HNWP, KUN). **YARLUNG ZANGBO-BRAHMAPUTRA**: **Cona**, S of Cona Town, 4000 m, 2011, *D. Y. Hong et al. H11010* (PE); **Jeba** Township, 3840 m, 2011, *X. F. Gao et al. 13902* (CDBI); **Phile-la**, 4270 m, 1914, *R. E. Cooper 1723* (BM); S slope of **Poshan**, 3800–3900 m, 1975, *Qinghai-Xizang Suppl. Exped. 75-1947* (paratypes of *C. xizangensis*: E, KUN, PE).

15. Codonopsis canescens Nannf., Svensk Bot. Tidskr. 34: 386. pl. 4. 1940; Hong, Lammers & Klein in Wu, Raven & Hong, Fl. China 19: 522. 2011; Hong, Monogr. *Codonopsis* Allied Gen. Camp. 96. 2015. Type: China, "Sikang: Taofu" (W Sichuan, Dawu), ca. 3100 m, 16. Sept. 1934, *H. Smith 12221* (holotype: UPS, n. v.; isotype: PE).

Roots carrot-shaped or elongated, 20–40 cm long, sometimes even longer,1–2.5 cm in diameter. Main stems usually many from one caudex, forming a clump, erect or ascending, 25–85 cm tall, leafy in middle part; branches numerous, mostly aggregated at lower or middle parts, leafy, fertile or sterile, densely white-villous. Leaves on main stem alternate, those on branches nearly opposite; petiole shorter than 4 mm; leaf blade ovate, broad-ovate or nearly cordate, 0.8–1.8 cm long, 0.5–1.2 cm broad, densely white-puberulent on both surfaces, margin entire, apex acute or obtuse. Flowers solitary, terminal on main stem and upper branches; pedicel 2–15 cm long. Calyx tube adnate to whole hypanthum, semiglobose, densely white hispidulous; calyx lobes ovate-lanceolate, 5–6 mm long, 2–5 mm broad, widest at middle, densely white-puberulent, margin entire; sinus between calyx lobes broad and obtuse. Corolla white, some with blue veins at base inside, some purple at base, bowl-shaped, 1.5–1.8 cm long, 2–2.5 cm across, shallowly lobed; corolla lobes deltoid, villous at apex. Stamens glabrous; filaments slightly dilated at base. Capsule semiglobose below, conical above, 1–1.7 cm long, 0.6–1 cm in diameter. Seeds ellipsoid, brown-yellow. Figure 47–34.

Phenology: flowering from June to September; fruiting from August to October.

Habitats: grassy slopes, sunny and stony river terraces, thickets or open forests; 2900–4500 m.

Distribution: **endemic**; Tangut and N Hengduan (Map 47–82).

TANGUT: Dêgê, Jangra, 3200 m, 1974. *Sichuan Plant Exped. 7026* (CDBI); Mengda Gou, 1957, *Y. W. Tsui 4513* (PE); **Garzê,** Sênkar, Baili, 3380 m, 1964, *J. D. Li 64–61* (PE, SM); Tuoba, 1977, *P. C. Kuo & W. Y. Wang 21204* (HNWP); **Jomda**, Aila Shan, 3400–3700 m, 2012, *H. J. Wang et al. JD–001* (PE); W SW of Jomda city, 3625 m, 2004, *D. E. Boufford et al. 31208* (A); **Nangqên,** Ji'nyinsib, 4200 m, 1965, *Y. C. Yang 1179* (KUN, PE); **Qamdo**, Garma, 3490 m, 2010, *Kham Exped. 10–1130* (PE); **Sêrtar**, Wungda, 3400 m, 1974, *Sichuan Plant Exped. 6576* (CDBI, PE); **Serxü,** Mabja, Kardê, 3500 m, 1979, *s. coll. 287* (SM); **Yushu**, Jiangxigou Forest Farm, 3800 m, 1971, *P. C. Kuo & W. Y. Wang 8392* (HNWP); Xiao Surmang, 1996, *T. N. Ho et al. 2276* (E, KUN, PE); **Zamtang**, Hongxin, 3600 m, 1975, *Sichuan Plant Exped. 9285* (CDBI, PE). **N HENGDUAN: Baiyü**, by Jinshajiang, 3157 m, 2010, *Kham Exped. 10–0650* (PE); **Dawu**, mountain near Larmasery 2900 m, *T. L. Dai et al. Dawu 64-1* (PE); mountain near the city, 3400 m, 2012, *D. Y. Hong & Q. Wang H12070* (PE); **Gonjo**, N of the city, 3650 m, 1991, *J. S. Yang 91-489* (KUN, PE); **Luhuo** (Zhaggo), Left bank of Xianshui River, Buddhist nunnery, 3300 m, 1961, *J. S. Ying 9402* (KUN, PE); Yiliang Zhaizi, 3100 m, 1964, *T. L. Dai et al. (Luhuo) 64-3* (PE, SM); **Markam**, Jiaolong 4500 m, 1976, *Qinghai-Xizang Exped. 11827* (KUN, PE).

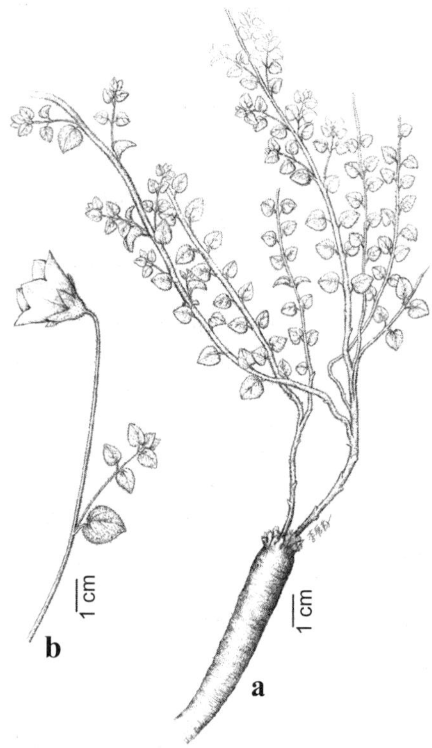

Figure 47–34: **Codonopsis canenscens** Nannf.: **a**, root, caudex, and lower part of two stems; **b**, upper part of stem. From Hong, *Monogr. Codonopsis Allied Gen. Camp.*, 2015.

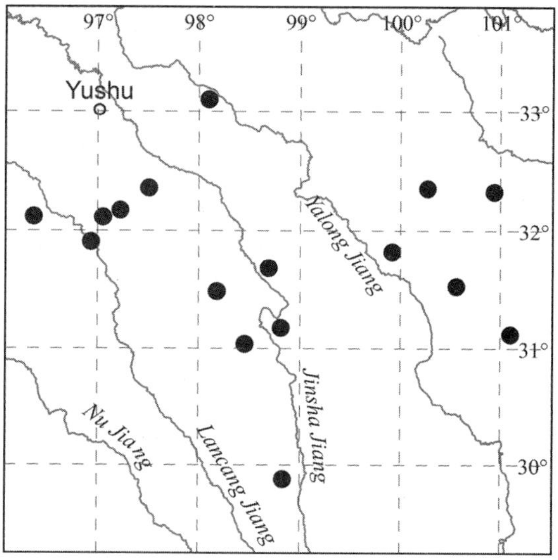

Map 47–82: **Codonopsis canescens** Nannf.

16. Codonopsis foetens Hook. f. & Thomson, J. Proc. Linn. Soc. Bot. 2: 16. 1857; Clarke in Hooker, Fl. Brit. India 3: 433. 1881; Hong, Monogr. *Codonopsis* Allied Gen. Camp. 99. 2015. Type: India (Sikkim), Kengna Lanch, 4570 m, 24 Jul. 1849, *J. D. Hooker* s. n.(lectotype designated by Hong, 2015: K; isolectotype: BM); India (Sikkim), Lachen, 4270 m, 15 Jul. 1849. *J. D. Hooker* s. n.(paratype: K); India (Sikkim), Samding, 4880 m, 11 Sept. 1849, *J. D. Hooker* s. n.(paratype: K)

Roots slender, 10–15 cm long, 1–1.5 cm in diameter. Caudex up to 15 cm long, abundantly branched. Stems several to many from one caudex, caespitose, ascending, 20–40 cm tall, very sparsely villous; branches many, aggregated in lower part, slender, sterile. Leaves on main stem alternate, mostly yellow, scalelike, only several upper ones green and ordinary, sessile, smaller than those on branches, those on branches opposite or nearly opposite, all ordinary leaves with petiole ca. 2 mm long; leaf blade cordate or cordate-ovate, 0.5–1.5 cm long, 0.5–1.5 cm broad, rather densely white hispidulous on both surfaces, margin entire, apex obtuse. Flowers solitary, rarely 2 or 3 in a cyme, terminal on main stem, more or less pendent. Calyx tube adnate to whole hypanthium, semiglobose, glabrous or very sparsely white-hispidulous; calyx lobes ovate-oblong or ovate-lanceolate, 6–12 mm long, 3.5–6.5 mm broad, rather densely hispidulous, margin entire or verrucose-denticulate, revolute; sinus between calyx lobes mostly rounded, rarely slightly truncate. Corolla blue or pale purple, rarely white, with dark purple veins, campanulate, 1.8–3 cm long; corolla lobes suborbicular, 5–7 mm long, mostly puberulent, apex obtuse or acute. Stamens glabrous; filaments slightly dilated at base. Capsule obconical at inferior part, conical at superior part, 1.5–2 cm long, 1.2–1.5 cm in diameter. Seeds not seen.

Distribution: widely distributed in the Pan-Himalaya (Map 47–83); also in other parts of Gansu of China.

1a. Leaf blade 5–9 mm long, 5–7 mm broad, rarely up to 15 mm long, 12 mm broad; calyx lobes often verrucose-serrate .... 16a. subsp. *foetens*

1b. Leaf blade 10–15 mm long, 10–15 mm broad; calyx lobes always entire .... 16b. subsp. *nervosa*

16a. Codonopsis foetens subsp. **foetens**

Leaf blade 5–9(–15) mm long, 5–7(–12) mm broad; calyx lobes often verrucose-serrate.

Phenology: flowering from July to September; fruiting in September and October.

Habitats: meadows, thickets, or alpine debris; 3650–4880 m.

Distribution: **endemic**; Sikkim & Darjiling, Bhutan, M Yarlung Zangbo, L Yarlung Zangbo, Yarlung Zangbo- Brahmaputra, and U Irrawaddy.

SIKKIM & DARJILING: Darjiling, Hahnak, 3960 m, 1938, *G. Ghose & Co.* s. n.(PE); **Sikkim**, Alpine region, 4260–4870 m, *J. D. Hooker* s. n.(P, PE); Chakung Chu, 3960 m, 1913, *R. E. Cooper 931* (E); Changu, 3660 m, 1913, *R. E. Cooper 528* (E); Chum-la-Chenchena, 4400 m, 1966, *K. Nishioka* s. n.(TI); Ewutsola, 2200 m, 1909, *W. W. Smith & G. H.Cave 1298* (B); Jamrachen Chu, 3960 m, 1909, *W. W. Smith & G. H. Cave 1693* (B, K); North Distr., Lasha Chhu, below Phaklung, 4420 m, 1996, *Edinburgh*

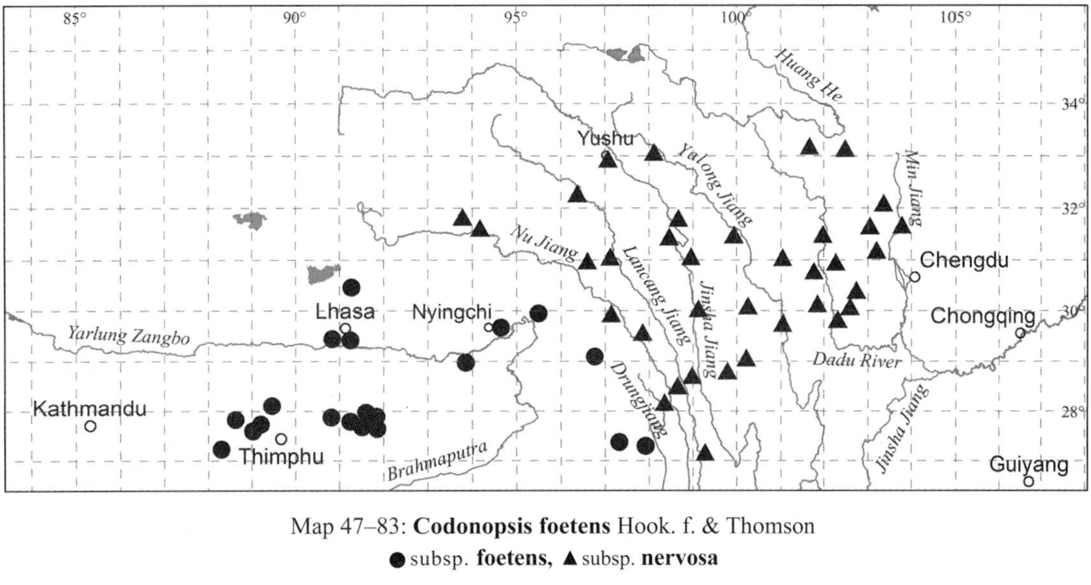

Map 47–83: **Codonopsis foetens** Hook. f. & Thomson

● subsp. **foetens,** ▲ subsp. **nervosa**

Exped. to N Sikkim (1996) 336 (E, TI); Nathui-la, 3960 m, *F. M. Baiby* s. n.(E) ; Shanka-la, 3960 m, 1913, *R. E. Cooper 352* (E). **BHUTAN**: **Shingbe**, Me La, 4270 m, 1933, *F. Ludlow & G. Sherriff 381* (BM); **Thimphu**, Yele-la, 4270 m, 1914, *R. E. Cooper 1840* (BM, E) ; **Upper Bumthang Chu**, Pangotang, Tsampa (Tolegong), 4420 m, 1949, *F. Ludlow et al. 19320* (BM). **Upper Kulong Chu**: Shingphel, Pemaling, near Tashi Yangtsi, 4200 m, 2000, *S.A. Bowes Lyon 13156* (E) ; **Upper Kuru Chu**, Singhi Dzong, Lhuntse, 4200 m, 2000, *S. A. Bowes Lyon 13092* (E) ; **Upper Mo Chu**, Valley SW of Sinchu-la, 4000 m, 1983, *C. Sargent 174* (E); Yala Pyala, 3350 m, 1938, *B. J. Gould 1360* (K). **M YARLUNG ZANGBO**: **Damdoi**: Weiyuanliang, 1985, *J. H. Tong 3025* (PE) ; **Lhasa**, Tang-la, 4270–4570 m, *F. Kingdon-Ward 5943* (K); Tuna to Khamba, 4880 m, *B. J. Gould 2353* (K). **Yadong**, near Chumbi, Pem La, 1892, *G. King's Collector* s. n.(B); Chumolari, 4880 m, 1912, *R. Lepcha 485* (E, G, K); Kangbu Spring, 4471 m, 2013, *PE-Tibet Exped. 2664* (PE); Lower Yadong, Zeli-la, 4200 m, 1974, *Qinghai-Xizang Exped. 74–2351* (KUN, PE); Pagri, 4300 m, 1974, *Qinghai-Xizang Exped.* 2854 (PE); Pagri to Heiwu, 1953, *P. C. Tsoong 5827* (PE). **L YARLUNG ZANGBO**: **Guxiang**, 4200 m, 1965, *J. S. Ying & D. Y. Hong 650852* (PE); **Yi'ong Farm**, pass of Mt Bala, 4000 m, 1983, *B. S. Li et al. 6697* (PE); **Mainling**, near Kaamula Lake, 4300 m, 1983, *B. S. Li et al. 05913* (PE); **Nyingchi**, Nambu-la, 4270–4570 m, 1924, *F. Kingdon-Ward 6097* (BM, K); Sheji-la, 4508 m, 2012, *D. Y. Hong et al. H12046* (PE). **YANGLUNG ZANGBO-BRAHMAPUTRA**: **Baishan Pass**, 4530 m, 2013, *Y. S. Chen et al. 13–0993* (PE); **Boshankou**, 4500–4600 m, 1975, *Qinghai-Xizang Suppl. Exped. 751537* (KUN, PE); **Boshankou**, N slopes, 4500 m, 1975, *C. Y. Wu et al.* 75–1007 (KUN, PE); **Chumolari**, 4880 m, 1912, *R. Lepca 485* (PE); W of **Cona** town, 4450 m, 2011, *D. Y. Hong et al. H11007*(PE); **Sangjiu** Village to **Ridong**, 4500 m, 2012, *FLPH Tibet Exped. 12–1425* (PE); **Shangzayü**, A'za to Rawu, 4100 m, 1973, *Qinghai-Xizang Exped. 73–1072* (KUN, PE)**Tawang**, 4150 m, 2002, *S. A. Bowes Lyon 15049* (E); **Yumen**, 4600 m, 2013, *Y. S. Chen et al. 13–0355* (PE); **Ze La**, ca. 4570 m, 1938, *F. Kingdon-Ward 14124* (BM); **Zayü**, Mingqi Cattle Farm, 3900 m, 1973, *Integrated Surv. Team 804* (PE); **U IRRAWADDY**: **Seinghku Valley**, 3960 m, 1926, *F. Kingdon-Ward 7568* (K).

16b. Codonopsis foetens subsp. **nervosa** (Chipp) D. Y. Hong, Novon 20(4): 422. 2010 & Monogr. *Codonopsis* Allied Gen. Camp. 102. 2015. ≡ *Codonopsis ovata* Benth. var. *nervosa* Chipp, J. Linn. Soc. Bot. 38: 385. 1908. ≡ *Codonopsis nervosa* (Chipp) Nannf., Acta Horti Gothob. 5: 26. pl. 13b. 1930. Type: China, Sichuan, Tachienlu (Kangding), 2740–4110 m, *A.E. Pratt 632* (lectotype designated by Hong, 2015: K; isolectotype: BM); W China, *E. H. Wilson 3984* (paratype: K)

Codonopsis macrantha Nannf., Notes Roy. Bot. Gard. Edinburgh 16: 157. fig. 1a, pl.229. 1931. ≡ *Codonopsis nervosa* (Chipp) Nannf. var. *macrantha* (Nannf.) L. T. Shen in Hong, Fl. Reipubl. Popularis Sin. 73(2): 57. 1983. ≡ *Codonopsis nervosa* (Chipp) Nannf. subsp. *macrantha* (Nannf.) D. Y. Hong & L. M. Ma in Li & Zhu, Fl. Sichuan. 10: 541. 1992. Type: China, Yunnan, Salween-Kiujiang Divide, 28°40′N, 98°15′E, Aug. 1919, *G. Forrest 18949* (holotype: E; isotypes: K, W).

Leaf blade mostly 1–1.5 (–2.4) cm long, 1–1.5 (–2) cm broad, calyx lobes always entire; stems sometimes densely hirsute, multiflowered. Figure 47–35: c & d.

Phenology: flowering in July and August; fruiting from August to October.

Habitats: grassy slopes, meadows with shrubs, or edges of forests; 3000–4700 m.

Distribution: Tangut, N and S Hengduan, and U Irrawaddy; also in other parts of SE Gansu and E Qinghai of China.

TANGUT: Aba (Ngawa), between upper Aba and Jigzhi, 1975, *Sichuan Plant Surv. 10498* (CDBI, PE); **Baxoi**, Demula Shan, 4500 m, 1980, *C. C. Ni et al. 1217* (PE) ; **Biru**, Nujiang, 3900 m, 1976, *D. D. Tao 11366* (KUN, PE) ; **Dêgê**, Damayan Gou, S. D., *S. X. Jia 148* (PE); Haizi Shan, 1951. *Y. W. Tsui 4983* (PE); **Garzê**, Dongge, 3600 m, 1960, *Medic. Plants Surv. 15503* (CDBI); S of Garzê, 4105–4350 m, 2005, *D. E. Boufford et al.*

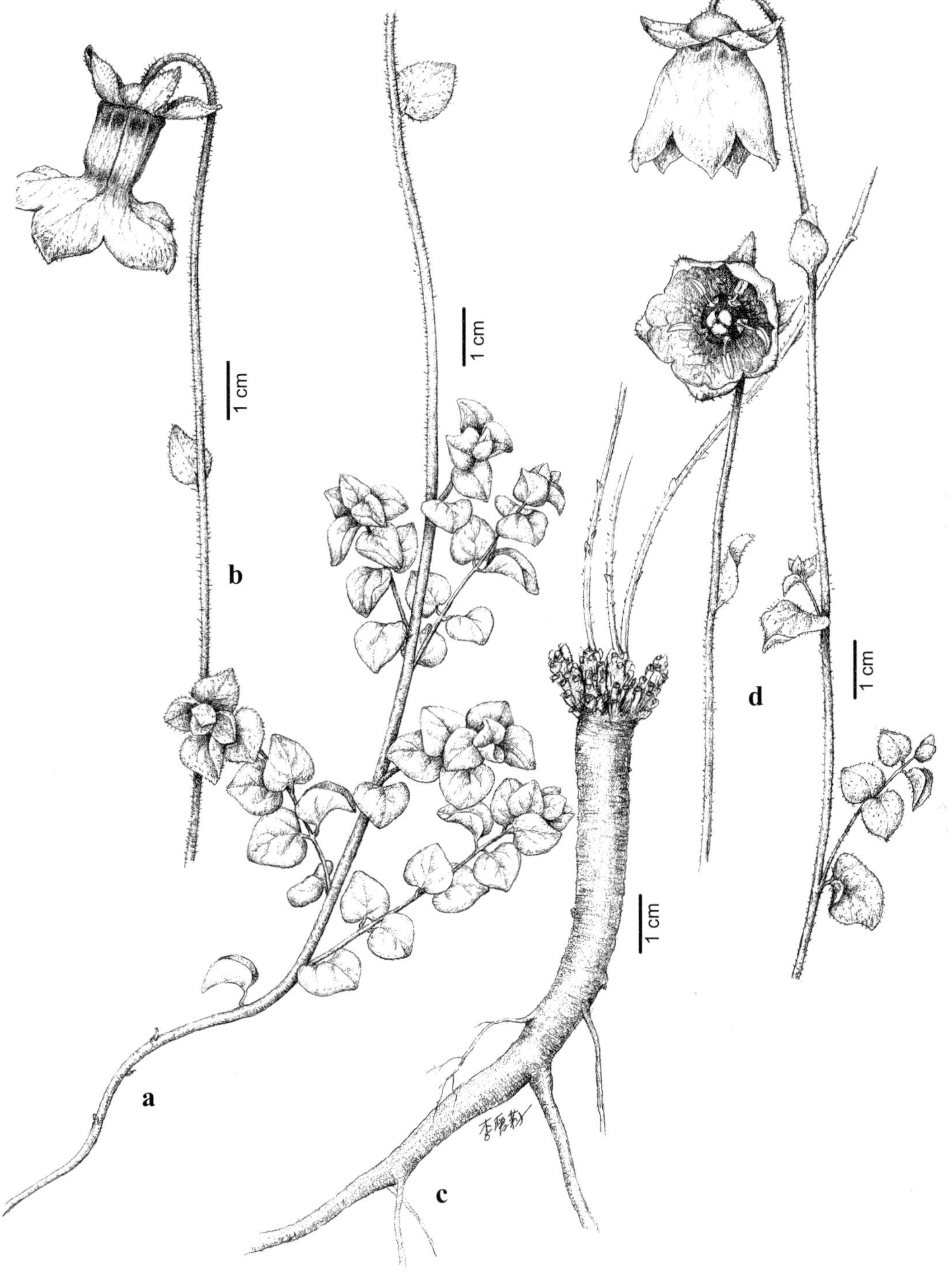

Figure 47–35: **a** & **b**, **Codonopsis bulleyana** Forrest ex Diels: **a**, lower part of stem; **b**, upper part of stem. **c** & **d**, **Codonopsis foetens** Hook. f. & Thomson subsp. **nevrosa** (Chipp) D. Y. Hong: **c**, root, caudex, and basal part of stems; **d**, middle and upper parts of stems. From Hong, *Monogr. Codonopsis Allied Gen. Camp.,* 2015.

34147 (A); **Gonjo**, Sadong, 2010, *Kham Exped. 10–1980* (PE); **Jomda**, Qinglongdong, 3750–4100 m, 1961. *s. coll. 1363* (KUN) ; **Nangqên**, Maozhuan, 4500 m, 1965, *Y. C. Yang 01416* (KUN, PE); Xiangda, 3800 m, 1965, *Y. C. Yang 01405* (KUN, PE); **Qamdo**, Dama La, 4400 m, 2010, *Kham Exped. 10–1150* (PE); Toba, 3700–4100 m, 1976, *Qinghai-Xizang Exped. 12712* (KUN, PE); **Riwoqê**, Riwoqê, 3800 m, 1976, *Qinghai-Xizang Exped. 12882* (KUN, PE); **Sogxian**, Gaqing Shan, 4100–4300 m, 1976, *Qinghai-Xizang Exped. 11131* (KUN, PE); Rongbo to Ya'ngamdo, Qala Shan, 3700–3800 m, 1976, *Qinghai-Xizang Exped. 11010* (PE); **Sêrxü**, N of Luoxu, 4800–4850 m, 2004, *D. E. Boufford et al. 31208* (A); **Yushu**, Baitang, Reshuigou, 4300 m, *s. coll. 0596* (PE); Go Qu, 3950 m, 1996, *T. N. Ho et al. 2253* (A, E, PE). **N HENGDUAN**: **Baiyü**, Azi , 4300 m, 1982, *Z. X. Tang et al.728* (PE); **Barkam**, Zhegu Shan, 1958, *C. L. Wu 33001* (PE); Zhuoke Ji, 3950–4150 m, 1997, *D. E. Boufford et al. 27906* (E); **Batang**, Yidun, by Haizi Lake, 4300 m, 1959, *S. Jiang & C. L. Jin 02736* (KUN, PE); **Danba**, Taiping, 3800 m, 1964, *T. L. Dai et al. 64-1* (PE, SM); **Dawu**, Gedaliangzi (pass), 3900 m, meadows, 1995, *D. Y. Hong et al. 95056* (PE); Yuko, 4300 m, 1974, *Sichuan Plant Surv. 5963* (CDBI, PE); **Heishui**: Shidiaoluo, 2900 m, 1959, *Heishui Group 1183* (KUN); Waboliangzi, 2300 m, 1959, *Sichuan Econ. Plant Surv. (Aba) 2622* (KUN); **Hongyuan**, Chazhenliangzi, 3700 m, 1989, *C. S. Cao 1472* (PE); Zamkou, 3400 m, 2007, *X. J. He et al. 145172* (SZ); **Jinchuan**, Anning, 3800 m, 1964, *J. D. Li et al.* 64–68 (PE, SM); **Litang**, Cogsum, 4000 m, 1964, *L. T. D. S 004* (PE); **Lixian**, s. loc., 3600 m, 1951, *T. C. Lee 2823* (SZ); **Maoxian**, Shaba, 4300 m, 1979, *Maowen Exped. 1050* (SM); **Markam**, Lawula, 3800 m, 1976, *Wu 4495* (KUN); **Wenchuan**, Balang Shan, 4300 m, 2007, *X. J. He et al. 146376* (SZ) & *153347* (SZ); Fengyi, 3850 m, 1963, *X. Y. Tang & S. J. Yue 63-44* (PE); **Xiaojin**, Lianghekou, Dashuigou, 3600 m, 1958, *S. S. Chang & Y. X. Ren 6189* (PE); Rilong, 3900 m, 1975, *Sichuan Plant Surv. 9700* (PE); **Yajiang**, Bailuogongka, 3900–4300 m, 1964, *Z. M. Xu & J. D. Li 64–59* (SM) ; Jianziwan Pass, 4456 m, 2010, *E. D. Liu et al. 2666* (KUN); **Zogang**, Nujiang Valley, 3700 m, 1975, *M. Zang 007* (KUN). **S HENGDUAN: Baoxing**, Daqiangpeng, 4000 m, 1954, *T. P. Soong 39085* (PE, SZ); Shizi Shan, 3000m, 1933, *T. T. Yü 2349* (PE); **Cawarong**, Himala, 3500 m, 1935, *C. W. Wang 65634* (PE); Nga La, 3800 m, 1935, *C. W. Wang 66061* (KUN, PE); **Daocheng**, Bowa Shan, 4570 m, 1982, *Sichuan Veget. Surv.* (*Daocheng*) *28967* (CDBI); Julong, Bowa Shan, 4600 m, 1973, *Sichuan Veget. Surv. 3608* (CDBI, KUN, PE); Shengmu, Gongba Shan, 4700 m, 1973, *Sichuan Plant Surv.* (*Daocheng*) *3637* (CDBI, KUN, PE); **Dêqên**, Mekong-Salween Divide, Doker-la, 3960–4270 m, 1913, *F. Kingdon-Ward 1100* (paratype of *C. macrantha*: E); Baima Shan, 4000 m, 1937, *T. T. Yü 9071* (KUN, PE); **Kangding**, Luba, 4030 m, 1981, *Q. Q. Wang 28618* (CDBI); Tagong, 4380 m, 2008, *D. Y. Chen & C. B. Wang 198113* (SZ); Tsipula, 4000 m, 1922, *H. Smith 4198* (E); Yulonggong, Gomba La, 3800–4000m, 1934, *H. Smith 10678* (PE); Zheduo La, 3600 m, 1934, *H. Smith 11080* (PE); Zheduo Shan, 4080 m, 2012, *D. Y. Hong & Q. Wang H12057* (PE); **Luding**, Moxi, 3200 m, 1980, *G. R. Xu 23371* (CDBI); **Tianquan**, Jinpeng Shan, 4400 m, 1978, *Tianquan Exped. 78–571* (SM); **Weixi**, Tungchuling, 4100 m, 1937, *T. T. Yü 9051* (KUN, PE); Yezhi, 3600 m, 1935, *C. W. Wang 68497* (PE); **Xiangcheng**, Wuming Shan, Maxionggou, 4200 m, 1981, *Qinghai-Xizang Exped. 4979* (CDBI, KUN, PE). **U IRRAWADDY**: **Drungjiang** (Upper Kiukiang Valley), Tsugum, 3800 m, 1938, *T. T. Yü 19887* (KUN, PE).

17. Codonopsis bulleyana Forrest ex Diels, Notes Roy. Bot. Gard. Edinburgh 5: 171. 1912; Tsoong, Contr. Inst. Bot. Nat. Acad. Peiping 3(3) : 92. fig. 8A. 1935 ; Hong, Monogr. *Codonopsis* Allied Gen. Camp. 104. 2015. Type: China, Yunnan, "eastern flank of the Lichiang Range" (Lijiang), 27°15′ N, 3300-3660 m, Jul. 1906, *G. Forrest 2731* (holotype: E).

Cyananthus mairei H. Lév., Cat. Pl. Yunnan 25. 1915. Type: China, Zochers du Io-ch'an (Yunnan, Qiaojia) , 3300 m, Aug. 1911, *E. E. Maire* s. n.(holotype: E).

Roots carrot-shaped, 15–30 cm long, ca. 2 cm in diameter. Main stem erect or ascending, 25–60 cm tall, sparsely white-villous; branches aggregated at lower part of main stem, sterile,

1–10 cm long, densely villous. Leaves alternate, those on branches nearly opposite; petiole 2-4 mm long, densely villous; leaf blade shallowly cordate, broad-ovate or ovate, 1–1.8 cm long, 0.8–1.4 cm broad, both surfaces white-hispidulous, base cordate or rounded, margin sinuous or obscurely crenate, or nearly entire, apex acute or obtuse. Flowers solitary, terminal on main stem, very rarely second flower terminal on an upper branch, slightly pendent; pedicel 4–8 cm long, usually sparsely villous. Calyx tube adnate to whole hypanthium, semiglobose, nearly glabrous; calyx lobes ovate, 8–13 mm long, 5-7 mm broad, strongly reflexed, both surfaces hispidulous at least towards apex, margin sinuous, revolute, apex acute. Corolla pale blue with purplish tube, tubular at lower half or 1/3, but suddenly spreading and becoming infundibular above, 1.8–2.8 cm long, shallowly lobed to 1/3; tube 0.5–1.2 cm wide, limb 2–2.8 cm wide; corolla lobes broad. Stamens glabrous. Capsule hemispheric in inferior part, conical and rostrate in superior part, 2–2.4 cm long, 1–1.5 cm in diameter. Seeds ellipsoid, ca. 1.5 mm long, brown-yellow. Figure 47–35: a & b.

Phenology: flowering from July to September; fruiting in September and October.

Habitats: grassy slopes or scrubs; 2800–4600 m.

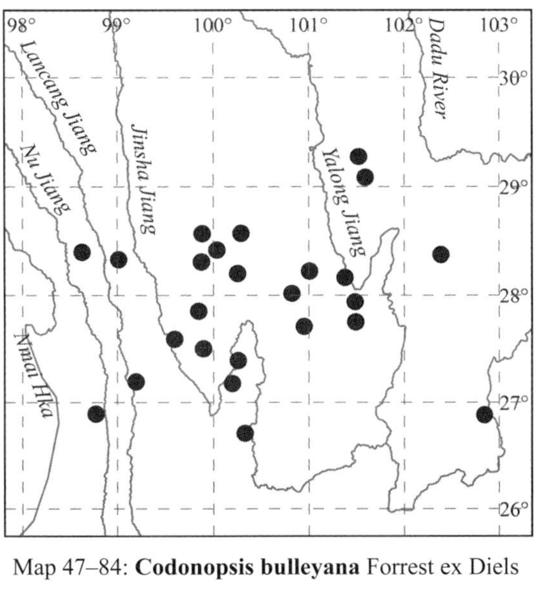

Map 47–84: **Codonopsis bulleyana** Forrest ex Diels

Distribution: S Hengduan (Map 47–84); also in NE Yunnan of China.

Notes: *Codonopsis bulleyana* differs from *C. foetens* subsp. *nervosa* mainly in corolla shape and colour. In the latter it is broad-campanulate and uniformly pale purple or pink, whereas in the former it is tubular for the lower half to 1/3, but then abruptly dilated into funnel-shaped and is purplish below, blue above. Their distribution ranges are different, with the latter in the north, while the former in the south, overlapping only in Dêqên, NW Yunnan, and in Daocheng, SW Sichuan. Furthermore, the former has broader calyx lobes, 5-7 mm broad vs. 3.5-6.5 mm broad in the latter.

S HENGDUAN: Daocheng, Gongga-ling to Luorong, 4100 m, forests, 1981, *Qinghai-Xizang Exped. 5461* (CDBI, KUN, PE); near Laocunkou, 4000 m, 1983, *K. Y. Lang et al. 2697* (KUN); **Dêgên,** Baima Shan, 4570–4880 m, 1913, *F. Kingdon-Ward 1095* (E); Mekong-Salween Divide, Doker-La, 3660-3960 m, 1917, *G. Forrest 14839* (E, K); Yunling, 4150–4300 m, 1981, *Inst. Bot. Hengduan Mts Exped. 3616* (PE); **Fugong,** Chiuchuan-Mekong Divide, 1922, *G. Forrest 23177* (E, K, W); **Heqing**, Siangshui River, Ganhaizi, 3800 m, 1929, *R. C. Ching 24300* (PE); **Jiulong**, 43 km N of Jiulong, 4188 m, 2010, *Kunming-Edinburgh Exped. to Sichuan 303* (E); Tanggo, 4100 m, 1980, *F. A. Liu & Q. Q. Wang 22852* (CDBI); **Muli,** Gibboh Shan, 3960 m, 1932. *J. F. Rock 24167* (E, K, PE); Kangwu, 3500 m, 1983, *Qinghai-Xizang Exped. 13173* (KUN, PE); Muli-Taoba, 4100 m, 2011, *D. Y. Hong et al. H11028* (PE); **Shangri-la** (Zhongdian), S side of Daxue Shan, 1998, *D. E. Boufford et al. 29167* (A, E); Haba Snow Range, 3800 m, 1981, *Inst. Bot. Hengduan Mts Exped. 2976* (PE); near Haba Hai, 4100 m, 2000, *E. von Raab-Straube et al. 242* (B); **Weixi**, Kangpu, 3500 m, 1935, *C. W. Wang 64664* (PE); Yezhi, 3600 m, 1935, *C. W. Wang 68655* (KUN, PE); **Xiangcheng**, Daxue Shan, 4200 m, 1981, *Qinghai-Xizang Exped. 3769* (CDBI, KUN, PE); **Xide**, Lianghekou, 3660 m, 1938, *T. K. Wang & T. S. Wen 655* (PE) *& 623* (PE); **Yanyuan**, Bailing, 3700 m, 1983, *Qinghai-Xizang Exped. 12730* (KUN, PE); Yuanbao, Dalu Shan, 3800 m, 1982, *Qinghai-Xizang*

Exped. 12233 (KUN, PE); **Yulong (Lijiang),** Nguluko, 2800 m, 1937, *T. T. Yü 15422* (KUN, PE); Yuhu, Xuesongcun, 3800 m, 1955, *K. M. Feng 21490* (PE, SZ); Yulong Snow Range, 4000 m, 1914, *Handel-Mazzett 4302* (G, WU).

18. Codonopsis thalictrifolia Wall. in Roxburgh, Fl. India 2: 106. 1824; Hooker & Thomson, J. Proc. Linn. Soc. Bot. 2: 16. 1857; Hong, Monogr. *Codonopsis* Allied Gen. Camp. 107. 2015. ≡ *Glossocomia thalictrifolia* (Wall.) Wall., Cat. 1297. 1828. ≡ *Wahlenbergia thalictrifolia* (Wall.) A. DC. in de Candolle, Prodr. 7: 425. 1839. ≡ *Campanula thalictrifolia* (Wall.) Spreng., Cura Post. 2(2): 77. 1825. Type: Nepal, Emodu ad Gossain Than, *Wallich 1297* [lectotype designated by Hong, 2015: K (K000814201); isolectotypes: BM, E, G, K (K000814200)].

Roots carrot-shaped, 15–20 cm long, 0.5–1 cm in diameter. Stems erect or ascending, main stem robust, 30–80 cm tall, glabrous or sparsely villous; branches sterile, many aggregated at base of main stem, 4–6 cm long, villous. Leaves on main stem alternate, those on branches nearly opposite; petiole 2–4 mm long, white-hirsute; leaf blade nearly orbicular, 3–19 mm long, 3–6 (–14) mm broad, both surfaces hirsute, base cordate or truncate, margin nearly entire or crenate, apex obtuse or acute. Flowers several or single, terminal on main stem and upper branches, pendent. Calyx tube adnate to whole hypanthium, semiglobose, 3–4 mm long, 6–10 mm wide, glabrous, rarely sparsely hispidulous; calyx lobes oblong or linear-lanceolate, broadest at middle, 5–11 mm long, 2–4 mm broad, hispidulous outside, margin entire, crenulate or denticulate, often revolute, apex obtuse; sinus between calyx lobes rounded or truncate. Corolla pale blue, tubular-campanulate, lower half tubular, upper half infundibular, 2.2–5.5 cm long, 1.5–4.3 cm wide, shallowly lobed, totally glabrous; corolla tube 1.8–4.2 cm long, 0.8–1.2 cm wide; corolla lobes triangular, 3–4 mm long. Filaments slightly dilated at base, glabrous; anthers villous at connective. Capsule semiglobose below, conical and rostrate above, ca. 1.5 mm long, 9 mm in diameter. Seeds ellipsoid, brown-yellow, smooth.

Distribution: **endemic**; U Yarlung Zangbo, C and E Nepal, Sikkim & Darjiling, M and L Yarlung Zangbo, and Yarlung Zangbo-Brahmaputra (Map 47–85).

Two allopatric subspecies are easily recognized

1a. Corolla mostly shorter than 3.3 cm long .... 18a subsp. *thalictrifolia*
1b. Corolla mostly over 3.5 cm long .... 18b. subsp. *mollis*

18a. Codonopsis thalictrifolia subsp. **thalictrifolia**

Corolla 2.2–4 cm long, but mostly shorter

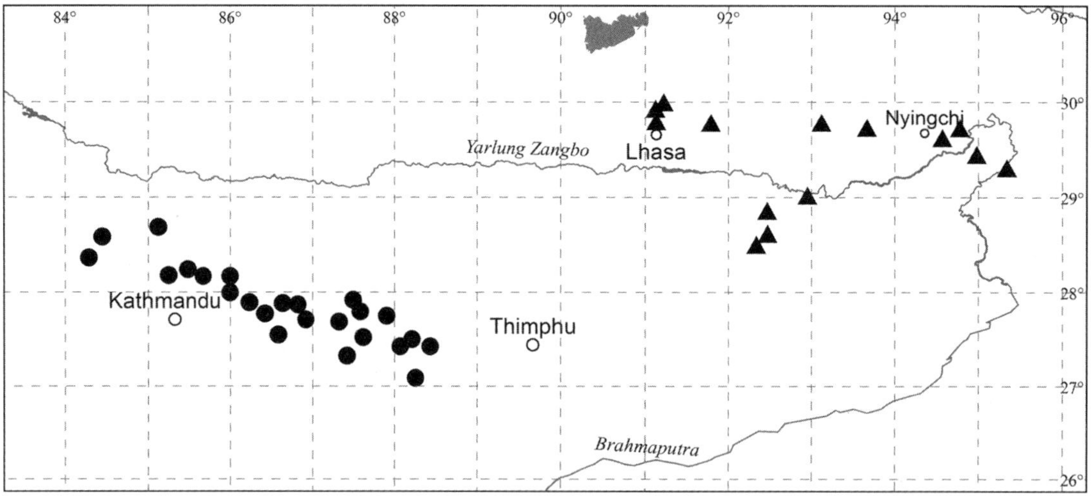

Map 47–85: **Codonopsis thalictrifolia** Wall.
● subsp. **thalictrifolia,** ▲ subsp. **mollis**

Figure 47–36: **a** & **b**, **Codonopsis thalictrifolia** Wall. subsp. **mollis** (Chipp) D. Y. Hong: **a**, lower part of stem; **b**, upper part of stem. **c**, **Codonopsis viridiflora** Maxim.: **a** & **b** from *Icon. Cormophyt. Sin.* 4, 1975; **c** from Hong, *Monogr. Codonopsis Allied Gen. Camp.*, 2015.

than 3.3 cm.

Phenology: flowering from June to August; fruiting in September and October.

Habitats: grassy slopes, alpine meadows with shrubs, or forests; 3400–5560 m.

Distribution: U Yarlung Zangbo, C and E Nepal, and Sikkim & Darjiling.

U YARLUNG ZANGBO: **Dinggyê**, Chengtanggou, 4151 m, 2013, *PE-Tibet Exped. 3516* (PE); Kelabei Shan, 4300–4800 m, 1990, *B. S. Li et al. 13129* (PE); Riwu, 3900 m, 2011, *X. F. Gao et al. 14470* (CDBI); **Gyirong**, Bahaogou, 4200 m, 1972, *Xizang Chinese Mater. Medic. & Herbs Exped. 699* (HNWP, PE); **Nyalam**, near the city, 4300 m, 2010, *D. Y. Hong & Q. Wang XZ 002* (PE); near Quxam, 3850 m, 1972, *Xizang Chinese Mater. Medic. & Herbs Exped. 1578* (PE). **C NEPAL: Bagmati,** Rasuwa, Laurebinayak, 3450–3950 m, 1995, *T. Hoshino et al. 9536089* (KATH); Rasuwa, Singum Gompa (3200 m)-Gosainkund (4200 m), 1972, *Kanai, Hara & Ohba 721857* (BM, E, KUN, TI); **Gandaki**, Manang, Bimtakothi, 3810 m, 1950. *D. G. Lowndes 1458 A* (BM, E); **Janakpur**, Dolakha, Gaurivinayak, 3810 m, 1965, *T. B. Shrestha & P. R. Shakya 3788* (G, KATH); Ramechhap, near Thare, 4150 m, 1985, *H. Ohba et al. 8570743* (E). **E NEPAL**: **Kosi**, Sankhuwasabha, Jaljale Pokhari, 4000 m, 1972, *J. F. Dobremez 1653* (BM, G); Sankhuwasabha, Thudam, 3400 m, 1972, *H. Kanai et al. 720698* (TI); **Mechi**, Taplejung, Kambachen, 4060 m, 1989, *KEKE 493* (E, K) ; Taplejung, Tamur Valley, Mewa Khola, Topke Gola, 4270 m, 1956, *J. D. A. Stainton 919* (BM, CAL, E, PE); **Sagarmatha,** Solukhumbu, Lumding Khola- Raule, 4270 m, 1964, *D. McCosh 412* (BM, E, KATH). **SIKKIM & DARJILING: Darjiling,** s. loc., 3960 m, 1838, *G. Ghose* s. n.(PE); **Sikkim**, Bikbari, 4270 m, 1983, *S. Macpherson 76* (E); Dongkung, 5560 m, 1972, *Pradhan, Norbu & Naku 23* (BM); West Distr., Jongri Pass, 4400 m, 1992, *D. G. Long et al.* 424 (E); Jongri, 4270 m, 1881, *G. Watt 5767* (E); Leonak, 3660 m, 1909, *W. W. Smith & G. H. Cave 1759* (BM); Thangshing to Lambi, Prek Chhu Valley, 4050 m, 1992, *D. G. Long et al. 575* (E); Zemu Valley, 3810 m, 1909, *W. W. Smith & G. H.Cave 145* (E).

18b. Codonopsis thalictrifolia subsp. **mollis** (Chipp) D. Y. Hong, Monogr. *Codonopsis* Allied Gen. Camp. 110. 2015. ≡ *Codonopsis mollis* Chipp, J. Linn. Soc. Bot. 38: 381. 1908. ≡ *Codonopsis thalictrifolia* Wall. var. *mollis* (Chipp) L. T. Shen in Hong, Fl. Reipubl. Popularis Sin. 73(2): 55. 1983. Type: China, Tibet, hills above Lhasa, Aug. 1904, *H. J. Walton* s. n. (lectotype designated by Hong, 2015: E; isolectotype: K). Corolla 3.1–5.5 cm long, but mostly longer than 3.5 cm. Figure 47–36: a & b.

Phenology: flowering from July to September; fruiting in September.

Habitats: thickets or meadows; 3400–4600 m.

Distribution: M and L Yarlung Zangbo, and Yarlung Zangbo-Brahmaputra.

M YARLUNG ZANGBO: **Lhasa,** 5 km N of Naiguodong, 4300 m, 1965, *Y. T. Chang & K. Y. Lang* 1696 (PE); **Lhünzhub,** Poindo, 4100–4400 m, 1972, *Xizang Chinese Mater. Medic. & Herbs Exped. 1917* (PE); Wulong, Gyironggou, 4550 m, 1972, *Xizang Chinese Mater. Medic. & Herbs Exped. 1963* (PE); **Maizhokunggar**, Maizhokunggar to Mi La, 4400 m, 2010, *D. Y. Hong & Q. Wang H10010* (PE); Rutog, Mila Pass, 4350 m, 2012, *FLPH Tibet Exped. 12–0484* (PE). **L YARLUNG ZANGBO: Gongbo'gyamda,** Numari, 1952, *P. C. Tsoong 7185* (PE); Pasum Lake, 3660 m, 1947, *F. Ludlow et al. 14140* (E); Sumdo, 4300 m, 2010, *D. Y. Hong & Q. Wang H10008* (PE); **Gyaca**, behind the Harbor, 3800–4300 m, 1972, *Xizang Chinese Mater. Medic. & Herbs Exped. 4535* (HNWP, PE); **Mainling**, Nyima La, 4110 m, 1947, *F. Ludlow et al. 14446* (E); Zêtanggang, 3800 m, 1972, *Xizang Chinese Mater. Medic. & Herbs Exped. 3896* (HNWP, PE); **Nyingchi**, Sheji-la, 4508 m, 2012, *D. Y. Hong et al. H12048* (PE); the valley behind College of Agriculture and Animal Husbandry, 3400-4200 m, 1983, *B. S. Li et al. 6296* (PE). **YARLUNG ZANGBO-BRAHMAPUTRA: Lhünzê**, Shirap, Char Chu, 3960–4270 m, 1936, *F. Ludlow & G. Sherriff 1972* (E); Xinggexi to Nyaimai, 4300 m, 1960, *G. X. Fu 622* (PE); Xoisar-202 high way, 4343 m, 2013, *Y. S. Chen et al. 13-0270* (PE); **Mêdog**, Nam La, 4270 m, 1924, *F.*

Kingdon-Ward 5995 (K, E).

19. Codonopsis viridiflora Maxim., Bull. Acad. Imp. Sci. Saint-Pétersbourg 27: 496. 1881; Komarov, Trudy Imp. S.-Peterburgsk. Bot. Sada 29(1): 116. 1908; Hong, Monogr, *Codonopsis* Allied Gen. Camp. 112. 2015. Type: China, Kansu, Yussun-chatyma Valley, 2740 m, *N. M. Przewalsky, 11/23. July 1880* (holotype: LE).

Codonopsis bicolor Nannf., Acta Horti Gothob. 5: 24, pl. 14. 1930. Type: China, Sichuan: Songpan, "Dongrergo, in prato alpino fruticoso, ca. 4200 m", 8 Aug. 1922, *H. Smith 3530* (lectotype designated by Hong, 2015: E; isolectotypes: BM, PE).

Roots carrot-shaped, 10–15 cm long, 1–3 cm in diameter. Stems 2 to several from one caudex, nearly erect, ascending, or scandent, sometimes twining apically, 0.3–2 m long, branched at lower part, sparsely hirsute or nearly glabrous. Leaves on main stem alternate, those on branches opposite or nearly opposite, and thus branches appearing a pinnate leaf; leaf blade broad-ovate to lanceolate, 1.5–3.5(–5) cm long, 1.3–3 cm broad, both surfaces hispidulous, base slightly cordate or rounded, margin crenate, apex obtuse. Flowers solitary, terminal on both main stem and branches, very rarely only on main stem; pedicel 6–15 cm long, glabrous or sparsely hirsute at top. Calyx tube adnate to whole hypanthium, semiglobose, glabrous; calyx lobes ovate to ovate-lanceolate, 10–15 mm long, 6–8 mm broad, sparsely hispidulous, margin ciliate, sparsely crenate or entire, apex obtuse. Corolla green or yellow-green, purplish only at base, or white, blue, greenish purple, or whitish blue with purple veins, often purple inside tube at base, campanulate, 1.7–2.5 cm long, 2–3 cm wide, lobed to 2/5–1/2, totally glabrous; corolla lobes triangular, 8–10 × 9–11 mm. Stamens glabrous; filaments slightly dilated at base. Capsule ca. 1.5 cm in diameter. Seeds ellipsoid, brown-yellow. Figure 47–36: c.

Phenology: flowering and fruiting from July to October.

Habitats: alpine meadows, thickets, forest margins, or forests; 2200–4130 m.

Distribution: L Yarlung Zangbo, Yarlung Zangbo-Brahmaputra, Tangut, N and S Hengduan (Map 47–86); also in S Ningxia, and in other parts of SE Gansu and E Qinghai of China.

L YARLUNG ZANGBO: Gongbo' gyamda, Pasum-la, 3960 m, 1938, *F. Ludlow et al. 6922* (BM); Pasum Lake, 3050–3350 m, 1924, *F. Kingdon-Ward 6102* (K); **Rawu**, just W of Rawu, 3950 m, 2000, *D. E. Boufford et al. 29688* (A). **YARLUNG ZANGBO-BRAHMAPUTRA: Zayü**, vicinity of Naxue, 1973, *Qinghai-Xizang Exped. 73–457* (KUN). **TANGUT: Baima,** Golog, 3400–3450 m, 1983, *W. Y. Wang 26814* (HNWP); Valley S of Baima, 3360 m, 2000,

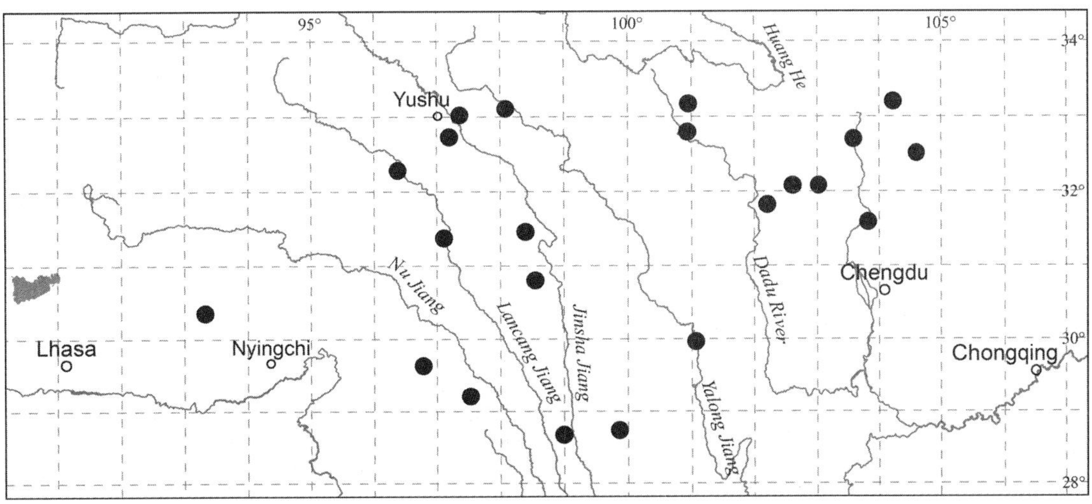

Map 47–86: **Codonopsis viridiflora** Maxim.

Sino-British Qinghai Alpine Garden Society Exped. 758 (E) & *761* (E); **Chindu**, Lochen-la, 3960 m, 1933, *F. Kingdon-Ward 12239* (BM); Xiwu, 3620 m, 1996, *T. N. Ho et al. 1865* (A, BM, E, PE); **Dêgê**, Damayanda Gou, s. d., *S. X. Jia* s. n.(PE); **Gonjo**, Wabar, 3800 m, 1976, *Qinghai-Xizang Exped. Veget. group 9524* (PE); **Jomda**, near the town, 3500 m, 1976, *Qinghai-Xizang Exped. 12535* (HNWP, KUN, PE); Tangpu, 3500 m, 1976, *Qinghai-Xizang Exped. 12443* (HNWP, PE); **Qamdo**, W of Toba, 3480–3550 m, 2004, *D. E. Boufford et al. 32524* (A); **Sêrxü**, N of Luoxu, 3780 m, 2005, *D. E. Boufford et al. 33676* (A) ; **Yushu**, Baitang, SE of Baitang, 3700 m, 1996, *T. N. Ho et al. 2009* (A, KUN). **N HENGDUAN: Barkam**, s. loc., 3000–4000 m, 1964, *J. D. Li & Z. M. Xu* (*Barkam*) *64–87* (PE, SM); **Heishui**, Guagu Shan, 3550 m, 1957, *X. Li 74028* (PE, SZ); Sergu, 3300 m, 1964, *J. D. Li et Z. M. Xu 64-91* (PE, SM); **Hongyuan**, Shuajinsi, 3500 m, 1957, *T. Y. Chang & H. F. Zhou 23594* (PE, SZ); **Jiuzhaigou** (Nanping), Xowa, 4100 m, 1979, *s. coll. 881* (SM); **Maoxian**, Sanlong, 1952, *Z. He 13702* (SZ); **Pingwu**, Huya, 3500 m, 1978, *Pingwu Exped. 500* (SM); **Songpan**, Yanchong, 3650 m, 1983, *Z. X. Tang 1584* (PE); **Yajiang**, No 4 Distr., 3720 m, 1974, *D. Q. Wang 5004* (PE). **S HENGDUAN**: **Dêqên**, Baima Shan, Shelong, 4130 m, 1937, *T. T. Yü 9191* (PE); Doker-la, 3000 m, *C. W. Wang 64907* (PE); **Xiangcheng**, Daxue Shan, 4297 m, 2012, *F. S. Yang et al. 063* (PE).

Sect. 2. Codonopsis

Codonopsis subg. *Eucodonopis* Kom. ser. *Volubiles* Kom., Trudy Imp. S. - Petersburgsk Bot. Sada 29: 104. 1908.

Roots all carrot-shaped. Stems all twining, rarely scandent, very occasionally also ascending. Ovary inferior, very rarely half-inferior to both calyx and corolla, very occasionally superior to calyx (*C. pilosula*). Fruit a capsule. Seeds numerous, wingless.

Eighteen species, distributed from N Pakistan eastwards to Korean Peninsula, and Taiwan of China; 17 species in the Pan-Himalaya (7 endemic). They grow from sea level in Korea Peninsula to 4300 m in the Himalaya.

20. Codonopsis pilosula (Franch.) Nannf., Acta Horti Gothob. 5: 29. 1930; Tsoong, Contr. Inst. Bot. Nat. Acad. Peiping 3(3): 99, 1935; Hong, Novon 20: 422. 2010 & Monogr. *Codonopsis* Allied Gen. Camp. 117. 2015. ≡ *Campanumoea pilosula* Franch., Nouv. Arch. Mus. Hist. Nat., ser. 2, 6: 72. 1883. Type: China, "Mont Pekinensis" (Beijing), cult. in "Ta-tchiao-cha" as medicinal plants, Jul. 1863, *A. David 2333* (lectotype designated by Hong, 2010: P).

Codonopsis modesta Nannf., Acta Horti Gothob. 5: 26. 1930. ≡ *Codonopsis pilosula* (Franch.) Nannf. var. *modesta* (Nannf.) L. T. Shen in Hong, Fl. Reipubl. Popularis Sin. 73(2): 41. 1983. Type: China, "Sze-ch'uan bor. (N Sichuan): Chún-ché", ca. 3200 m, 24 Aug. 1922, *H. Smith 4304* (lectotype designated by Hong, 2010: BM; isolectotypes: E, PE).

Roots carrot-shaped or fusiform-cylindrical, often branched, 15–50 cm long, 1–3 cm in diameter. Stems twining, glabrous, 1–2 m long, branched, lateral branches 15–30 cm long, branchlets 1–5 cm long, sterile or fertile. Leaves on main stem and branches alternate, those on branchlets opposite; petiole 0.5–4 cm long, sparsely hispidulous; leaf blade ovate or narrowly ovate, 1–8.5 cm long, 0.8–6 cm broad, adaxially green, abaxially gray-green, both surfaces sparsely or densely hispidulous, densely or sparsely appressed hirsute or villous, rarely glabrous, base cordate or nearly cordate to rounded, margin crenate to entire, apex obtuse or acute. Flowers solitary and terminal on branches, pedicellate. Calyx from nearly totally free from hypanthium to calyx tube adnate to hypanthium up to middle; calyx lobes lanceolate or elliptic, 1–2.2 cm long, 0.5–1 cm broad, margin nearly entire or slightly waved, apex obtuse or acute; sinus between calyx lobes narrow and pointed. Corolla yellow-green, with purple spots inside, campanulate, 1.8–2.3 cm long, lobed to 1/5–1/4; corolla lobes deltoid, apex acute. Filaments glabrous, slightly dilated at base. Capsule semiglobose in inferior part, conical at superior part. Seeds ovoid, brown-yellow.

Distribution: Tangut, N and S Hengduan (Map 47–87); also widely distributed in Korea Peninsula, the Far East of Russia, and other parts

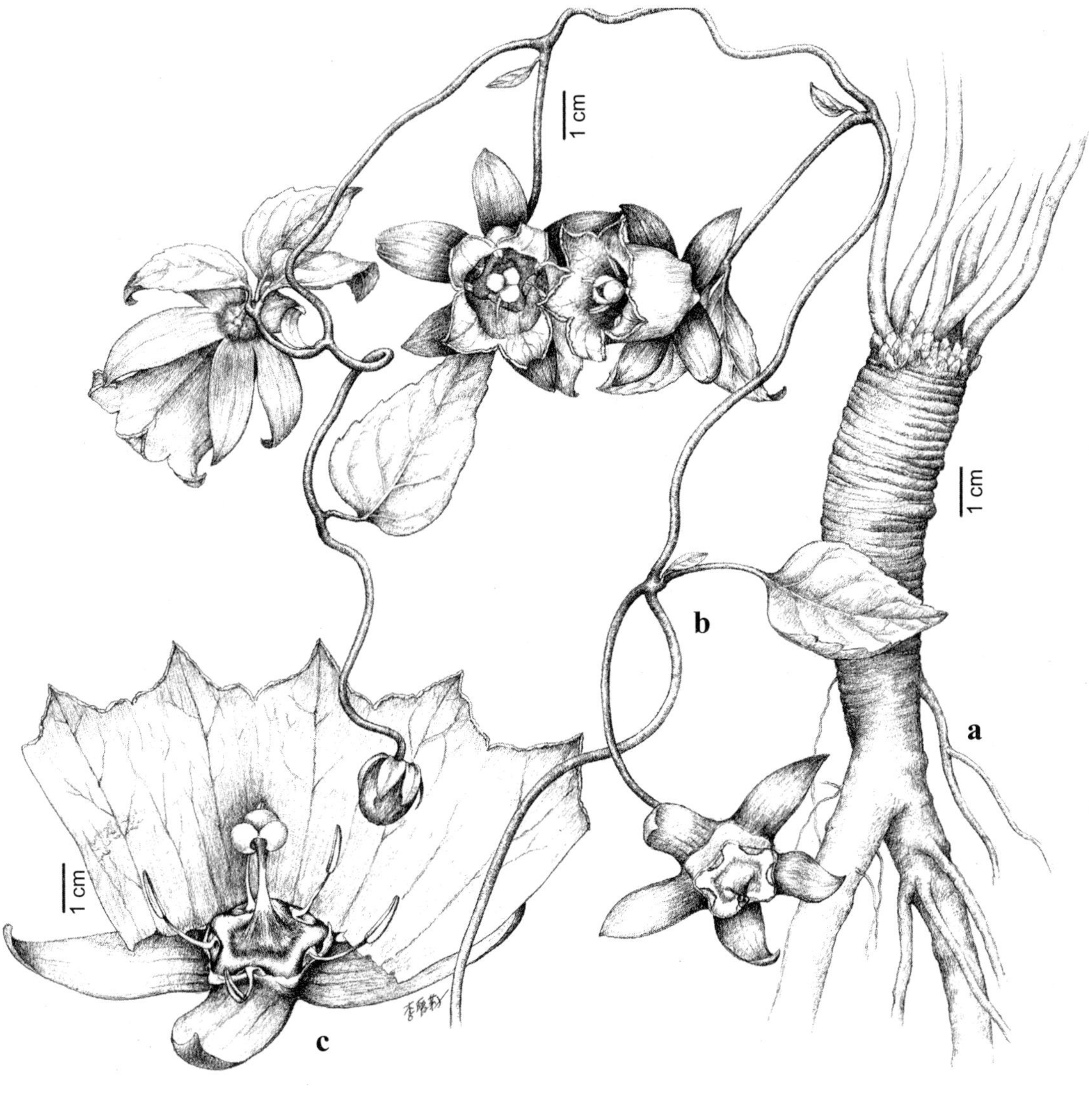

Figure 47–37: **Codonopsis pilosula** (Franch.) Nannf. subsp. **pilosula**: **a**, root, caudex, and basal part of stems; **b**, upper part of twining stem; **c**, a flower with corolla opened, showing calyx, stamens, disc, style, and stigma. From Hong, *Monogr. Codonopsis Allied Gen. Camp.*, 2015.

of China. Three subspecies; two in our region, while subsp. *tangshen* in Central China.

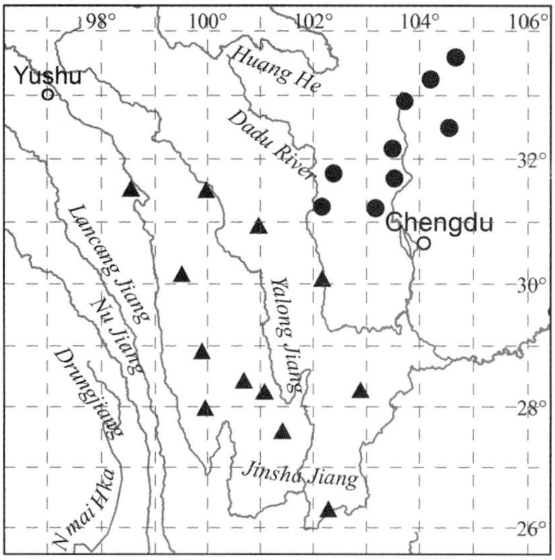

Map 47–87 **Codonopsis pilosula** (Franch.) Nannf.
● subsp. **pilosula,** ▲ subsp. **handeliana**

1a. Leaf blade usually hirsute on both surfaces . ... 20b. subsp. *handeliana*
1b. Leaf blade hispidulous, rarely glabrous .... 20a. subsp. *pilosula*

20a. Codonopsis pilosula subsp. **pilosula**

Leaf blade 1–6.5 cm long, 0.8–5.5 cm broad, mostly shallowly cordate or truncate, less often rounded at base, both surfaces hispidulous, rarely glabrous. Adnate part of calyx longer than 3 mm; calyx lobes 1.4–1.8 cm long. Corolla 2–2.3 cm. **Chromosomes**: 2n=16. Figure 47–37.

Phenology: flo-wering from June to September; fruiting in September and October.

Habitats: grasses, thickets, edges of woods, or woods; 2000–3600 m.

Distibution: N Hengduan; widely distribution in N and NE China, Korea Peninsula, and the Far East of Russia.

N HENGDUAN: Barkam: Somang, 2910 m, 1957, *T. Y. Chang & H. F. Zhou 22578* (SZ); Zhuokeji, 2800 m, 1957, *T. Y. Chang & H. F. Zhou 23431* (SZ); **Heishui,** near Luhua, 2500 m, 1957, *X. Li & J. X. Zhou 73872* (SZ); Naizigou, 3200 m, 1957, *X. Li 73280* (SZ); **Lixian**, Dabanzhao, 1956, *W. K. Hu 53–952* (SZ); Miyaluo, forests, 2900 m, 1964, *T. L. Dai et al. 64-1* (PE); Zhegu Shan, 3300 m, 1957, *T. Y. Chang & H. F. Zhou 23862* (SZ); **Maoxian**, Huangcaoping, 2450 m, 1964, *T. L. Dai et al.* (*Maowen*) *64-7* (PE); Wangzhou Shan, 2000 m, 1964, *T. L. Dai et al.* (*Maowen*) *64-1* (PE); **Jiuzhaigou** (Nanping), Jiuzhaigou, 2200 m, 1986, *C. Y. Wu et al. 552* (KUN); **Songpan**, near Balang, 3300 m, 1937, *K. T. Fu 1721* (PE); Hongyankuan, 3000 m, 1959, *Sichuan Econ. Plant Surv. 1563* (PE); **Wenchuan**, Longxi, 1963, *S. J. Yue 63-27* (PE); **Wenxian**, Zhong'an, 2000 m, 1959, *Z. Y. Zhang 6745* (HWNP); **Xiaojin**, Shanong, 3000 m, *Sichuan Plant Exped. 9712* (CDBI).

20b. Codonopsis pilosula subsp. **handeliana** (Nannf.) D. Y. Hong & L. M. Ma in Li & Zhu, Fl. Sichuan. 10:532. 1992; Hong, Monogr. *Codonopsis* Allied Gen. Camp. 125. 2015. ≡ *Codonopsis handeliana* Nannf. in Handel-Mazzetti, Symb. Sin. 7(4): 1078. 1936. ≡ *Codonopsis pilosula* var. *handeliana* (Nannf.) L. T. Shen in Hong, Fl. Reipubl. Popularis Sin. 73(2): 41. 1983. Type: China, Yunnan, "Dshungdien" (Zhongdian), 3500–3600 m, 12 Aug. 1914, *Handel-Mazzetti 4777* (lectotype designated by Hong, 2015: WU; isolectotype: W)

Leaf blade 1–3 cm long, 0.8–2.5 cm broad, both surfaces usually hirsute. Adnate part of calyx 3 mm long, or even longer; calyx lobes 1.5–2 cm long. Corolla 2–2.8 cm long.

Phenology: flowering in July and August; fruiting in September.

Habitats: forests, thickets, or margins of woods; at an altitude of 2300–3900 m.

Distribution: Tangut, N and S Hengduan; also in NE Yunnan of China.

TANGUT: Jomda, Jomda Town, 3380–3400 m, 2010, *X. F. Gao et al. 12581* (CDBI). **N HENGDUAN: Dawu**, Gaqag, 3900 m, 1961, *S. Jiang 9633* (PE); Wesheng, 3100 m, 1974, *Sichuan Plant Surv. 5808* (CDBI, PE). **S HENGDUAN: Daocheng**, Mengzi, 3500 m, 1971, *Z. G. Liu 288* (CDBI); Gongga-ling, 3300 m, 1973, *Sichuan Veget. Surv. 2475* (CDBI); **Huili**, Deyang, 1976, *Sichuan Veget. Surv. 12098* (CDBI); **Kangding**, Chungo Valley, ca. 2900 m, 1934. *H. Smith 11086* (PE); Yala, 1951, *W. K. Hu & Z. He 10652* (PE); **Muli**, Kumatian, 2900 m, 1937, *T. T. Yü 7730* (KUN, PE); Yazui Forest Farm, 3100 m, 1978, *Sichuan Plant Surv. 7356* (CDBI); **Shangri-la** (Zhongdian), outside Alpine Bot. Gard., 3300 m, 2010, *D. Y. Hong et al. H10017* (PE); Zhongdian Valley, 3300 m, 1939, *K. M. Feng 1962* (KUN, PE); **Xiangcheng**, No. 4 Distr., Qianjin, 4000 m, 1972, *Sichuan Plant Exped. 1428* (CDBI). **Yanyuan**: Mianba, 2300–2420 m, *F. G. Zhu 002* (PE).

21. Codonopsis hemisphaerica P. C. Tsoong ex D. Y. Hong, Plant Divers. Resour. 36(3): 288. fig. 2. 2014. Type: China, Sichuan, Baoxing, Puxigou, 2500 m, 21 Jul. 1954., *T. P. Soong 39208 A* (holotype: PE; isotype: SZ).

Roots carrot-shaped, up to 2.5 cm in diameter, branched or not. Stems over 1 m, twining, glabrous. Leaves on main stem alternate; petiole 1–2.7 cm long, hispidulous; leaf blade ovate or ovate-lanceolate, 2.5–5 cm long, 1.5–3 cm broad, both surfaces densely white-puberulent, base cordate to rounded, margin entire; leaves on branches similar, but smaller. Flowers solitary, terminal on main stem and branches; pedicel 0.5–4.5 cm long, glabrous. Calyx tube hemispheric, adnate to hypanthium up to top, glabrous; calyx lobes triangular-lanceolate, broadest at base, 8–10 mm long, 4–6 mm broad, glabrous, reflexed from anthesis, margin entire; sinus between calyx lobes narrow and pointed. Corolla white, campanulate, glabrous, 2.8 cm long, lobed to middle; corolla lobes triangular. Stamens glabrous; filaments as long as anthers, slightly dilated at base. Style as high as stamens. Figure 47–38.

Phenology: flowering in July and August.

Habitats: thickets or forests; 1000–2500 m.

Distribution: N and **S** Hengduan (Map 47–88); also in other parts of Sichuan of China (Ebian, Hongxi, and Leibo).

Notes: This newly published species resembles *Codonopsis pilosula* very closely, particularly in habit and leaves, and many of the specimens cited below were originally identified as that species. However, it differs from the latter in having corolla white (vs. yellow-green or pale purple), deeply divided with corolla lobes as long as tube, ovary inferior to both calyx and corolla, and calyx lobes entirely reflexed from anthesis onwards.

N HENGDUAN, Pingwu: Sier Township, 2300 m, 1958, *H. L. Tsiang 11051* (PE, SZ); **Wenchuan**, Fengyi, 1800 m, 1963, *S. J. Yue & X. Y. Tang 63–13* (PE). **S HENGDUAN: Baoxing**, Ganyang Xiaogou, 2400 m, 1936, *K. L. Chu 3770* (P, SZ); Tú-pa-kou, 2130 m, 1939, *S. Y. Hu 1260* (A, K, SZ); Yongxing, 2450 m, 1954, *T. P. Soong 39324* (SZ); **Ganluo**, Sizu, 1959, *s. coll. 4433* (PE, SZ).

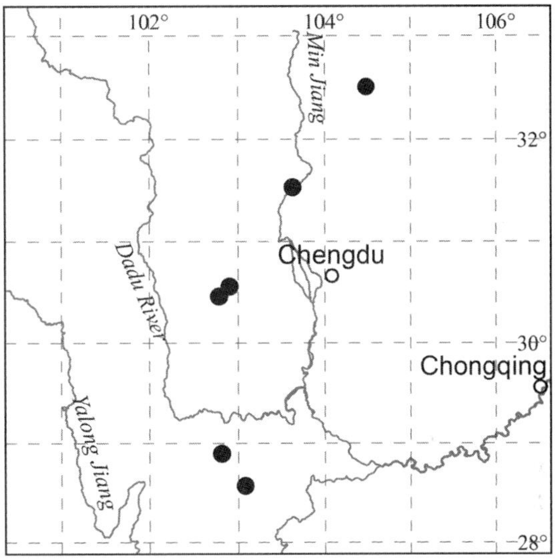

Map 47–88: **Codonopsis hemisphaerica** P. C. Tsoong ex D. Y. Hong

22. Codonopsis bomiensis D. Y. Hong, Plant Divers. Resour. 36(3): 285. fig. 1. 2014. Type : China, Tibet, Bomi, Guxiang, 3200 m, 30 Jul. 1965, *J. S. Ying & D. Y. Hong 650812* (holotype: PE; isotypes: PE).

Codonopsis rotundifolia Benth. var. *angustifolia* Nannf. ex Fletcher, Bot. Mag. 167: tab. 131. 1950. Type: China, Tibet, Mainling County, Doshong, 2900 m, 28 Jul. 1938, *F. Ludlow, G. Sherriff & G. Taylor 5377A* (holotype: BM).

Roots carrot - shaped, 10–35 cm long, 0.5– 3 cm in diameter. Caudex short, up to 1.5 cm long.

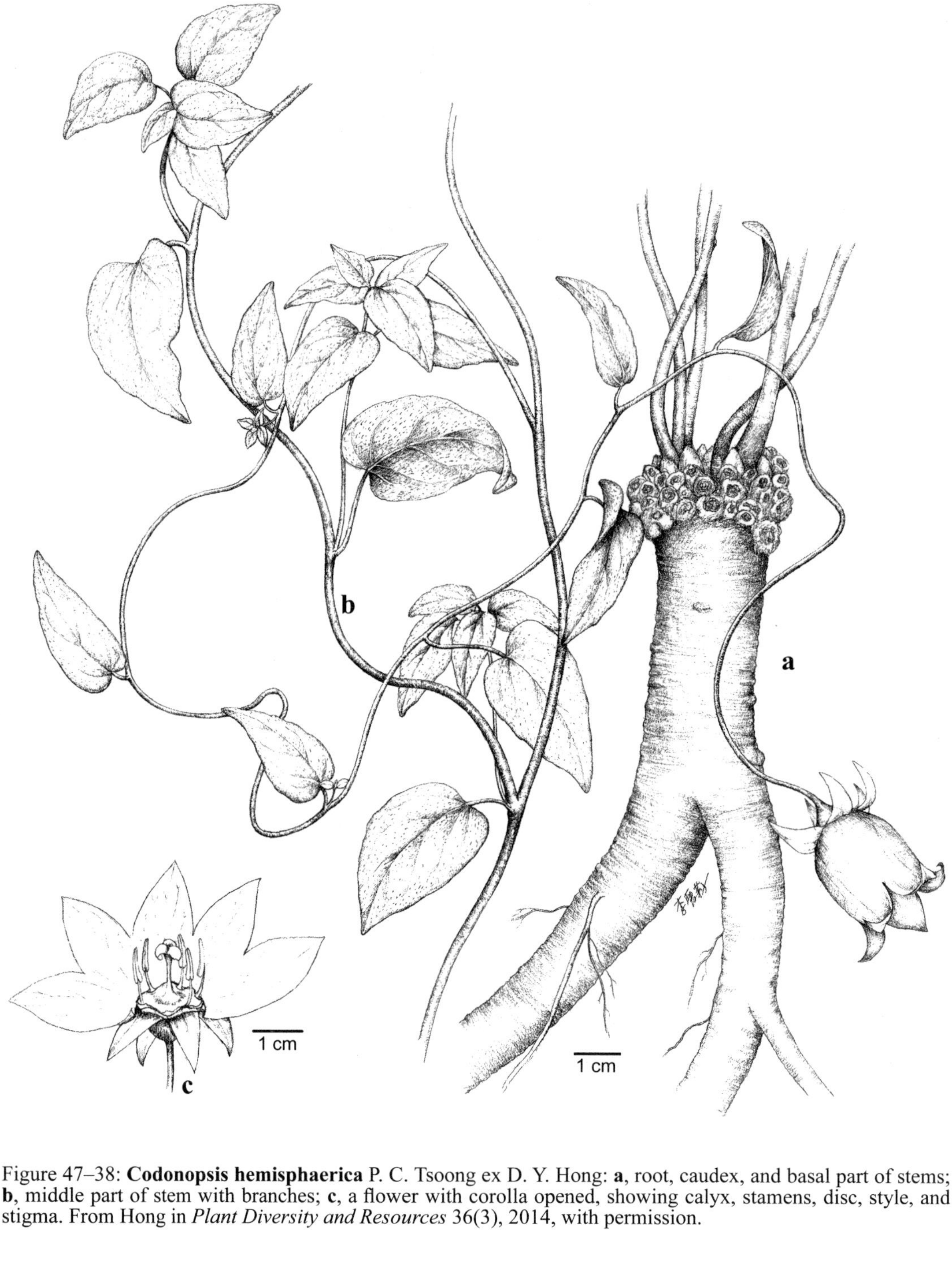

Figure 47–38: **Codonopsis hemisphaerica** P. C. Tsoong ex D. Y. Hong: **a**, root, caudex, and basal part of stems; **b**, middle part of stem with branches; **c**, a flower with corolla opened, showing calyx, stamens, disc, style, and stigma. From Hong in *Plant Diversity and Resources* 36(3), 2014, with permission.

Stems several from one caudex, twining, or less frequently ascending, glabrous or sparsely villose, branched, long branches fertile or sterile; short branches found above long branches and mostly fertile. Leaves alternate; petiole 5– 12 mm long; leaf blade ovate or ovate-lanceolate, 2–12 cm long, 1–4.5 cm broad, adaxially sparsely hispidulous, abaxially sparsely or densely hirsute, base cuneate or obtuse, margin serrate, crenulate, rarely lobed or entire, apex acute or short - acuminate. Flowers terminal on main stem or also on branches; pedicel glabrous, up to 10 cm long. Calyx adnate to hypanthium up to middle, calyx lobes and corolla inserted at the same level; calyx tube semiglobose, glabrous; calyx lobes lanceolate to linear-lanceolate, 1.4–2 cm long, 0.5–1 cm broad, glabrous, apex acute or obtuse, entire or denticulate; sinus between lobes narrow and pointed. Corolla pale yellow-green, with purple veins, campanulate, 2– 3 cm long, 1.5–2 cm wide; corolla lobes deltoid, acute, 5–7 mm long, ciliate. Stamens glabrous; filaments slightly dilated at base. Capsule hemispheric in inferior part, conical at superior part, ca. 1.5 cm long, 1.5 cm in diameter. Seeds ellipsoid, brown, ca. 0.8 mm long. **Chromosomes**: 2n = 16. Figure 47–39.

Phenology: flowering in July and August; fruiting in September and October.

Habitats: forests or scrubs; 2700– 4000 m.

Distribution: **endemic**; confined to L Yarlung Zangbo (Map 47–89).

Notes: *Codonopsis bomiensis* has been confused with *C. levicalyx* and *C. rotundifolia*. However, *C. levicalyx* is conspecific with *C. henryi*, and *C. bomiensis* differs distinctly from *C. henryi* in having calyx lobes glabrous, upright, and larger, 1.4 –2 cm long, and the corolla campanulate, 1.5–2 cm wide, and with ciliate lobes. It differs from *C. rotundifolia* by its hairy stems and leaves, and its crenulate or serrate leaf blade.

L YARLUNG ZANGBO: Qomdo, 3300 m, 1979, *R. F. Huang 778* (HNWP); vicinity of **Tongmai**, 1960 m, 1965, *P. G. Xiao et al. 2251* (KUN); **Yi'ong**, Yi'ong Farm, 3500 m, 1983, *XZ Plant Resources Exped. 3621* (XZ); Zhamo, 2700 m, 1965, *Y. T. Chang & K. Y. Lang 653* (PE); **Mainling**, Bagar, 3200 m, 1974, *Qinghai-Xizang Exped. 74-1947* (KUN, PE); Gongga Harbor, 3200 m, 22. July 1972, *Xizang Chinese Mater. Medic. & Herbs Exped: 4155* (HNWP, PE); Gyala, 2835 m, 1938, *F. Ludlow, G. Sherriff & G. Taylor 5377* (paratype of *C. roundifolia* var. *angustifolia*: BM); Hongwei Lumbering Yard, 3100 m, 1972, *Xizang Chinese Mater. Medic. & Herbs Exped. 3635* (HNWP, PE); Lusha, 2900 m, 1947, *F. Ludlow et al. 13292* (BM); Tsangbo Valley, 2740–3050 m, 1924, *F. Kingdon-Ward 6030* (BM, E); **Nyingchi**, Naixi, 3100 m, 1965, *Y. T. Chang & K. Y. Lang 1187* (KUN, PE); Nambu-la, 3810 m, 1938, *F. Ludlow et al. 6971* (BM); Shoga Dzong, 3200 m, *F. Ludlow et al. 14156* (BM).

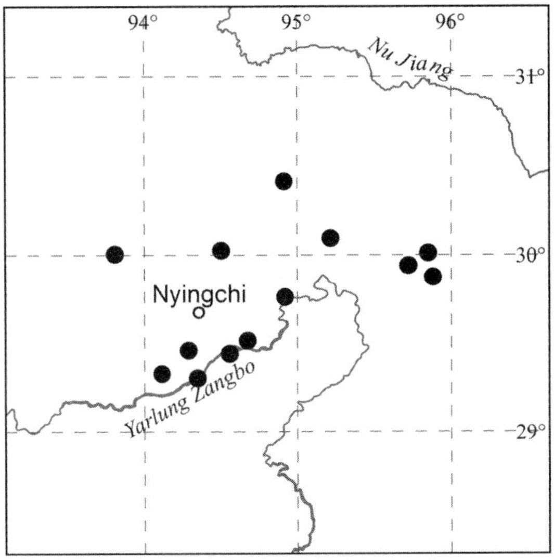

Map 47–89: **Codonopsis bomiensis** D. Y. Hong

23. Codonopsis deltoidea Chipp, J. Linn. Soc. Bot. 38:387. 1908; Tsoong, Contr. Inst. Bot. Nat. Acad. Peiping 3(3): 94. fig. 8B. 1935; Hong, Monogr. *Codonopsis* Allied Gen. Camp. 131. 2015. Type: "China: Mt Omi (Sichuan, Emei)", Aug. 1904, *E. H. Wilson 5035* [lectotype designated by Hong, 2015: K (K000575333); isolectotypes: BM, K (K000575334 & K000575335), PE]; loc. eodem, 1980 m, Oct. 1903, *E. H. Wilson 3988* (paratypes: K, PE)

Roots carrot-shaped or cylindrical, more or less branched, 15–30 cm long, 1–1.5 cm in diameter. Stems twining, over 1 m long, sparsely

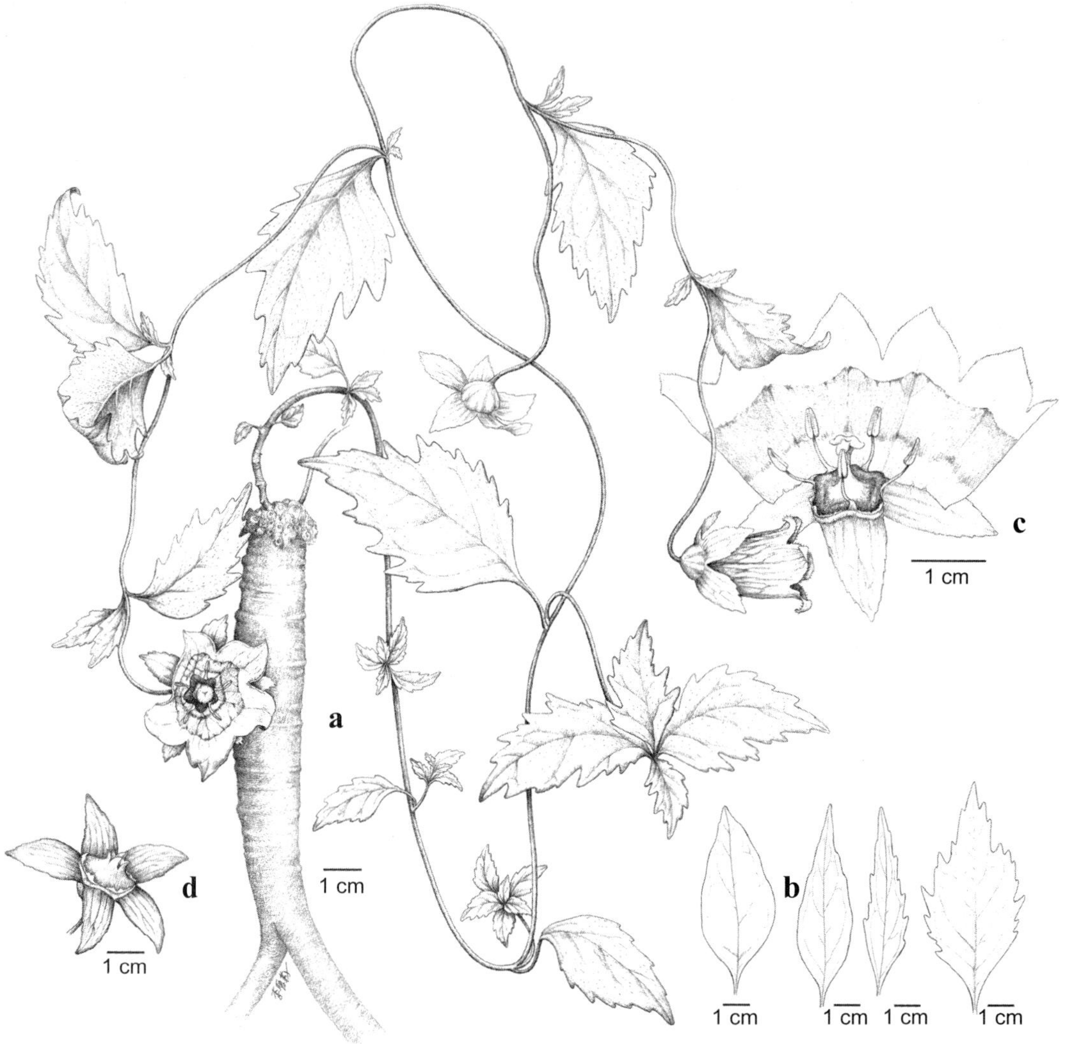

Figure 47–39: **Codonopsis bomiensis** D. Y. Hong: **a**, habit; **b**, different forms of leaves. **c**, a flower with corolla opened, showing calyx, stamens, disc, style, and stigma; **d**, capsule. From Hong in *Plant Diversity and Resources* 36(3), 2014, with permission.

villous, branches and branchlets sterile or with a terminal flower. Leaves on main stem alternate; petiole 1.5–8 cm long, villous or hirsute; leaf blade deltoid-ovate or broad-ovate, 3–9.5 cm long, 4–10 cm broad, sparsely puberulent or hirsute, adaxially green, abaxially gray-green, sparsely puberulent, hirsute along veins, base shallowly cordate, truncate, rounded or broad-cuneate, margin dentate, apex acuminate; leaves on branches opposite, but petiole shorter, leaf blade ovate to narrowly ovate. Flowers solitary or sometimes in cyme, terminal on main stem and branches; pedicel less than 2 cm long, villous. Calyx tube adnate to whole hypanthium, semiglobose, glabrous or villous along ribs; calyx lobes ovate, 1.1–1.3 cm long, 0.5–0.7 cm broad, margin verrucose-serrate, often white villous-ciliate, apex acute. Corolla pale yellow-green, with purple veins, rarely white, broad-tubular, 2–2.5 cm long, 1–1.5 cm wide; corolla lobes 1/5–1/4 as long as corolla length, deltoid. Filaments slightly dilated at base, glabrous. Capsule semiglobose below, short-conical above, 1.2–1.4 cm long, 0.8–1 cm in diameter, fruiting calyx lobes 1.4–1.9 cm long, 0.8–1.2 cm broad. Seeds ovoid, smooth, yellow-white. Figure 47–40.

Phenology: flowering and fruiting from July to October.

Habitats: forests, margins of forests, or scrubs; 1000–2600 m.

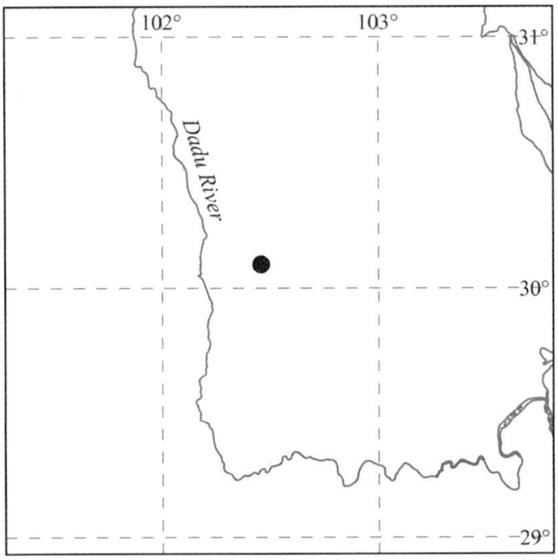

Map 47–90: **Codonopsis deltoidea** Chipp

Distibution: S Hengduan, confined to Tianquan (Map 47–90); also in other parts of Sichuan in China (Ebian, Emei, and Hongya).

S HENGDUAN: Tianquan, Yinchang, Niujinggou, 2600 m, 1953, *H. L. Tsiang 37516* (PE, SZ).

24. Codonopsis rotundifolia Benth. in Royle, Ill. Bot. Himal. 254. t. 62. 1834; Hooker, Bot. Mag. tab. 4942. 1856; Hong, Novon 20: 423. 2010 & Monogr. *Codonopsis* Allied Gen. Camp. 134. 2015. ≡ *Wahlenbergia rotundifolia* (Benth.) A. DC. in de Candolle, Prodr. 7: 425. 1839. ≡ *Glossocomia rotundifolia* (Benth.) Rupr., Bull. Cl. Phys. - Math. Acad. Imp. Sci. Saint - Pétersbourg 15: 210. 1857. Type: Bentham's tab. cited.

Codonopsis rotundifolia Benth. var. *grandiflora* Hook. f., Bot. Mag. 13: tab. 5018. 1857. Type: Hooker's tab. cited.

Codonopsis lurida Lindl., Edwards's Bot. Reg. 25 (Misc.): 82. 1839. ≡ *Glossocomia lurida* (Lindl.) Lindl., Edwards' Bot. Reg. 28: pl. 3. 1842. ≡ *Wahlenbergia lurida* (Lindl.) G. Manetti, Cat. Pl. Hort. Medic.: 106. 1842. Type: Edwards' plate cited (1842).

Codonopsis nepalensis H. Hara, J. Jap. Bot. 53(5):138. fig.2. 1978. Type: C Nepal, Seti Khola, Annapurna Himal., 3200 m, 5 Aug. 1954. *Stainton, Sykes & Williams 6687* (holotype: BM; isotype: E).

Codonopsis longifolia D. Y. Hong, Acta Phytotax. Sin. 18(2): 245. fig. 1.4. 1980. Type: China, S Tibet, Nyalam, near Quxam, 3600–3700 m, 27 Aug. 1972, *Xizang Chinese Mater. Medic. & Herbs Exped. 1563* (holotype: PE).

Roots carrot-shaped, ca. 18 cm long, 1 cm in diameter. Stems over 1 m long, twining, branched, glabrous or sparsely villous. Leaves alternate; petiole 0.5–2.5 cm long, sparsely hirsute or glabrous; leaf blade from deltoid-orbicular to ovate-lanceolate, 3–9 cm long, 2–6 cm broad, both surfaces nearly glabrous or sparsely hirsute along veins, base mostly cordate, less often truncate or rounded, margin dentate or entire, apex short-acuminate to obtuse. Flowers nearly

Figure 47–40: **Codonopsis deltoidea** Chipp: **a**, root, caudex, and basal part of a stem; **b**, upper part of stem; **c**, a flower with corolla opened, showing calyx, stamens, disc, style, and stigma; **d**, capsule with persistent calyx. From Hong, *Monogr. Codonopsis Allied Gen. Camp.*, 2015.

leaf-opposed, in cyme, or solitary and terminal; pedicel glabrous or villous, 2–20 cm long. Calyx adnate to hypanthium up to middle; calyx tube glabrous; calyx lobes broad-elliptic or ovate-lanceolate, 1.5–3 cm long, 0.5–1.5 cm broad, upright, both surfaces glabrous or hispidulous, margin minute-serrate to entire, glabrous or ciliate. Corolla green, greenish white, pale blue or pinkish white, light purple, purple inside, campanulate, 1.8–3 cm long, lobed to 1/3–2/5, glabrous. Filaments glabrous, slightly dilated at base. Capsule globose in inferior part, conical in superior part, 1.5–2.5 cm long, 1.2–2 cm in diameter. Seeds oblong, brown, ca. 1.5 mm long. **Chromosomes**: 2n=16. Figure 47–41.

Phenology: flowering from July to September; fruiting in August and September.

Habitats: meadows, scrubs, or forests; 1830–3800 m.

Distribution: N Pakistan, Jammu & Kashmir, U Ganga & Indus, U Yarlung Zangbo, W and C Nepal, (Map 47–91); also in Punjab of India.

Netes: Overall examination of a large number of specimens in BM, CAL, E, G, K, KATH, P, PE, and W showed that the indumentum on stems, leaves and calyx lobes, and the shape of leaf base varied gradually from the west (N Pakistan) to the east (Nepal and S Tibet). Therefore, it is considered unjustified to divide this group into different species or even subspecies.

N PAKISTAN: **Azad Kashmir**, Neelum Valley, 1999, *U. H. Zargar786445* (KUH); **Baltistan**, Baltal, 2740–3050 m, 1912–1913, *R. R. Stewart* s. n. (KUH); Kishenganga Valley, road to Nanga Parbat via Gangabal Lake, 1950 m, 1939, *I. D. Stewart 17810* (PE); Kishenganga Valley, Kerau to Reshua, 1680 m, 1939, *R. R. & I. D. Stewart 17663* (KUH); Surgun, 3200–4000 m, 1953, *F. Schmid 540* (G); **Hazára**, Kagan, 1899, *Inayat* s. n.(CAL); Kagan Valley, Shogran, 1960, *S. M. A. Kazmi* s. n. (KUH); Murree Hills, 1907, *R. R. Stewart* s. n.(K). **JAMMU & KASHMIR**: **Nara Wag**, Wangal Valley, 1940, *P. W. Piufold 248* (BM); **Pahalgam**, 1913, *M. A. Evershed* s. n.(BM); **Poonch**, 2130 m, 1955, *Jan Mohd J11* (BM); **Sonamarg**, 2590 m, 1940, *F. Ludlow & G. Sherriff 7903* (BM); **Tangmarg**, 1830 m, 1956, *O. Polunin 56/348* (B, BM, E,); Mt. Trunkhal, 1969, *O. Polunin 9633* (BM). **U GANGA & INDUS:** above **Barua**, 1890, *J. H. Lace 377* (CAL, E); **Chamba**, Tarlok Nath, 2743 m, 1930, *Koelz 1078* (P); **Upper Chenab Valley**, 2440 m, 1879, *R. Ellis 375* (CAL, K); upper Chenab, Dida stream, 2440 m, 1879, *B. H. Baden-Powell 216* (CAL); Jalauri Pass, Rulu, 1830–2130 m, *Edgeworth 5043* (K); Kulu, 2130 m, 1887, *G. Watt 9684* (E); **Lahul**, Gondla, 1941, *N. L. Bor 9943* (K); **Manali**, 3050 m, 1941, *N. L. Bor 15583* (K); **Pangi**, Gindi, left bank of Chenab River, near Soar, 2740 m, 1878, *s. coll.256 H* (E); Pangi, Kilar, W of Darwas, 2590 m, 1897, *J. H. Lace 1643* (E); **Sangla**, Baspa Valley, Shimla Hill State, 3660 m, 1939, *G. Sherriff 7485* (BM, E); **Shimla** Hills, 1885, *J. R. Drummond 22628* (E, K); Deoban, 2740 m, 1898, *J. S. Gamble*

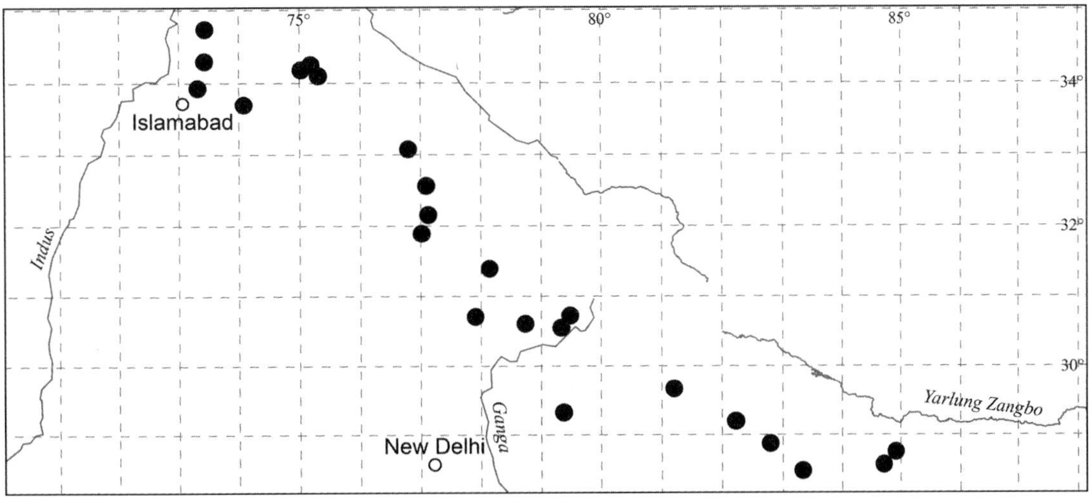

Map 47–91: **Codonopsis rotundifolia** Benth.

Figure 47–41: **Codonopsis rotundifolia** Benth.: **a**, upper part of stem; **b**, a flower with calyx removed and corolla opened, showing stamens, disc, style, and stigma. From Hong, *Monogr. Codonopsis Allied Gen. Camp.*, 2015.

27239 (K); between Josimath and Badrinath, 3300 m, 1975, *F. Billiet & J. Leonard 6565* (E, K); **Tehri-Garhwal**, Ganga Valley, near Derali, 2440–2740 m, 1883, *J. F. Duthie 609 (b)* (G). **W NEPAL**: **Karnali**, Bamara, 2750 m, 1981, *N. P. Manandhari & D. P. Joshi 6879* (KATH); Dolpa, Tarakot, Bheri River, ca. 3350 m, 1952, *Polunin, Sykes & Williams 2440* (BM, E); Jumla, 2440 m, 1952, *Polunin, Sykes & Williams 4401* (BM, E, G); **Seti**: Bajhang, Thogundanda, 3000 m, 1972, *M. S. Bista & D. P. Joshi 728* (KATH). **C NEPAL**: **Dhawalagiri**, Baglung, Dogadi Khola, 3660 m, 1954, *Stainton, Sykes & Williams 3797* (BM, CAL, PE); Mustang, Larjung (S of Tukuche), 1954, *Polunin, Sykes & Williams 1975* (BM, E); Phagure Dhuri: 3810 m, 1954, *Polunin, Sykes & Williams 3416* (BM, E); **Gandaki**, Gorkha, 3320 m, 1994, *M. Suzuki et al. 9485289* (E, KATH).

25. Codonopsis reflexa D. Y. Hong, Plant Divers. Resours. 36(3): 289. fig.3. 2014. Type: Nepal, Nr. Dogadi Khola (Dhawalagiri Zone, Baglung Distr.), 3960 m, 11. Aug. 1954, *Stainton, Sykes & Williams 3822* (holotype: PE).

Plants glabrous throughout except for filaments and style. Stems twining, over 1 m, branched from upper parts. Leaves all alternate; petiole 1.5–4 cm long; leaf blade ovate-lanceolate, 2.5–7.5 cm long, 1.2–4 cm broad, base cordate, margin serrate or crenate, apex acute or obtuse; those on branches much smaller and rounded at base. Flowers terminal on main stem and major branches. Calyx adnate to whole hypanthium; calyx tube obdeltoid; calyx lobes ovate-lanceolate, 1.4–1.6 cm long, 0.6–0.8 cm broad, broadest below middle, apex acuminate, strongly reflexed. Corolla dark blue with reddish ring near base, funnel-shaped, 2.1–4.7 cm long, ca. 2–4 cm broad, deeply divided to over 3/5 of length; corolla lobes ovate-lanceolate, 1.3–2.9 cm long, apex acute. Filaments dilated at base, dilated part villose-ciliate. Stigma 3- or 2-fid; stigma lobes broad-elliptic. Figure 47–42

Phenology: flowering in August and September.

Habitats: climbing upon shrubs or tall herbs; 3500–4000 m.

Distribution: **endemic**; confined to Central Nepal (Map 47–92).

Notes: *Codonopsis reflexa* is rather similar to *C. rotundifolia,* but it differs distinctly from that species in having corolla deeply divided to over 3/5 of overall length, calyx lobes reflexed, and filaments that have the dilated part villose-ciliate.

C NEPAL: **Dhawalagiri**, Mustang, Kali Kandaki Tal (SW Tukuche), 3500 m, 1967, *H. J. Lange 61* (paratype: B).

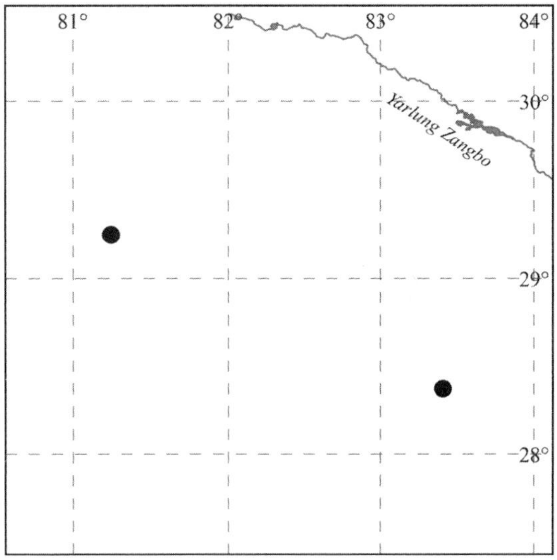

Map 47–92: **Codonopsis reflexa** D. Y. Hong

26. Codonopsis henryi Oliv., Hooker's Icon. Pl. 20: pl. 1967. 1891; Komarov, Trudy Imp. S.-Peterburgsk. Bot. Sada 29(1): 109. 1908; Hong, Monogr. *Codonopsis* Allied Gen. Camp. 136. 2015. Type: "China, Prov. Hupeh (Hubei), Fang Distr. (Fangxian CO)", 1889, *A. Henry 6651* [lectotype designated by Hong, 2015: K (K000575336); isolectotypes: BM, CAL, K (K000575337), LE, P, PE]

Codonopsis levicalyx L. T. Shen, Acta Phytotax. Sin. 13(3): 55. pl. 5. fig. 11a & b. 1975. Type: China, Sichuan, Nanping (Jiuzhaigou), Lingjiang, Xingfu, Baowujia, 2900 m, 13 Oct. 1964, *X. Y. Tang, J. Z. Wu et al. (N) 64017 (*non *6407)* (holotype: PE).

Roots carrot-shaped. Stems over 1 m long, twining, branches less than 3 cm long, with 2–4 leaves, sterile or with a terminal flower, sparsely

Figure 47–42: **Codonopsis reflexa** D. Y. Hong: **a**, upper parts of stems; **b**, corolla opened; **c**, a stamen, ventral view, showing dilated and hairy filament base. From Hong in *Plant Diversity and Resources* 36(3), 2014, with permission.

puberulent or glabrescent at internodes. Leaves on main stem alternate, rarely opposite, those on branches nearly opposite; petiole 0.5–2 cm long, puberulent; leaf blade ovate-lanceolate or lanceolate, 3–16 cm long, 1–7 cm broad, adaxially green, puberulent, abaxially gray-green, appressed-puberulent, base decurrent or cuneate, margin conspicuously dentate, apex acuminate. Flowers solitary, terminal on branches; pedicel ca. 1 cm long, puberulent. Calyx tube adnate to hypanthium up to middle, semiglobose, puberulent; calyx lobes reflexed from anthesis, triangular, 6–9 mm long, 3–4 mm broad, apex acute, puberulent and ciliate; sinus between calyx lobes broad and obtuse. Corolla campanulate, or more or less tubular-campanulate, 1.5–3 cm long, 1–1.2 cm wide; corolla lobes triangular, glabrous. Stamens glabrous; filaments slightly dilated at base. Capsule ca. 1.5 cm in diameter. Figure 47–43: a & b.

Phenology: flowering and fruiting from July to September.

Habitats: grassy slopes or thickets; 2300–3800 m.

Distribution: N Hengduan (Map 47–93); also in W Hubei and SE Gansu of China.

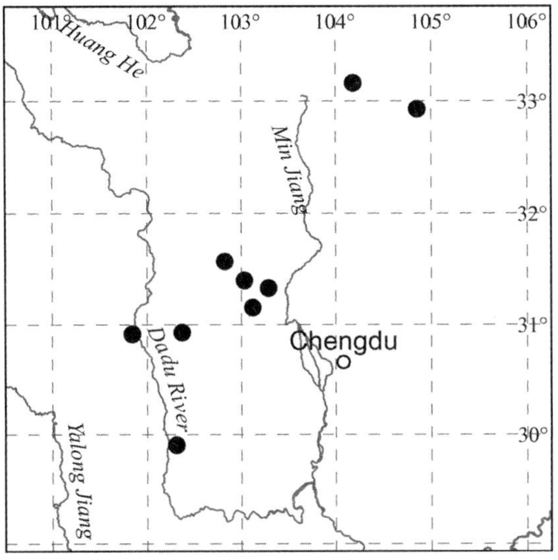

Map 47–93: **Codonopsis henryi** Oliv.

N HENGDUAN: Danba, Dongnai, 2800 m, *Sichuan Medic. Plant Resources Surv. 20518* (paratype of *C. levicalyx*: WCU); **Lixian**, Luoyonggou, 3800 m, 1964. *T. L. Dai et al. 64-11* (PE); Miyaluo, 1958, *C. L. Wu 33604* (PE); **Wenchuan**, s. loc., 1961, *X. Y. Tang & S. J. Yue 63-01* (PE) ; **Xiaojin**, Rilong, 3100 m, 1975, *Sichuan Plant Surv. 9685* (PE).

27. Codonopsis affinis Hook. f. & Thomson, J. Proc. Linn. Soc. Bot. 2: 12. 1857; Komarov, Trudy Imp. S. -Peterburgsk. Bot. Sada 29(1): 107. 1908; Hong, Monogr. *Codonopsis* Allied Gen. Camp. 138. 2015. Type: India (Sikkim), Lachen, 2740 m, 3 Aug.1849, *J. D. Hooker* s. n.(lectotype designated by Hong, 2015: K); loc. eodem, Lachen, 2130–2740 m, 3 Aug. 1849, *J. D. Hooker* s. n.(paratype: K); India (Sikkim), Lachung, 1830 m, 2 Oct. 1849, *J. D. Hooker* s. n.(paratypes: K, PE, W).

Codonopsis affinis Hook. f. & Thomson var. *burmanica* C. B. Clarke in Hooker, Fl. Brit. India 3: 431. 1881. Type: Myanmar, s. loc., *Griffith 3439* (holotype: K)

Roots much thickened, carrot-shaped, branched. Stems twining, over 2 m long, branched, nearly glabrous or sparsely white-pubescent. Leaves on main stem and major branches alternate, those on branchlets nearly opposite; petiole 1–6 cm long, pubescent; leaf blade ovate or ovate-oblong, 2.5–16 cm long, 1–9 cm broad, adaxially green, sparsely white-pubescent, abaxially gray-green, densely white-pubescent, base deeply cordate to rounded, margin entire to shallowly crenate, apex short-acuminate. Flowers solitary, terminal, or leaf-opposed; pedicel 1–4 cm long, hispidulous. Calyx tube adnate to nearly whole hypanthium, semiglobose, scaberulose; calyx lobes distant from each other, linear-oblong or linear-lanceolate, 8–17 mm long, 2–3 mm broad, puberulent on both surfaces, margin waved, apex acute or slightly obtuse; sinus between calyx lobes truncate. Corolla yellow-green at lower and middle parts, purple-red at upper part, broad-campanulate, 1.3–2 cm long, 1–1.6 cm wide, shallowly 5-lobed, apically hirsute outside; corolla lobes triangular, often recurved. Filaments slightly dilated at base, glabrous. Capsule semiglobose or nearly globose below, conical and rostrate above, 1–1.5 cm in diameter, rostrum ca. 4 mm long. Seeds ellipsoid or ovoid, yellow-brown, smooth. Figure 47–43: c.

Figure 47–43: **a** & **b**, **Codonopsis henryi** Oliv.: **a**, root, caudex, and basal part of stems; **b**, a section of stem, showing reflexed calyx lobes and tubular-campanulate corolla. **c**, **Codonopsis affinis** Hook. f. & Thomson: a section of stem, showing linear calyx lobes and campanulate corolla. From Hong, *Monogr. Codonopsis Allied Gen. Camp.*, 2015.

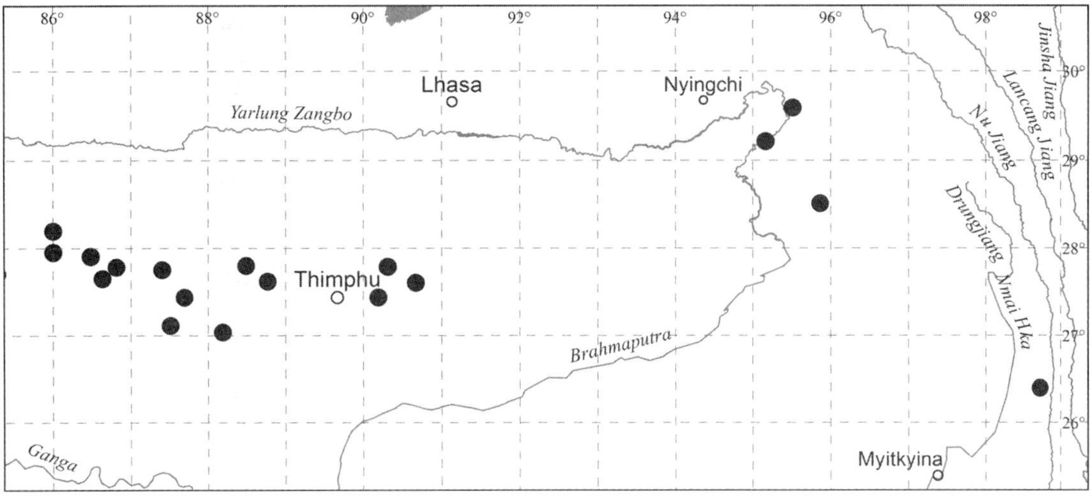

Map 47–94: **Codonopsis affinis** Hook. f. & Thomson

Phenology: flowering and fruiting from July to October.

Habitats: forests; 2300–3400 m.

Distribution: **endemic (?)**; U Yarlung Zangbo, C and E Nepal, Sikkim & Darjiling, Bhutan, Yarlung Zangbo-Brahmaputra, and S Hengduan (Map 47–94).

Notes: The type (*Griffith 3439*) of *C. affinis* var. *burmanica* is from Myanmar, but there is no indication of the precise locality, and thus it is uncertain whether the species is endemic to the Pan-Himlalaya or not.

U YARLUNG ZANGBO: **Nyalam**, near Quxam, 3200 m, 1972, *Xizang Chinese Mater. Medic. & Herbs Exped. 1579* (HNWP, PE); near Zham, 3200 m, 1972, *Xizang Chinese Mater. Medic. & Herbs Exped. 1215* (HNWP, PE). **C NEPAL**: **Janakpur**, Dolakha, Khare Khola, Patale Pokhar, (4000 m)-a pass (4200 m)- Phedi Kharka (2100 m), 1983, *H. Ohba et al. 8332037* (KATH). **E NEPAL: Kosi**, Dhankuta, 2100 m, 1963, *H. Hara et al. 6300417* (TI); Sankhuwasabha, lower Kashwa Khola, above Hedangha, ca. 2300 m, 1981, *C. Grey-Wilson et al. 4398* (K); Sankhuwasabha, Pamya, 2743 m, 1966, *Banerjee & P. R. Sakya 5609* (KATH); Terhathum, near Chauki, 2700 m, 1989, *KEKE 64* (E, K); Terhathum, Tinjure Danda, 2740 m, 1967, *Williams & Stainton 8407* (BM, K); **Mechi**, Taplejung, Chauke (2700 m) –a pass (3850 m)- Tute (2300 m), 1977, *H. Ohashi et al. 774168* (TI) & *775515* (TI); **Sagarmatha**, Solukhumbu, Namche Bazar to Toktok, Monjo, 2780 m, 2005, *DNEP 3 BX 288* (E). **SIKKIM & DARJILING: Darjiling**, Phalut, 3600 m-Ratho Chu 2100 m-Raman 2400 m, 1972, *H. Kanai et al. 721440* (BM, TI); Senchal-Tiger Hill-Ghum, 2400 m, 1969, *H. Hara et al. 69715* (TI); **Sikkim**, Bijan, 1888, *King's Collector* s. n.(CAL); Humbo, 3350 m, 1875, *C. B. Clarke 25551 (K)* & *25605* (K); Jumbok, 3050 m, 1870, *C. B. Clarke 12897A* (CAL, K); Kulhart, 2130 m, 1870, *C. B. Clarke 12897* (K) & *13003* (K); Lachen, 2440 m, 1909, *Ribu & Rhomoo 3066* (CAL); Lachung, 2740 m, 1911, *Ribu & Rhomoo 5533* (E); Rathangoong, 1908, *Ribu 118* (CAL); Thumbo, 3350 m, 1875, *C. B. Clarke 25560* (CAL). **BHUTAN**: **Chendebi**, 3050 m, 1949, *F. Ludlow et al. 17075* (BM, E); **Gichha**, 2440 m, 1914, *R. E. Cooper 3152 b* (E); **Gasa** Dzong, 3050 m, 1949, *F. Ludlow et al. 17402* (BM). **YARLUNG ZANGBO-BRAHMAPUTRA: Mêdog**, 29°22′ N, 95°08′ E, 2000 m, 2003, *X. F. Gao et al. 7855* (CDBI); 2003, *X. F. Gao et al. 8071* (CDBI); **Zayü**, Luhit F. D., Dreyi to Tezu, 1957, *R. S. Rao 10907* (CAL). **S HENGDUAN: Lushui**, 3660 m, 1926, *G. Forrest 26973* (K).

28. Codonopsis viridis Wall. in Roxburgh, Fl. India 2 : 103. 1824; Komarov, Trudy Imp. S.-Peterburgsk. Bot. Sada 29(1): 106. 1908; Hong, Monogr. *Codonopsis* Allied Gen. Camp. 140. 2015. ≡ *Campanula viridis* (Wall.) Spreng., Cura Post. 78. 1825. ≡ *Glossocomia viridis* (Wall.) Rupr., Bull. Cl. Phys.

- Math. Acad. Imp. Sci. Saint - Petersbourg 15: 210. 1857. ≡ *Wahlenbergia viridis* (Wall.) A. DC. in de Candolle, Prodr. 7: 424. 1839. Type: Nepal, *Wallich 1298* [lectotype designated by Hong, 2015: K (K000781220); isolectotypes: E, G, K (K000781221)].

Codonopsis griffithii C. B. Clarke in Hooker, Fl. Brit. India 3: 431. 1881. *Codonopsis sp.* Griffith, Notul. Pl. Asiat. 4: 281. 1854 & Icon. Pl. Asiat. t. 482. 1854. Type: Griffith's tab. cited.

Stems twining, over 1 m long, glabrous, branched. Leaves alternate; petiole 0.5–2 cm long, puberulent; leaf blade ovate to lanceolate, 2–11 cm long, 1–6 cm broad, adaxially hispidulous, abaxially densely hispidulous or hirsute, base cordate to rounded, margin minute-serrate or nearly entire, apex acute to acuminate. Flowers in cyme, extra-axillary, almost leaf-opposed; pedicel 2.5–7 cm long, glabrous at lower and middle part, pubescent at top. Calyx tube adnate to whole hypanthium, semiglobose, glabrous or sparsely pubescent; calyx lobes linear to linear-lanceolate, 10–20 mm long, 2–3 mm broad, reflexed, both surfaces hispidulous entirely or only towards apex, with one or two verrucose denticles on each side; sinus between calyx lobes broad and truncate. Corolla pale greenish white, flushed and spotted crimson, purple inside, or tube white, lobes deep purple, campanulate, 2–3.2 cm long, 2–3 cm wide, lobed to 1/3 of overall length, pubescent, less frequently glabrous outside; corolla lobes deltoid, sometimes recurved. Filaments glabrous, slightly dilated at base. Capsule globose in inferior part, conical in superior part, 1–2 cm long, 1.2–1.6 cm in diameter. Seeds oblong, brownish-yellow, ca.0.7 mm long. Figure 47–44.

Phenology: flowering and fruiting from September to November.

Habitats: scrubs on open slopes, or edges of forests; 1200–3000 m. [*L. Dhevoj 503* (in BM and E) from Nepal, Goketey, 1930 records 3960–4270 m].

Distibution: U Ganga & Indus, W, C and E Nepal, Sikkim & Darjiling, Bhutan, and Yarlung Zangbo-Brahmaputra (Map 47–95); also in Meghalaya of India.

U GANGA & INDUS: Pithoragarh, **Kuthi**, 2130 m, *R. Strachey & J. E. Winterbottom No3* (BM, K, P, PE); **Ralam Valley**, 1900, *Inayat 24631* (CAL); **Tehri Garhwal**, Rampur, 1938, *Kirah Rain 8857* (E); near Khathee, 2130 m, *Madden's 374* (E, K). **W NEPAL: Seti**, Bajhang, 1660–2230 m, 1991, *M. Suzuki et al. 60714* (E). **C NEPAL**: **Bagmati**, Dhading, S of Ganesh Himal., Tiru Danda, 1830 m, 1965, *J. D. A. Stainton 5099* (BM); between Khangjung and Syarpagaon, north side of Langtang River, 2200 m, 1966, *D. H. Nicolson 2409* (BM); Rasuwa, 1900 m, 2001, *D. G. Long et al. 113* (E); Grang, 1800 m, 1971,

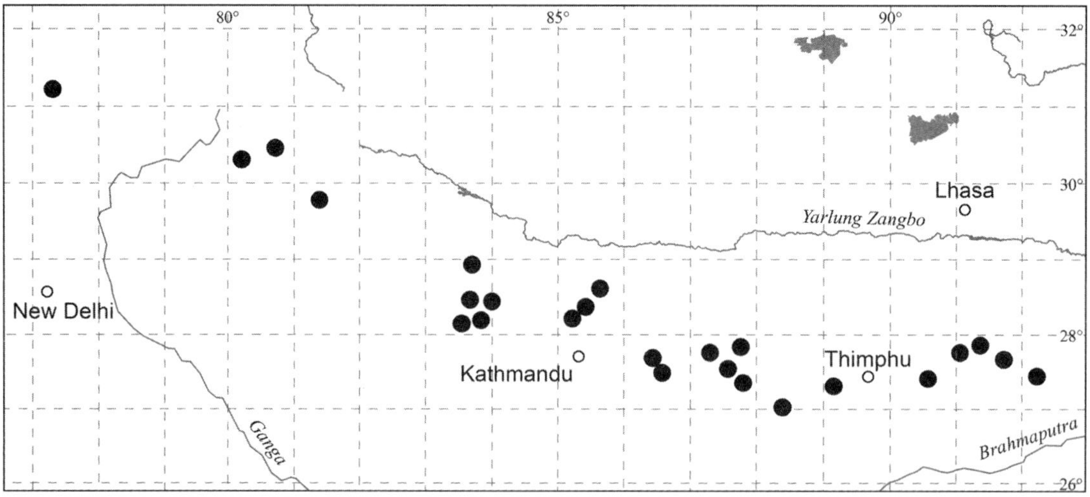

Map 47–95: **Codonopsis viridis** Wall.

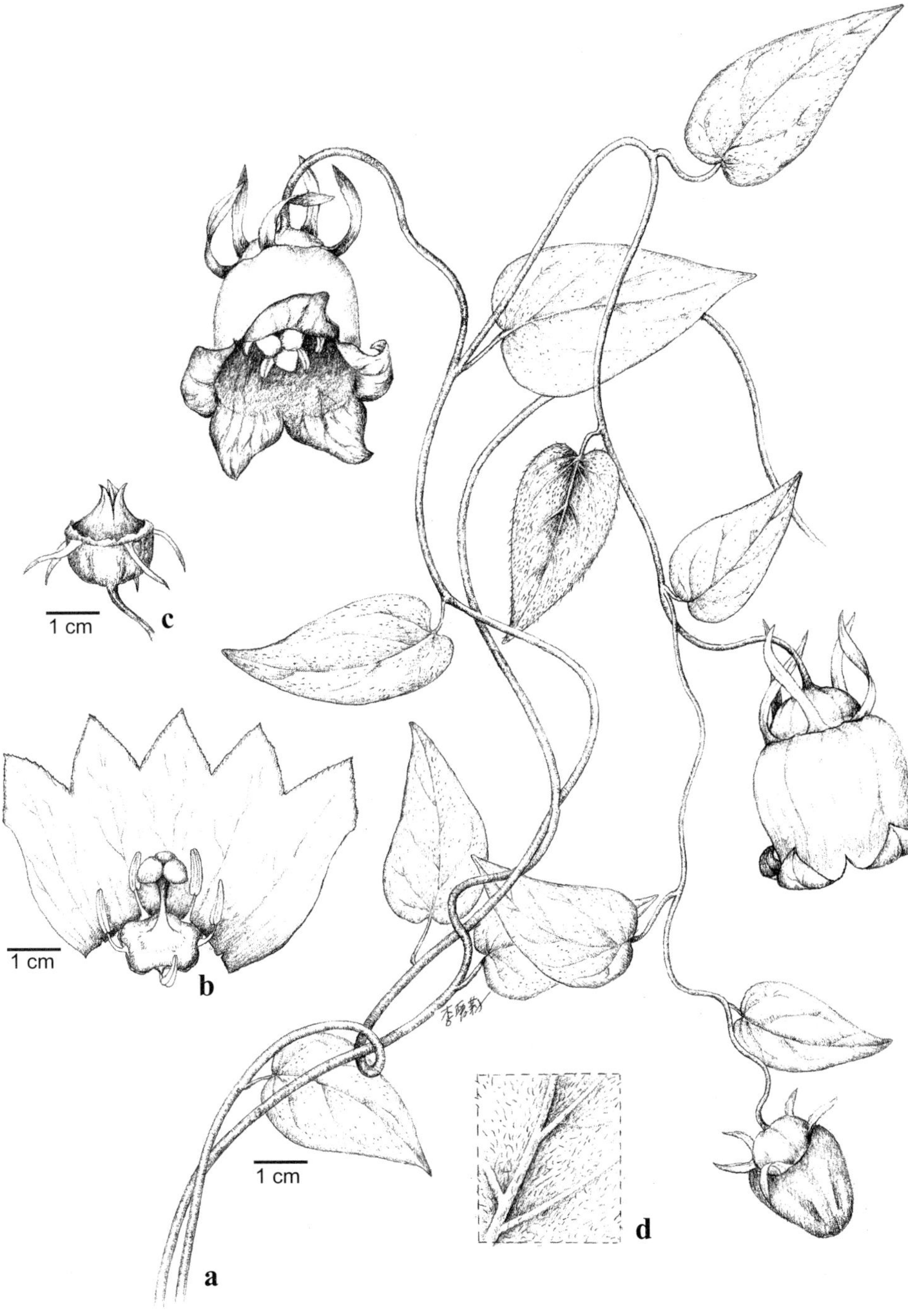

Figure 47–44: **Codonopsis viridis** Wall., the type species of the genus *Codonopsis*: **a**, upper parts of twining stems; **b**, a flower with calyx removed and corolla opened, showing stamens, disc, style, and stigma; **c**, 3-loculicidal capsule with persistent calyx, **d**, abaxial surface of leaf blade. From Hong, *Monogr. Codonopsis Allied Gen. Camp.*, 2015.

J. F. Dobremez 925 (BM); Thade, 2000 m, 1977, *N. P. Manandhar & Party 387* (KATH). **Dhawalagiri**: Mustang, S of Tukucha, Taglung, Kali Gandaki, 2740 m, 1954, *Stainton, Sykes & Williams 7929* (BM); Lete-Larjung Cobang, 1985. *D. P. Joshi et al. 131/85* (KATH); Myagdi, Muna, 1830 m, 1954, *Stainton, Sykes & Williams 4224* (BM, CAL, PE); Parbat, E of Gorepani, ca. 1800 m, 1973, *C. Grey-Wilson & Phillips 879* (BM, K); **Gandaki**, Kaski, Lumle, 1539 m, 1985, *D. P. Joshi et al. 11/85* (KATH); Mardi Khola: 1680 m, 1954, *Stainton, Sykes & Williams 8450* (BM, G); **Janakpur**, Dolakha, Yarsa-Jiri, 1980 m, 1964, *Banerjee, Shrestha & Upadhyay 2879* (KATH); Ramechhap, 1800–2700 m, 1985, *H. Ohba et al. 853195* (E, TI); **Kathmandu**, Phulchoke, 1965, *A. D. Schilling 691* (K, TI). **E NEPAL**: **Kosi**: Sankhuwasabha, above Milke Bhanjang, ca. 3000 m, 1981, *C. Grey-Wilson 4556* (K); Terhathum, between Chauki and Terhathum, 2200 m, 1981, *C. Grey-Wilson 4689* (BM, K); **Mechi**: Panchthar, Yektin-Birwa, 1963, *H. Kanai et al. 6300427* (TI); Yektin-Akasay-Batasay, 1963, *H. Hara et al. 6300428* (TI); Taplejung, Tapletok, 2300 m, 1971, *J. F. Dobremez 1388* (BM); Taplejung, Yamphudin, 2400 m, 1989, *KEKE 899* (E, K); Diorali, 1900 m, 1970, *H. Kanai & T. B. Shrestha 672722* (TI); **Sagarmatha**, Solukhumbu, 2350–2800 m, 1985, *H. Ohba et al. 8581347* (BM, E). **SIKKIM & DARJILING: Darjiling**, Victoria Fall, 1980 m, 1874, *J. S. Gamble 2335 A* (K). **BHUTAN**: **Tobrang**, Trashiyangsi Valley, 2290 m, 1934, *F. Ludlow & G. Sherriff 917* (BM, E); **Tongsa**, 1830 m, 1914, *R. E. Cooper 2073* (BM, E); **Wamlong**, Menthong La, 1974, *R. Nawang & S. Tshering* s. n.(E); **Upper Mo Chu**, between Tamji and Goen Gaza, 1984, *I. W. J. Sinclair & D. G. Long 4962* (E). **YARLUNG ZANGBO-BRAHMAPUTRA: Dirang** Dzong, 1520–1830 m, 1938, *F. Kingdon-Ward 14218* (BM); between **Gyipu** and Kapteng, Nyam Jang Chu, 2130 m, 1938, *F. Ludlow et al. 7063* (BM); **Muktur**, Tawang Chu, S Tisel, 2440 m, 1936, *F. Ludlow & G. Sherriff 2834* (BM).

29. Codonopsis farreri J. Anthony, Notes Roy. Bot. Gard. Edinburgh 15: 181. pl. 216. 1926; Shen & Hong in Hong, Fl. Reipubl. Popularis Sin. 73(2): 45. 1983; Hong, Monogr. *Codonopsis* Allied Gen. Camp. 143. 2015. Type: Myanmar, western flank of the Chimili, N'Maikha-Salween Divide, 26°23′ N, 98°48′ E, 3350–3650 m, Aug. 1924, *G. Forrest 24885* (lectotype designated by Hong, 2015: E; isolectotypes: K, P, W); loc. eodem, 26°24′ N, 98°48′ E, 3650 m, Jun. 1925, *G. Forrest 26973* (paratypes: E, K, P, W); Myanmar, Chimili Valley, the uppermost alpine woodlands, 3350 m, 31. Jul. 1919, *Farrer 1144* (paratype: E).

Codonopsis farreri J. Anthony var. *grandiflora* S. H. Huang, Acta Bot. Yunnan. 6: 394. 1984. Type: China, Yunnan, Fugong, 3100–3600 m, 2 Aug. 1979, *Q. Ling 791720* (holotype: KUN).

Codonopsis gombalana C. Y. Wu in Wu & Li, Yunnan Trop, Subtrop. Fl. Res. Rep. 1: 81. pl. 17. 1965. Type: China, Yunnan, Gongshan, Cikai, Heipu Shan, bamboo thickets, 9 Oct. 1940, *K. M. Feng 8295* (holotype: KUN; isotypes: PE).

Stems 0.5–5 m long, twining, rarely ascending or scandent, branched. Leaves on main stem and major branches alternate or nearly opposite, those on branchlets opposite or pseudo-verticillate; petiole 1–1.5 cm long; leaf blade ovate, lanceolate or elliptic, 0.5–4 cm long, 0.5–1.5 cm broad, both surfaces glabrous, base cuneate or rounded, margin entire, apex acute or slightly obtuse. Flowers solitary, terminal; pedicel 5–6 cm long. Calyx tube adnate to whole hypanthium, glabrous; calyx lobes linear-lanceolate, 2–2.6 cm long, 0.3–0.5 cm broad, margin nearly entire, apex acute or slightly obtuse. Corolla pale green or greenish yellow, with brown-red or purple reticulate veins and with a dark purple patch at base of each of five main veins, campanulate, 3.3–5.5 cm long, ca. 3 cm wide; corolla tube 2–4 cm long; corolla lobes broad-triangular, obtuse, 1.2–1.6 cm long, 1.2–2 cm wide. Stamens glabrous; filaments slightly dilated at base. Figure 47–45.

Phenology: flowering from June to September

Habitats: mostly in *Rhododendron* and bamboo thickets, less frequently in forests or meadows, often on granites, less often on limestones; 3000–3700 m.

Figure 47–45: **Codonopsis farreri** J. Anthony: **a**, upper part of stem; **b**, a flower with corolla opened, showing calyx, stamens, disc, style, stigma, and a dark purple patch at the base of each of five main veins on corolla; **c**, style and stigma. From Hong, *Monogr. Codonopsis Allied Gen. Camp.*, 2015.

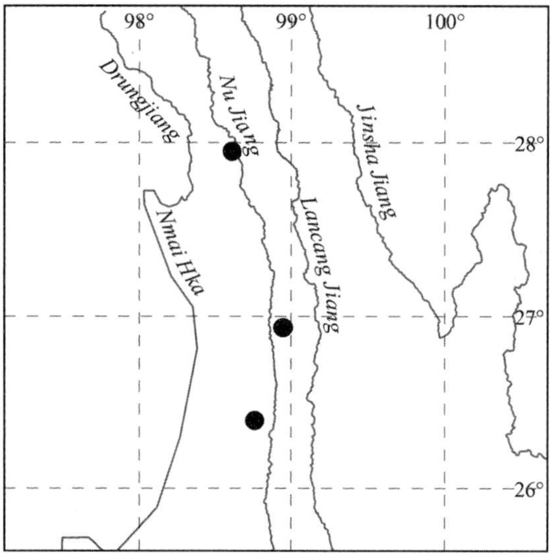

Map 47–96: **Codonopsis farreri** J. Anthony

Distribution: **endemic**; confined to S Hengduan and U Irrawaddy (Gaoligong Shan) (Map 47–96).

S HENGDUAN: Fugong, Lishadi, 3560 m, 2005, *Gaoligong Shan Biodiversity Survey 27015* (A, CAS); Lumadeng, 3650 m, 2005, *Gaoligong Shan Biodiversity Survey 28643* (A, CAS); **Gongshan**,; Salween-Kiukiang Divide, Lunguailaka, 3600 m, 1938, *T. T. Yü 20265* (paratypes of *C. gombalana:* E, KUN, PE); Cikai, Heipu Pass, 3490 m, 2006, *Gaoligong Shan Biodiversity Survey 32059* (A, CAS).

30. Codonopsis benthamii Hook. f. & Thomson, J. Proc. Linn. Soc. Bot. 2: 14. 1857; Hong, Novon 20: 420. 2010 & Monogr. *Codonopsis* Allied Gen. Camp. 143. 2015. Type: India, Sikkim, Lachen, 2740 m, 31 Jul. 1849, *J. D. Hooker* s. n.[lectotype designated by Hong 2010: K (K000814213)]; Sikkim, Lachen, 2750–3050 m, 3 Aug. 1849, *J. D. Hooker* s. n.[paratype: K (K000814212)].

Codonopsis macrocalyx Diels, Notes Roy. Bot. Gard. Edinburgh 5: 170. 1912. Type: China, Yunnan, "on the eastern flank of the Lichiang Range" (Lijiang), 27°20′N, 3050–3350 m, Sept. 1906, *G. Forrest 3008* (lectotype designated by Hong, 2010: E; isolectotype: K).

Codonopsis macrocalyx Diels var. *parviloba J.* Anthony, Notes Roy. Bot. Gard. Edinburgh 15: 183. 1926. Type: China, Yunnan, "on the eastern flank of the Tali range" (Dali), 25°40′N, 2740–3050 m, Jul. 1906. *G. Forrest 3854* (lectotype designated by Hong 2010: K; isolectotype: E).

Codonopsis macrocalyx Diels var. *coerulescens* Hand.-Mazz., Akad. Wiss. Wien Math.-Naturwiss. Kl. Anz. 61 (20): 169. 1924. Type: China, Yunnan, Mekong-Salween Divide ("Schöndsu-la"), 28°4′N, 4025 m, 2 Aug. 1916, *Handel-Mazzetti 9618* (lectotype designated by Hong, 2015: WU; isolectotype: E).

C. pianmaensis S. H. Huang, Acta Bot. Yunnan. 6: 393. fig. 1. 1984. Type: China, W Yunnan,Lushui County, Pianma, 3500 m, 17 Aug. 1964. *S. K. Wu 8428* (holotype: KUN).

Roots carrot-shaped or cylindrical, 20–30 cm long, 1.5–2.5 cm in diameter. Stems 0.3–2 m long, ascending, scandent, sometimes twining, branched, branches sterile or fertile, sparsely puberulent. Leaves alternate or those on branches nearly opposite; petiole 1–6 cm, sparsely villous; leaf blade broad-ovate to ovate-lanceolate, 3–10.5 cm long, 1.5–7 cm broad, adaxially green, sparsely puberulent, abaxially gray-white, puberulent, base shallowly cordate to cuneate, margin irregularly pinnatifid to shallowly lobed, or dentate, crenate, rarely nearly entire, apex acute or obtuse. Flowers solitary, terminal; pedicel 1–8 cm long, villous. Calyx tube adnate to whole hypanthium, glabrous or sparsely villous; calyx lobes ovate or lanceolate, 1–2.5 cm long, 0.4–0.5 cm broad at anthesis, glabrous, rarely sparsely hispidulous, margin crenate, ciliate; sinus between calyx lobes narrow and pointed. Corolla yellow, green, yellow-green, or rarely purple, pale brown-red at base, tubular, slightly dilated above, 1.8–4 cm long, 1.5–2 cm wide, shallowly lobed for 1/6–1/7 of overall length, glabrous; corolla lobes triangular. Filaments entirely villous, dilated at base; connectives enveloping anthers, villous at back. Capsule 1.2–1.6 cm in diameter; inferior part hemispheric, fruiting calyx lobes up to 2.5 cm long, 1.5 cm broad. Seeds ovoid, brown-yellow. Figures 47–46: a–c.

Phenology: flowering & fruiting from July to October.

Habitats: grassy slopes, forest margins, scrubs, on granites or marbles; 2100–4300 m.

Distribution: E Nepal, Sikkim & Darjiling, Bhutan, M Yarlung Zangbo, Yarlung Zangbo-Brahmaputra, S Hengduan, and U Irrawaddy (Map 47–97); also in SW Yunnan (Yongde) of China.

E NEPAL: Kosi, Sankhuwasabha, 2540-3500 m, 1988, *M. Suzuki et al. 8820505* (E, KUN); Bakang, above Sekidim, ca. 2800 m, 1981, *C. Grey-Wilson et al. 4070* (BM, K); **Mechi**, Taplejung, Topke Gola – Shewaden, 2700 m, 1977, *H. Ohashi et al. 772671* (TI) ; **Sagarmatha**, Solukhumbu, 2880 m, 1997, *M. Wakabayashi et al. 9730115* (E). **SIKKIM & DARJILING: Sikkim,** Chola Range, 2440 m, 1892, *G. A. Gammie 1281* (CAL) ; Gangtok, 1877, *G. King 4439* (CAL); Lachen, 2440 m, 1909, *Ribu & Rhomoo 3016* (CAL); Lachung, 1879, *G. King's collector* s. n.(CAL); Laghep, 3350 m, 1913, *R. E. Cooper 509* (E); Selu Valley, 3050 m, 1892, *G. A. Gammie* s. n.(CAL); Yakla, 3050 m, 1869, *C. B. Clarke 10224* (CAL, K); Zemu Valley, 2440 m, 1909, *W. W. Smith & G. H. Cave 986* (B, CAL, E, K, P). **BHUTAN**: **Omta Tso**, 3810 m, 1949, *F. Ludlow et al. 19839* (BM); **Rinchen Chu**, 2900 m, 1937, *F. Ludlow & G. Sherriff 3527* (BM, E); **Rip-la**, 2900 m, 1949, *F. Ludlow et al. 20886* (BM); **Sengor**, between Sengor and Sheridrang, 2980 m, 1979, *A. J. C. Grierson & D. G. Long 2543* (E); **Thimphu**, 3210 m, 1984, *I. W. J. Sinclair & D. G. Long 5566* (E, K); **Upper Kulong Chu,** 2700 m, 2000, *S. Bowes Lyon 13179* (E); **Upper Mo Chu**, 2003, *S. Bowes Lyon 15127* (E). **M YARLUNG ZANGBO: Yadong**, Asang Monitory station, 3000 m, 2013, *PE-Tibet Exped. 2945* (PE); Chumbi Valley, 1877. *G. King 4270* (CAL); vicinity of Chumbi Tang, 3860 m, 1974, *Qinghai-Xizang Exped. 74–2509* (KUN). **YARLUNG ZANGBO-BRAHMAPUTRA**: **Dirang** Dzong, 2130–2440 m, 1938, *F. Kingdon-Ward 14228* (BM, PE); **Lage**, Duoxiang-la, 3000 m, 1992, *H. Sun et al. 0788* (KUN); Arza of **Shangzayü** to Rawu, 3900 m, 1973, *Qinghai-Xizang Exped. 73–1069* (KUN, PE). **S HENGDUAN: Bijiang**, Biluoxue Shan 4000 m, 1934, *H. T. Tsai 58236* (KUN, PE, SZ); **Cawarong** (Tsa-wa-rung), Chi-na-tung, 2800 m, 1935, *C. W. Wang 65204* (PE); E of Salween River, N of Alulaka, Muli-la, 2740 m, 1932, *J. F. Rock 22359* (E, PE); **Dali**, 3050 m, 1912, *G. Forrest 8954* (E, K); **Dêqên**, Doker-la, 3800–4150 m, 1915, *Handel-Mazzetti 8121* (W, WU); Junzila, 3600 m, 1938, *T. T. Yü 23257* (KUN, PE); Mekong-Salween Divide, E of Yuragan, Nake-la, 3810 m, 1932, *J. F. Rock 23127* (E, PE); Slaza, 3800 m, 1938, *T. T. Yü 22796* (KUN, PE); **Fugong**, Lumadeng, 3650 m, 2005, *Gaoligong Shan Biodiversity Survey 28673* (A, CAS); **Gongshan**, Binzhongluo, 2880 m, 2006, *Gaoligong Shan Biodiversity Survey 32891* (A, CAS); Champu-tung, Lung-pan-la, 3000 m, 1935, *C. W. Wang 67453* (KUN, PE); Cikai, 3220 m, 2006. *Gaoligong Shan Biodiver-*

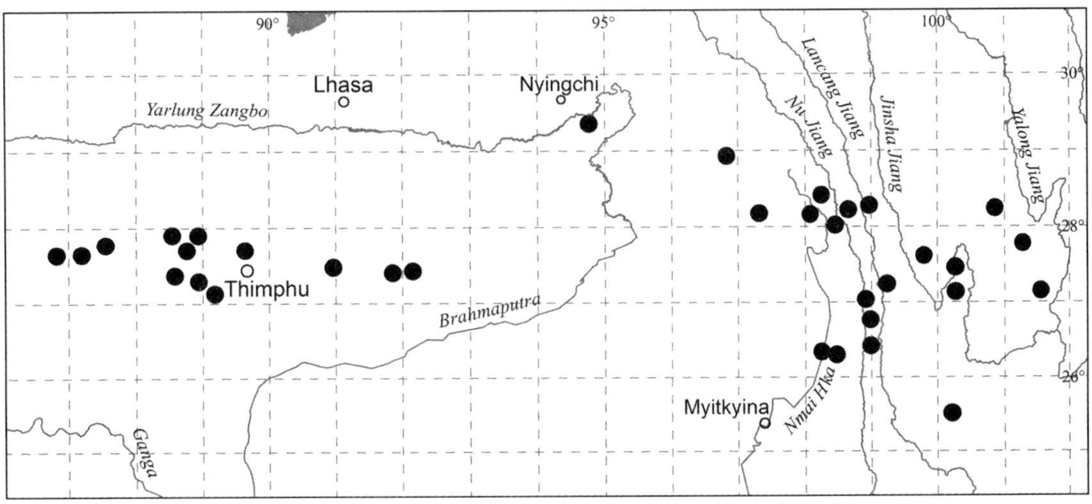

Map 47–97: **Codonopsis benthamii** Hook. f. & Thomson

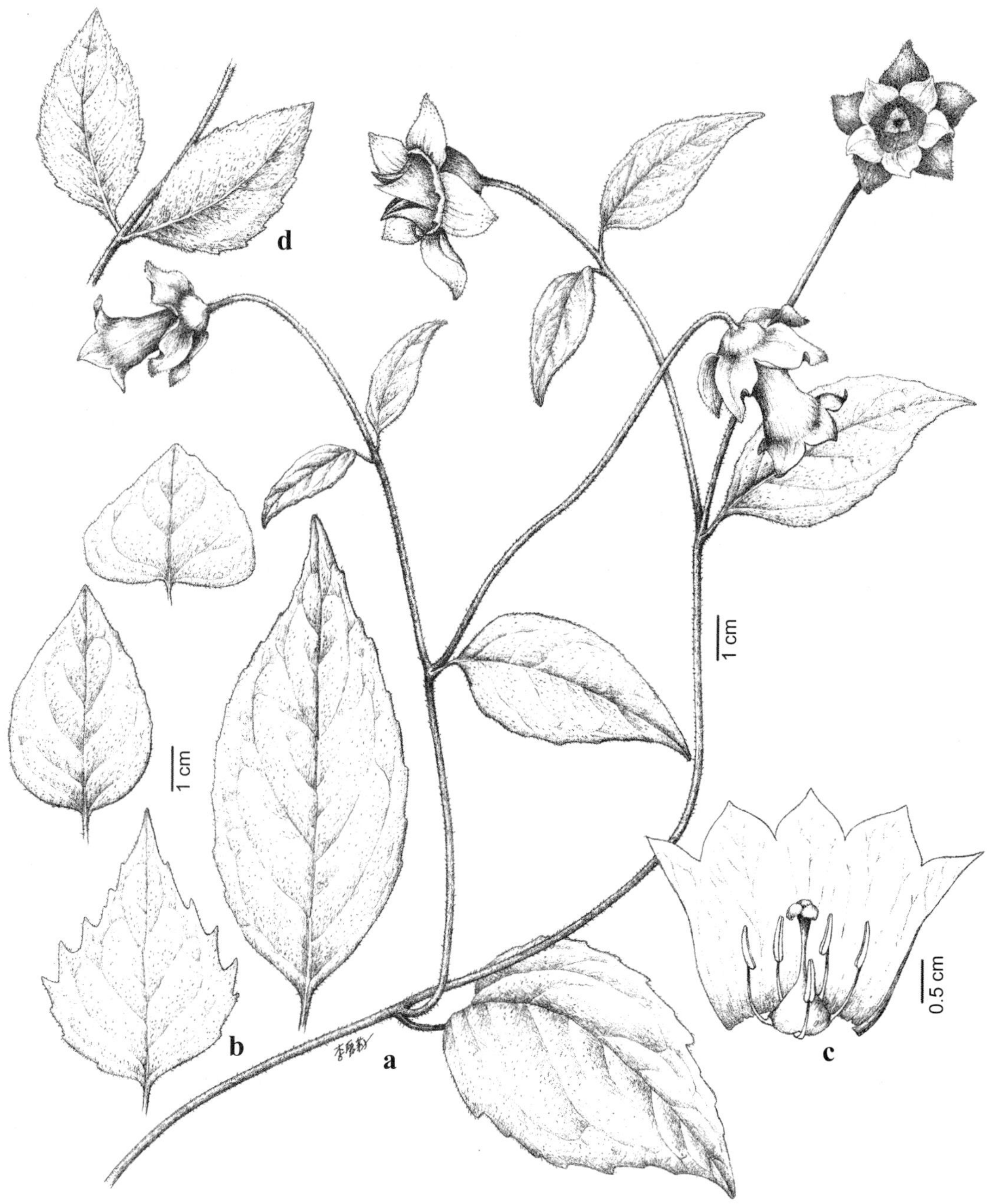

Figure 47–46: **a–c**, **Codonopsis benthamii** Hook. f. & Thomson: **a**, upper part of stem with a mature capsule and 3 flowers with tubular corolla; **b**, different forms of leaves; **c**, a flower with calyx removed and corolla opened, showing stamens and upper part of pistil. **d**, **Codonopsis tubulosa** Kom. subsp. **tubulosa**: a section of stem with opposite leaves. From Hong, *Monogr. Codonopsis Allied Gen. Camp.*, 2015.

sity Survey 34115 (A, CAS); Salween-Kiukiang Divide, Tsukuai, 3600 m, 1938, *T. T. Yü 20708* (E, KUN, PE); **Lushui**, 3050–3350 m, 1924, *G. Forrest 24723* (paratypes of *C. macrocalyx* Diels var. *parviloba* J. Anthony: E, K, W); **Muli**, 10 km S of Muli, 3050 m, 1924, *F. Kingdon-Ward 4853* (E); Wo-chio, 2800 m, 1937, *T. T. Yü. 7597* (KUN, PE); **Shangri-la** (Zhongdian), Chi-ren, 3200 m, 1937, *T. T. Yü 12402* (KUN, PE); Wengshui, 4000 m, 2010, *Kham Exped. 10–3273* (PE); Xiaozhongdian, 3800 m, 1939, *K. M. Feng 1886* (KUN, PE); **Weixi**, Yezhi, 3600 m, 1935, *C. W. Wang 68519* (PE); Kangpu, 3500 m, 1935, *C. W. Wang 64533* (PE); **Yanbian**, Zemulong, 3700–3800 m, 2002, *X. F. Gao et al. 3443* (CDBI, PE) ; **Yulong (Lijiang)**, 3100 m 1914, *C. Schneider 2302* (K, PE); 1914, *G. Forrest 12797* (E, PE). **U IRRAWADDY**: **Chaungmaw Hka** Valley, 2440–2740 m, *F. Kingdon-Ward 3284* (E); **Chimili Valley**, 3050–3350 m, 1919, *RA 1219* (E); **Di Chu Valley**, 3350 m, 1926, *F. Kingdon-Ward 7193* (K); **N' Maikha-Salween Divide**, 1931, *G. Forrest 29709* (E); **Ridong**, 1982, *Qinghai–Xizang Exped. 10111* (KUN, PE); **Seinghku Valley**, 3960 m, 1926, *F. Kingdon-Ward 7531* (K).

31. Codonopsis tubulosa Kom., Trudy Imp. S.-Peterburgsk. Bot. Sada 29(1): 112. tab. 2. fig. 3. 1908; Tsoong, Contr. Inst. Bot. Nat. Acad. Peiping 3(3): 103. fig. 13. 1935; Hong, Monogr. *Codonopsis* Allied Gen. Camp. 149. 2015. Type: China, Yunnan, Mengtze (Mengzi), Gran Mts, 2130 m, *A. Henry 10167* (lectotype designated by Hong, 2015: K; isolectotypes: A, E, LE, PE).

Codonopsis accrescenticalyx H. Lév., Cat. Pl. Yunnan 24. 1915; Anthony, Notes Roy. Bot. Gard. Edinburgh 15: 182. 1926, pro syn. Type: China, Yunnan, "pâturages des montagnes à Tcha-ho", 2600 m, Jul. 1912, *E. E. Maire s. n.*(holotype: E).

Roots carrot-shaped, simple or branched below middle, 10–20 cm long, 0.5–2 cm in diameter. Stems usually 1–3 m long, twining, rarely ascending, occasionally erect, branched; branches sterile or with a terminal flower, nearly glabrous or sparsely villous. Leaves opposite, tending to alternate distally; petiole 1–17 mm long, puberulent; leaf blade ovate to lanceolate, 3–12 cm long, 1–6 cm broad, adaxially hispidulous, abaxially sparsely to usually densely white-puberulent, base cuneate or obtuse, margin crenate or nearly entire, occasionally deeply incised, apex acute or obtuse. Flowers terminal; pedicel 1–6 cm long, villous. Calyx tube adnate to whole hypanthium, semiglobose, nearly glabrous to densely villous; calyx lobes ovate-lanceolate, ovate to broadly ovate-deltoid, 1–1.8 cm long, 0.5–1.4 cm broad, adaxially glabrous, abaxially sparsely villous, margin crenate and verrucose-toothed, ciliate-villous. Corolla greenish yellow, base often purple, tubular, slightly dilated above, lobed to ca. 1/6, 2–3.7 cm long, 1–1.6 cm wide, glabrous; corolla lobes glabrous or very minute-ciliate. Filaments dilated at base, ciliate above; connective densely villous. Capsule globose below, conical above, 1.5–2.2 cm long, 1–1.3 cm wide. Seeds ovoid, smooth, brown-yellow.

Distribution: Yanglung Zangbo-Brahmaputra, N and S Hengduan, and U Irrawaddy (Map 47–98); also in NE Guangxi and Guizhou of China, and the other part of Myanmar (Chin Hills).

1a. Corolla 2–3.7 cm long, corolla lobes glabrous; petiole 1–11 mm long .... 31a. subsp. *tubulosa*
1b. Corolla 1.8–2.2 cm long , corolla lobes sparsely minute-ciliate; petiole 10–17 mm long .... 31b. subsp. *vadsea*

31a. Codonopsis tubulosa subsp. **tubulosa**

Stems usually 1–3 m long, twining, rarely ascending. Petiole 1–11 mm long. Corolla 2–3.7 cm wide, glabrous. Figure 47–46: d.

Phenology: flowering from July to September; fruiting in September and October.

Habitats: shrubby slopes, forests, or meadows; 1400–3600 m.

Distribution: N and S Hengduan; also in NE Guangxi and Guizhou of China, and N Myanmar (Chin Hills).

N HENGDUAN: **Wenchuan**, Wolong, Dengshengou, 2700 m, 2007, *X. J. He et al. 150018* (SZ) & *150076* (SZ); **Xiaojin**, Rilong,

3100 m, 1975, *Sichuan Plant Surv. 9685* (CDBI). **S HENGDUAN: Baoxing**, Ganyang, 4000 m, 1936, *K. L. Chu 3558* (PE, SZ); **Butuo**, Tuojo, 2600 m, 1959, *Sichuan Econ. Plant Surv.* (*Liang*) *5926* (CDBI, KUN, PE); **Dali**, Cangshan, Zhonghesi to Xiangshiya, 1946, *T. N. Liou 21228* (PE); Chienchuan-Mekong Divide, 3350–3660 m, 1922, *G. Forrest 21991* (E, K); E side of Diancang Shan, 1984, *Sino-Amer. Bot. Exped. 999* (B, E, PE); Huadianba, ca. 3000 m, 1984, *K. Iwatsuki et al. 1441* (KUN); **Dechang**, 15 km from Dashan to the city, 2300 m, 2010, *E. D. Liu et al. 2558* (KUN); **Eryuan**, Fengyu to Liantie Pass, 2946 m, 2010, *E. D. Liu 2750* (KUN); **Ganluo**, Azhai, 2600 m, 2008, *X. J. He & C. S. Cao 184624* (SZ); Haitang, 21. July 1959, *Sichuan Econ. Plant Surv. (Liang) 4292* (PE); **Heqing**, Songqui, Ma'er Shan, 3500 m, 1936, *R. C. Ching 24086* (KUN); **Huidong**, Lunan, 1959, *S. K. Wu 1038* (KUN); Sunjiaqing, sunny slopes, 16. June 1959, *S. K. Wu 680* (KUN); **Huili**, Mt Longdschu-schan, 1914, *Handel-Mazzetti 5206* (W, WU); **Lanping**, 3000 m, 1933, *H. T. Tsai 53774* (KUN, PE,SZ); **Meigu**, s. loc., 1959, *Sichuan Econ.Plant Surv. (Liang) 1652* (CDBI, KUN, PE); **Mianning**, Tuowu, 2800 m, 1978, *Mianning Exped. 468* (SM); **Miyi**, Puwei Township, 2500 m, 1958, *Z. He 11305* (SZ); **Muli**, Mt. Gibboh, 3050 m, 1928, *J. F. Rock 16935* (E); 2740 m, 1905, *F. Kingdon-Ward 4563* (E, K); **Puge**, Tuomugou, Huangzhu Shan, 2500 m, 1959, *Sichuan Econ. Plant Surv. (Liang) 5487* (CDBI, PE); **Shangri-la (Zhongdian)**, near Xiaozhongdian, 3300 m, 1939, *K. M. Feng 1916* (KUN, PE); **Tianquan**, Erlang Shan, *H. L. Tsiang 37516* (PE, SZ); **Weixi**: s. loc., s. d. *G. Forrest 30505* (PE); **Xichang**, 2900 m, 1932, *T. T. Yü 1293* (PE); **Xide**, Mianshan, 2400 m, 2008, *X. J. He & C. S. Cao 195663* (SZ); **Yanbian**, Dapingzi, 2800 m, 1978, *Yanbian Exped. 338* (SM); **Yangbi**, Yangbi River basin, 3050–3660 m, 1922, *J. F. Rock 6210* (E); Zhongshan, 1984, *K. Iwatsuki et al. 874* (KUN); **Yanyuan**, Yanjing, Mt Yuanbao, 2600–3000 m, 1964, *T. L. Dai et al. 64-13* (PE); **Yuexi**, Bao'an, 2700 m, forests, 13. July 1959, *Sichuan Econ.Plant Surv. (Liang) 3866* (CDBI, KUN, PE); **Yulong (Lijiang)**, Laschiba, 1939, *R. C. Ching 21881* (KUN, PE); Ludian, Lidiping, Machang, 3000 m, 1956, *P. I. Mao 166* (KUN, PE, SZ); **Zhaojue**, Jiefang Gou, 2810 m, 1979, *Zhaojue Surv. 956* (SM). **U IRRAWADDY: Tengchong** (Tengyueh), s. loc., *E. B. Howell 23* (E, K).

31b. Codonopsis tubulosa subsp. **vadsea** (S. S. Dash & A. A. Mao) D. Y. Hong, Monogr. *Codonopsis* Allied Gen. Camp. 153. 2015. ≡ *Codonopsis vadsea* S. S. Dash & A. A. Mao, Nelumbo 53: 17. fig. 1. 2011. Type: Yarlung Zangbo-Brahmaputra region, Kurung Kumey District, Mt Vadse, 3800 m, 4 Aug. 2010, *A. A. Mao & S. S. Dash 31846* (holotype: CAL, n. v.; isotype: CAL, n. v.).

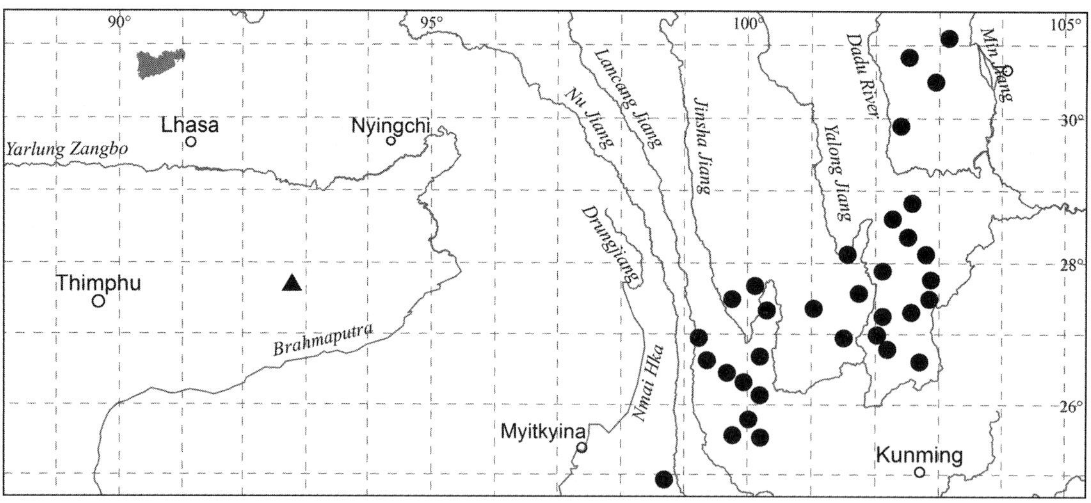

Map 47–98: **Codonopsis tubulosa** Kom.
● subsp. **tubulosa**, ▲ subsp. **vadsea**

Stems 0.5–1 m long, ascending or erect. Petiole 1–1.7 cm long. Corolla 1.8–2.2 cm long; corolla lobes very minutely ciliate.

Phenology: flowering from July to September.

Habitats: subalpine to alpine grassy slopes; 3700–4000 m.

Distribution: **endemic**; confined to Kurung Kumey in Yarlung Zangbo-Brahmaputra.

32. Codonopsis microtubulosa Z. T. Wang & G. J. Xu, Acta Phytotax. Sin. 31(2): 184. fig. 1. 1993; Hong, Lammers & Klein in Wu, Raven & Hong, Fl. China 19: 516. 2011, pro syn; Hong, Monogr. *Codonopsis* Allied Gen. Camp. 153. 2015. Type: China, Sichuan, Nanping (Jiuzhaigou), 2500 m, 23 Jul. 1986, *Z. T. Wang & B. Y. Yü 867206* (holotype: CPU; isotype: PE).

Herbs perennial, pilose throughout. Roots carrot-shaped, 8–21 cm long, 0.5–1 cm in diameter, simple or sometimes branched below middle. Caudex often short, 2–4 mm long, 6–10 mm in diameter. Stems 1 to several from one caudex, scandent or twining, 45–90 cm long, branched from base, less often simple. Leaves alternate, but those at lower and middle parts of stem usually opposite; petiole 2–12 mm long; leaf blade cordate-ovate, 1.1–3.6 cm long, 0.6–3.3 cm broad, base shallowly to deeply cordate, margin crenate or dentate, apex obtuse. Flowers solitary, terminal on both main stem and branches; pedicel 1.1–2.9 cm long. Calyx tube adnate to whole hypanthium, semi-globose, nearly glabrous; calyx lobes oblong-lanceolate, 6–8 mm long, 2–4 mm broad, both surfaces glabeous, margin entire, villous-ciliate, apex obtuse. Corolla tubular, 1–1.5 cm long, glabrous, lobed to1/4 – 1/5. Figure 47–47.

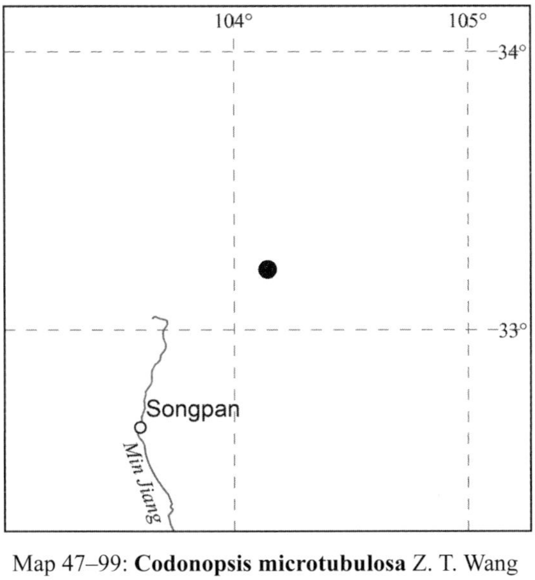

Map 47–99: **Codonopsis microtubulosa** Z. T. Wang & G. J. Xu

Phenology: flowering in July and August.

Habitats: ca. 2500 m.

Distribution: N Hengduan (Map 47–99); also in Chongqing (Chengkou) of China.

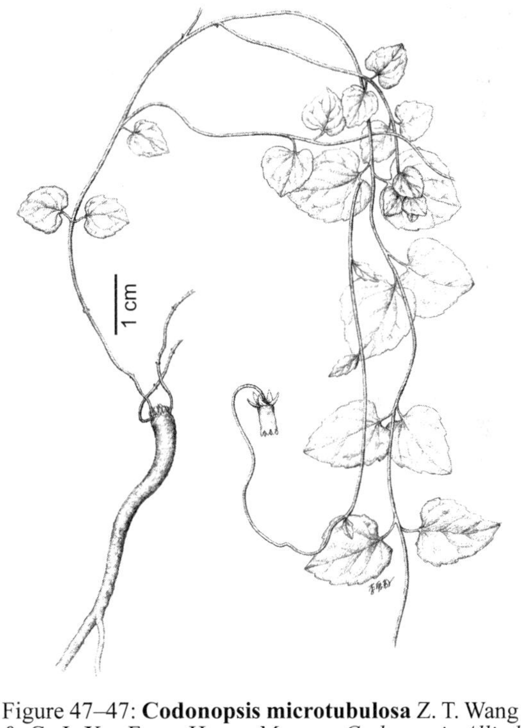

Figure 47–47: **Codonopsis microtubulosa** Z. T. Wang & G. J. Xu. From Hong, *Monogr. Codonopsis Allied Gen. Camp.*, 2015.

Notes: We reduced this species as a synonym of *C. pilosula* in *Fl. China* (Hong et al., 2011). Recently we examined two collections in the herbarium P, *Farges 1366* and *Farges* s. n., from Chengkou of Chongqing in China. The former has four individuals and the latter has 30 individuals. Thus, one of the collections came from a rather large population and therefore the material is adequate for judging the status of *C. microtubulosa*. It is a very distinct species, differing from all the other species in having long pilose hairs, and from *C. benthamii* and *C. tubulosa*, which are also of tubular corolla and twining stems, in having much smaller corolla only 1–1.5 cm long (vs. 1.8–4 cm long, 1–1.8 cm broad).

33. Codonopsis subglobosa W. W. Smith, Notes Roy. Bot. Gard. Edinburgh 8: 108. 1913; Anonymous, Icon. Cormophyt. Sin. 4: 381. fig. 6176. 1975; Hong, Monogr. *Codonopsis* Allied Gen. Camp. 154. 2015. Type: China, Yunnan, "the eastern flank of the Lichiang Range" (Lijiang), 27°20′N, 2740 m, Sept. 1910, *G. Forrest 6684* (lectotype designated by Hong, 2015: E; isolectotypes: BM, K).

Roots carrot-shaped or cylindrical, 30–50 cm long, 1.5–8 cm in diameter. Caudex short or rarely up to 40 cm long. Stems twining, often over 2 m long, branched, sparsely white-hirsute. Leaves on main stem and branches alternate, those on branchlets nearly opposite; petiole 0.5–2 cm long, sparsely white-hirsute; leaf blade broad-ovate to ovate-lanceolate, 0.5–3 cm long, 0.5–2.5 cm broad, adaxially hispidulous, abaxially hirsute, base shallowly cordate to rounded, margin slightly waved or shallowly crenate, apex obtuse or acute. Flowers solitary, terminal on branchlets or leaf-opposed; pedicel hirsute. Calyx tube adnate to nearly whole hypanthium, semiglobose, sparsely white-hirsute on ribs, rarely glabrous; calyx lobes distant to each other, nearly orbicular or rhomboid-ovate, 9–13 mm long, 6–8 mm broad, abaxially white-hirsute, rarely glabrous, margin serrate and ciliate, often revolute. Corolla pale yellow-green, dark red-purple at apex, globose-campanulate, 2–2.5 cm long; corolla lobes broad-triangular. Filaments slightly dilated at base. Stigma 3-fid; each stigma segment 2-lobed. Capsule semiglobose below, conical or rostrate above. Seeds ellipsoid or ovoid, smooth, yellow-brown. Figure 47–48: a.

Phenology: flowering from July to September; fruiting from September to November.

Habitats: stony meadows or scrubs; 2400–4100 m.

Distribution: **endemic**; Tangut, N and S Hengduan (Map 47–100).

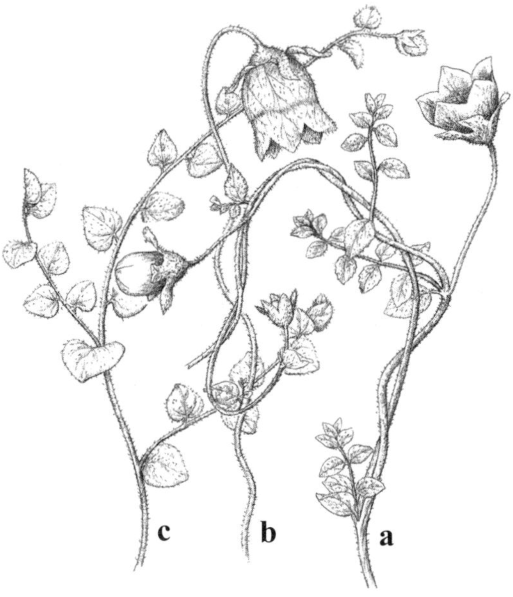

Figure 47–48: **a**, **Codonopsis subglobosa** W. W. Smith: upper part of twining stem. **b** & **c**, **Codonopsis bragaensis** Grey-Wilson: upper part of stems. **a** from *Icon. Cormophyt. Sin.* 4, 1975; **b** & **c** modified from Grey-Wilson, 1990.

TANGUT: Jomda, Gamtog to Gyorxung, 1951, *S. X. Jia 284* (PE). **N HENGDUAN: Baiyü**, Samar-Garyi, 3110 m, 2010, *Kham Exped. 10–0552* (PE); **Batang**, Yarigong, 3200 m, 1964, *J. D. Li 64-37* (PE); Rongguo Shan, 3800 m, 1964, *J. D. Li et al. 64-46* (PE); Zongza, 3000 m, 1964, *J. D. Li 64-44* (PE, SM); **Danba**, Taiping, Heifending, 3000 m, 1964, *T. L. Dai et al.* (*Danba*) *64-8* (PE, SM); **Dawu**, Qingdaqiao, 3240 m, 1964, *T. L. Dai* (*Qianning*) *64-5* (PE, SM); Wari, Muru, 3500 m, 1974, *Sichuan Plant Surv. 5901* (CDBI, PE); **Litang**, Litang Qu Valley, 4022 m, 2010, *Kunming-Edinburgh Exped. to Sichuan 183* (E). **S HENGDUAN**: **Daocheng**, Gonggaling, near temple, 3100 m, 1937, *T. T. Yü 12887* (KUN, PE); Julong, 3900 m, 1973, *Sichuan Veget. Surv. 2178* (CDBI, KUN, PE); Shengmu, Gêwa Village, 3500 m, 1964, *J. D. Li & Z. M. Xu 64-53* (PE, SM); **Dêqên**, Baima Shan, 3600 m, 1976, *Wu 4364* (KUN); vicinity of Dêqên town, 3300 m, 1986, *H. Sun & Z. G. Qian 710* (KUN); **Dêrong**, Deep Gorge of Ding Qu, 2600–2900 m, 2004, *D. E. Boufford et al. 31033* (A); **Heqing**, Heqing Valley, 2440 m, *G. Forrest 83* (E); **Jianchuan**, Jianchuan to Yangcheng, 2300 m, 1962,

C. C. Lu s. n. (KUN); **Jiulong**, Tanggo, 3240 m, 1980, *Sichuan Plant Surv. 28280* (CDBI); **Kangding**, s. loc., 1893, *J. A. Sourie 129* (G, K, P); **Muli**, Ker, 2900 m, 2011, *D. Y. Hong et al. H11035* (PE); Liziping, Baicaoping, 2750 m, 2011, *D. Y. Hong et al. H11029* (PE); Wa'ertsa, 3100 m, 1937, *T. T. Yü 14794* (E, PE); **Shangri-la** (Zhongdian), Dongwang, 2800 m, 1988, *J. S. Yang 88-118* (KUN, PE); **Xiangcheng**, Qianjin, 3800 m, 1972, *Z. G. Liu 1484* (CDBI); Reda, 3450–3650 m, 1998, *D. E. Boufford et al. 28793* (E); **Yajiang**, s. loc., 2800 m, 1964, *J. D. Li & Z. M. Xu 64-60* (SM); **Yanyuan**: Bailing Shan, 1986, *Z. T. Wang 869-383* (PE); Yanyuan to Kwapi, 2900–3200 m, 1914, *Handel-Mazzetti 5456* (E, K, W, WU); **Yulong** (Lijiang), Baisa, 1939, *R. C. Ching 30602* (KUN, PE); between Dali and Lijiang, 2800–2900 m, 1916, *Handel-Mazzetti 10058* (G. W, WU); Laschiba, 1939, *R. C. Ching 21374* (KUN, PE); Xuesongcun, 3000 m, 1937, *T. T. Yü 15601* (KUN, PE).

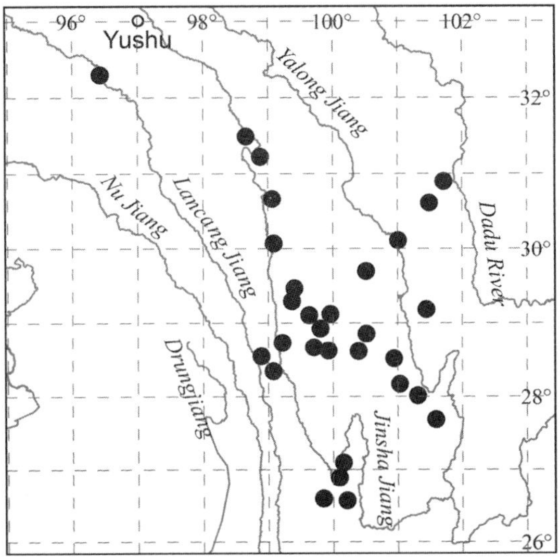

Map 47–100: **Codonopsis subglobosa** W. W. Smith

34. Codonopsis bragaensis Grey-Wilson, Plantsman, 12 (2): 99. fig. in p. 72. 1990; Hong, Monogr. *Codonopsis* Allied Gen. Camp. 156. 2015. Type: Nepal, Gandaki Zone, Manang Distr., "Marsyandi Valley, slopes of Annapurna 3 to the S. W. of Braga, c. 3,800 m," Jul. 1989, *C. Grey-Wilson et al.* s. n.(holotype: K).

Stems procumbent, twining above, ca. 1 m long, branched with leafy branches from lower half, unbranched or with few-leaved branches from upper half, densely pubescent. Leaves alternate, but those near base of stems opposite: petiole less than 2 mm long; leaf blade ovate to ovate-lanceolate, 1–2.9 cm long, 0.9–2.4 cm broad, pubescent on both surfaces, base rounded to slightly cordate, margin entire, apex obtuse to subobtuse. Flowers solitary, terminal on main stem, occasionally a lateral flower found in axil of uppermost leaf. Calyx tube adnate to whole hypanthium, pubescent; calyx lobes lanceolate-triangular, 0.9–1.2 cm long, margin entire but rather undulate. Corolla white flushed with mauve, with blue veins, campanulate, 1.6–2.4 cm broad, covered with white setae; corolla lobes somewhat spreading, broad-triangular, ca. 6 mm long. Stamens glabrous; filaments and anthers each ca. 5 mm long. Figure 47–48: b & c.

Phenology: flowering in July.

Habitats: low scrubs of *Berberis*, *Rosa sericea* and *Potentilla fruticosa*; ca. 4000 m.

Distibution: **endemic**; C Nepal (Map 47–101).

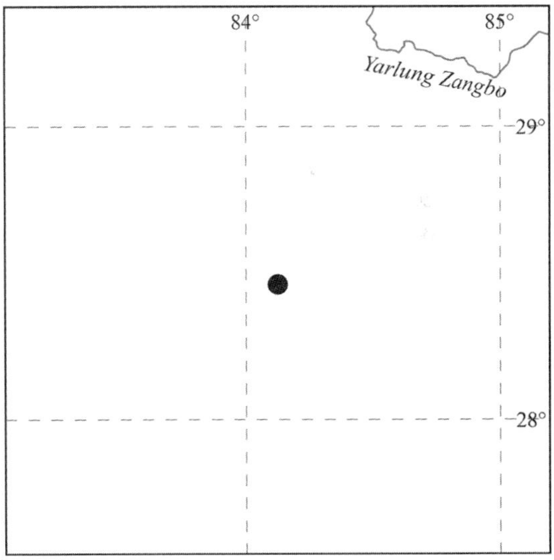

Map 47–101: **Codonopsis bragaensis** Grey-Wilson

Notes: *Codonopsis bragaensis* was locally abundant in the type locality, but it has not been found elsewhere so far. The species is distinct from all the others in having entirely pubescent calyx, lanceolate-triangular calyx lobes, and corolla which is densely covered with white setae and with characteristic coloration.

35. Codonopsis cordifolioidea P. C. Tsoong, Contr. Inst. Bot. Nat. Acad. Peiping 3(3): 93. fig. 7 & pl. 12. 1935; Hong, Lammers & Klein in Wu, Raven & Hong, Fl. China 19: 515. 2011; Hong, Monogr. *Codonopsis* Allied Gen. Camp. 157. 2015. Type: China, Yunnan, Shangpa County (Fugong), woods, 8 Oct. 1933. *H. T. Tsai 54814* [holotype: PE(00032074); isotypes: PE (00032071 & 00032075)].

Plants glabrous, or nearly glabrous, or only leaves sometimes very sparsely hispidulous. Stems up to 5 m long, twining, with a few very short branches. Leaves on main stem widely spaced, alternate, ca. 10 cm distant between two leaves; petiole 3–9 cm long; leaf blade broad-ovate, 4–10 cm long, 3–7 cm broad, adaxially green, abaxially gray-green, base nearly cordate, with a square sinus, margin nearly entire, apex short-acuminate or acute; leaves on branches usually only 2 and opposite, petiole less than 1 cm, similar to those on main stem in shape, but sinus not square. Flowers solitary, nearly opposite to leaves; pedicel 3–6 cm long. Calyx tube adnate to hypanthium up to middle, semiglobose; calyx lobes deltoid-lanceolate, 7–10 mm long, 5–6 mm wide, entire, apex acuminate; sinus between calyx lobes narrow. Corolla purple, dark blue or white but purple-red inside throat and limb, campanulate, 1.7–1.8 cm long, 1–1.2 cm wide, lobed to middle; corolla lobes lanceolate-deltoid. Filaments dilated at base. Capsule semiglobose in lower half, rostrate above, 1.3–1.8 cm in diameter. Seeds ellipsoid, brown. Figure 47–49: a & b.

Phenology: flowering and fruiting from July to October.

Habitats: forests; 1400–2200 m.

Distribution: **endemic**; S Hengduan and U Irrawaddy (Map 47–102).

S HENGDUAN: Fugong, s. loc., 1933, *H. T. Tsai 56554* (paratypes: PE); Lishadi, 1830 m, 2005, *Gaoligong Shan Biodiversity Survey 26972* (CAS). **U IRRAWADDY**: **Drungjiang**, 1750 m, 1982, *Qinghai-Xizang Exped. 9562* (KUN, PE); Drungjiang, 1600 m, 1990, *Drungjiang Invest. Team 1446* (CAS); Gameilai River, 1400 m, 1990, *Drungjiang Exped. 1446* (KUN).

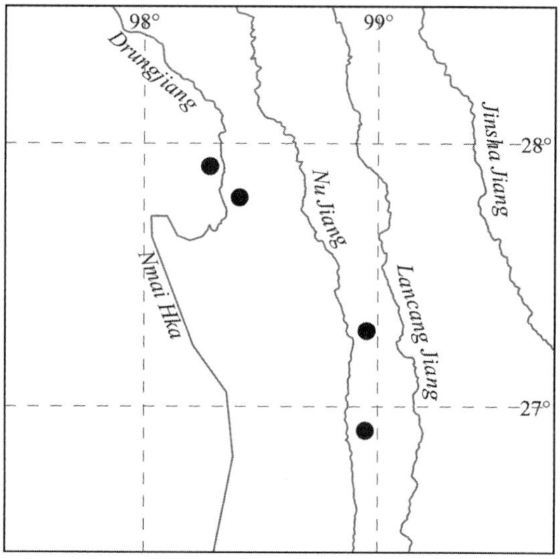

Map 47–102: **Codonopsis cordifolioidea** P. C. Tsoong

36. Codonopsis micrantha Chipp, J. Linn. Soc. Bot. 38: 382. 1908; Tsoong, Contr. Inst. Bot. Nat. Acad. Peiping 3(3): 97. fig. 10. 1935; Hong, Monogr. *Codonopsis* Allied Gen. Camp. 158. 2015. Type: China, Yunnan, Yunnanfu (Kunming), *P. Ducloux 513* (holotype: K).

Campanumoea violifolia H. Lév., Cat. Pl. Yunnan 24. 1915. Type: China, Yunnan, "haies - plaine de Tche-Hay", 2900 m, Aug. 1911, *E. E. Maire s. n.* (holotype: E ; isotype : E)

Melothria violifolia H. Lév., Cat. Pl. Yunnan, 65. 1916. Type: China, Yunnan,"haies de la plaine á Ta-Kiao", 2900 m, Aug. 1912. *E. E. Maire s. n.* (holotype: E).

Roots long-cylindrical, 20–30 cm long, 0.5–1 cm in diameter. Stems over 1 m long, twining, branched, sparsely villous or nearly glabrous. Leaves alternate, those on branches opposite; petiole 2–5 cm long; leaf blade cordate, 2–6 cm long, 2.4–4 cm broad, adaxially glabrous or sparsely white-hirsute, abaxially white-hirsute or nearly glabrous, base deeply cordate, basal broad and nearly square sinus, margin crenate or serrate, apex obtuse or acute. Flowers solitary, terminal or leaf-opposed; pedicel 1–3 cm long, glabrous. Calyx tube adnate to hypanthium up to near top, semiglobose; calyx lobes

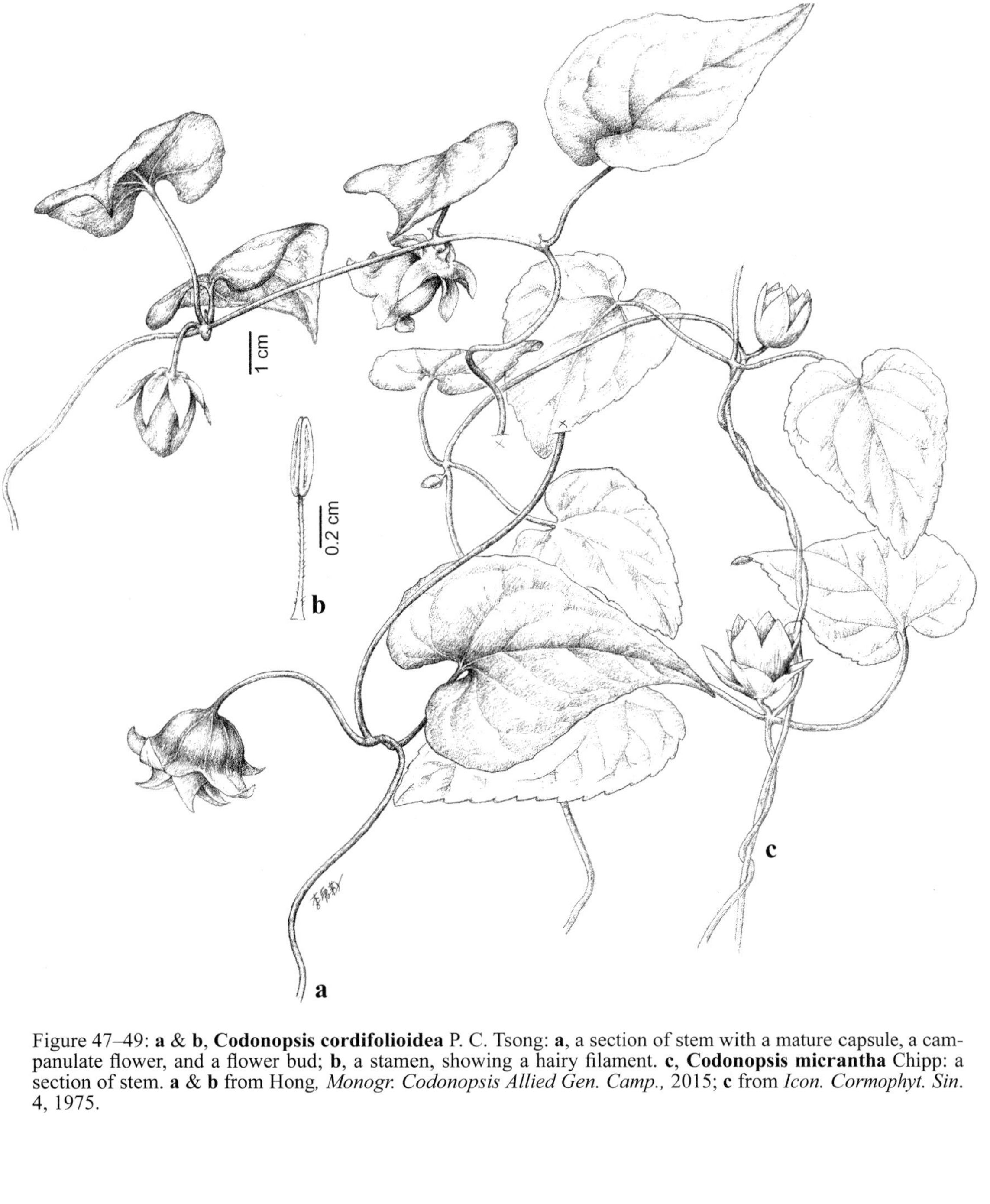

Figure 47–49: **a** & **b**, **Codonopsis cordifolioidea** P. C. Tsong: **a**, a section of stem with a mature capsule, a campanulate flower, and a flower bud; **b**, a stamen, showing a hairy filament. **c**, **Codonopsis micrantha** Chipp: a section of stem. **a** & **b** from Hong, *Monogr. Codonopsis Allied Gen. Camp.*, 2015; **c** from *Icon. Cormophyt. Sin.* 4, 1975.

triangular-lanceolate, 1–1.4 cm long, 0.4–0.6 cm broad, glabrous or sparsely ciliate, apex acute; sinus between calyx lobes narrow and pointed. Corolla white or yellowish white, campanulate, 7–9 mm long, divided to middle; corolla lobes triangular, glabrous or ciliate. Stamens glabrous; filaments slightly dilated at base. Capsule semi-globose below, conical and rostrate above, 8–10 mm in diameter, superior part very short, 2.5–3.5 mm long. Seeds ovoid, slightly compressed, short-tailed, ca. 1 mm long, 0.5 mm in diameter, brown-yellow. Figure 47–49: c.

Phenology: flowering and fruiting from July to October.

Habitats: scrubs or open forests; 1300–2900 m.

Distribution: S Hengduan (Map 47–103); also in C and E Yunnan and SW Guizhou of China.

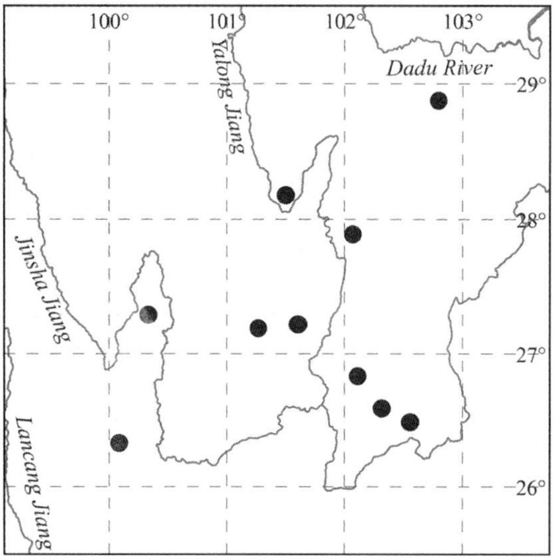

Map 47–103: **Codonopsis micrantha** Chipp

S HENGDUAN: Baidiao, 2300 m, 1959, *S. K. Wu (Sichuan Econ. Plant Surv.) 2700* (KUN); **Eryuan**: Niujie, Huoyan Shan, 1975, *B. Y. Qiu 61065* (KUN); **Ganluo**, s. loc., 1800 m, 1979, *s. coll. 867* (SM); **Huidong**, Matou Shan, 2200 m, 1961, *T. L. Dai & S. P. Tang 11843* (SM); **Huili**, Huili town, 1950 m, 1914, *Handel-Mazzetti 5231* (E, K, LE, W, WU); **Miyi**, Nanba, 2300 m, 1959, *S. K. Wu (Sichuan Econ.Plant Surv.) 454* (KUN); **Xichang**, Luoji Shan, 1978, *N. C. Guo 445* (SM); **Yanbian**, Hongbao, 2400 m, 2002, *X. F. Gao et al.3615* (CDBI); Qinghe, 1300 m, 2002, *X. F. Gao et al. 3939* (CDBI, PE); Yankou, 2220 m, 2002, *X. F. Gao et al. 3630* (CDBI); **Yulong** (Lijiang), Yulong Snow Range, Baishuihe, 2600 m, 1942, *K. M. Feng 9072* (KUN, PE).

Sect. 3. Leptocodon (Hook. f. & Thomson) D. Y. Hong, Monogr.*Codonopsis* Allied Gen. Camp. xx. 2015. ≡ *Codonopsis* subg. *Leptocodon* Hook. f. & Thomson, Ill. Himal. Pl. t. 16, A. 1855. ≡ *Leptocodon* (Hook. f. & Thomson) Lem., Ill. Hort. 3. Misc. 49. 1856. Type: *Codonopsis gracilis* Hook. f. & Thomson.

Herbaceous vines, strongly fetid. Leaves alternate, those on branches often opposite or nearly opposite. Flowers solitary, extra-axillary, less frequently axillary or leaf-opposed. Calyx tube entirely adnate to hypanthium, broad-obconical. Corolla long-tubular. Stamens alternate with 5 free, lamellar glands. Ovary semi-inferior, with superior part narrowly conical, over 3 times longer than inferior part, 3-locular; stigma 3-fid, segments clavate. Capsule apically 3-loculicidal. Seeds many.

Two species, in the Himalaya and adjacent regions; both in the Pan-Himalaya (1 endemic). They grow in thickets and forests at middle altitudes from 1100 to 3350 m.

37. Codonopsis gracilis Hook. f. & Thomson, Ill. Himal. Pl. t. 16, A. 1855; Grey-Wilson, Plantsman 12(2): 81. 1990; Hong, Monogr. *Codonopsis* Allied Gen. Camp. 168. 2015. ≡ *Campanumoea gracilis* (Hook. f. & Thomson) G. Nichols., Ill. Dict. Gard. 1: 259. 1884. ≡ *Leptocodon gracilis* (Hook. f. & Thomson) Lem., Ill. Hort. 3: Misc. 49. 1856. Type: India, Sikkim, Lachung, 1830–2130 m, 15 Aug. 1849, *J. D. Hooker* s. n. [lectotype designated by Hong, 2015: K (K000781242)]; Sikkim, Lachen, 1830–2440 m, May 1849, *J. D. Hooker* s. n. [paratypes: BM, K (K000781244)].

Plants glabrous throughout, rarely sparsely villous on young leaves and pedicel. Stems slender, over 1 m long, branched. Leaves alternate, rarely those on branchlets opposite, long-petiolate; leaf blade ovate-orbicular, 0.9–2.5 cm long, 0.9–2.5 cm broad, membranous, adaxially green, abaxially gray, base truncate or obtuse, margin sinuate-crenate, apex obtuse. Flowers solitary, pendent; pedicel slender, 1–5 cm long, straight or curved. Calyx glabrous; calyx tube 2–3 mm long; calyx lobes reflexed, narrowly oblong, 5–8 cm long, 1–2 mm broad, base attenuated, stalk-like, margin with 1 or 2 pairs of callose serratures. Corolla blue, violet or purple, tubular, 2.8–3.5 cm long, shallowly lobed; corolla tube ca. 4 mm wide; limb 1–1.5 cm wide; corolla lobes triangular, 4–5 mm long. Filaments ca. 2 cm long. Style ca. 2 cm long. Capsule semiglobose or obconic, inferior part 4–5 mm long; superior part conical or ovoid, 8–10 mm long. Seeds red-brown, linear-oblong, smooth, 1.5–1.8 mm long. Figure 47–50: a–c.

Phenology: flowering and fruiting from August to October.

Habitats: forests or thickets; 1100–2600 m.

Distribution: U Yarlung Zangbo, E Nepal, Sikkim & Darjiling, S Hengduan, and U Irrawaddy (Map 47–104); also in W Myanmar and other parts of Yunnan in China.

U YARLUNG ZANGBO: Dinggyê, Chengtang, Gamagou, 2426 m, 2013, *PE-Tibet Exped. 3217* (PE). **E NEPAL**: **Kosi**, Arun Valley, Hatiar, 1830 m, 1956, *J. D. A. Stainton 1437* (BM); Dhankuta, 1100 m, 1963, *H. Hara et al. 6300441* (BM, K); Sankuwasabha, 2020 m, 1998, *S. Noshino et al. 9830059* (E) ; Terhathum, Chittre, 2200 m, 1972, *J. F. Dobremez* (*DBR NEP*) *1474* (BM);

Figure 47–50: **a–c**, **Codonopsis gracilis** Hook. f. & Thomson: **a**, a branch with terminal flower; **b**, a calyx lobe; **c**, a flower with calyx removed and corolla opened, showing five lamellar glands alternate with stamens, and pistil. **d**, **Codonopsis hongii** Lammers: a section of stem with a flower, showing a calyx lobe and a part of leaf blade showing hirsute abaxial surface. **a** from Tsoong in *Contr. Inst. Bot. Nat. Acad. Peiping* 3(3), 1935; **b** & **c** from *Icon. Cormophyt. Sin.* 4, 1975; **d** from Hong in *Acta Phytotax. Sin.* 18: fig. 2: 1 & 2, 1980.

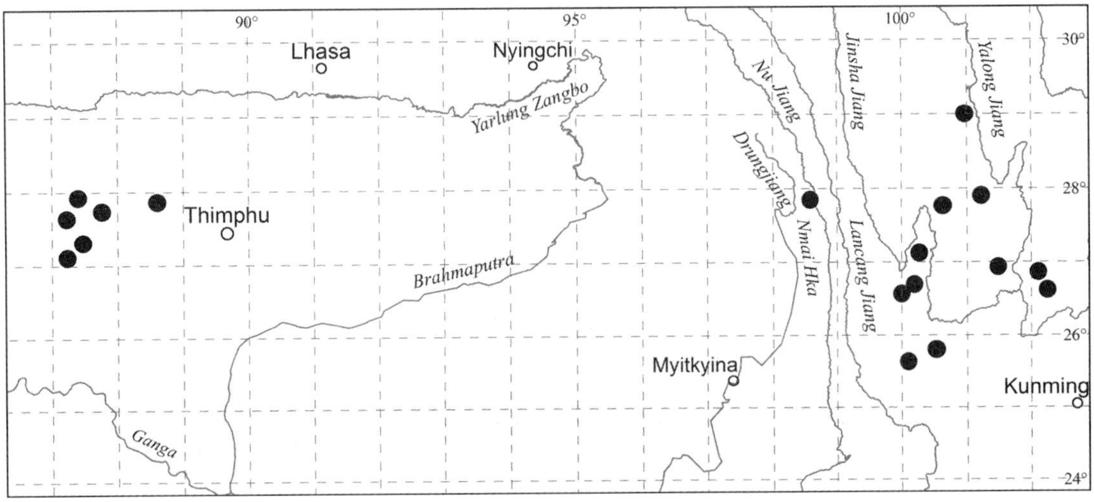

Map 47–104: **Codonopsis gracilis** Hook. f. & Thomson

Mechi, Taplejung, Ghunsa Khola, 2650 m, 1990, *KEKE 273* (E, K). **SIKKIM & DARJILING: Sikkim**, Choongthung, 1903, *Prain* s. n.(E); North Distr. 1930 m, 1966, *Edinburgh Exped. to N Sikkim (1966) 249* (E). **S HENGDUAN: Binchuan**, Jizhu Shan, 2000 m, 1963, *Jinsha-jiang Team 63–6660* (KUN); Pianjing to Shazi-jie, 1946, *T. N. Liou 21716* (PE); **Chabulang**, Waerzhai, 2700 m, 1983, *Qinghai-Xizang Exped. 13642* (PE); **Dali**, Dali to Tengchong, 2440–2740 m, 1903, *G. Forrest 987* (E); W of Dali, Mekong watershed, on the way to Youngchang and Tengchong, 1230–2440 m, 1922, *J. F. Rock 6835* (E); **Heqing**, Lianping, 2600 m, 1940, *R. C. Ching 23755* (KUN, PE); **Huili**, s. loc., 2400 m, 1958, *Z. He et al. 11532* (SZ); **Miyi**, Pingshan, Longganzi, 2300 m, 1978, *s. coll. 393* (SM); **Muli**, Milikigon, 2600 m, 1937, *T. T. Yü 14165* (A, KUN, PE); Taoba, 2200 m, 1978, *C. S. Cao et al. 8195* (SZ);**Yanbian**, s. loc., 2500 m, 1932, *T. T. Yü 1659* (A, PE); **Yulong (Lijiang)**, Jiazi, Baishuihe, 1900 m, 1955, *K. M. Feng 21540* (KUN, PE); W of Lijiang, near Ngaza, 1923, *J. F. Rock 10579* (A). **U IRRAWADDY**: **Drungjiang**, Meiliwang, 1800 m, 1991, *Drungjiang Exped. 1642* (KUN).

38. Codonopsis hongii Lammers, Novon 11: 67. 2001; Hong, Monogr. *Codonopsis* Allied Gen. Camp. 170. 2015. ≡ *Leptocodon hirsutus* D. Y. Hong, Acta Phytotax. Sin. 18: 246. fig. 2: 1–2, 1980. ≡ *Codonopsis hirsuta* (D. Y. Hong) K. E. Morris & Lammers, Novon 9: 387. 1999, non (Hand.-Mazz.) D. Y. Hong & L. M. Ma (1992). Type: China, Tibet, Zayü, Shangzayü, 2500 m, 9 Aug. 1973, *Qinghai-Xizang Exped. 73–1014* (holotype: PE).

Roots slender, not thickened. Stems up to 5 m long, hirsute on young parts. Leaves on main stem alternate, those on branchlets almost opposite, all leaves long-petiolate; leaf blade broad-ovate or nearly orbicular, 1.5–3 cm long, 1.5–2.5 cm broad, abaxially villous, rarely extremely sparsely villous, base truncate or rounded, margin gross-dentate, apex rounded. Flowers solitary, nearly leaf-opposed; pedicel slender, over 3 cm long, glabrous. Calyx tube hemispheric, 10-ribbed; calyx lobes ovate, overlapping each other at anthesis, separated at fruiting, upright, 5–7 cm long, 3–4 mm broad, hirsute abaxially and on margin, base cordate or nearly truncate, margin sinuate-crenate. Inferior part of capsule semiglobose, 8–10 mm in diam.; superior part long-conical, ca. 10 mm long. Seeds ellipsoid, ca. 1 mm long. Figure 47–50: d.

Phenology: flowering from July to September; fruiting from September to Nobember.

Habitats: mixed forests, thickets, or grasslands, on granites; 1660–3350 m.

Distribution: **endemic**; Bhutan, L Yarlung Zangbo, Yarlung Zangbo-Brahmaputra, S Hengduan, and U Irrawaddy (Map 47–105).

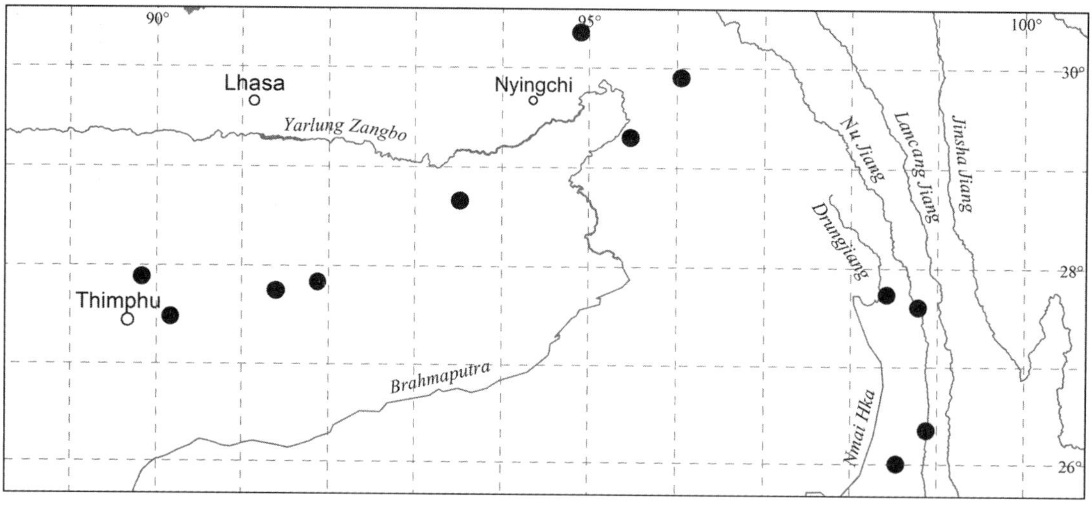

Map 47–105: **Codonopsis hongii** Lammers

BHUTAN: **Chendebi**, Deuchung, Khoma Chu, 2290 m, 1949, *F. Ludlow et al. 21479* (BM, E); Kancham Runakla, 1860 m, 1914, *R. E. Cooper 3264* (BM, E); **Tobrang**, between Trashiling and Chendebi, 2290 m, 1949, *F. Ludlow et al. 17059* (BM); Tashiling, Jangsa, 2590 m, 1914, *R. E. Cooper 2059* (E) ; Trashiyangsi Valley, 2440 m, 1934, *F. Ludlow & G. Sherriff 905* (BM, E); **Upper Mo Chu**, Goen Gaza - Gaza Zam, 2450 m, 1984, I. *W. J. Sinclair & D. G. Long 4980* (E, K). **L YARLUNG ZANGBO: Qomdo**, 2800 m, 2011, *D. Y. Hong et al. H11019* (PE); **Tangmai**, 2150 m, 1980, *C. C. Ni et al. 1649* (PE); **Yi'ong**, 2400 m, 11. July 1965, *J. S. Ying & D. Y. Hong 65–1272* (paratype: PE). **YARLUNG ZANGBO-BRAHMAPOTRA**: **Cona**, **Mama-Tea Farm**, 2532 m, 2011, *D. Y. Hong et al. H11009* (PE); **Lhunzê**, Zara-Maja, 2780 m, 2013, *Y. S. Chen et al. 13–0678* (PE); **Mêdog**, 52K to 80K, 3000m, 2012, *FLPH Tibet Exped. 12–1898* (PE). **S HENGDUAN**: **Gongshan**, Lazao, Salween Valley, 1940, *K. M. Feng 8673* (PE). **U IRRAWADDY**: Western flank of the **Chimi-li**, N' Maikha- Salween Divide, 3050–3350 m, 1924, *G. Forrest 24881* (E, K); **Drungjiang,** 1660 m, 2004, *Gaoligong Shan Biodiversity Survey 21262* (CAS, E); **Hpimaw Hill**, 1980 m, 1919, *RA 1289* (E); **Ngawchang Valley** above Kang Fang, 1220–1530 m, 1914, *F. Kingdon-Ward 1887* (E); Kang Fang, 1530 m, 1924, *F. Kingdon-Ward 3683* (E).

Sect. 4. Campanumoea (Bl.) D. Y. Hong, Monogr. *Codonopsis* Allied Gen. Camp. 172. 2015. ≡ *Campanumoea* Bl., Bijdr. Fl. Nedia India 13: 726. 1826. Type: *Codonopsis javanica* (Bl.) Hook. f. (*Campanumoea javanica* Bl.).

Roots carrot-shaped. Stems twining. Leaves alternate or opposite, leaf blade deeply cordate. Calyx epigynous or hypogynous; corolla always epigynous. Fruit a berry. Plants not fetid.

Two species, widely distributed in E and SE Asia from E Nepal to Indonesia, Philippines and Japan, both in our region (1 endemic). They grow in thickets, forests, and forest margins (thickets or herbs) at altitudes from 250 to 3200 m.

39. Codonopsis javanica (Bl.) Hook. f., Ill. Himal. Pl. 116. tab. 16B. 1855; Moeliono & Tuyn in van Steenis, Fl. Males. Ser. 1. 6: 119. 1960; Hong, Monogr.*Codonopsis* Allied Gen. Camp. 172. 2015. ≡ *Campanumoea javanica* Bl., Bijdr. 727. 1826. ≡ *Campanula javanica* (Bl.) D. Dietr., Syn. Pl. 1: 758. 1839. Type: Java Occid., Mons Megamondong, *Blume* s. n. (holotype: P).

Roots thickened, carrot-shaped. Stems twining, multi-branched, glabrous, up to 3 m long. Leaves opposite, rarely alternate, long-petiolate; leaf blade cordate or cordate-ovate, 2.6–8 cm long, 2–7.5 cm broad, glabrous or sometimes abaxially sparsely villous, margin dentate or crenate, or rarely entire. Flowers solitary, axillary, glabrous throughout. Calyx free from hypanthium, cleft nearly to base; calyx lobes lanceolate or ovate, rarely narrowly triangular, 0.8–2 cm long. Corolla white or yellow-green, purple or reddish inside, epigynous, campanulate, cleft to middle. Stamens 5. Stigma 4- or 5-fid; ovary 5-locular. Berry purple, or greenish white suffused with red, globose, 0.7–2.6 cm in diameter. Seeds irregular in shape, surface reticulate.

Distibution: Sikkim & Darjiling, Bhutan, Yarlung Zangbo-Brahmaputra, S Hengduan, and U Irrawaddy (Map 47–106); also in Indonesia, Japan, Laos, Thailand, Vietnam, and S China, E India, and S Myanmar.

1a. Corolla 1.5–3.2 cm long; berry violet, 1.5–2.6 cm in diameter ...... 39a. subsp. *javanica*
1b. Corolla 0.6–1.7 cm long; berry greenish white suffused with red, 0.7–1.5 cm in diameter .... 39b. subsp. *japonica*

39a. Codonopsis javanica subsp. **javanica**

Leaf blade 3.5–8 cm long, 2.2–7.5 cm broad. Calyx lobes lanceolate, 1–2 cm long. Corolla 1.5–3.2 cm long. Berry violet, 1.5–2.6 cm in diameter. Figure 47–51: a & b.

Phenology : flowering and fruiting from May to November.

Habitats: forests, thickets, or forest margins ; 270–3200 m.

Distribution: Sikkim & Darjiling, Bhutan, Yarlung Zangbo-Brahmaputra, S Hengduan, and U Irrawaddy; also widely distributed in other parts of SE Asia: S China, Meghalaya and Nagaland of India, Indonesia, Laos, S Myanmar, Thailand, and Vietnam.

SIKKIM & DARJILING: Darjiling, Rishap, 1070 m, 1870, *C. B. Clarke 12388 A* (K); Runguet, 1220 m, 1881, *J. S. Gamble 9766* (K); **Sikkim**, Lechong, 910 m, 1874, *C. B. Clarke 684* (K). **BHUTAN**: **Chungkar**, 2130 m, 1938, *F. Ludlow et al. 6782* (BM); **Punakha**, Gichha, 2440 m, 1914, *R. E. Cooper 2705* (E); **Shali**, Trashiyangsi Valley, 1980 m, 1934, *F. Ludlow & G. Sherriff 1053* (BM, E). **YARLUNG ZANGBO- BRAHMAPUTRA: Ukhrul**, 1520–1830 m, 1948, *F. Kingdon-Ward 17895* (BM, PE). **S HENGDUAN: Dali**, s. loc., 1946, *T. N. Liou 14230* (KUN); **Fugong**, Lumadeng, 1780 m, 2005, *Gaoligong Shan Biodiversity Survey 26999* (A, CAS); Pihe to Gongshan, 1980, *J. S. Yang* s. n.(KUN); **Gongshan**, Cikai, 1500 m, 2002, *Gaoligong Shan Biodiversity Survey 15876* (CAS); **Longyang**, summit of Gaoligong Shan, 2050 m, 2000, *H. Li et al. 13114* (CAS); Lisushan to Huangcaoping, 1941, *H. C. Wang 792* (PE); Mangkuan, 1350 m, 1998, *H. Li et al. 10616* (CAS); **Lushui**, Yaojiaping to Jiangqiao, 2000 m, 1964, *S. K. Wu 8506* (KUN); **Weixi**, by Lancang Jiang, 1940, *K. M. Feng 8923* (KUN, PE); **Yangbi**, Pingpo, 1946, *T. N. Liou 22521* (PE); **Yulong** (Lijiang), Yulong Snow Range, 2600 m, 1942, *K. M. Feng 9072* (PE); **Yunlong**, Caojian, 2300 m, 1993, *Q. E. Yang 9305* (PE); Caojian, 2406 m, 2006, *A. H. Li et al. SCSB-A-000172* (KUN). **U IRRAWADDY**: **Chaungmaw Hka**, 910–1220 m, 1911, *F. Kingdon-Ward 3491* (E); **Hpimaw,** Ngawchang Valley 1530 m, 1923, *F. Kingdon-Ward 3667* (E); Ngawchang Valley, 1220 m, 1919, *R. F. 1273* (E); Nyitadi, 2130 m, 1920, *R. F. 1838* (E); **N'Maikha-**

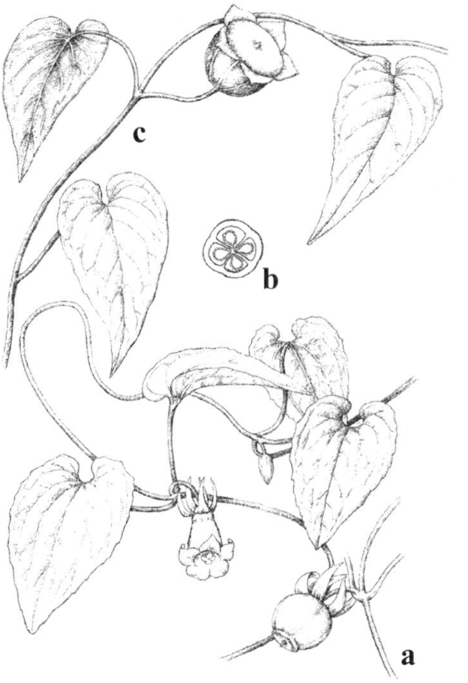

Figure 47–51: **a** & **b**, **Codonopsis javanica** (Bl.) Hook. f. subsp. **javanica**: **a**, a section of twining stem with a berry and a flower; **b**, cross section of 4-locular ovary. **c**, **Codonopsis inflata** Hook. f. & Thomson: a section of stem with a berry opposite leaf. All from *Fl. China* 19, 2011.

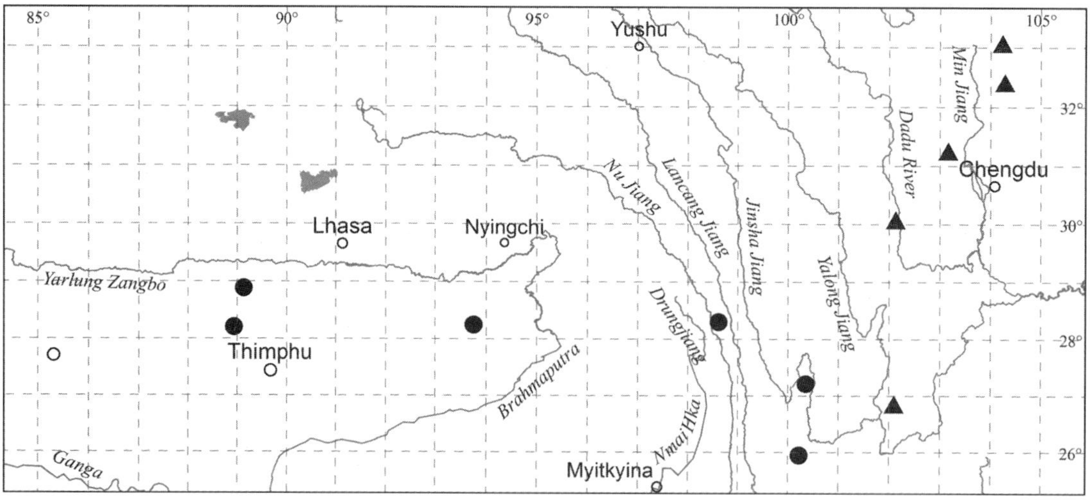

Map 47–106: **Codonopsis javanica** (Bl.) Hook. f.

● subsp. **javanica**, ▲ subsp. **japonica**

Salween Divide, Nwar-Hka Valley, 1220–1830 m, 1914, *F. Kingdon-Ward 1927* (E); **Satawgaw**, s. loc., 1924–1925, *G. Forrest 24969* (E, K, PE) & *25394* (E, K, PE); **Tengchong**, Guangming, 2000 m, 1980, *S. T. Li 80–378* (KUN); Jietou, N of Datang, 1950 m, 1998, *H. Li et al. 11037* (CAS); Mangkuan, 1525 m, forests, 2003, *Gaoligong Shan Biodiversity Survey 19074* (CAS); Qushi, 1930–2120 m, 1960, *W. Q. Yin 60–1034* (KUN); Sanyuan, 1560 m, 1998, *H. Li et al. 11431* (CAS, E); Wuhe, 1900 m, 2003, *Gaoligong Shan Biodiversity Survey 17960* (CAS).

39b. Codonopsis javanica subsp. **japonica** (Makino) Lammers, Bot. Bull. Acad. Sin. (Taipei) 33(3): 285. 1992. ≡ *Campanumoea javanica* Bl. var. *japonica* Makino, Bot. Mag. (Tokyo) 22: 155. 1908. ≡ *Campanumoea javanica* subsp. *japonica* (Makino) D. Y. Hong, Fl. Reipubl. Popularis Sin. 73(2): 71. 1983. ≡ *Campanumoea japonica* Maxim., Bull. Acad. Imp. Sci. Saint-Pétersbourg 12: 67. 1867, non Sieb. ex Morr., Belgique Hort. 13: 337. 1863. ≡ *Campanumoea maximowiczii* Honda, Bot. Mag. (Tokyo) 50: 389. 1936. Type: Japan, Nagasaki, 1863, *Maximowicz* s. n.(holotype: K).

Leaf blade 2.6–8 cm long, 2.2–6.5 cm broad. Calyx lobes ovate or lanceolate, or rarely narrowly triangular, 0.8–1.2(–1.7) cm long. Corolla 0.6–1.3 (–1.7) cm long. Berry greenish white suffused with red, 0.7–1.5 cm in diameter.

Phenology: flowering in August and September; fruiting from September to November.

Habitats: thickets or forest margins; 250–2000 m.

Distribution: N and S Hengduan; also widely distributed in E and C China (N Guangdong, N Guangxi, NE Yunnan, Taiwan, and northwards to S Anhui, S Zhejiang, Hubei, and other parts of Sichuan), and Japan (northwards to Honshu).

N HENGDUAN: Lixian, Laisugou, 3100 m, *D. P. He 46460* (SZ) ; **Pingwu**, Yaodu, 650 m, 1937, *K. T. Fu 2359* (PE). **S HENGDUAN: Miyi,** s. loc., 1958, *Z. He et al. 11285* (SZ); **Tiangquan**, Jiaojiping, 1175 m, 1959, *Sichuan Econ. Plant Surv. (Ya) 521* (PE); Zhongzhen, 1951, *W. K. Hu & Z. He 11799* (PE).

40. Codonopsis inflata Hook. f. & Thomson, Ill. Himal. Pl. t. 16, C. 1855; Hong, Monogr. *Codonopsis* Allied Gen. Camp. 180. 2015. ≡ *Campanumoea inflata* (Hook. f. & Thomson) C. B. Clarke in Hooker, Fl. Brit. India 3: 436. 1881. Type: India, Sikkim, 1520–2130 m, *J. D. Hooker* s. n. [lectotype designated by Hong, 2015: K (K000814187); isolectotype: CAL]; Sikkim, Lachung, *J. D. Hooker* s. n. [paratype: K (K000814190)].

Roots slightly thickened, branched. Stems twining, almost glabrous. Leaves alternate; petiole 3–6 cm long; leaf blade ovate to ovate-lanceolate, 4–10 cm long, 2–5 cm broad, adaxially glabrous or sparsely puberulent, abaxially glaucous, puberulent, base deeply cordate, margin entire or sinuate, apex acuminate. Flowers solitary, almost leaf-opposed, or in terminal and few-flowered cyme; pedicel 2–3 cm long. Calyx adnate to hypanthium up to top; calyx tube obdeltoid-conical, puberulent; calyx lobes inserted at same level as corolla, ovate-triangular, ca. 1 cm long, margin often revolute. Corolla pale yellow or pale green, often purple at margin and along veins, urceolate, 2.5–3 cm long, cleft to 1/3. Stamens 5; filaments glabrous. Style glabrous; stigma 3-fid; ovary 3-locular. Berry yellow-green, globose, 1.2–1.5 cm in diameter. Seeds ovoid-globose, surface reticulate. Figure 47–51: c.

Phenology: flowering and fruiting from August to November

Habitats: evergreen broad-leaved forests or herb communities at forest margins; 900–2750 m.

Distibution: **endemic**; E Nepal, Sikkim & Darjiling, Bhutan, and Yarlung Zangbo-Brahmaputra (Map 47–107).

E NEPAL: **Mechi**, Mai Pokhari, 1963, *H. Hara et al. 6300416* (BM, TI). **SIKKIM & DARJILING: Darjiling,** Jospokri, 2440 m, 1912, *G. H. Cave* s. n.(E); Kurseong, 1370 m, 1845, *J. S. Gamble 538A* (K); Labdah, 1530 m, 1918, *G. H. Cave* s. n.(E); Rungbee, 1875, *C. B. Clarke 24788B* (K); **Sikkim,** Chateng, 2440 m, 1917, *G. H. Cave* s. n.(E); Gangtok, Penlong La, 2100 m, 1964, *H. Hara & K. C. Pradhan* s. n.(TI). **BHUTAN: Dumsong**, 1830 m, 1875, *C.*

B. Clarke 26420C (K). **YARLUNG ZANGBO-BRAHMAPUTRA: Baibung**, 1500 m, 1992, *H. Sun et al.1117* (KUN); **Dagmo**, 1900 m, 1993, *H. Sun et al. 4267* (KUN); Dagmo to Gutang, 1700 m, 1974, *Qinghai-Xizang Exped. 74–4988A* (KUN, PE); **De'ergong**, 1600 m, 1974, *Qinghai-Xizang Exped. 74–4445* (KUN, PE); **Gelin**, 2000 m, 1992, *H. Sun et al. 2668* (KUN); **Hanmi**, 1100 m, 1992. *H. Sun et al. 0489* (KUN); **Luhit Valley**, 1220 m, 1922, *F. Kingdon-Ward 7695* (K); Therolian, Luhit Valley, 1220 m, 1950, *F. Kingdon-Ward 20283* (BM); **Mêdog** Distr., 1974, *Qinghai-Xizang Exped. 1588* (KUN, PE); **Mêdog** Town, 1256 m, 2012, *D. Y. Hong et al. H12001* **(PE)**; Xianaba to Dagmo, 1700–1800 m, 1982, *B. S. Li & S. Z. Cheng 01358* (PE); **Xiran**, 900 m, 1992, *H. Sun et al. 1517* (KUN); **Zangbo Gorge** (Gampe-la), 1520–1830 m, 1924, *F. Kingdon-Ward 6371* (K).

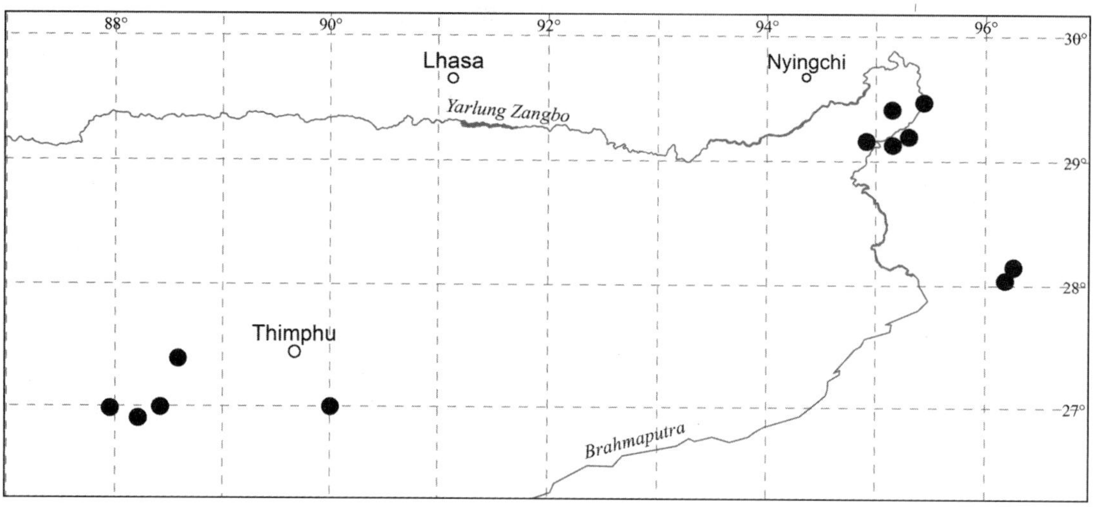

Map 47–107: **Codonopsis inflata** Hook. f. & Thomson

5. HIMALACODON D. Y. Hong & Q. Wang

Hong & Wang in Wang et al., J. Syst. Evol. 52(5): 548. 2014; Hong, Monogr. *Codonopsis* Allied Gen. Camp. 181. 2015. ≡ *Codonopsis* subg. *Obconicicapsula* D. Y. Hong, Acta Phytotax. Sin. 18(2): 245. 1980. Type: *Himalacodon dicentrifolius* (C. B. Clarke) D. Y. Hong & Q. Wang (*Wahlenbergia dicentrifolia* C. B. Clarke).

Roots carrot-shaped, branched. Stems erect, multi-branched, each branch bearing a terminal flower. Calyx tube adnate to hypanthium up to top, ovary almost entirely inferior, obconical; calyx lobes and corolla inserted at same level. Corolla campanulate, shallowly 5-lobed. Disc or glands lacking. Stamens entirely glabrous; filaments slightly dilated at base. Ovary 3-locular, ovules many; style clavate, densely glandular-hairy; stigma 3-fid, stigma lobes nearly triangular, broadest at base. Fruit a loculicidal capsule, obconical. Seeds many, narrowly ovoid, compressed. Plants not fetid. Chromosomes 2n=14.

Recent references: Hong & Pan, 2012, Pollen morphology of the platycodonoid group (Campanulaceae s. str.) and its systematic implications. *J. Integr. Plant Biol.* 54: 773–789; Wang et al., 2013, Molecular phylogeny of the platycodonoid group (Campanulaceae s. str.) with special reference to the circumscription of *Codonopsis*. *Taxon* 62: 498–504; Wang et al., 2014, Phylogenetic analyses reveal three new genera. *J. Syst. Evol.* 52(5): 541–550.

Himalacodon is the most closely related to *Pankycodon* D. Y. Hong & X. T. Ma. They are paraphyletic in our molecular trees (Wang et al., 2013 & 2014). However, they can not be clumped into a single genus not only phylogenetically, but also morphologically. In *Himalacodon* leaves on main stem are alternate, seeds narrowly ovoid, slightly compressed, wingless, and hypanthium is as high as ovary, whereas in *Pankycodon* leaves are all opposite, seeds lunate and strongly compressed with a wide wing and hypanthium is higher than ovary by 1–2 mm. Wang et al. (2014) provides more information about the genus and its allies.

Monospecific, **endemic**; U Yarlung Zangbo, Nepal, and Sikkim & Darjiling.

1. Himalacodon dicentrifolius (C. B. Clarke) D. Y. Hong & Q. Wang in Wang et al., J. Syst. Evol. 52 (5): 549. 2014; Hong, Monogr. *Codonopsis* Allied Gen. Camp. 183. 2015. ≡ *Wahlenbergia dicentrifolia* C. B. Clarke in Hooker, Fl. Brit. India 3: 430. 1881. ≡ *Campanopsis dicentrifolia* (C. B. Clarke) Kuntze, Revis. Gen. Pl. 2: 379. 1891. ≡ *Codonopsis dicentrifolia* (C. B. Clarke) W. W. Smith, Rec. Bot. Surv. India 4: 388. 1913. Type: India, Sikkim, Phulloot, 3350 m, 27 Oct. 1875, *C. B. Clarke 25722 A* (holotype: K).

Plants totally glabrous. Roots carrot-shaped, more or less branched, up to 40 cm long, 3 cm long. Stems erect or pendulous from rocks, 30–60 cm tall; branched at upper part. Leaves mostly alternate, those on branches nearly opposite; petiole 2–6 mm long; leaf blade ovate or ovate-oblong, 1.1–4.5 cm long, 0.7–2.2 cm broad, abaxially purple or red, base cuneate or rounded, margin entire or sinuous, sometimes slightly revolute, apex acute or obtuse. Flowers solitary, terminal on main stem and branches; pedicel 0.5–3 cm long. Calyx tube adnate to hypanthium nearly to top, obconical, 5–7 mm long, 7–10 mm wide; calyx lobes nearly linear, 7–10 cm long, ca. 1 mm broad, margin entire, apex acuminate; sinus between calyx lobes broad and obtuse. Corolla blue, blue-purple or pale blue, campanulate, 2.5–3.5 cm long, 5-lobed to ca. 1/3; corolla lobes broad-ovate, 0.8–1.3 cm long, 1–1.5 cm broad, apex acute. Filaments slightly dilated below, slender above, 3–4 mm long. Capsule shortly conical in superior part, obconical or obdeltoid-conical in inferior part, obscurely 10-ribbed, 1.2–1.5 cm long, 1–1.2 cm wide. Seeds ovoid, compressed, wingless. **Chromosomes**: 2n=14. Figure 47–52.

Phenology: flowering and fruiting from August to October.

Habitats: rocks by broad-leaved forests or in thickets: 2700–4100 (–5200) m.

Distribution: **endemic**; U Yarlung Zangbo, W, C and E Nepal, and Sikkim & Darjiling (Map 47–108).

U YARLUNG ZANGBO: **Gyirong**, Gyirong Town, near Todang Fall, 3100–3300 m, 1990, *B. S. Li et al. 13685* (PE); **Nyalam**, Quxam, 3200 m, 2011, *D. Y. Hong et al. H11004* (PE); near Quxam, 3300 m, 1972, *Xizang Chinese Mater. Medic. & Herbs Exped. 1598* (HNWP, PE); **Tingri**, Rongxar, 2700 m, 1959, *s. coll. 749* (PE). **W NEPAL**: **Seti**, Bajura / Bajhang, 2190–3350 m, 1991, *M. Suzuki et al. 9170642* (E); Barpak, 4570–5200 m, 1928, *L. Dheuoj 52* (BM); Pongsing, 1830–4220 m, 1929, *L. Dhevoj 93* (BM); **Tsumje**, 3500 m, 1953, *S. Nakao* s. n. (TI). **C NEPAL**: **Bagmati**, S of Gurjakhani, 3650 m, 1954, *Stainton, Sykes & Williams 3864* (BM, E, G, P, PE); Rasuwa, 3500 m, 1994, *F. Miyamoto et al. 9420197* (KUN); Rasuwa, Chilime Kharka-Chilimagean, ca. 3350 m, 1949, *O. Polunin 1454* (BM, E); **Dhawalagiri**, Myagdi - Baglung, Jalja La, 3250 m, 1996, *M. Mikage et al. 9684221* (BM) ; **Gandaki**, Manang, around Bimtang, 3500 m, 1994, *M. Suzuki et al. 9470377* (E, KATH); **Janakpur**, Dolakha, S of Rupte, 3200 m, 1974, *J. H. de Haas 2556* (BM). **E NEPAL**: **Kosi**, Sankhuwasabha, Milke Danda, Gidde, 3450 m, 1991, *H. Ohba et al. 9120118* (BM, TI); Terhathum, Arun Valley, Maghang Khola, E of Num, 3350 m, *J. D. A. Stainton 1725* (BM, E); Bhe-

Figure 47–52: **Himalacodon dicentrifolius** (C. B. Clarke) D. Y. Hong & Q. Wang: **a**, root and basal part of two stems; **b**, upper part of stem; **c**, a flower with corolla opened, showing calyx, stamens, style, and stigma; **d**, cross section of ovary showing three locules. From Wang et al., *J. Syst. Evol.* 52(5), 2014, with permission.

ding, 3650–3960 m, 1930, *L. Dhevoj 0335* (BM, E); Chauki, 3000 m, 1972, *J. F. Dobremez 1546* (BM, G); **Mechi**, Taplejung, Hile Kharka (3530 m) – Mane Bhanjyang (3490 m), 1999, *Y. Omori et al. 9950022* (E); **Sagarmatha**, Solukhumbu, Tashing Dingma (3300 m) – Thuli Kharka (4100 m), 1997, *M. Wakabayashi et al. 9715260* (E). **SIKKIM & DARJILING**: **Sikkim**, Gantong, 3650–3960 m, 1913, *R. E. Cooper 748* (E); Saudekbu, 1908, *Ribu 371* (E).

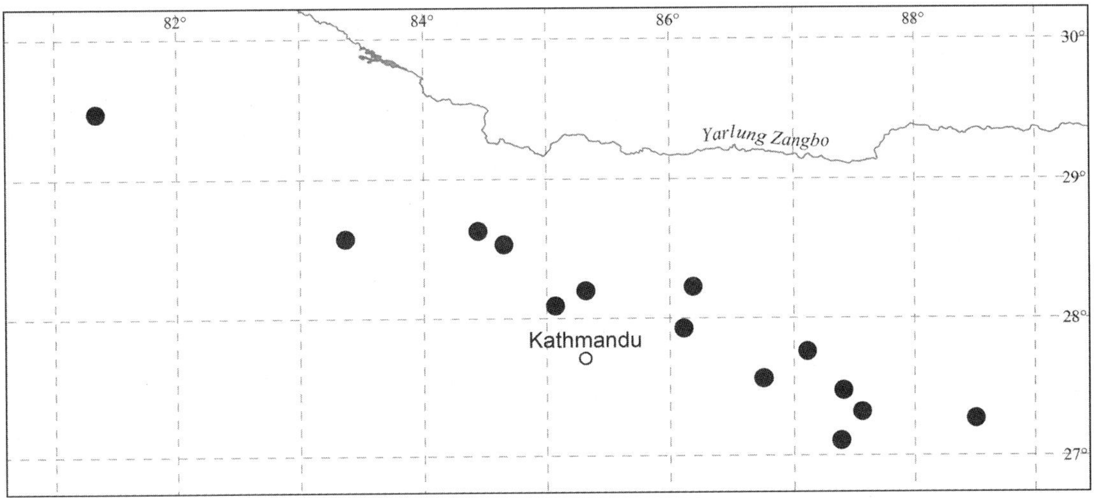

Map 47–108: **Himalacodon dicentrifolius** (C. B. Clarke) D. Y. Hong & Q. Wang

6. **PANKYCODON** D. Y. Hong & X. T. Ma

Hong & Ma in Wang et al., J. Syst. Evol. 52(5) : 549. 2014; Hong, Monogr. *Codonopsis* Allied Gen. Camp. 184. 2015. Type: *Pankycodon purpureus* (Wall.) D. Y. Hong & X. T. Ma (*Codonopsis purpurea* Wall.)

Roots attenuate. Stems pendulous from rocks or trees. Leaves opposite. Flowers solitary, terminal on main stem and branches. Hypanthium free from ovary at apex. Calyx tube adnate to whole hypanthium, i. e. up to base of corolla, (4- or) 5-lobed. Corolla campanulate, (4- or) 5-lobed to near middle. Stamens (4 or) 5, glabrous; filaments slightly dilated at base. Ovary nearly totally inferior to both calyx and corolla, 3-locular; ovules many; style clavate, densely glandular-hairy above; stigma 3-fid, lobes deltoid-lanceolate at early anthesis, later lanceolate. Fruit a capsule, 3-loculicidal. Seeds many, lunate, strongly compressed, widely winged. Plants not fetid. Chromosomes 2n=14.

Recent reference: Wang et al. 2014, Phylogenetic analyses three reveal new genera. *J. Syst. Evol.* 52(5): 541–550.

Monospecific, in the Pan-Himalaya and adjacent regions (SW Yunnan of China and Meghalaya of India).

1. Pankycodon purpureus (Wall.) D. Y. Hong & X. T. Ma in Wang et al., J. Syst. Evol. 52(5): 549. 2014; Hong, Monogr. *Codonopsis* Allied Gen. Camp. 186. 2015. ≡ *Codonopsis purpurea* Wall. in Roxburgh, Fl. India 2: 105. 1824. ≡ *Campanula purpurea* (Wall.) Spreng., Cura Post. 78. 1825. ≡*Wahlenbergia purpurea* (Wall.) A. DC. in de Candolle, Prodr. 7: 425. 1839. Type: Nepal, Gossain Than, *Wallich 1299* (lectotype designated by Hong, 2014: K; isolectotypes: BM, E, G, LD, P, W).

Plants glabrous throughout. Roots attenuate, yellow-white, smooth. Stems several from one caudex, few-branched, 30–80 cm long. Leaves opposite or sometimes alternate towards apex; petiole 1–10 mm long; leaf blade ovate or lanceolate, 2–12 cm long, 1.5–5 cm broad, base rounded or cuneate, margin nearly entire or sinuous, apex acute or obtuse. Flowers terminal or leaf-opposed at top. Calyx tube adnate to whole hypanthium, semiglobose-obconical; calyx lobes (4 or) 5, triangular-lanceolate, 1–2 cm long, 0.6–1 cm broad, margin entire,apex acute. Corolla dark red-purple, campanulate or funnelform-campanulate, 2–3.5 cm long, (4- or) 5-lobed to middle; corolla lobes triangular. Stamens glabrous; filaments ca. 6 mm, slightly dilated at base. Capsule obovoid-conical below, conical above, 2–3 cm long, 1.5–2 cm wide; valves 5–8 mm long. Seeds lunate, strongly compressed, wide-winged, yellow, ca. 2 mm long. **Chromosomes**: 2n=14. Figure 47–53

Phenology : flowering and fruiting Sept–October

Habitats: grassy slopes, scrubs, often pendent from rocks, sometimes epiphytic on trunks in forests; 2000–3300 m.

Distribution: U Ganga & Indus, U Yarlung Zangbo, C and E Nepal, Sikkim & Darjiling, Bhutan, Yarlung Zangbo-Brahmaputra, and S Hengduan **(**Map 47–109**);** also in SW Yunnan (Jindong) of China and Meghalaya of India.

U GANGA & INDUS: **Kumaon**, Lohashal, 1830 m, s. d. *R. Strachey & J. E. Winterbottom 1* (K); **Nainital**, *M. K. Wall 1901* (BM). **U YARLUNG ZANGBO: Nyalam,** between Zham and Friendship Bridge, 2000 m, 1972, *Xizang Chinese Mater. Medic. & Herbs Exped. 1308* **(HNWP, PE);** below Zham, 2066 m, 2011, *D. Y. Hong et al. H11003* **(PE). C NEPAL**: **Bagmati**, between Dhunche and Singdum Gompa, 1900m, 1972, *H. Kanai et al. 723166* (BM, KUN, TI); Grang, E side of Trisuli River, 1800 m, 1966, *D. H. Nicolson 2357* (BM, KATH); Langdeng, 3050 m, 1931, *K. H. Sharma E126* (BM); Langtang Khola, 2440–2900 m, 1949, *O. Polunin 1805* (BM, E, G, P); Rasuwa, 2600 m, 2001, *F. Shrestha & N. Joshi 152/2001* (KATH); Rasuwa, near Bokejhund, 1950 m, 1977, *N. P. Manandhar & Party 389* (KATH); **Gandaki**, Kaski, 2050 m, 1988,

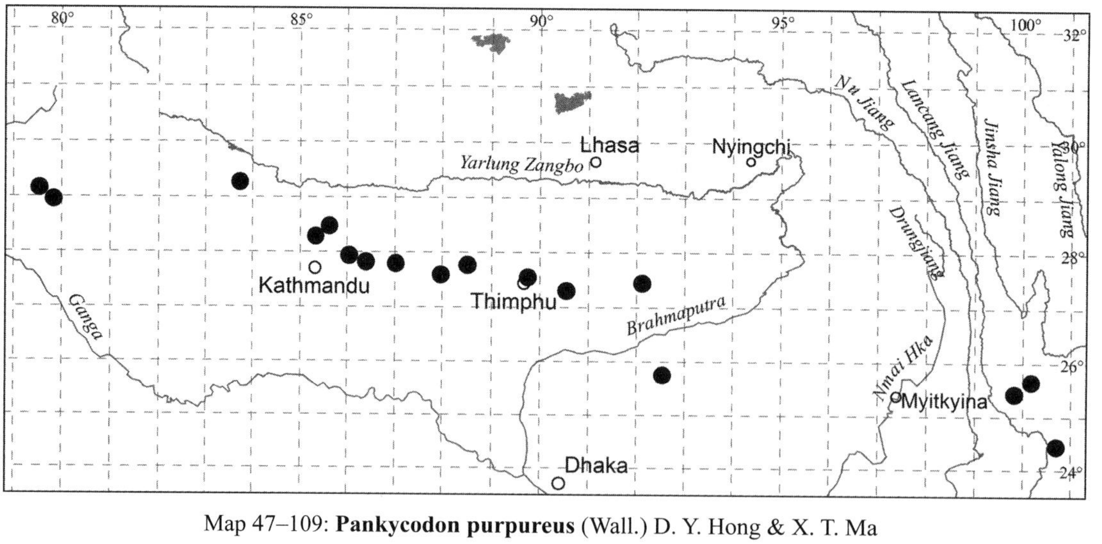

Map 47–109: **Pankycodon purpureus** (Wall.) D. Y. Hong & X. T. Ma

Figure 47–53: **Pankycodon purpureus** (Wall.) D. Y. Hong & X. T. Ma: **a**, root, caudex, and basal part of two stems; **b**, upper part of stem; **c**, longitudinal section of flower, showing hypanthium higher than ovary; **d**, stigma with upper part of style. From Wang et al., *J. Syst. Evol.* 52(5), 2014, with permission.

M. Suzuki et al. 8860475 (BM); Kaski Distr., 1900 m, 1988, *M. Suzuki et al. 10831* (KATH); Marsiandi Valley, Dharamsala, 2440 m, 1950, *D. G. Lowndes 1481* (BM, CAL, E); **Janakpur**, Dolakha, 1120–2030 m, 2007, *S. Noshiro et al 20710072* (E); Dolakha, Kalingchok, Thala to Tale Bisauna, ca. 2500 m, 1970, *H. Kanai et al. 672793* (KATH, TI); **Kathmandu**, Phulchoke, SW of Kathmandu, 2440 m, 1965, *A. D. Schilling 601* (K); Thankot, 12 km W of Kathmandu, 1830 m, 1956, *K. de B. Codrington 1* (BM). **E NEPAL**: **Mechi**, Taplejung, Shewaden-Papung, ca. 2200 m, 1977, *H. Ohashi et al. 773971* (TI); Taplejung, Tamur Valley, Mewa Khola, 2440 m, 1956, *J. D. A. Stainton 1305* (BM, E); **Sagarmatha**, Solukhumbu, 1950–3100 m, 1997, *M. Wakabayashi et al. 9730094* (E). **SIKKIM & DARJILING**: **Sikkim**, Cheungtong, 1830 m, 1913, *R. E. Cooper 444* (E). **BHUTAN: Dotena,** Thsiubu Chu, 1949, *F. Ludlow et al. 17535* (BM); **Tongsa,** 1980 m, 1914, *R. E. Cooper 2071* **(E);** Mangde Chu, 1980 m, 1949, *F. Ludlow et al. 19558* **(BM, E). YARLUNG ZANGBO-BRAHMAPUTRA: Senge** Dzong, 2130-2440 m, 1938, *F. Kingdon-Ward 14180* (BM). **S HENGDUAN: Dali,** 2740–3050 m, 1910, *G. Forrest 6818* (BM, E); **Yangbi**, Meixi, Malutang, 2380 m, 1963, *Zhongdian Exped. 63-4236* (KUN).

7. **PSEUDOCODON** D. Y. Hong & H. Sun

Hong & Sun in Wang et al., J. Syst. Evol. 52(5): 546. 2014; Wang & Hong, Phytotaxa 204 (1): 60. 2015; Hong, Monogr. *Codonopsis* Allied Gen. Camp. 190. 2015. ≡ *Codonopsis* subg. *Pseudocodonopsis* Kom., Trudy Imp. S. -Peterburgsk. Bot. Sada 29(1): 102. 1908. Type: *Pseudocodon convolvulaceus* (Kurz) D. Y. Hong & H. Sun (*Codonopsis convolvulacea* Kurz).

Herbs mostly twining, much less frequently ascending. Roots globose or ellipsoid, but in *Pseudocodon petiolatus* they elongated, attenuate and constricted at middle part. Stems single from caudex. Leaves alternate, rarely opposite. Flowers solitary, less frequently in cyme of 2–4, terminal. Calyx tube entirely adnate to hypanthium; calyx lobes 5, inserted at same level as corolla, entire. Corolla 5-cleft to near base, rotate. Stamens 5; filaments short, base strongly dilated, dilated part semiorbicular and densely lanate at periphery adaxially. Ovary nearly entirely inferior to both calyx and corolla, 3-locular; stigma 3-fid, stigma lobes ovate. Fruit a 3-loculicidal capsule. Seeds many, wingless. Plants not fetid. Chromosomes 2n=14.

Recent references: Hong & Pan, 2012, Pollen morphology of the platycodonoid group (Campanulaceae s. str.) and its systematic implications. *J. Integr. Plant Biol.* 54: 773–789; Wang *et al.*, 2013, Molecular phylogeny of the platycodonoid group (Campanulaceae s. str.) with special reference to circumscription of *Codonopsis*. *Taxon* 62: 498–504; Wang et al., 2014, Phylogenetic analyses reveal three new genera. *J. Syst. Evol.* 52(5): 54–550; Hong, 2015, A Monograph of *Codonopsis* and Allied Genera (Campanulaceae). Science Press & Academic Press; Wang & Hong, 2015, Taxonomic revision of the genus *Pseudocodon* (Campanulaceae) based on character analysis and molecular phylogeny. *Phytotaxa* 204 (1): 49–64.

The genus *Pseudocodon* differs distinctly from *Codonopsis* in a number of characters. Corolla is cleft to near the base and rotate (vs. shallowly lobed and campanulate or tubular); disc is absent (vs. always present with various forms); filaments are always abruptly dilated at base into a semiorbicular and densely lanate lamella, rarely ciliate (vs. slightly dilated and glabrous or villous-ciliate); roots are globose, less frequently ellipsoid, rarely elongated (vs. carrot-shaped or nearly columnar, rarely long-ovoid); pollen is of short colpi, with colpus length/polar axis length 0.2–0.3 (vs. ≥0.7), smooth colpus membrane (vs. spinulose) and high sexine spinules ≥2 μm (vs. ≤ 1 μm). Furthermore, in our molecular phylogenetic tree (Wang et al., 2013 & 2014) it forms a monophyletic clade with *Himalacodon* D. Y. Hong & Q Wang and *Pankycodon* D. Y. Hong & X. T. Ma, but separated from the clade consisting of the core *Codonopsis*.

The taxonomy of *Pseudocodon* is problematic. In the genus only *P. rosulatus* has been stable in taxonomic status and is distinct from all the other species. *P. petiolatus* D. Y. Hong is newly described species. The rest species form the *Codonopsis convolvulacea* Kurz complex, and the number of species in the complex has been controversial among authors, from one to nine species.

Eight species, extending from Nepal eastwards to SW China via Bhutan and N Myanmar; all in the Pan-Himalaya (5 endemic).

1a. Basal leaves rosulate, persistent; stems erect or ascending, rarely procumbent, never twining .... 8. *P. rosulatus*
1b. Basal leaves absent; stems nearly always twining.
  2a. Stems and leaves more or less white-hirsute; leaves at least partially opposite .... 4. *P. hirsutus*
  2b. Stems and leaves glabrous; leaves all alternate
    3a. Leaf blade filiform or linear, less than 5 (–9) mm wide, mostly aggregated at lower part of stems; stems less than 1 m long .... 5. *P. graminifolius*
    3b. Leaf blade cordate-ovate to linear-lanceolate, 0.4–8 cm wide, evenly distributed, stems usually over 1 m long.
      4a. Calyx lobes strongly reflexed; roots elongate, attenuated, with a constriction at middle part; petiole 1.5–2.5 cm, nearly as long as leaf blade .... 7. *P. petiolatus*
      4b. Calyx lobes upright; roots globose or ellipsoid; petiole 0.2–6(12) cm long, much shor-ter than leaf blade.
        5a. Leaf blade retroserrate, margin sparsely setose .... 6. *P. retroserratus*
        5b. Leaf blade entire, crenate, dentate or serrate, margin without setae.
          6a. Petiole 1–3 cm long, as long as or nearly as long as leaf blade; leaf blade dentate, mostly cordate, less often rounded at base; corolla with a wide dark red-purple ring at base .... 3. *P. grey-wilsonii*
          6b. Petiole usually from absent to 1 cm, rarely up to 6 (12) cm long, much shorter than leaf blade; leaf blade entire, serrate or dentate, mostly cuneate or rounded, very rarely cordate at base; corolla without a wide colourful ring at base.
            7a. Leaf blade membranous, dentate or serrate, very rarely entire; calyx lobes mosty 3–7 mm long, 1.5–3 mm broad, less often 8–16 mm long, 3–6 mm broad .... 2. *P. vinciflorus*
            7b. Leaf blade papery, mostly entire, less often serrate; calyx lobes 6–18 mm long, 2.2–10 mm broad .... 1. *P. convolvulaceus*

1. Pseudocodon convolvulaceus (Kurz) D. Y. Hong & H. Sun in Wang et al., J. Syst. Evol. 52(5): 548. 2014; Wang & Hong, Phytotaxa 204(1): 61. 2015. ≡ *Codonopsis convolvulacea* Kurz, J. Bot. 11: 195. 1873; Chipp, J. Linn. Soc. Bot. 38: 389. 1908, p. p., excl. syn.; Anthony, Notes Roy. Bot. Gard. Edinburgh 15: 177. 1926, p. p., excl. syn. & vars. *limprichtii*, *hirsuta*, & *pinifolia*; Tsoong, Contr. Inst. Bot. Nat. Acad. Peiping, 3(3): 90. fig. 6. 1935, p.p., excl. var. *pinifolia*; Shen & Hong in Hong, Fl. Reipubl. Popularis Sin. 73(2): 67. 1983, p.p., quoad type var. & var. *forrestii*. Type: China, Yunnan, Mengzi, 1520–1830 m, s. d., *A. Henry 9425* (neotype designated by Hong, 2010: PE; isoneotypes: E, K, LE).

Roots ovoid-globose or ovoid, 2.5–5 cm long. Stems twining or nearly ascending, usually with a few branches, up to more than 1 m long, glabrous. Leaves alternate or sometimes opposite, evenly distributed along stems; petiole from absent to 6 (–12) cm long; leaf blade linear-lanceolate to cordate-ovate, 2–10 cm long, 0.4–8 cm broad, base cuneate, rounded or slightly cordate, margin entire or serrate, apex obtuse, acute or acuminate. Flowers solitary, terminal on main stem and branches, sometimes in cyme of 2–4; pedicel 2–12 cm long, glabrous. Calyx tube adnate to whole hypanthium, obconical, 3–7 mm long, glabrous, 10-ribbed; calyx lobes usually lanceolate, less frequently subulate or long-ovate, 6–18 mm long, 2.2–10 mm broad, glabrous, apex acuminate or acute; sinus between calyx lobes narrow, pointed or slightly obtuse. Corolla pale blue or blue-purple; corolla lobes elliptic, 1–5 cm long, 0.6–2 cm broad, apex acute. Filaments 1–3 mm long. Superior part of capsule short-conical, valves ca. 4 mm long; inferior part obconical, 1–1.6 mm long. Seeds numerous, oblong, ca. 1.5 mm long, brown-yellow.

Distribution: Yarlung Zangbo-Brahmaputra, S Hengduan, and U Irrawaddy (Map 47–110); also in S Myanmar and other parts of China (W Guizhou, S Sichuan, and C & NE Yunnan).

1a. Petiole up to 3 mm long; calyx lobes 6–10 mm long, 2.2–4 mm broad; corolla lobes 1–3 cm long .... 1a. subsp. *convolvulaceus*

1b. Petiole 0.3–6 (12) cm long; calyx lobes 10–18 mm long, 2.5–10 mm broad; corolla lobes 1.8–5 cm long .... 1b. subsp. *forrestii*

1a. Pseudocodon convolvulaceus subsp. **convolvulaceus**

Leaves nearly sessile or with petiole up to 3 mm long; leaf blade lanceolate to linear-lanceolate, 2–7 cm long, 0.5–1.5 cm broad, papery, entire, base rounded to cuneate. Calyx lobes 6–10 mm long, 2.2–4 mm broad. Corolla lobes 1–3 cm long. Figure 47–54.

Phenology: flowering in August and September; fruiting from September to November.

Habitats: grassy slopes, thickets, or forests, on granites; 1100–2500 m.

Distribution: Yarlung Zangbo-Brahmaputra, S Hengduan, and U Irrawaddy; also in S Myanmar and other parts of Yunnan in China.

YARLUNG ZANGBO-BRAHMAPUTRA: Tembang, 1830–2130 m, 1938, *F. Kingdon-Ward 13943* (BM). **S HENGDUAN**: **Longyang**, Lujiang, Bawan Village, 1800 m, 2003, *Gaoligong Shan Biodeversity Survey 18391* (CAS). **U IRRAWADDY**: **Tengchong**, Qushi, Longkou Village, 1998. *H. Li et al. 11205* (CAS); Sanyun, Pingdi Village, 1335 m, 2003, *Gaoligong Shan Biodiversity Survey 18057* (CAS); Volcanic mouth, 1800 m, 1980, *S. T. Li 80–342* (PE).

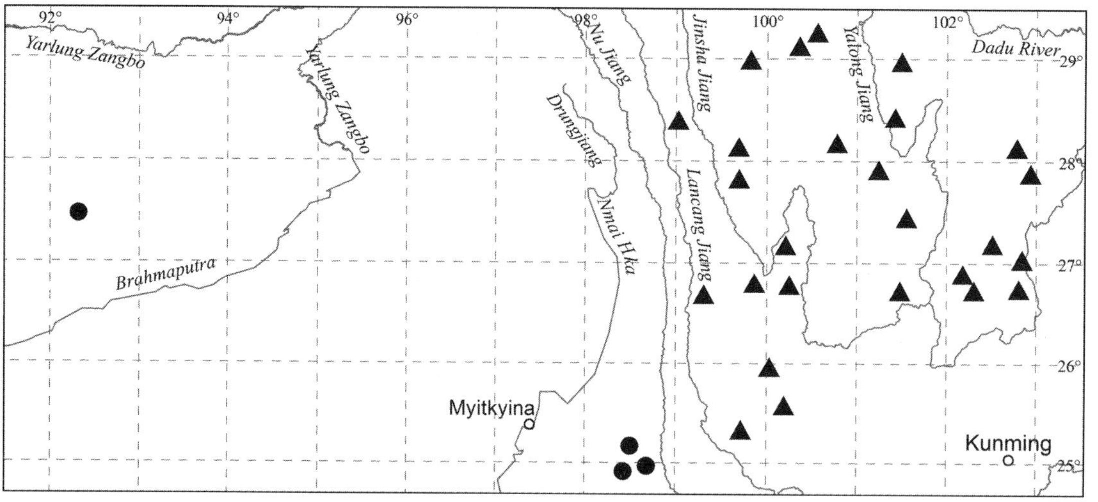

Map 47–110: **Pseudocodon convolvulaceus** (Kurz) D. Y. Hong & H. Sun
● subsp. **convolvulaceus**, ▲ subsp. **forrestii**

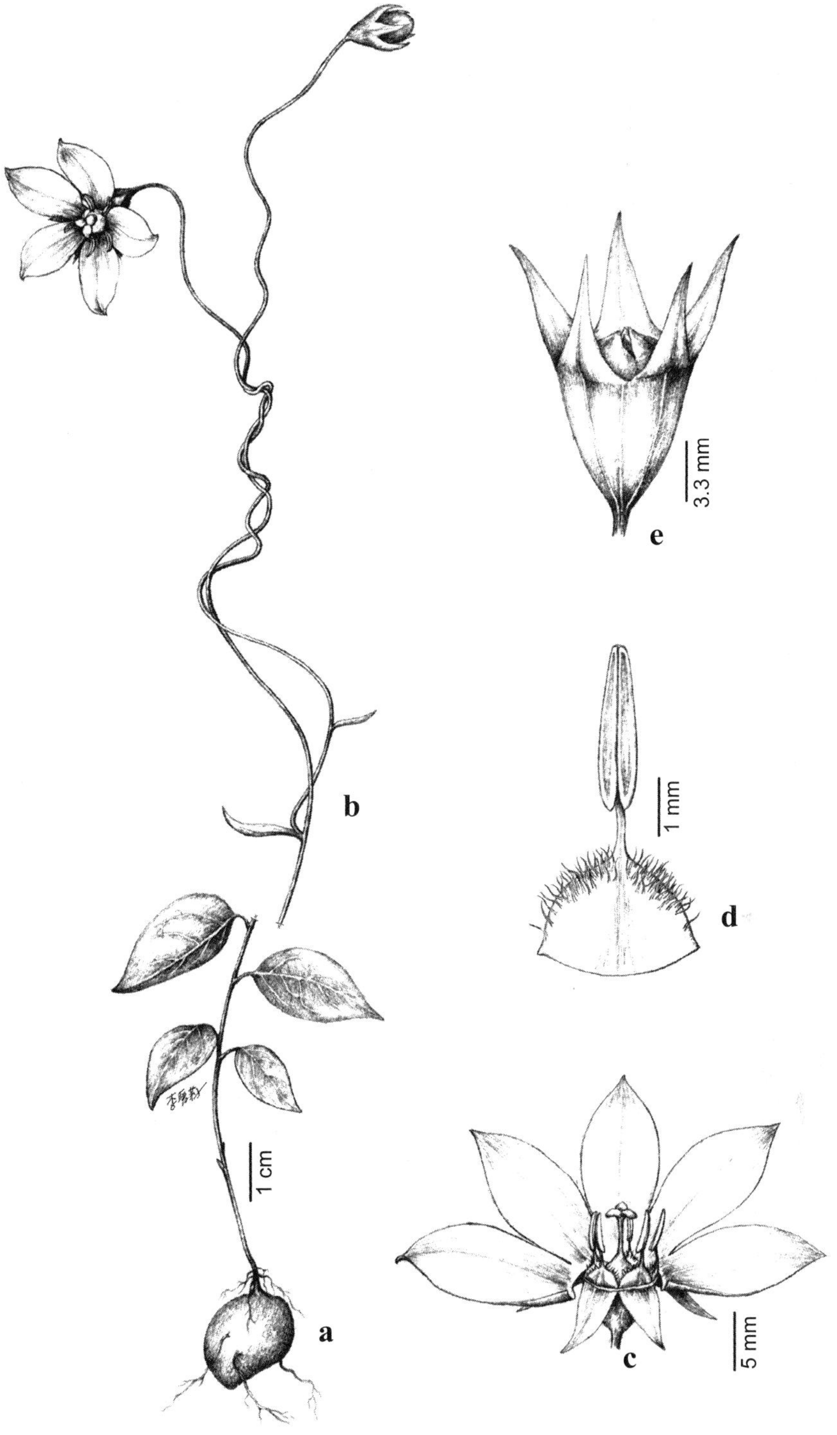

Figure 47–54: **Pseudocodon convolvulaceus** (Kunz) D. Y. Hong & H. Sun subsp. **convolvulaceus**: **a**, root and lower part of stem; **b**, upper part of stem; **c**, a flower with corolla opened, showing calyx, stamens, style, and stigma; **d**, ventral side of stamen, showing much dilated and hairy filament base; **e**, capsule. From Wang et al, in *J. Syst. Evol.* 52(5), 2014, with permission.

1b. Pseudocodon convolvulaceus subsp. **forrestii** (Diels) D. Y. Hong in Wang & Hong, Phytotaxa 204(1): 61. 2015. ≡ *Codonopsis forrestii* Diels, Notes Roy. Bot. Gard. Edinburgh 5: 171. 1912. ≡ *Codonopsis convolvulacea* subsp. *forrestii* (Diels) D. Y. Hong & L. M. Ma in Li & Zhu, Fl. Sichuan. 10: 546. 1992. ≡ *Codonopsis convolvulacea* var. *forrestii* (Diels) Ballard, Bot. Mag. 162: tab. 9581. 1939. Type: China, Yunnan, on the banks of Yangtze, 1520–2130 m, Sept. 1904, *G. Forrest 48* (lectotype designated by Hong in Wang & Hong, 2015: E; isolectotype: PE).

Codonopsis limprichtii Lingelsh. & Borza, Repert. Spec. Nov. Regni Veg. 13: 391. 1914; Lammers & Klein, Bot. Stud. (Taipei) 51: 555, 2010. p. p., excl. syn. ≡ *Codonopsis convolvulacea* Kurz var. *limprichtii* (Lingelsh. & Borza) J. Anthony, Notes Roy. Bot. Gard. Edinburgh 15: 178. 1926. Type: China, Yunnan: "Talifu (Dali), an schattigen Plätzen unter Kiefern vor Ai-ning-po, bei Pu-peng", 2500 m, 21 Aug. 1913. *Limpricht 851* (holotype: WRSL, n. v. ; isotype: WU).

Codonopsis mairei H. Lév., Cat. Pl. Yunnan, 24. 1915. Type: China, Yunnan, "paturages des mont. à Ta-hai », 3200 m, Jul. 1911, *E. E. Maire s. n.* (holotype :E).

Codonopsis efilamentosa W. W. Smith, Notes Roy. Bot. Gard. Edinburgh 8: 107. 1913; Lammers & Klein, Bot. Stud. (Taipei) 51: 559. 2010. ≡ *Codonopsis convolvulacea* Kurz var. *efilamentosa* (W. W. Smith) L. T. Shen in Hong, Fl. Reipubl. Popularis Sin. 73(2): 69. 1983. Type: China, Yunnan, "the eastern flank of the Lichiang Range" (Lijiang), 27°30'N, 3050 m, Jul. 1910, *G. Forrest 6258* (holotype:E).

Leaves all petiolate with petiole 0.3–6(12) cm; leaf blade cordate-ovate, ovate-lanceolate to linear-lanceolate, 2–11 cm long, 0.4–8 cm broad, papery, entire or serrate, base cordate to cuneate. Calyx lobes 10–18 mm long, 2.5–10 mm broad. Corolla lobes 1.8–5 cm long. **Chromosomes: 2n=14.**

Phenology: flowering from July to September; fruiting from September to November.

Habitats: forests, thickets, grassy slopes, or edges of forests; 1500–4000 m.

Distribution: S Hengduan; also in other parts of Myanmar and in W Guizhou, S Sichuan, and C & NE Yunnan of China.

S HENGDUAN: Butuo, Shaluo, 2400 m, 1979, *s. coll. 711* (SM); **Dali**, Cangshan, 2500 m, 1914, *C. Schneider 3777* (G, PE); Dali Range, 1924, *G. Forrest 28101* (E, PE); Huadianba, 2900–3100 m, 1984, *K. Iwatsuki et al. 984* (KUN); **Chabulang** to Wanglang, 1983, *Qinghai-Xizang Exped. 14111* (KUN, PE); **Daocheng**, Riwa, Lamuge, 3900 m, 1973, *s. coll. (Dao) 2417* (KUN); valley E of Daocheng, 3800 m, 2010, *Kunming & Edinburgh Exped. to Sichuan 244* (E); **Dêqên**, Yinmin Ganchong, 2700 m, *S. B. Lan 142* (PE); **Heqing**, Huangping, 2600 m, 1963, *Jinshajiang Exped. 6565* (KUN, PE); Lianping, Nanxilu, 2800 m, 1929, *R. C. Ching 23859* (KUN, PE); **Huidong**, Baisha, 3000 m, 1959, *Sichuan Econ. Plant Surv. (Xi) (59) 774* (KUN); Tangtang, 2100 m, 1978, *Huidong Exped. 543* (SM); **Huili**, Yimen, Chahe, 1958, *Z. He et al. 11619* (SM, SZ); **Jianchuan**, Chünchuan-Mekong Divide, 3350–3660 m, 1922, *G. Forrest 22275* (E, K, W); **Jiulong**, Tanggo, 3200 m, *Y. B. Yang 23013* (CDBI); **Lanping**, 2800 m, 1933, *H. T. Tsai 53707* (KUN, PE, SZ); **Meigu**, Bapu, 2100 m, 1979, *s. coll. 540* (SM); **Miyi**, Wanqiu, 1300 m, 1963, *Miyi Working Team 194* (CDBI); **Muli**, Bo'ao, 2200 m, 1978, *C. S. Cao et al. 8316* (CDBI, SZ); Muli-Yongning, 3100–3500 m, 1915, *Handel-Mazzetti 7267* (WU); **Ningnan**, Xinjian, 2000 m, 1978 *Ningnan Exped. 517* (SM); **Puge**, Jingyingsuo, 1959, *Sichuan Econ. Plant Surv. (Pu) 5452* (KUN); Ju'andahuo Shan, 3400 m, 1979, *Puge Exped. 579* (SM); **Shangrila** (Zhongdian), Lamaso, 3100 m, 1937, *T. T. Yü 12644* (KUN, PE); 4 km NW of Zhongdian town, 3400 m, 2000, *E. von Raab-Straube, M. Smalla & H. Sun 417* (B); near Zhongdian Alpine Bot. Gard., 3350 m, 2010, *D. Y. Hong et al. H10015* (PE); **Xiangcheng**, No.4 Distr., 3400 m, 1981, *Qinghai-Xizang Exped. 4555* (PE); **Yanbian**, Yankou, 2800 m, 2001, *C. M. Tan P315 B* (SZ); **Yanyuan**, Bailing, 3000 m, 1983, *Qinghai-Xizang Exped. 12902* (KUN, PE); **Yongping**, between Dali and Baoshan, 2200 m, 1997,

Gaoligong Shan Exped. (1997) 8539 (E); **Yulong** (Lijiang), Hutiaoxia, 1900 m, 1987, *H. Ohba et al. 99* (KUN); Ma-on-Shan, 1939, *R. C. Ching 21314* (KUN, PE); Shudi, 1500–1750 m, 1987, *K. Iwatsuki et al. 1893* (KUN); Tamichung, 1939, *R. C. Ching 21575* (KUN, PE); Xuesongcun, 3500 m, 1989, *T. J. Zhang 0028* (PE); Yulong Snow Range, 1922, *J. F. Rock 5476* (E, PE); **Zhaojue**, Poluo, 2420 m, 1979, *s. coll. 893* (SM).

2. Pseudocodon vinciflorus (Kom.) D. Y. Hong in Wang et al., J. Syst. Evol. 52(5): 548. 2014 & in Wang & Hong, Phytotaxa 204(1): 61. 2015. ≡ *Codonopsis vinciflora* Kom., Trudy Imp. S.-Peterburgsk. Bot. Sada 29(1): 103. 1908. ≡ *C. convolvulacea* var. *vinciflora* (Kom.) L. T. Shen in Hong, Fl. Reipubl. Popularis Sin. 73(2): 68. 1983. ≡ *C. convolvulacea* subsp. *vinciflora* (Kom.) D. Y. Hong in Wu, Fl. Xizang. 4: 582. 1985. Type: China, "Se-czuan (Sichuan) inter Va-sy-kou et Za-li in valle fl. Siao-tzin-che", 15 Jul. 1893, *G. N. Potanin s. n.* (holotype: LE).

Codonopsis forrestii Diels var. *heterophylla* C. Y. Wu in Wu & Li, Yunnan Trop. Subtrop. Fl. Res. Rep. 1: 80. pl. 30. fig. 1. 1965. Type: China, Yunnan, Dêqên, Hongpo, 3240 m, 3 Sept. 1937, *T. T. Yü 10173* (holotype: KUN; isotypes: PE).

Herbs twining, glabrous throughout except filaments and style. Roots globose. Stems from 20 cm to over 1 m long. Leaves all alternate; petiole 3–12 mm long; leaf blade membranous, ovate to lanceolate, rarely linear-lanceolate, 1.5–9.5 cm long, 0.3–3.3 cm broad, base rounded to broad-cuneate, margin usually dentate, rarely entire, apex acute; those on branches much smaller. Flowers solitary, terminal only on main stem or also on upper branches; pedicel 1.5–20 cm long. Calyx lobes subulate or subulate-lanceolate, 3–16 mm long, 1.5–6 mm broad. Corolla blue-purple, 2.5–5 cm across; corolla lobes elliptic or oblong. Ovary mostly obconical, less often obovoid; style slightly longer than stamens. Fruit usually obconical or obconical-ovoid, 1–1.3 cm long, 0.6–0.8 cm wide.

Distribution: **endemic**; M and L Yarlung Zangbo, Tangut, N and S Hengduan (Map 47–111).

1a. Calyx lobes 3–7 mm long, 1.5–3 mm broad; leaf blade 1.5–4 cm long, 0.3–2 cm broad .... 2a. subsp. *vinciflorus*

1b. Calyx lobes 8–16 mm long, 3–6 mm broad; leaf blade 3–9.5 cm long, 0.7–3.3 cm broad .... 2b. subsp. *dianchuanicus*

2a. Pseudocodon vinciflorus subsp. **vinciflorus**

Leaf blade 1.5–4 cm long, 0.3–2 cm broad; calyx lobes 3–7 mm long, 1.5–3 mm broad. Figure 47–55: a.

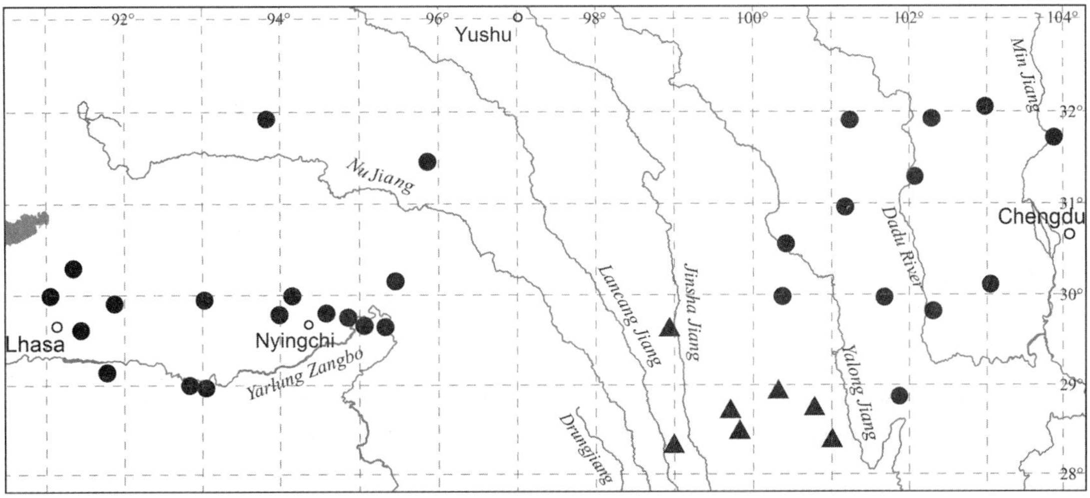

Map 47–111: **Pseudocodon vinciflorus** (Kom.) D. Y. Hong
● subsp. **vinciflorus**, ▲ subsp. **dianchuanicus**

Figure 47–55: **a**, **Pseudocodon vinciflorus** (Kom.) D. Y. Hong subsp. **vinciflorus**: upper part of stem. **b**, **Pseudocodon grey-wilsonii** (J. M. H. Shaw) D. Y. Hong: upper part of stem. From Hong, *Monogr. Codonopsis Allied Gen. Camp.*, 2015.

Phenology: flowering from June to September; fruiting in September and October.

Habitats: thickets, forests, or grassy slopes; 2600–4600 m.

Distribution: M and L Yarlung Zangbo, Tangut, N and S Hengduan.

M YARLUNG ZANGBO: **Lhasa**, mountain near Kartosi, 4600 m, 1965, *Y. T. Chang & K. Y. Lang 1995* (KUN, PE); hills N of Lhasa, 3960 m, 1943, *F. Ludlow & Sherriff 9872* (BM, E); **Lhünzhub,** Songpan to Poindo, 4175 m, 2012, *FLPH Tibet Exped. 12–0422* (PE); **Maizhokunggar**, between the city and Mila Pass, 4150 m, 2010. *D. Y. Hong & Q. Wang H10001* (PE); **Nêdong**, Gyêrba, 3805–3864 m, 2012, *FLPH Tibet Exped. 12–0756* (PE). **L YARLUNG ZANGBO: Bomi**, Yuren, 3300 m, 1973, *Qinghai-Xizang Exped. 73–1340* (KUN); **Gongbo'gyamda**, W of Bahe Zeng, by highway 318, 3335 m, 2000, *D. E. Boufford 30064* (A); near the city, S of Nyang Qu, 3465 m, 2010, *D. Y. Hong & Q. Wang H10003* (PE); Pasum Lake, 3660 m, 1947, *F. Ludlow et al. 14141* (BM, E); **Gyaca**, Lhasoi, 3400 m, 2011, *D. Y. Hong et al. H11012* (PE); **Mainling,** Gyala, 2840 m, 1938, *F. Ludlow 7545* (BM); near Huilong, 3100 m, 2011, *D. Y. Hong et al. H11013* (PE); between Tripe and Gyala, 2890 m, 1938, *F. Ludlow et al. 5333* (BM, E); Tsangbo Gorge, 2740 m, 19. July 1924, *F. Kingdon-Ward 5958* (BM, E); **Nangxian**, Jyêmdong to Laiyi Village, *FLPH Tibet Exped. 12–1123* (PE); Kyim Dzong, 3510 m, 1936, *F. Ludlow & G. Sherriff 1938* (BM); **Nyingchi**, Naixi, 3040 m, 1965, *Y. T. Chang & K. Y. Lang 1076* (PE); Serji La, W slope, 3074 m, 2008, *Liu JQ–08XZ–225* (KUN). **TANGUT**: **Dêngqên**, Damdoi, 2800 m, 1961, *s. coll. 1665* (KUN); **Sogxian**, Rongbo to Rabo, 3800–3900 m, 1976, *D. D. Tao 10995* (KUN); **Zamtang**, 2990–3050 m, 2007, *D. E. Boufford et al. 38874* (KUN). **N HENGDUAN: Barkam**, Dalangjiao, 2650 m, 1957, *H. F. Zhou & T. Y. Chang 23003* (SZ); **Dawu**, Bonglong, 3500 m, 1979, *s. coll. 146* (SM); **Heishui**, Waboliangzi, 2500 m, 1963, *X. Y. Tang & Z. L. Kuen 63011* (PE); **Jinchuan**, Anning, Muma Shan, 2800 m, 1958, *X. Li 78087* (PE, SZ); **Litang**, Lapo, 3900 m, 1973, *Sichuan Veget. Surv. 3692* (CDBI, PE); **Maoxian**, Sanlong, 1952, *Z. He & Z. L. Zhou 13063* (SZ); **Xinlong**, Longlaxi, 3400 m, 1974, *Z. S. Qin 6347* (CDBI, PE). **S HENGDUAN**: **Jiulong**, Sanyanlong, 3500 m, 1980, *Jiulong Exped. 299* (SM); **Kangding**, Tongolo, 1893, *J. A. Soulie 2760* (P); Zierpo, 2800 m, 1981, *Z. J. Zhao et al. 115211* (SZ); **Luding**, Erlang Shan, 2600 m, 1959, *Sichuan Econ.Plant Surv.* (*Garzê*) *0010* (PE); **Tianquan**, Erlang Shan, Ganhaizi, 2500 m, 1953, *W. K. Hu 35419* (SZ).

2b. Pseudocodon vinciflorus subsp. **dianchuanicus** D. Y. Hong & Q. Wang in Wang & Hong, Phytotaxa 204(1): 62. 2015. Type: China, Yunnan, Dêqên, Tswalung-la, 2600 m, thickets, Sept. 1937, *T. T. Yü 10286* (holotype: PE; isotype: KUN).

Leaf blade 3–9.5 cm long, 0.7–3.3 cm broad. Calyx lobes 8–16 mm long, 3–6 mm broad.

Phenology: flowering from July to September; fruiting in September and October.

Habitats: sunny grassy slopes or forests, but mostly climbing up thickets; 2700–3900 m.

Distribution: confined to N and S Hengduan.

N HENGDUAN: Markam, Chubarong, 3000 m, 2010, *Kham Exped. 10–2401* (PE). **S HENGDUAN: Daocheng**, Gongga-ling, 3100 m, 1937, *T. T. Yü 12881* (KUN, PE); Riwa, 3300 m, 1981, *Qinghai-Xizang Exped. 4291* (CDBI, PE); **Dêqên**, Baima Shan, 3000 m, 1935, *C. W. Wang 69784* (PE); Huangpu, 3240 m, 1937, *T. T. Yü 10172* (KUN); Mabam Pass to **Donglang**, 3100–3200 m, 1983, *Qinghai-Xizang Exped. 14525* (KUN, PE); **Muli**, Bo'ao, Dingdonghai, 2800 m, 1983, *Qinghai-Xizang Exped. 13657* (KUN); Lea-Lang-Keng-din, 2800 m, 1937. *T. T. Yü 7524* (KUN, PE); **Xiangcheng**, Rawu, 3500–3800 m, 1981, *Qinghai-Xizang Exped. 4127* (CDBI, KUN, PE); Shangyou, 3600 m, 1973, *Sichuan Veget. Surv. 3231* (PE).

3. Pseudocodon grey-wilsonii (J. M. H. Shaw) D. Y. Hong in Wang et al., J. Syst. Evol. 52(5): 548. 2014 & in Wang & Hong, Phytotaxa 204(1): 62. 2015. ≡ *Codonopsis nepalensis* Grey-Wilson, Plantsman, 12(2): 99. 1990, non Hara (1978). ≡ *Codonopsis grey-wilsonii* J. M. H. Shaw, New Plantsman, 3(2): 93. 1996. ≡ *Codonopsis convolvulacea* subsp.

grey-wilsonii (J. M. H. Shaw) D. Y. Hong, Novon 20(4): 421. 2010. Type: Nepal, Kali Gandaki Valley, W of Marpha, 3500 m, 6 Sept. 1973, *Grey-Wilson & Phillips 777* (holotype: K; isotype: K).

Herbs twining, glabrous throughout except filaments and style. Root tubers globose or ovoid. Stems over 1 m long. Leaves all alternate; petiole 1–3 cm long; leaf blade ovate to ovate-lanceolate, 2.5–4.5 cm long, 1.5–2.5 cm broad, base cordate to rounded, margin dentate, apex obtuse to acute; leaves on sterile branches much smaller. Flowers solitary and terminal on main stem and fertile branches; pedicel up to over 10 cm long. Calyx lobes lanceolate or ovate-lanceolate, 1–1.8 cm long, 0.4–0.7 cm broad. Corolla blue-purple, very rarely white, often with a wide dark red-purple ring at base, 3–6.5 cm wide; corolla lobes ovate, apex acute-obtuse. Filaments ca. 1.3 cm long; anthers as long as filaments. Ovary obconical or obovoid; style nearly as long as stamens or slightly longer; stigma lobes ovate. Fruit obconical or obovoid. Figure 47–55: b.

Phenology: flowering and fruiting from August to October.

Habitats: thickets or forests; 2100–4300 m.

Distribution: **endemic**; U Yarlung Zangbo, W, C, and E Nepal, Bhutan, and Yarlung Zangbo-Brahmaputra (Map 47–112).

Notes: The species is distinctly different from *P.vinciflorus* in having calyx lobes much larger and lanceolate or ovate-lanceolate, petiole much longer, and corolla with a wide dark-purple ring inside at the base. Furthermore, our recent molecular data indicate that three accessions from S Tibet form a clade sister to a clade of the accessions from N of the Himalaya. Therefore, *P. grey-wilsonii* is a distinct species, which is isolated from *P. vinciflorus* by the Himalaya.

U YARLUNG ZANGBO: Gyirong, near Gyirong Town, 2150 m, 1981, *C. C. Ni et al. 2180* (PE); Gyirong Town, 3100–3130 m, 2010, *PE-Xizang Exped. 0185* (PE); Sale, Labi-Sale, 3000 m, 1990, *B. S. Li et al. 13337* (PE); Zaxiang, 2900 m, 1990, *B. S. Li et al.13207* (PE). **Nyalam**: Zham to Lixin, 2800 m, 1990, *B. S. Li & H. Li 14007* (PE); Zham, 2100 m, 2011, *D. Y. Hong et al. H11005* (PE); Zham, Lixin, 2492 m, thickets at edges of forests, 19. August 2010, *PE-Xizang Exped. 1500* (PE). **W NEPAL**: ; **Karnali**, Dolpa, Tarakot, 2700 m, 1973, *Einarsson, Skärby & Wetterhall 3674* (BM, LD); Dolpa, Kaigaun, 2800 m, 1973, *Einarsson, Skärby & Wetterhall 3756* (BM, LD). **C NEPAL**: **Bagmati**, Dhading, near Nyilo, 3350 m, 1973, *D. Lichter 120* (KATH); Rasuwa, Kyangchin Gomba, 3600 m, 1971, *J. F. Dobremez 1024* (G); Langtang: 3660 m, 1970, *J. D. A. Stainton 6675* (BM, E); **Dhawalagiri**, Baglung, near Dogadi Khola. 3960 m, 1954, *Stainton, Sykes & Williams 3822* (BM, G);

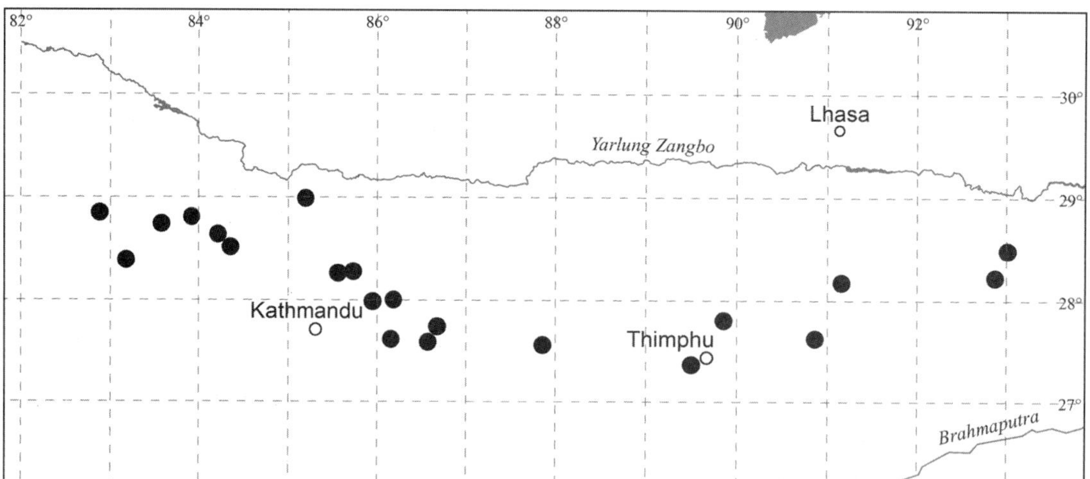

Map 47–112: **Pseudocodon grey-wilsonii** (J. M. H. Shaw) D. Y. Hong

Mustang, Tukucha, Samargaon, 3960 m, 1954, *Stainton, Sykes & Williams 7245* (BM); Myagdi, Tukucha, Kali Gandaki, 3200 m, 1954, *Stainton, Sykes & Williams 7455* (BM, P); **Gandaki**, Manang, Chame, 2590 m, 1950, *D. G. Lowndes 1401* (BM, E, G); **Janakpur**, Dolakha, Prés de Thari, 2100 m, 1954, *A. Zimmerman 1230* (G). **E NEPAL: Mechi**, Tamur Valley, Thapabu Khola, 2290 m, 1956, *J. D. A. Stainton 1213* (BM); **Sagarmatha**, Solukhumbu, 2660 m, 1997, *M. Wakabayashi et al. 9715296* (E). **BHUTAN**: **Bumthang**, 2900 m, 1949, *F. Ludlow et al. 19647* (E) & *19674* (BM, E); **Ha**, Hatö, 2870 m, 1949, *F. Ludlow et al. 19620* (BM, E); **Hatey**, s. loc., 2900 m, 1938, *B. J. Gould 1541* (K); **Paro**, Belala, 2740 m, 1937, *F.Ludlow & G. Sherriff 3568* (BM); **Thimphu**, Duke-la, 2740 m, 1914, *R. E. Cooper & A. K. Bulley 2665* (E). **YARLUNG ZANGBO-BRAHMAPUTRA: Lhakang**, 3810–4110 m, 1933, *F. Ludlow & G. Sherriff 506* (BM); **Chayul-Charms** Rv., 3960–4270 m, 1936, *F. Ludlow & G. Sherriff 2431* (BM, E); 3660 m, 1936, *F. Ludlow & G. Sherriff 2714* (BM); **Chayul Chu**, 3350 m, *F. Ludlow & G. Sherriff 2423* (BM).

4. **Pseudocodon hirsutus** (Hand.-Mazz.) D. Y. Hong in Wang et al., J. Syst. Evol. 52(5): 548. 2014 & Monogr. *Codonopsis* Allied Gen. Camp. 203. 2015. ≡ *Codonopsis limprichtii* var. *hirsuta* Hand.- Mazz., Akad. Wiss. Wien Math.-Naturwiss. Kl. Anz. 61: 169. 1924. ≡ *Codonopsis convolvulacea* var. *hirsuta* (Hand.- Mazz.) Nannf. in Handel- Mazzetti, Symb. Sin. 7(4): 1077. 1936. ≡ *Codonopsis hirsuta* (Hand.-Mazz.) D. Y. Hong & L. M. Ma in Li & Zhu, Fl. Sichuan. 10: 546. 1992. Type: China, "Prov. Setschwan austro-occ.: Ad terram abruptam calceam reg. temperatae supra vicum Lodjiahosan pr. oppidum Yenyüen"(Sichuan, Yanyuan), 3000 m, 30 Sept. 1914, *Handel-Mazzetti 5423* (lectotype designated by Hong, 2015: WU; isolectotype: W).

Roots globose or ovoid-globose, 1–2 cm in diameter. Stems up to 1 m long, decumbent below, twining only at upper part, hirsute at leafing section, very rarely glabrous. Cauline leaves nearly opposite or alternate, aggregated at lower part, sessile or with extremely short petiole up to 8 mm long; leaf blade ovate or narrowly oblong, 1.5–8 cm long, 0.3–3.2 cm broad, adaxially glabrous, rarely sparsely hirsute, abaxially densely hirsute, very rarely glabrous, base cuneate or rounded, margin entire or sinuous-crenulate, often revolute, apex acute or obtuse. Flowers solitary, terminal. Calyx tube adnate to hypanthium up to near top, obconical, 4–7 mm across, glabrous; calyx lobes subulate or linear-lanceolate, 6–18 cm long, 2–6 mm broad, apex acuminate or acute, entire. Corolla pale blue or blue-purple; corolla lobes elliptic, 1–3.5 cm long, 0.5–1.6 cm broad. Capsule short-conical at superior part, obconical at inferior part. Seeds numerous, brown-yellow, smooth. **Chromosomes**: 2n=14. Figure 47–56: a.

Phenology: flowering in August and September; fruiting in October.

Habitats: open thickets, waste fields, sides of fields, or sunny grassy slopes; calcareous soils; 2400–3400 m.

Distribution: S Hengduan (Map 47–113); also in SE Yunnan of China.

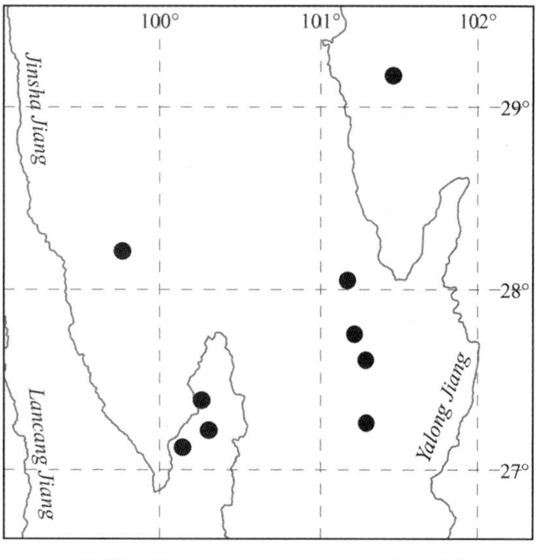

Map 47–113: **Pseudocodon hirsutus** (Hand.-Mazz.) D. Y. Hong

S HENGDUAN: Jiulong, Tanggo, 3200 m, 1979, *Q. Q. Wang 21134* (CDBI); **Liziping,** Baicaoping Village, 2800 m, 2011, *D. Y. Hong et al. H11034* (PE); **Muli,** No. 1 Distr., Ninglang

Figure 47–56: **a**, **Pseudocodon hirsutus** (Hand. -Mazz.) D. Y. Hong: a section of stem and upper part of stem. **b**, **Pseudocodon graminifolius** (H. Lév.) D. Y. Hong: plant. **c** & **d**, **Pseudocodon retroserratus** (Z. T. Wang & G. J. Xu) D. Y. Hong: **c**, a leaf showing retroserrate and setose margin; **d**, flower. From Hong, *Monogr. Codonopsis Allied Gen. Camp.*, 2015.

to Eya, 2700–2900 m, 1983, *Qinghai-Xizang Exped. 14838* (PE); **Shangri-la** (Zhongdian), Hong Shan, 3180 m, 1995, *Roy. Bot. Gard. Edinburgh Exped. To Dêqên Pref. 131* (E); **Yanbian**, Yankou, Dawan, 2657 m, 2011, *D. Y. Hong et al. H11043* (PE) ; **Yanyuan**, Yangjuan, 2002, *2002 UW Biodiversity 2002–137* (SZ); Zhongxing Village, 2951 m, 2011, *D. Y. Hong et al. H11023* (PE); **Yulong (Lijiang)**, Ganhaizi, 2900 m, 2010, *D. Y. Hong et al. H10027* (PE); Xuesongcun to Qiaomaidi, 2800–3200 m, 2010 , *D. Y. Hong et al. H10031* (PE); below Xuesongcun, 2550 m, 2010, *D. Y. Hong et al. H10036* (PE).

5. **Pseudocodon graminifolius** (H. Lév.) D. Y. Hong in Wang et al., J. Syst. Evol. 52(5): 548. 2014 & Monogr. *Codonopsis* Allied Gen. Camp. 204. 2015. ≡ *Codonopsis graminifolia* H. Lév., Cat. Pl. Yunnan 24. 1915. Type : China, Yunnan, s. loc., 2600 m, Jul. 1911, *E. E. Maire s. n.* [lectotype designated by Hong in Wang & Hong, 2015: E (E00265634); isolectotypes: E (E00265635), P].

Codonopsis limprichtii Lingelsh. & Borza var. *pinifolia* Hand.-Mazz., Akad. Wiss. Wien Math.- Naturwiss. Kl. Anz. 61: 170. 1924. ≡ *Codonopsis convolvulacea* Kurz var. *pinifolia* (Hand.-Mazz.) Nannf. in Handel-Mazzetti, Symb. Sin. 7(4): 1077. 1936. ≡ *Codonopsis pinifolia* (Hand.-Mazz.) Grey-Wilson, Plantsman 12(2): 89 & 76. fig. A. 1990, comb. invalid. Type: China, Yunnan, Yangtze-Mekong Divide, Weixi, 2800–3250 m, 22 Sept. 1916, *Handel-Mazzetti 10053* (lectotype designated by Hong, 2015: WU; isolectotypes: E, W); Yunnan, Lidjiang, Yulung-schan, 3350 m, 31 Jul. 1914, *Handel-Mazzetti 4370* (paratypes: W, WU).

Plants glabrous throughout except filaments. Roots ovoid or oblong, 0.8–1.5 cm in diameter. Stems up to 1 m long, decumbent below, twining only at upper part, sometimes nearly ascending, simple or branched. Cauline leaves alternate, usually aggregated at lower part, sessile, pinoid, linear, rarely linear-lanceolate, 2–10.5 cm long, 0.1–0.5(–0.9) cm broad, base cuneate, margin entire, often revolute, apex acuminate. Flowers solitary, terminal mostly only on main stem, sometimes also on branches. Calyx tube adnate to hypanthium up to top, obconical; calyx lobes long-triangular, 4–10 mm long, 2–3 mm broad, entire. Corolla blue or purple, often with a red ring at base; corolla lobes elliptic, 0.8–2.6 cm long, 0.4–1.2 cm broad. Capsule obconical, ca. 6 mm in diameter. at inferior part, short-conical at superior part. Seeds brown-yellow. Figure 47–56: b.

Phenology: flowering from July to September; fruiting in September and October.

Habitats: open grassy slopes, thickets, or forests; 1500–3300 m.

Distribution: N and S Hengduan (Map 47–114); also in W Guizhou, S Sichuan (Jinyang), and C and NE Yunnan of China.

N HENGDUAN: Litang, s. loc., 3600 m, 1979, *Litang General Surv. 004* (SM). **S HENGDUAN: Binchuan**, Jizu Shan, 1946, *T. N. Liou 22183* (PE); Xiayang to Huxi, 1946, *T. N. Liou 21512* (PE); **Butuo**, Jiaojihe, 1500 m, 1959, *Sichuan Econ. Plant Surv. (Liang) 5684* (PE); **Dali**, Cangshan, 2740–3050 m, 1913, *G. Forrest 11516* (BM, E, PE); Lanfeng Shan, half way to top, 1944, *H. C. Wang 4449* (PE); **Dechang**, s. loc., 1800 m, 1978, *Hu 356* (SM); **Heqing**, Baiyan, Shaxi, 2600 m, 1940, *R. C. Ching 24456* (PE); Huangping, 1700 m, 1990, *X. F. Gao 518* (KUN); **Huidong**, s. loc. 1978, *Huidong Exped. 432* (SM); **Huili**, s. loc., 2600 m, 1932, *T. T. Yü 1485* (PE); **Jianchuan**, Shizhong Shan, 1975, *B. Y. Qiu 61092* (CDBI, PE); Yuhua, 2680 m, 2010, *D. Y. Hong et al. H10013* (PE); **Lanping**: Biluo Shan, 3050 m, 1964, *S. K. Wu 8932* (KUN); **Liziping**, Baicaoping, 2600 m, 2011, *D. Y. Hong et al. H11024* (PE); **Mianning**, Maoniu Shan, 3000 m, 1960, *J. S. Ying 4332* (PE); **Miyi**: s. loc., 2300 m, 1978, *s. coll. 239* (SM); **Muli**, 5 km, S of Muli, 3050 m, 1927, *F. Kingdon-Ward 4864* (E); **Ningnan**, s. loc., 2000 m, 1978, *Ningnan Exped. 639* (SM); **Shangri-la** (Zhongdian), Haba, 2700 m, 1962, *Zhongdian Exped. 2114* (KUN); **Xichang** (His-Chang), Lushan, 1978, *Y. W. Zhu 362* (SM); **Xide**, Hongmo, 2000 m, 1979, *s. coll. 827* (SM); **Yangbi**, Zhongshan, 2400 m, 1984, *K. Iwatsuki et al. 1135* (KUN); **Yanbian**, Dapingzi, 2800 m, 1978, *Yanbian Exped. 329* (SM); Yankou, Dawan, 2657 m, 2011, *D.*

Y. Hong et al. H11042 (PE); **Yanyuan**, Dalinsi 2600–2700 m, 1965, *F. G. Zhu 006* (PE); **Yulong (Lijiang)**, Ganhaizi, 2900 m, 2010, *D. Y. Hong et al. H10028* (PE); Licking Valley, 1910, *G. Forrest 6277* (BM, E, PE); Tiejiashan, 2600 m, 2010, *D. Y. Hong et al. H10025* (PE); Tong Shan, 27°20′N, August 1914, *G. Forrest 13134* (BM, E, PE); Xuesongcun to Qiaomaidi, 3100 m, 2010, *D. Y. Hong et al. H10032* (PE); **Zhaojue**, s. loc., 1979, *s. coll. 927* (SM).

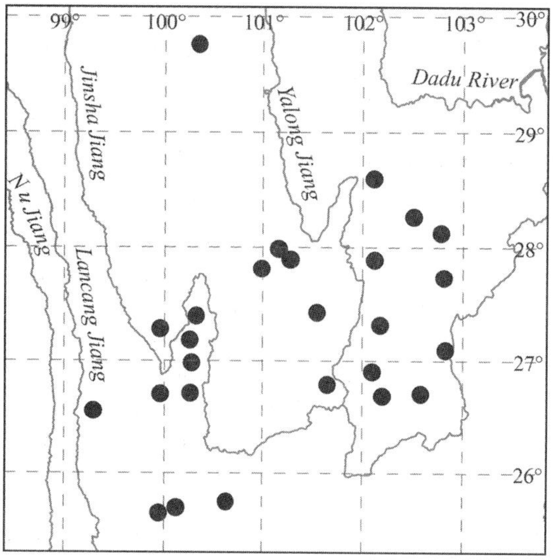

Map 47–114: **Pseudocodon graminifolius** (H. Lév.) D. Y. Hong

6. Pseudocodon retroserratus (Z. T. Wang & G. J. Xu) D. Y. Hong & Q. Wang in Wang & Hong, Phytotaxa 204(1): 62. 2015. ≡ *Codonopsis retroserrata* Z. T. Wang & G. J. Xu, Acta Phytotax. Sin. 31(2): 186. fig. 2. 1993. Type: China, Sichuan, Yanyuan, 2400 m, 5 Sept. 1986, *Z. T. Wang 869382* (holotype: CPU; isotype: PE).

Plants glabrous throughout except leaf margin and ciliate filament base. Roots globose or short-conical, 2–3 cm long, 1.5–2.2 cm in diameter. Stems twining, mostly over 1 m long. Leaves alternate, nearly sessile or with petiole ca. 2 mm; leaf blade papery, ovate or ovate-lanceolate, 1.5–4 cm long, 1–3 cm broad, base shallowly cordate or rounded, margin retroserrate, sparsely white-setose, apex acute. Flowers solitary, terminal only on main stem, or also on upper short branches; pedicel 5–30 cm long. Calyx tube entirely adnate to hypanthium, obconical, 5–7 mm long; calyx lobes lanceolate, 8–12 mm long, 3–5 mm broad. Corolla pale blue or violet; corolla lobes elliptic, 2.2–2.5 cm long, 1–1.5 cm broad, apex obtuse. Figure 47–56: c & d.

Phenology: flowering in August and September.

Habitats: thickets; 2400–2600 m.

Distribution: **endemic**; confined to S Hengduan (Map 47–115).

Notes: *Pseudocodon retroserratus* (*Codonopsis retroserrata*) was treated as a synonym of *Codonopsis convolvulacea* subsp. *forrestii* in *Fl. China* (Vol. 19), but it is a distinct species with retroserrate and white-setose leaf margin. Our recent molecular analysis shows that it forms an independent clade in the genus (Wang & Hong, 2015).

S HENGDUAN: **Muli**, near the town, 2600 m, 1937, *T. T. Yü 7669* (KUN, PE); Taoba, 2600 m, 2011, *D. Y. Hong et al. H11038* (PE).

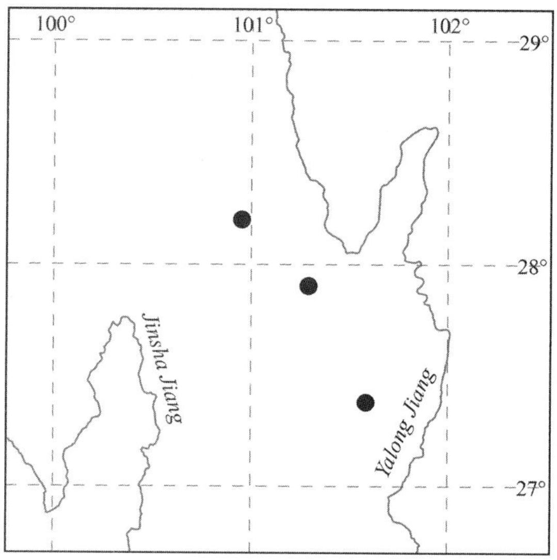

Map 47–115: **Pseudocodon retroserratus** (Z. T. Wang & G. J. Xu) D. Y. Hong & Q. Wang

7. Pseudocodon petiolatus D. Y. Hong & Q. Wang in Wang & Hong, Phytotaxa 204(1): 62. 2015. Type: China, Sichuan, Muli, 913 Farm, Zamin, 2600 m, thickets, 23 Aug. 1983, *Qinghai-Xizang Exped. 13331* (holotype: PE; isotype: PE).

Plants glabrous throughout except ciliate filament base. Roots elongated and attenuated with a constriction at middle part, over 15 cm long, 1.3 cm in diameter. Stems twining, slender, over 1 m long. Leaves alternate; petiole 1.5–2.5 cm long; leaf blade cordate to ovate, 2–3.5 cm long, 1.5–2.5 cm broad, base cordate, truncate to rounded, margin entire or shallowly crenulate, apex obtuse to acute. Flowers solitary, terminal on main stem. Calyx tube adnate entirely to hypanthium, obtriangular; calyx lobes strongly reflexed even since anthesis, linear-lanceolate, 7–9 mm long, 3–3.5 mm broad. Corolla blue; corolla lobes elliptic, ca. 2 cm long, 1 cm broad. Filaments 4 mm long, base strongly dilated into horizontally rectangular, ciliate; anthes 4 mm long. Style 2 mm long; stigma lobes elliptic- oblong, ca. 4 mm long, 2.5 mm wide. Figure 47–57: a–c.

Phenology: flowering in August.

Habitats: thickets; ca. 2600 m.

Distribution: **endemic**; confined to S Hengduan (Muli) (Map 47–116).

Notes: *Pseudocodon petiolatus* differs distinctly from all the other species of the genus in having roots elongated with a constriction at middle part, petiole 1.5–2.5 cm long, nearly as long as leaf blade, calyx lobes strongly reflexed even since anthesis, and stigma lobes large.

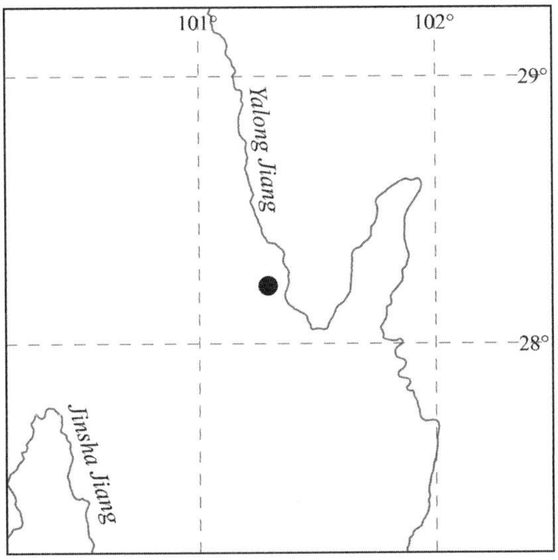

Map 47–116: **Pseudocodon petiolatus** D. Y. Hong & Q. Wang

8. Pseudocodon rosulatus (W. W. Smith) D. Y. Hong in Wang et al., J. Syst. Evol. 52(5): 548. 2014 & Monogr. *Codonopsis* Allied Gen. Camp. 209. 2015. ≡ *Codonopsis rosulata* W. W. Smith, Notes Roy. Bot. Gard. Edinburgh 13: 157. 1921. Type: China, Sichuan, Muli, 28°12′N, 3350 m, among heavy grasses, Aug. 1918, *G. Forrest 16856* (holotype: E)

Plants glabrous throughout except stamens. Roots globose or ovoid, 0.5–1.5 cm long, 0.5–1 cm in diameter. Caudex 2–4 cm long, less than 1 mm in diameter, with 2–4 alternate and very small scales. Stems single, erect, ascending, or more or less decumbent, 10–50 cm tall, simple or with 2–3 branches. Basal leaves rosulate; petiole 0.1–3 cm long; leaf blade broad-ovate, ovate or elliptic, 0.5–7 cm long, 0.4–5 cm broad, base slightly cordate or rounded, margin crenate, apex obtuse; cauline leaves scale-like, lanceolate or linear-lanceolate, 1–5 mm long, 0.5–2 mm broad, base rounded or cuneate, margin sinuous-crenate, or nearly entire, apex acute. Flowers mostly 1–3, rarely up to 8, terminal on main stem and branches; pedicel 4–15 cm long. Calyx tube adnate to hypanthium up to near top, obconical, 2–5 mm long, calyx lobes ovate-triangular, 3–6 mm long, 1.5–3 mm broad, 1 vein visible, apex acute, sinus between calyx lobes narrow and pointed. Corolla pale blue, blue or blue-purple; corolla lobes ovate or broad-ovate, 7–14 mm long, 5–7 mm broad, apex acute or obtuse. Capsule obconical below, slightly truncate-obtuse above, ca. 1.2 cm long, 0.6 cm in diameter. Seeds compressed, ellipsoid, brown-yellow. Figure 47–57: d.

Phenology: flowering and fruiting in September and October.

Habitats: grassy slopes, open oak-pine forests, or edges of forests; 2600–3400 m.

Distribution: **endemic**; confined to S Hengduan (Map 47–117).

S HENGDUAN: Liziping, 3180 m, 1978, *C. S. Cao et al. 8390* (CDBI, SZ); **Muli**, Qing'o Forest Farm, 3400 m, *S. K. Wu 2598* (CDBI, KUN); **Yanbian**, Dapingzi, 2700 m, 1978, *Yanbian Exped. 372* (SM); Yankou, Dawan, 2700 m, 2002, *X. F. Gao & Y. L. Peng 3669* (PE); Yankou,

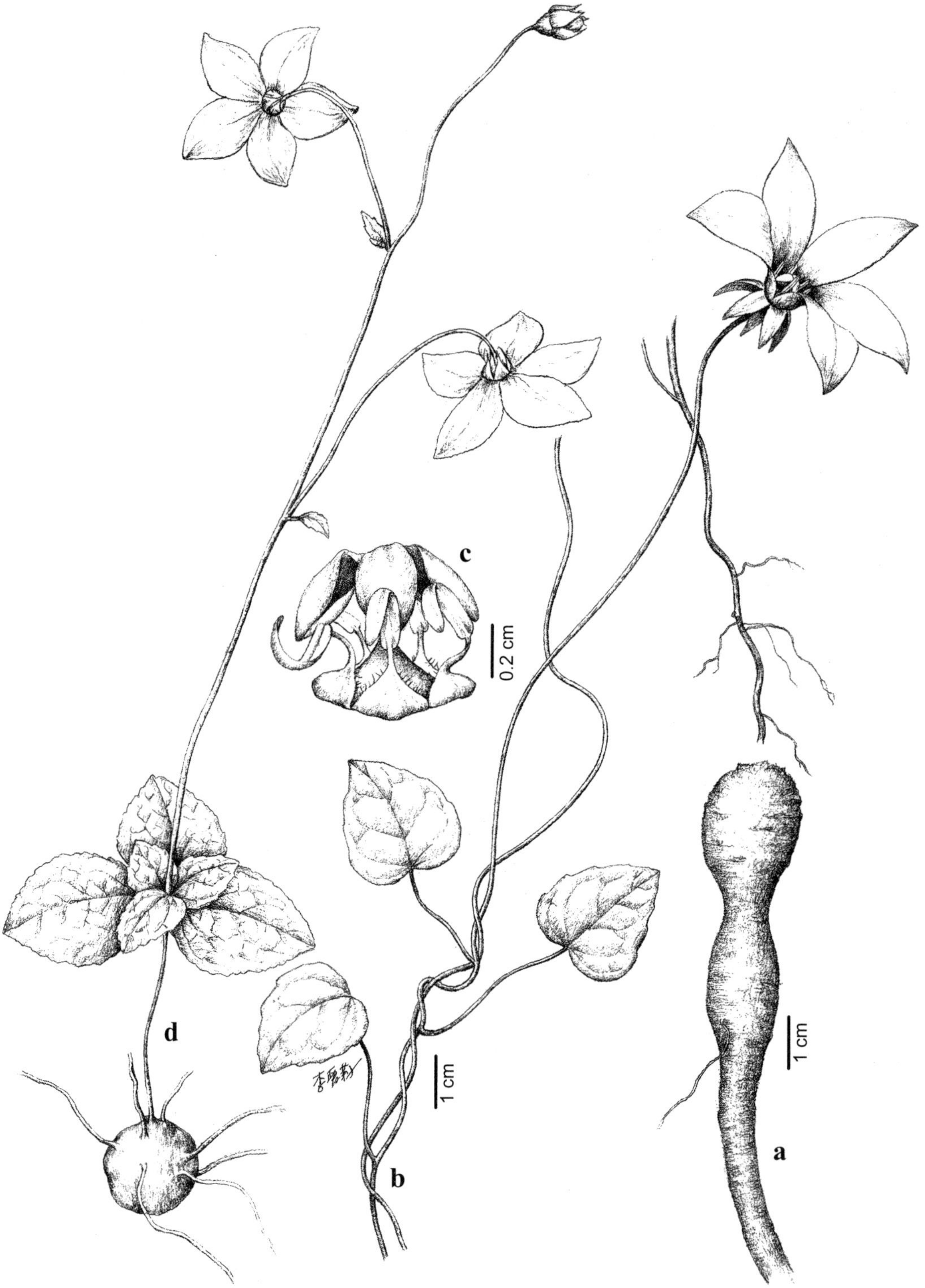

Figure 47–57: **a–c**, **Pseudocodon petiolatus** D. Y. Hong: **a**, root and lower part of stem; **b**, a section of stem and upper part of stem; **c**, a flower with calyx and corolla removed, showing stamens, style, and stigma. **d**, **Pseudocodon rosulatus** (W. W. Smith) D. Y. Hong: plant. From Hong, *Monogr. Codonopsis Allied Gen. Camp.*, 2015.

Dawan, 2657 m, 2011, *D. Y. Hong et al. H11039* (PE) ; **Yanyuan**, s. loc., 3000 m, 1914, *Handel-Mazzetti 5422* (W, WU).

Trib. 2. Wahlenbergieae Endl., Gen. Pl. 513. 1838; Hong & Wang, J. Syst. Evol. 53(3): 207. 2015. Type: *Wahlenbergia* Schrad. ex Roth

Wahlenbergioideae (Endl.) Kolak., Bot. Zhurn. 72(12): 1574. 1987. p. p.

Fruit a capsule, dehiscent by apical (above the insertion of calyx lobes) valves. Pollen porate, with poroid apertures. Chromosome x =7, 8, 9, 17.

Sixteen genera, all in Africa except for *Wahlenbergia*, which is widely distributed in Africa and Australia, with some species in Asia and Europe; one genus in our region.

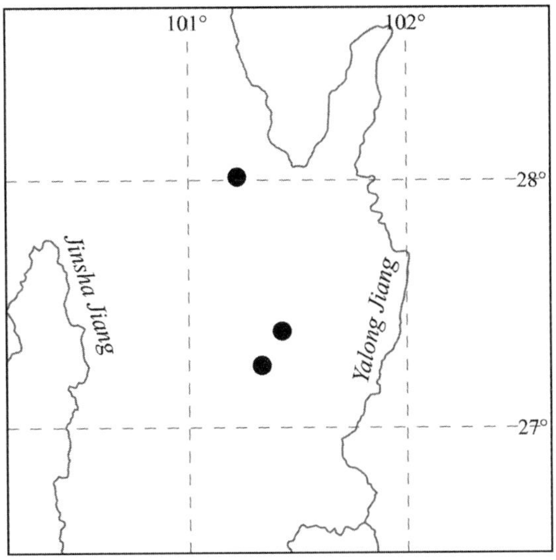

Map 47–117: **Pseudocodon rosulatus** (W. W. Smith) D. Y. Hong

8. **WAHLENBERGIA** Schrad. ex Roth

Roth, Nov. Pl. Sp. 399. 1821, nom. cons. ≡ *Campanula* sect. *Wahlenbergia* (Schrad. ex Roth) D. Dietr. (1839). Type: *Wahlenbergia elongata* (Willd.) Schrad. ex Roth [*Campanula elongata* Willd. = *Wahlenbergia campanuloides* (Delile) Vatke].

Cephalostigma A. DC. (1830).

Herbs perennial or annual, rarely subshrubs or shrubs. Leaves alternate, rarely opposite, usually sessile. Flowers small to medium-sized, sessile or pedicellate, terminal or axillary, solitary or in thyrses, panicles, or fascicles. Calyx lobes typically 5. Corolla campanulate or infundibular, lobed, cleft, or divided; corolla lobes typically 5. Stamens typically 5; filaments free and distinct, dilated at base, dilated part triangular, ciliate; anthers coherent. Ovary inferior, (2–) 3 (–5)-locular; stigma (2–) 3 (–5)-fid, segments linear. Fruit a capsule, apically loculicidal. Seeds many or few.

Two hundred and sixty species; mainly in the Southern Hemisphere, particularly in S Africa and Australia, extending northwards to western Europe, South and East Asia, and northern South America; three species (one endemic) in the Pan-Himalaya.

1a. Leaves all or mostly opposite; flowers usually solitary and terminal; plants glabrous .... 2. *W. peduncularis*
1b. Leaves all alternate; flowers usually 2–several in a cyme; plants hirsute, rarely entirely glabrous.
2a. Perennials with a carrot-shaped tap root; plants very sparsely hirsute, rarely entirely glabrous; corolla cleft to 1/3–2/3, lobes oblong .... 1. *W. marginata*
2b. Annuals; plants rather densely hirsute; corolla cleft to near base, lobes linear ..... 3. *W. hirsuta*

1. Wahlenbergia marginata (Thunb.) A. DC., Monogr. Camp. 143. 1830; Nannfeldt, Acta Horti Gothob. 5: 31. 1930; Hong, Fl. Reipubl. Popularis Sin. 73(2): 29. 1983. ≡ *Campanula marginata* Thunb. in Murray, Syst. Veg. (ed. 14): 211. 1784. Type: Japan, not seen.

Wahlenbergia dehiscens (Roxb. ex Wall.) A. DC., Monogr. Camp. 145. 1830. ≡ *Campanula dehiscens* Roxb. ex Wall., As. Res. 12: 571. t. 6, 1819. Type: Nepal, *Wallich 1294* (holotype: K).

Wahlenbergia agrestis (Wall.) A. DC., Monogr. Camp. 145. 1830. ≡ *Campanula agrestis* Wall. in Roxburgh, Fl. India 2: 97. 1824. Type: Nepal (Napalia), *Wallich 1292* [**lectotype here designated**: K (K000781159); isolectotype: K].

Herbs perennial, rarely annual. Roots elongate, attenuate. Stems erect or ascending, 10–50 cm tall, branched at base, glabrous or sparsely hirsute below. Leaves alternate, mostly at lower part of stem, lowermost ones usually caducous, others sessile or with petiole up to 7 mm long; leaf blade spatulate, oblanceolate, elliptic, or linear, 1–4 cm long, 0.2–0.8 cm broad, glabrous or sparsely hirsute, margin sinuate, sparsely serrulate, or entire. Flowers terminal and sometimes axillary, upright; pedicel slender, up to 20 cm long. Calyx tube obovoid or conical, glabrous; calyx lobes triangular or subulate. Corolla blue, broadly campanulate, 3–12 mm long, cleft to

Figure 47–58: **a–d**, **Wahlenbergia marginata** (Thunb.) A. DC.: **a**, plant; **b**, flower; **c**, longitudinal section of flower; **d**, cross section of ovary. Redrawn by Ai-Li LI from *Icon. Cormophyt. Sin.* 4, 1975.

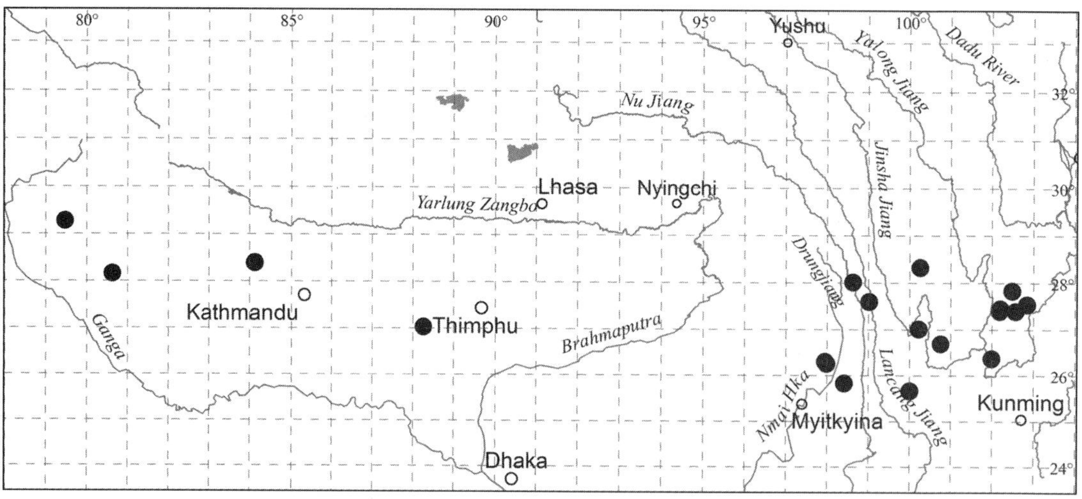

Map 47–118: **Wahlenbergia marginata** (Thunb.) A. DC.

1/3–2/3; corolla lobes obovate, oblong, or narrowly oblong. Capsule obconical, obovoid, or subglobose, 3–8 mm long, 3-locular. Seeds yellow-brown or dark brown, ellipsoid, slightly compressed, 0.3–0.5 mm long, smooth. **Chromosomes**: 2n=18, 36, 54, 64, 72, and ca. 90. Figure 47–58

Phenology: flowering and fruiting from February to May.

Habitats: waste areas, fields, open or grassy slopes; below 2800 m.

Distribution: U Ganga & Indus, Nepal, Sikkim & Darjiling, S Hengduan, and U Irrawaddy (Map 47–118); also in Indonesia, Japan, Korea Peninsula, Laos, Malaysia, Philippines, Sri Lanka, and Vietnam, and other parts of China, India, and Myanmar; also in Australia, Papua New Guinea; naturalized in the Hawaiian Islands and North America.

Notes: The types of the two synonyms are from Nepal, but I have not seen any additional specimen from Nepal.

U GANGA & INDUS: s. loc., 1844, *Edgeworth 241* (K) & 242 (K); **Loodhaua** (?), 1844, *Edgeworth 243* (K); **Doddabetta**: Nilgiris, 1973, *C. Townsend & T. P. Ramamoorthy 134* (K); **Kumaon**: 1200 m, *R. Strachey & J. E. Winterbottom 11* (K); **Upper Gangetie Plain**, Kheri, 1898, *Inufat 22217* (K); **Banda**: Border of Jhil, 1901, *A. S. Beil 184* (K). **SIKKIM & DARJILING: Darjiling**, 1830 m, 1881, *J. S. Gamble 9562* (K). **S HENGDUAN**, **Butuo**, Tuojue, 2475 m, 1976, *Vegetation Group (Butuo) 14142* (PE); **Decang**, Yinlu Commune, 2150 m, *Exped. to Decang 11979* (PE); **Daliang Shan**, 2500 m, 1934, *T. T. Yü 3854* (PE); **Gongshan**, Salween Valley, Dara, 1750 m, 1938, *T. T. Yü 19192* (PE); **Huili**, Lixi, 2000 m, 1976, *Vegetation Exped. to Huili 12452* (PE); **Meigu**, Houbutuo, 1959, *Sichuan Econ. Plant Surv. (Liang 59) 5959* (PE); **Puge**, near the city, 1600 m, 1959, *C. T. Kuan 7314* (PE); **Weixi**, Kangpu, 2500 m, 1935, *C. W. Wang 64331* (PE); **Yangbi**, Jinzhansanchang, 1940. *R. C. Ching 22338* (PE); **Yongshen**, 1960. *S. Jiang 6192* (PE); **Yuexi**, Zhongsuoba, 1700 m, 1959, *s. coll. 3445* (PE); **Yulong** (Lijiang), Xuesongcun, 2850 m, 2010, *D. Y. Hong et al. H10033* (PE); **Zhaojue**, Wanchang, 2980 m, 1982, *Z. X. Tang et al. 184* (PE). **U IRRAWADDY**: **N'Maikha-Salween Divide**, 1931, *G. Forrest 29706* (PE); Upper Burma, 1887, *C. B. Collett* s. n. (K).

2. Wahlenbergia peduncularis (Wall. ex A. DC.) Hook. f. & Thomson, J. Proc. Linn. Soc. 2: 22. 1857; Clarke in Hooker, Fl. Brit. India 3: 430. 1881; Haridasan & Mukherjee in Hajra & Sanjappa (eds), Fasc. Fl. India 22: 114. 1996. ≡ *Campanula peduncularis* Wall. ex A. DC. in de Candolle, Prodr. 7: 483, 1839. Type: India, Kumaon, *Wallich 7139* (holotype: K).

Herbs perennial, creeping, glabrous throughout except for filaments. Roots carrot-shaped, up to 5 mm in diameter. Caudex sometimes branched. Stems slender, up to 15 cm long. Leaves opposite, or upper ones alternate; petiole 2–4 mm long; leaf blade ovate, elliptic or lanceolate, 0.6–1.8 cm long, 0.3–0.9 cm broad, base rounded, margin crenulate or entire, apex obtuse. Flowers solitary, terminal; pedicel 1.5–3.5 cm long. Calyx tube obconical; calyx lobes linear-lanceolate, ca. 5 mm long, 2 mm broad, entire. Corolla pale lilac, lobed to near base, 5–7 mm long; corolla lobes lanceolate. Filaments dilated at base, dilated part villous-ciliate. Ovary 3-locular; stigma 3-fid. Capsule obconical, ca. 5 mm long, 4 mm in diameter.

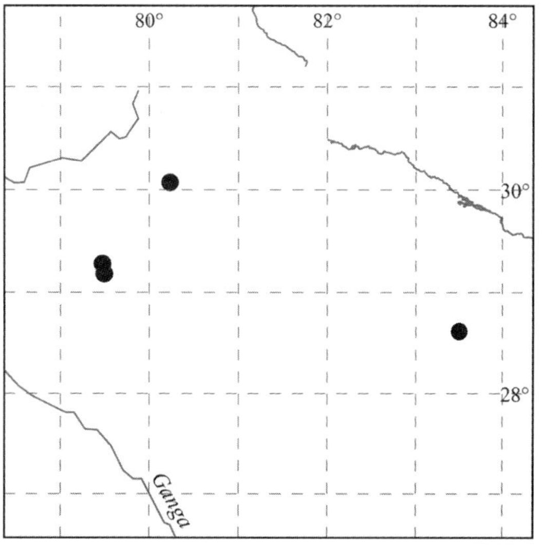

Map 47–119: **Wahlenbergia peduncularis** (Wall. ex A. DC.) Hook. f. & Thomson

Phenology: flowering in April and May.

Habitats: sandy terraces, near irrigation channel; 600–2700 m.

Distribution: **endemic**: U Ganga & Indus and C Nepal (Map 47–119).

U GANGA & INDUS: Kumaon, Numsiari (Gorifst), 610 m, 1910, *N.Gill 487* (E, K); Kumaon, Chaur, 2040 m, *R. Strachey & J. E. Winterbottom 8* (K). **C NEPAL**: **Dhawalagiri**, Mayangdi Khola, 1100 m, 1954, *Stainton, Sykes & Williams 2540* (E).

3. Wahlebergia hirsuta (Edgew.) Tuyn in van Steenis, Fl. Males. Ser. 1. 6: 113 (in note). 1960. ≡ *Cephalostigma hirsutum* Edgew., Trans. Linn. Soc. London 20: 81. 1846. Type: India, Himachal Pradesh, Banasar, 1220–1520 m, 1844, *M. P. Edgeworth 252* (**lectotype here designated**: K; isolectotype: BM).

Herbs annual. Roots attenuate, up to 2 mm in diameter. Stems erect, 3–10 cm tall, simple or multibranched, diffuse, hirsute. Leaves alternate, sessile; leaf blade elliptic or ovate, 0.5–3.5 cm long, 0.3–1.5 cm broad, adaxially glabrous, abaxially hirsute or hispidulous along veins, base cuneate, margin entire or sparsely serrate, apex obtuse or acute. Flowers in a terminal cyme of 2-several; pedicel 0.5–1.5 cm long, sparsely hispidulous. Calyx tube hemispherical, hirsute or nearly glabrous, 2–3 mm in diameter; calyx lobes narrowly triangular, 1.5–2 mm long, ca. 0.5 mm broad, sparsely hirsute-ciliate. Corolla blue, pale blue or whitish, slightly longer than or as long as calyx lobes, lobed to near base; corolla lobes linear. Filaments dilated below, dilated part villous. Style exserted; stigma 3-fid. Capsule hemispherical, ca. 3 mm in diameter. **Chromosomes**: 2n=42.

Phenology: flowering and fruiting from June to November.

Habitats: forests, open thickets, or cultivated fields; 800–1850 m.

Distribution: Jammu & Kashimir and U Ganga & Indus (Map 47–120); also in Africa: Cameroon, Congo, Ethiopia, Nigeria, Oman, Tanzania, and Uganda.

JAMMU & KASHIMIR: **Kashmir**, Basasli, 1370 m, 1876, *C. B. Clarke 31560* (BM, K). **U GANGA & INDUS**: **Shimla**, 1520 m, 1885, *H. Collett 654* (BM, K); loc. eodem. 1830 m, 1884, *J. R.Drummond 2751* (BM, K); **Dalhousie**, 1220 m, 1874, *C. B. Clarke 23122* (K).

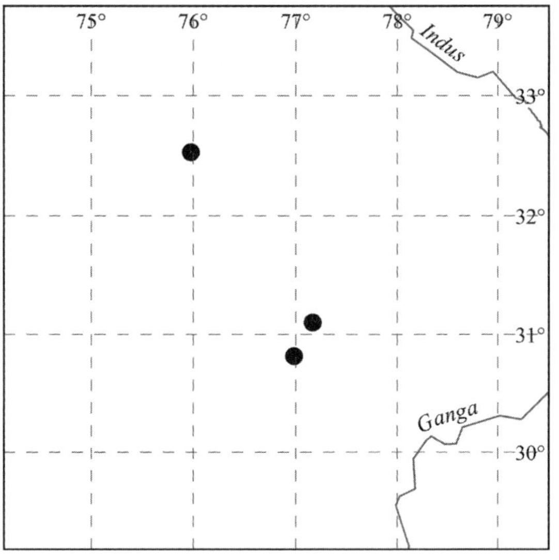

Map 47–120: **Wahlebergia hirsuta** (Edgew.) Tuyn.

Trib. 3. Campanuleae Durmort., Fl. Berg. 58. 1827; Hong & Wang, J. Syst. Evol. 53(3): 207. 2015. Type: *Campanula* L.

Fruit a capsule dehiscent by lateral pores or valves (below the insertion of calyx lobes) or dry and indehiscent. Pollen porate with ponoid apertures. Chromosomes x=6, 7, 8, 9, 10, 11, 13, 15, 17.

Twenty seven genera, cosmopolitan, but mostly in the area from the Mediterranean to western Asia; five genera in the Pan-Himalaya.

9. CAMPANULA L.

Linnaeus, Sp. Pl. 1: 163. 1753. Type: *Campanula latifolia* L. (designated by Hitchc. & M. L. Green, 1929).

Neocodon Kolak. & Serdyuk.

Herbs perennial, with long horizontal rhizomes or with short caudex and thickened, more or less fleshy roots, less often annual. Basal leaves sometimes rosulate; cauline leaves all alternate. Flowers solitary and terminal, sometimes also in upper axils and inflorescence raceme-like, or in cymes; cymes paniculate or capitulate, latter subtended by an involucre. Calyx adnate to hypanthium; calyx lobes 5, sometimes with an appendage between them. Corolla campanulate, tubular-campanulate, or infundibular, sometimes nearly rotate, 5-lobed. Filaments dilated at base; anthers coherent or rarely connivent. Disc absent. Ovary inferior, 3–5-locular. Fruit a capsule, poricidal by lateral pores, crowned by persistent calyx lobes. Seeds many, ellipsoid, smooth.

About 420 species, throughout the Arctic and N temperate regions, most abundant from the Mediterranean to the Caucasus; 27 species (12 endemic) in the Pan-Himalaya.

1a. Basal leaves rosulate, persistent at anthesis; cauline leaves mostly at base, upper ones, if present, usually linear, sessile or nearly so; capsule poricidal above middle (subg. 2. **Rapunculus**).

2a. Plants glabrous, with horizontal rhizomes; stems simple.

3a. Calyx tube narrowly cylindrical; calyx lobes filiform, longer than corolla . 25. *C. aristata*

3b. Calyx tube obovoid or obconic; calyx lobes subulate or narrowly triangular, shorter than corolla .... 27. *C. immodesta*

2b. Plants variously hairy, without horizontal rhizomes; stems simple or branched.

4a. Stems villous; calyx sparsely villous, rarely glabrous; capsule 4–8 mm long .... 24. *C. calcicola*

4b. Stems glabrous; calyx glabrous; capsule 9–20 mm long.

5a. Lower cauline leaves spatulate to ovate, 1–3 cm long; flowers pendent .... 23. *C. crenulata*

5b. Lower cauline leaves ovate to linear, 0.5–1.5 cm long; flowers upright .... 25. *C. chrysospleniifolia*

1b. Basal leaves usually wilted at anthesis; cauline leaves many and evenly distributied; capsule poricidal at base (subg. 1. **Campanula**).

6a. Calyx with appendages between calyx lobes.

7a. Corolla 3–6.5 cm long, lantern-like; leaves up to 7 cm long .... 18. *C. punctata*

7b. Corolla 0.8–2 cm long, campanulate or tubular; leaves less than 3 cm long.

8a. Corolla 0.8–1 cm long; style exserted beyond corolla lobes.

9a. Plants up to 45 cm tall, caespitose or not; stems glabrous to sparsely hirsute, white .... 19. *C. leucoclada*

9b. Plants ca. 10 cm tall, caespitose; stems hirsute, not white .... 20. *C. polyclada*

8b. Corolla 1.3–2 cm long; style included.

10a. Leaf blade 4–6 mm broad; corolla campanulate, glabrous .... 21. *C. staintonii*

10b. Leaf blade 5–15 mm broad; corolla tubular, hairy outside .... 22. *C. leucantha*

6b. Calyx without appendages.

11a. Flowers in a head of 2 to several, neither pedunculate, nor pedicellate, heads aggregated in a large compound head .... 2. *C. glomeratoides*

11b. Flowers solitary or in a lax inflorescence.

12a. Leaf blade 7–13 cm long, 3–7 cm broad; flowers 3–5.5 cm long; stems 0.7–2 m tall, glabrous .... 1. *C. latifolia*

12b. Leaf blade much smaller; flowers 4–2.8 cm long; stems up to 70 cm tall, hairy or glabrous.

13a. Annual herbs; rosulate basal leaves often present at anthesis; each branch with compound cymes .... 17. *C. dimorphantha*

13b. Perrenial herbs; basal leaves often absent at anthesis; branches with a single cyme or solitary flower.

14a. Corolla rotate or nearly rotate, cleft for 3/4–5/6 of overall length; calyx lobes glabrous on both surfaces; flowers solitary and terminal.

15a. Leaf blade obovate-elliptic or elliptic, sparsely hispidulous; calyx tube hairy .... 3. *C. nakaoi*

15b. Leaf blade linear, glabrous; calyx tube glabrous .... 7. *C. rotata*

14b. Corolla campanulate or tubular-campanulate, cleft for 1/2 or slightly over 1/2 of overall length; calyx lobes hairy; flowers usually in a cyme or solitary.

16a. Flowers only setose; corolla 3–8 mm long (corolla $\leq$ 8 mm long in *C. yunnanensis*, but calyx lobes in that species with a pair of teeth).

17a. Leaf blade lanceolate to linear; calyx lobes setose only on veins .... 5. *C. tenuissima*

17b. Leaf blade ovate to ovate-lanceolate; calyx-lobes densely setose .... 6. *C. tristis*

16b. Flowers with other types of hairs; corolla mostly longer.

18a. Cauline leaves linear, 1–6 cm long, with length/width ratio over 5 .... 4. *C. sylvatica*

18b. Cauline leaves elliptic or ovate, with length/width ratio less than 2.

19a. Cauline leaves less than 5 mm long; stems many from one caudex, caespitose; corolla lobed to near middle .... 8. *C. microphylloides*

19b. Cauline leaves much larger; stems single or several from one caudex; corolla mostly more shallowly lobed.

20a. Leaves and calyx green, hispid, not tomentose and gray-white; plants erect or ascending, 20–70 cm tall.

21a. Calyx lobes 2–4 mm long, less than 1 mm broad, with one pair of verrucose denticles ; corolla lobes slightly longer than tube .... 10. *C. mekongensis*

21b. Calyx lobes 2–8 mm long, 1.5–3 mm broad, mostly entire; corolla lobes ca. 1/2 as long to ca. as long as tube .... 9. *C. pallida*

20b. Leaves and calyx tomentose and gray-white; plants prostrate or ascending, less than 35 cm long.

22a. Calyx lobes elliptic or oblong, widest above base, not at base, entire; abaxial surface of leaf blade and calyx-lobes covered with mixed white hispidulous and hispid hairs .... 14. *C.argyrotricha*

22b. Calyx lobes widest at base; leaf blade and calyx lobes glabrous or mostly covered with one kind of hairs.

23a. Calyx lobes triangular, with a pair of teeth; corolla up to 8 mm long .... 15. *C. yunnanensis*

23b. Calyx lobes subulate, entire; corolla usually longer (8–10 mm long but tubular in *C. alsinoides*).

24a. Corolla tubular, 2–3(–5) mm in diameter; stems very slender, filiform .... 13. *C. alsinoides*

24b. Corolla campanulate, wider; stems not filiform.

25a. Corolla rotate above, lobed to over middle; calyx lobes always entire; style 1.3–1.5 cm long, much longer than corolla .... 16. *C. chinensis*

25b. Corolla campanulate, lobed to 1/3, at most to 2/5; calyx lobes sometimes toothed; style slightly longer than corolla.

26a. Style slightly longer than corolla; corolla 1–1.8 cm long; calyx lobes 3–5 mm long, sometimes with a pair of teeth...................... 11. *C. cana*

26b. Style much shorter than corolla; corolla 1.3–2.8 cm long; calyx lobes 5–7 mm long, always entire .... 12. *C. cashmeriana*

Subg. 1. Campanula

Basal leaves absent or wilted at anthesis; cauline leaves evenly distributed, upper ones sessile or petiolate. Flowers many, in cymes aggregated into a raceme, spike, or capitulum. Capsule poricidal toward base.

About 280 species, throughout the Arctic and N temperate regions of the Old World, extending southwards into Macronesia, tropical E Africa, the Arabian Peninsula, Indochina, and eastwards into the Aleutian Islands; 27 species (11 endemic) in the Pan-Himalaya.

1. Campanula latifolia L., Sp. Pl. 165. 1753; Hooker & Thomson, J. Proc. Linn. Soc. Bot. 2: 22. 1857; Sims, Curtis's Bot. Mag. 52: tab. 2553. 1825; Clarke in Hooker, Fl. Brit. India 3: 439. 1881. Type: "Habitat in Angliae, Sueciae montosis, sepibus." s. n. (lectotype designated by Haridasan & Mukherjee in Hajra & Sanjappa, Fasc. Fl. India 22: 47. 1996: LINN, n. v.).

Herbs perennial. Rhizomes horizontal. Stems erect, 0.6–2 m tall, glabrous. Leaves alternate, lower ones with petiole up to 2.5 cm long, upper ones sessile; leaf blade broad-lanceolate to ovate, 5–14 cm long, 2–7 cm broad, adaxially nearly glabrous, abaxially hirsute along veins, margin dentate to serrate. Flowers usually solitary, sometimes more in axils of upper leaves forming a raceme; pedicel 0.3–3.5 cm long, often curved, with or without a pair of bracteoles. Calyx tube obovoid-conical, glabrous but sparsely puberulent at base; calyx lobes subulate, 11–15 mm long, 2–3.5 mm broad, reflexed since anthesis, margin hispidulous-ciliate, entire or denticulate. Corolla dark purple-blue or violet, narrow-campanulate, 3–5.5 cm long, corolla lobes ca. 2/3 as long as tube, acute. Filaments 4–6 mm long, slightly dilated at base, dilated part ciliate. Style included; stigma 3–fid, segments linear. Capsule ca. 1 cm long. **Chromosomes**: 2n = 34.

Phenology: flowering and fruiting from June to October.

Habitats: meadows; 2400–3800 m.

Distribution: N Pakistan, Jammu & Kashmir, U Ganga & Indus, U Yarlung Zangbo, and W Nepal (Map 47–121); also in Armenia, Azerbaijan, Georgia, Iran, Russia, Turkey; Europe.

Notes: The unvalid name *Campanula brunonis* Wall., Cat. 1285, belongs here.

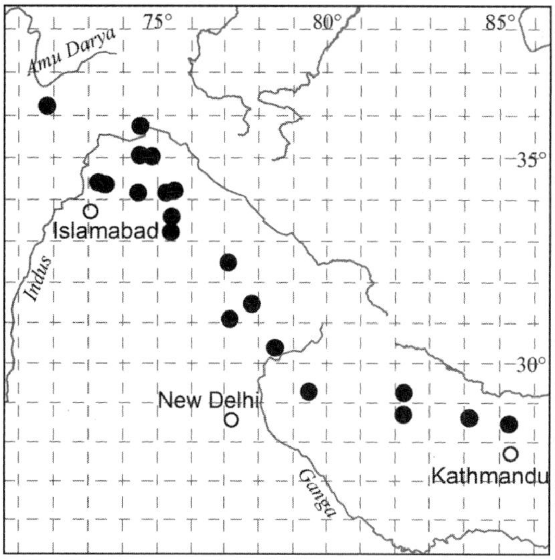

Map 47–121: **Campanula latifolia** L.

N PAKISTAN: **Azad Kashmir**, Muzaffarabad, Sinnai Leepa Valley, 1986, *M. Qaiser & R. Y. Hashnu 7784* (KUH); **Baltistan**, Astore, on way to Dumail, 2003, *S. W. Khan et al.531* (KUH); Kishenganga, Shönthar Valley, 2440 m, 1935, *F. Ludlow & G. Sherriff 1500* (E); Kishenganga Valley and road to Nanga Parbat, below Khel, 1830 m, 1939, *R. R. & I. D. Stewart 17795* (KUH); **Giligit**, Gorai-Gilgit Road, 2900 m, 1928, *B. B.Osmaston 24* (K); **Hazara**: Kagan Valley, 3110 m, 1896, *Inayat 19892* (K); **Swat**, Bishigram, ca. 2740 m, 1953, *R. R. Stewat & A. Rahman 24992* (BM). **JAMMU & KASHMIR**: **Amarnath**, 3050 m, 1977, *J. D. A. Stainton 7952* (E); Kashmir, **Desu**, 2590 m, 1936, *M. K. Timins 182* (BM); **Gulmarg**, 2590 m, 1919, *H. H. Rich 1213* (K); near Gulmarg, 2440–2740 m, 1892, *J. F. Duthie* s. n. (E); **Liddar**, 3350 m, 1956, *O. Pulunin 56–494* (BM). **U GANGA & INDUS**: **Bassahir-Kunawar**, 1885, *J. R. Drummond 22369* (K); **Lahul**, Sissu, 3200 m, 1938, *N. L. Bor 9298* (K); Sissu, 3140 m, 1941, *N. L. Bor 11911* (E, K) & *11912* (K); **Shimla**, Pangi, Bushahr, 1934, *N. Parmanand 861* (K); Shimla Hill, 3660 m, 1939, *Sherriff 7484* (BM, E); **Tihri-Garhwal**, 1883, *J. F. Duthie 603* (BM); **Upper Chenab Valley**, 2440 m, 1879, *R. Ellis 361* (K). **U YARLUNG ZANGBO**: **Gyirong**: Gyirong Town, vicinity of Naixia Village, 2013, *PE-Tibet Exped. 3977* (PE). **W NEPAL**: **Bheri**, Jajarkot, 3500 m, 1973, *Einarsson, Skärby & Wetterhall 3660* (BM); near Jajarkot, Bhei River, 3350 m, 1952, *Polunin, Sykes & Williams 2462* (BM, E); **Karnali**, SE of Jumla, Kunvigaon, 2740 m, 1952, *O. Polunin et al. 4864* (E); Pishang, 3810 m, 1931, *K. N. Sharma E. 50* (BM); Ranmagaon, above Ranmagaon, 3350 m, 1954, *Stainton, Sykes & Williams 3336* (BM, E).

2. Campanula glomeratoides D. Y. Hong, Acta Phytotax. Sin. 18: 247. fig. 2: 3–5. 1980; Hong, Fl. Reipubl. Popularis Sin. 73(2): 84. pl. 12: 3–4. 1983; Hong *et al.* in Wu, Raven & Hong, Fl. China 19: 532. 2011. Type: China, Tibet, Bomi, 2700 m, Jul. 1965, *Y. T. Chang & K. Y. Lang 548* (holotype: PE).

Plants annual (?), white hispid throughout. Rhizomes unknown. Stems erect, robust, up to 60 cm tall. Lower cauline leaves winged-petiolate; leaf blade oblanceolate or linear-oblong; middle and upper cauline leaves sessile; leaf blade lanceolate or ovate-lanceolate, up to 9 cm long, 1 cm broad, margin dentate. Flowers sessile, 2 to several in sessile capitula aggregated into a terminal involucrate compound capitulum; involucral bracts ovate-lanceolate to orbicular, forming a series continuous with upper cauline leaves, 1 or 2 in capitula with several flowers, but absent in capitula with only 2 flowers. Calyx tube obovoid or obconic; calyx lobes ligulate, ca. 4 mm long, apex obtuse. Corolla blue, tubular-campanulate, ca. 1.5 cm long. Filaments ca. 3 mm long; anthers ca. 4 mm long. Style slightly longer than corolla. Fruit not seen. Figure 47–59: a & b.

Phenology: flowering in July.

Habitats: meadows; ca. 2700 m.

Distribution: **endemic**; confined to L Yarlung Zangbo (Map 47–122).

L YARLUNG ZANGBO: **Bomi**, Zhamo,

Figure 47–59: **a** & **b**, **Campanula glomeratoides** D. Y. Hong: **a**, lower and upper parts of plant, showing a flowering capitulum; **b**, flower. **c–e**, **Campanula rotata** D. Y. Hong: **c**, upper part of stem; **d**, flower, showing stamens and style; **e**, a stamen. **a** & **b** from *Fl. Reipubl. Popularis Sin.* 73(2), 1983; **c–e** from Hong in *Phytotaxa*, 227(2), 2015, with permission.

2700 m, 1973, *Qinghai-Xizang Exped. 73–1461* (paratypes: KUN, PE).

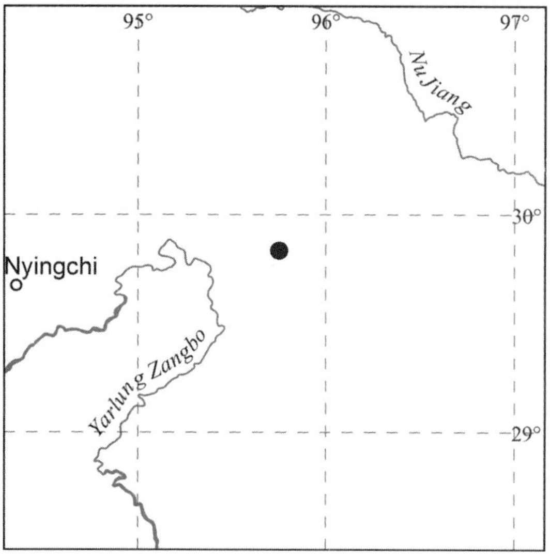

Map 47–122: **Campanula glomeratoides** D. Y. Hong

3. Campanula nakaoi Kitam., Acta Phytotax. Geobot. 15: 108. 1954; Hong, Fl. Reipubl. Popularis Sin. 73(2): 84. pl. 15–1. 1983; Type: Nepal, circa Shiar Khola, prope Tsumje, Khar, inter segetis, 3200 m, 5 Jul. 1953, *S. Nakao* s. n. (holotype: KYO, n. v.); Jsumje, 3400 m, 18 Jul. 1953, *S. Nakao* s. n. (paratype: KYO, n. v.).

Plants perennial. Rhizomes slender and long. Stems prostrate, ascending or erect, simple but sometimes branched, usually dark purple, less frequently straw-coloured, up to 40 cm long, nearly glabrous or sparsely retrorse-hispid. Middle cauline leaves largest, obovate-elliptic or elliptic, 1.2–3.5 cm long, 0.4–1.2 cm broad, apex obtuse; lower cauline leaves gradually attenuate into short petiole; upper cauline leaves sessile; all leaves sparsely serrate, hispidulous near margin or along veins. Flowers terminal, solitary, pendulous. Calyx tube obconic, densely hirsute or granulate-glandular; calyx lobes subulate, 5–10 mm long, 1.5–2 mm broad, both surfaces glabrous, margin serrate and hispidulous-ciliate. Corolla blue or blue-purple, broadly campanulate, sometimes nearly rotate, 1.4–2 cm long, glabrous; corolla lobes up to 4 times longer than tube. Style ca. 1 cm long. Figure 47–60: a.

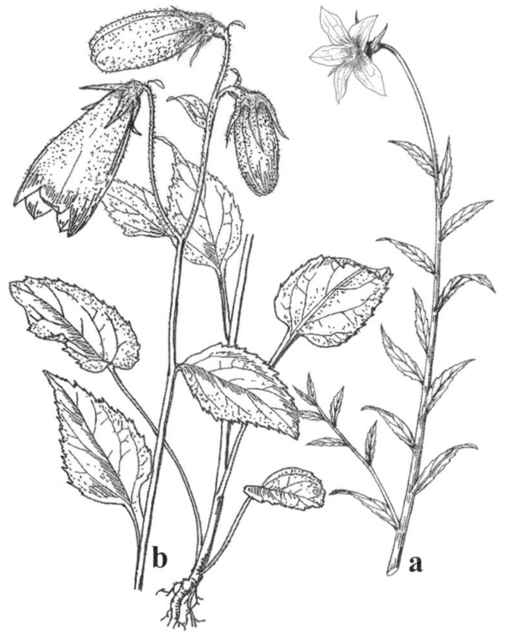

Figure 47–60: **a**, **Campanula nakaoi** Kitam.: **a**, upper part of stem. **b**, **Campanula punctata** Lam.: lower and upper parts of plant. **a** from *Fl. Reipubl. Popularis Sin* 73(2), 1983; **b** redrawn by Ai-Li LI from *Icon. Cormophyt. Sin.* 4, 1975.

Phenology: flowering from July to September; fruiting in October.

Habitats: margins of pine forests, pine forests, or thickets; 2800–3700 m.

Distribution: **endemic**; confined to U Yarlung Zangbo, C Nepal, and E Nepal (Map 47–123).

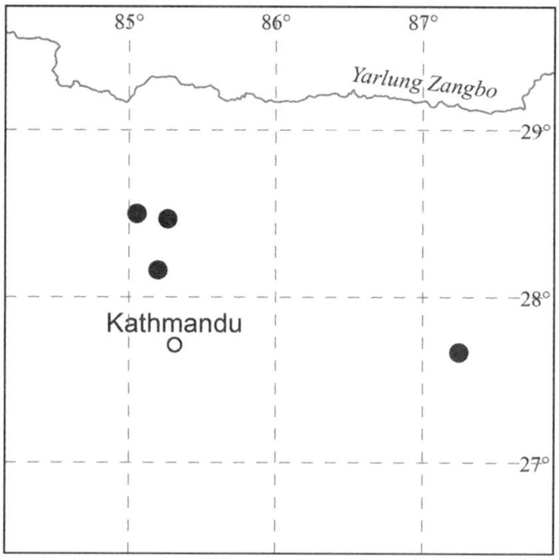

Map 47–123: **Campanula nakaoi** Kitam.

U YARLUNG ZANGBO: **Gyirong**, Gyirong Town, 3700 m, 2013, *PE-Tibet Exped. 3835* (PE); Gyirong Gou, 3141 m, 2013, *PE-Tibet Exped. 4159* (PE). **C NEPAL**: **Bagmati**, Dhading, Mailung Khola, 3510 m, 1965, *J. D. A. Stainton 5126* (E). **E NEPAL**: **Kosi**, Sankhuwasabha, 3720 m, 1991, *EMAK 416* (E).

4. Campanula sylvatica Wall. in Roxburgh, Fl. India 2: 97. 1824; Hooker & Thomson, J. Proc. Linn Soc. Bot. 2: 22. 1857; Nasir in Nasir & Ali, Fl. Pakistan 155: 13. fig. 3: E–F. 1984; Haridasan & Mukherjee in Hajra & Sanjappa, Fasc. Fl. India 22: 51. 1996. Type: Nepal, *Wallich 1293* (**lectotype here designated**: K [K000814433]; isolectotypes: E, K, W).

Herbs perennial. Roots attenuate. Stems erect, 20–40 cm tall, 0.5–1.5 mm in diameter, poorly branched, sparsely hispidulous. Basal and cauline leaves sessile, linear, 1–6 cm long, 0.1–0.5 cm broad, both surfaces shortly setose. Flowers solitary and terminal, or in a cyme of 2–4; pedicel 0.5–1.3 cm long, glabrous or hispidulous. Calyx tube semiglobose, hispidulous; calyx lobes subulate, 5–8 mm long, 1.5–2 mm broad, entire, hispidulous. Corolla mauvish blue, tubular-campanulate, 0.9–1.6 cm long, glabrous; corolla lobes ca. 1/2 as long as tube. Style included. Capsule semiglobose, ca. 5 mm long, 5 mm broad.

Phenology: flowering and fruiting from April to August.

Habitats: 1300–2800 m.

Distribution: U Ganga & Indus, W, C, and E Nepal, and Bhutan (Map 47–124); also in Punjab of India.

U GANGA & INDUS: **Almora**, 1620 m, *R. Strachey & J. E. Winterbottom 10* (K, P); **Kumaon**: s. loc., 1845, *T. Thomson s. n.* (K). **W NEPAL**: **Karnali**, Kalikot, Tila River, 1680 m, 1952, *O. Polunin et al. 1926* (E, P); Karnali Valley, between Badalkot and Sika, 1524 m, 1952, *O. Polunin et al. 3970* (E); Mugu, 2380 m, 2008, *C. A. Pendry et al. JRS A 107* (E); **C NEPAL**: **Bagmati**, Kavrepalanchok, Magna Deorali, 1980 m, 1964, *D. McCosh 04* (E); Bhragu, 2130 m, 1949, *O. Polunin 094* (E); Bhote Kosi, 1830 m, 1974, *J. D. A. Stainton 6963* (E). **BHUTAN**: **Chalimay-la**, 2130 m, 1914, *R. E. Cooper* 1504 (E); **Drugye** Dzong, Paro Chu, 2440–2740 m, 1949, *F. Ludlow et al. 16217* (E); **Thimphu**, Thimphu Chu, 7 km N of Thimphu, 2600 m, 1975, *A. J. C. Grierson & D. G. Long 603* (E); **Wangdu**, Samten Gang-Wangdu Phodrang, 1370–2180 m, 1938, *B. J. Gould 794* (K); Inharbages, 2000 m, 1967, *H. Kanai et al. 1704* (E, KUN).

5. Campanula tenuissima Dunn, Kew Bull. 1924: 385. 1924; Nasir in Nasir & Ali, Fl. Pakistan 155: 16. fig. 4C. 1984. Type: Pakistan, Kashmir, Jhelum Valley Road, 762 m, rock crevices, 28–29 Aug. 1919, *R. R. Stewart 5104* (holotype: RAW, n. v.; isotypes: E, K, KUH).

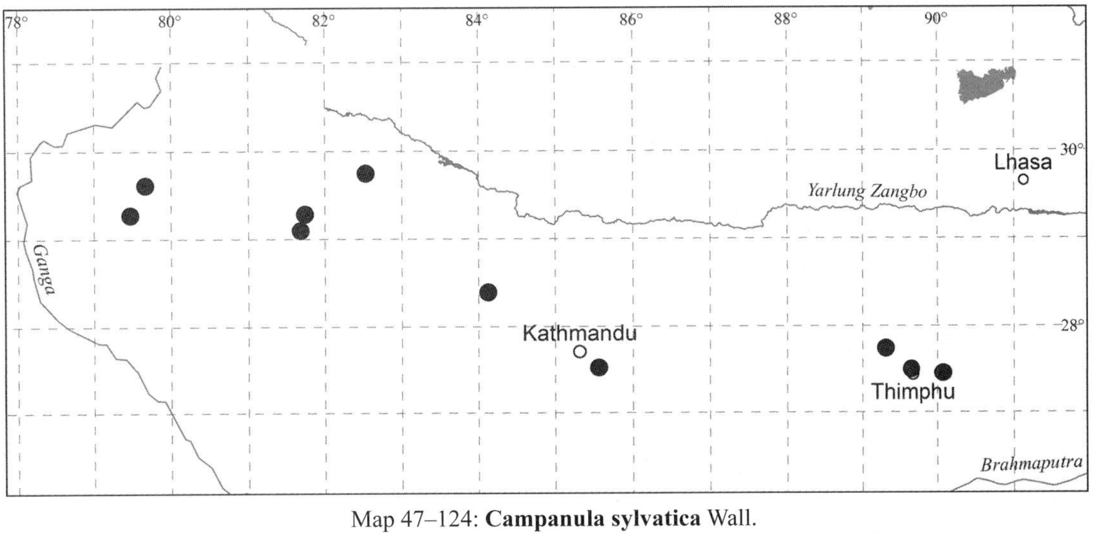

Map 47–124: **Campanula sylvatica** Wall.

Herbs perennial. Roots attenuate. Stems many, caespitose, 20–30 cm tall, 0.5–1 mm in diameter, pubescent to strigose. Leaves sessile or attenuated into a short petiole; leaf blade narrowly elliptic, 0.5–1.5 cm long, 0.2–0.4 cm broad, both surfaces strigose, margin entire to dentate, apex acute. Flowers terminal on branches, in a cyme of 2 to several or solitary; pedicel hirsute. Calyx setose along ribs; calyx tube 3–4 mm long, base attenuate; calyx lobes linear or subulate, 4–6 mm long, 1–1.5 mm broad. Corolla white, bluish or rose pink, campanulate, 5–8 mm long. Capsule subglobose, attenuate toward base, 4–5 mm long, 3–4 mm in diameter.

Phenology: flowering and fruiting from August to October.

Habitats: rocks; 750–2500 m.

Distribution: **endemic**; N Pakistan and Jammu & Kashmir (Map 47–125).

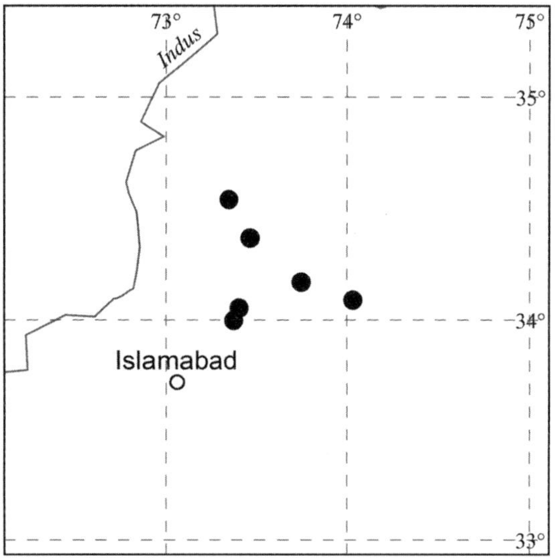

Map 47–125: **Campanula tenuissima** Dunn

N PAKISTAN: **Azad Kashmir**, Muzaffarabad, Leapa Valley, 1979, *B. Ahmad & Dilawre 332* (ISL); Jhelum Valley Road, 760 m, 1917, *R. R. & I. D. Stewart 3731* (KUH); **Hazara**, Mandri, Kagan Valley, 1896, *Inayat 19894* (K); Murree Hills, Changla-Nathia Goli, 2450 m, 1949, *R. R. Stewart* s. n. (KUH); Murree, SE of Changla, 1914, *E. M. Saunders s. n.* (K); Hazara, near Dunga Gali, 1975, *M. N. Chaudhari & M. A. Siddiqi 80* (ISL) & *100* (ISL). **JAMMU & KASHMIR**: Uri, 1220 m, 1934, *R. R. Stewart* s. n. (KUH);

6. Campanula tristis Kitam., Acta Phytotax. Geobot. 16: 132. 1956 & Fl. Afghan. 381. tab. 93. 1960; Nasir in Nasir & Ali, Fl. Pakistan 155: 17. fig. 4D. 1984. Type: Afghanistan, Nuristan, Wama to Trokikorl, 1500 m, *Kitamura* s. n. (holotype: KYO, n. v.).

Herbs perennial. Stems caespitose, erect, 15–20 cm tall, much branched, white- hirsute. Leaves alternate; petiole 3–8 mm long, white-hirsute; leaf blade oblong to ovate, 1–2.5 cm long, 0.3–0.8 cm broad, those on branches sessile and smaller, adaxially hirsute, abaxially pubescent, base cuneate, margin crenate, serrate to dentate, apex acute or obtuse. Flowers terminal on main stem and branches, in cymes or solitary; pedicel 2–10 mm long, white-hirsute. Calyx tube semiglobose white-hirsute; calyx lobes triangular to ovate, 2–4 mm long, 1 mm broad, setose along midrib and margin. Corolla tubular, 3–4 mm long, sparsely hispidulous outside; lobes narrowly ovate, 1/2 as long as to as long as tube. Style shorter than corolla. Capsule subglobose, 3–4 mm long, 3–4 mm wide.

Phenology: flowering and fruiting from June to September.

Habitats: 2130–2740 m.

Distribution : N Pakistan (Map 47–126); also in other part of Afghanistan (Nuristan).

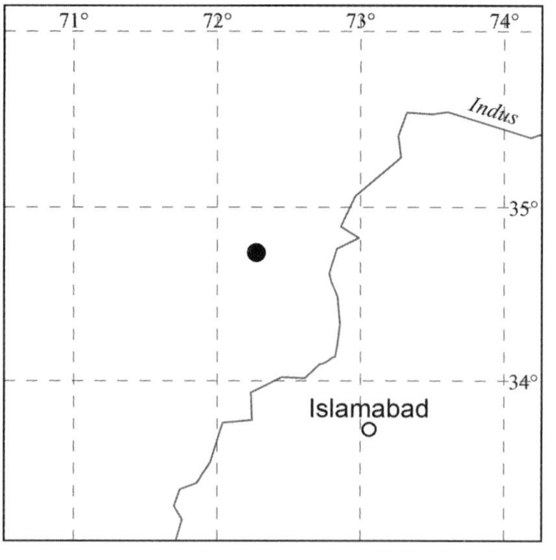

Map 47–126: **Campanula tristis** Kitam.

N PAKISTAN: **Swat**, between Barikot and Mingora, 1965, *K. H. Rechinger 30491* (K, W).

7. Campanula rotata D. Y. Hong, Phytotaxa 227(2): 196. fig. 2. 2015. Type: Tibet (Xizang), Gongbo' gyamda, Bahe Town, Dianka Village, 29°59′N, 93°41′E, meadows by forests, 3350 m, 23 Sept. 2012, *FLPH Tibet Exped. 12-2125* (PE).

Perennial herbs. Stems caespitose, ascending or procumbent, 30–70 cm tall, branched, glabrous or sparsely puberulent. Leaves alternate, sessile, linear or linear-elliptic, 3–9 cm long, 0.2–0.6 cm broad, both surfaces glabrous, margin usually sparsely serrate, sometimes entire. Flowers terminal on main stem and branches, pendulous; pedicel 2–6 cm long, glabrous. Calyx tube obconical, glabrous, 10-ribbed; calyx lobes subulate, 7–10 mm long, ca. 1 mm broad at base, glabrous, with 1–2 pairs of verrucose denticles. Corolla pale purple, rotate; corolla tube 2 mm long; corolla lobes ovate, ca.11 mm long, 6 mm broad. Filaments ca. 3 mm long, base much dilated, semicircular and villous-ciliate; anthers connivent into a tube, ca. 6 mm long. Style ca. 18 mm long; stigma 3-fid. Figure 47–59: c–e.

Phenology: flowering in September.

Habitats: meadows; ca. 3350 m.

Distribution: **endemic**; confined to L Yarlung Zangbo (Map 47–127). So far no additional collection has been found.

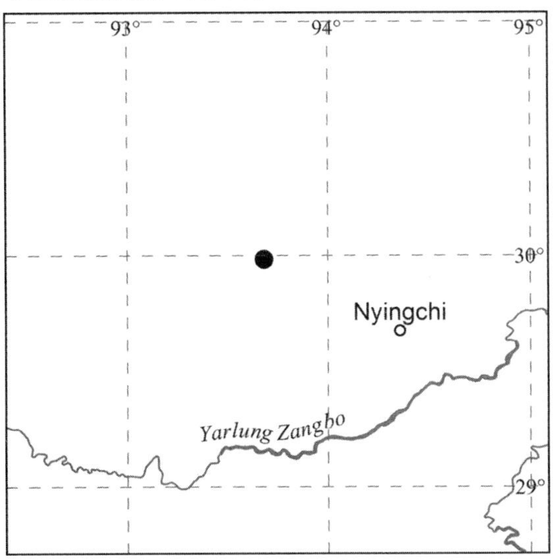

Map 47–127: **Campanula rotata** D. Y. Hong

Notes: This newly discovered species is characterized by its rotate corolla, which is lobed for 5/6 of the overall length, with a corolla tube only ca. 2 mm long, linear or linear-elliptic leaves, denticulate calyx lobes, and by the connivent, not just coherent, anthers. The deeply cleft corolla makes the systematic position of this species intermediate between *Campanula* and *Asyneuma*.

8. Campanula microphylloidea D. Y. Hong, Phytotaxa 227(2): 199. fig. 3. 2015. Type: China, Tibet, Lhozhag County, Lhakan, 3150 m, rock crevices, Aug. 2013, *Y. S. Chen et al. 13–1387* (holotype: PE).

Herbs perennial. Roots attenuate, ca. 3 mm across. Stems caespitose, many from one caudex, 10–17 cm tall, upper parts branched, puberulent. Leaves alternate, sessile; leaf blade up to 5 mm long, 2.5 mm broad, both surfaces densely puberulent, base cuneate, margin entire or with 1–2 pairs of teeth, apex rounded or acute. Flowers terminal on main stem or both on main stem and branches, solitary. Calyx tube obconical, densely puberulent, 10-ribbed, five ribs entering calyx lobes stronger than others; calyx lobes broad-triangular, ca. 3.5 mm long, 2 mm broad, densely

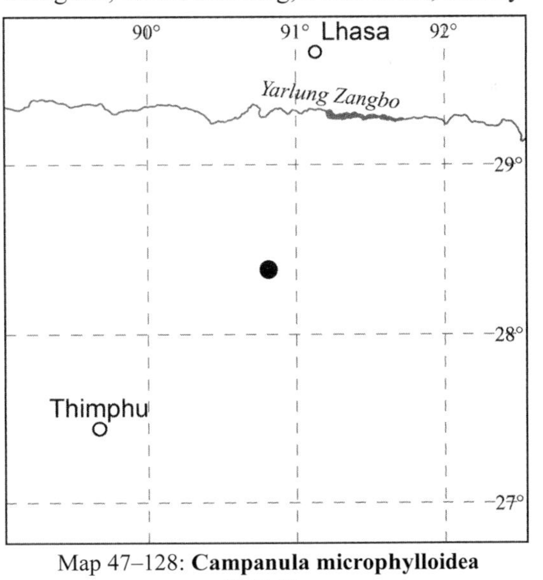

Map 47–128: **Campanula microphylloidea** D. Y. Hong

puberulent, entire, apex acute. Corolla pale blue, tubular-campanulate, ca. 14 mm long, lobed to near middle, puberulent outside; corolla lobes ca.

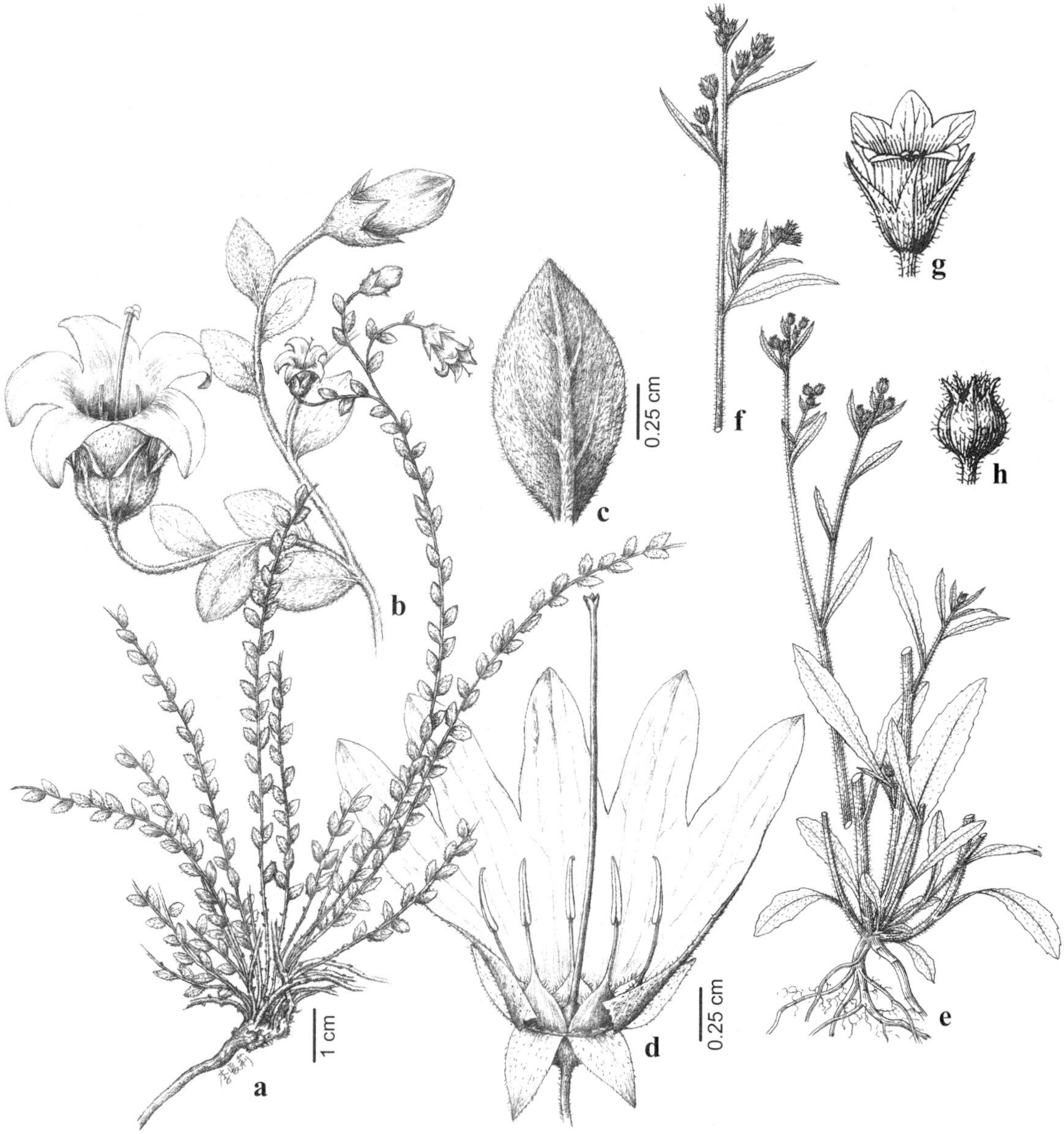

Figure 47–61: **a–d**, **Campanula microphylloidea** D. Y. Hong: **a**, plant; **b**, upper part of stem; **c**, abaxial view of leaf; **d**, a flower with corolla opened. **e–h**, **Campanula dimorphantha** Schweinf.: **e**, lower part of plant and upper part of branch; **f**, upper part of branch; **g**, a normal flower; **h**, fruit developed from a cleistogamous flower. **a–d**, from Hong in *Phytotaxa*, 227(2), 2015, with permission; **e–h** from *Fl. Reipubl. Popularis Sin.* 73(2), 1983.

6.5 mm long, lanceolate. Stamens ca. 6 mm long, glabrous. Style ca. 16 mm long, exserted. Figure 47–61: a–d.

Phenology: flowering in August.

Habitats: rocky crevices; ca. 3100 m.

Distribution: **endemic**; confined to M Yarlung Zangbo (Lhozhag) (Map 47–128). Known only from the type specimen.

Notes: This newly described species resembles mostly *Campanula cana* Wall., but differs distinctly from the latter in having much smaller and sessile leaves, solitary flowers, entire calyx lobes, and the plants entirely covered with a single type of hairs: densely puberulent.

9. Campanula pallida Wall., Asiat. Res. 13: 375. 1820; Haridasan & Mukherjee in Hajra & Sanjappa, Fasc. Fl. India 22: 49. 1996; Hong, Lammers & Klein in Wu, Raven & Hong, Fl. China 19: 533. 2011. Type: Napaul (C Nepal), *Gardner* s. n. anno1818 (in *Herb. Wallich*) [lectotype desighated as 'type' by Nasir, Fl. Pakistan 155: 20. 1984: BM (BM000521740); isolectotype: K].

Campanula colorata Wall. in Roxburgh, Fl. India 2: 101. 1824. Type: Nepal, *Wallich 1287* (**lectotype here designated**: K; isolectotype: E).

Campanula colorata var. *tibetica* Hook. f. & Thomson, J. Proc. Linn. Soc. Bot. 2: 23. 1857. ≡ *Campanula pallida* Wall. var. *tibetica* (Hook. f. & Thomson) H. Hara, J. Jap. Bot. 50: 1975. **syn. nov.** Type: India, Jammu & Kashmir, Chenab Valley, 1520–2440 m, 17 Jun. 1848, *T. Thomson* s. n. (holotype: K).

Campanula ramulosa Wall. in Roxburgh, Fl. India 2: 100. 1824. Type: Nepal, *Wallich 1286* (holotype: K).

Perennials. Roots carrot-shaped, sometimes slightly thicker than stems. Stems single, sometimes 2 or rarely several from a single caudex, ascending or erect, up to 60 cm tall, hirsute. Lower leaves winged-petiolate, upper leaves sessile; leaf blade elliptic, rhombic-elliptic, or oblong, 1–4 cm long, 0.3–1.5 cm broad, abaxially densely hirsute or hispid only along veins, adaxially appressed-hispid, margin denticulate or almost entire, apex acuminate, acute, or obtuse. Flowers pendent, terminal, sometimes in cymes. Calyx tube obconic or campanulate, hispid; calyx lobes deltoid, narrowly triangular, or subulate, 2–8 mm long, 1.5–3 mm broad, abaxially hispid entirely or only along veins, margin entire, rarely serrulate. Corolla purple, blue-purple, or blue, tubular-campanulate, 4–15 mm long; corolla lobes ca. 1/2 as long to about as long as tube. Style less than 2/3 as long as corolla. Capsule obconic, obovoid, or globose. Seeds golden brown, ellipsoid, compressed. **Chromosomes**: 2n=24, 28. Figure 47–62: a & b.

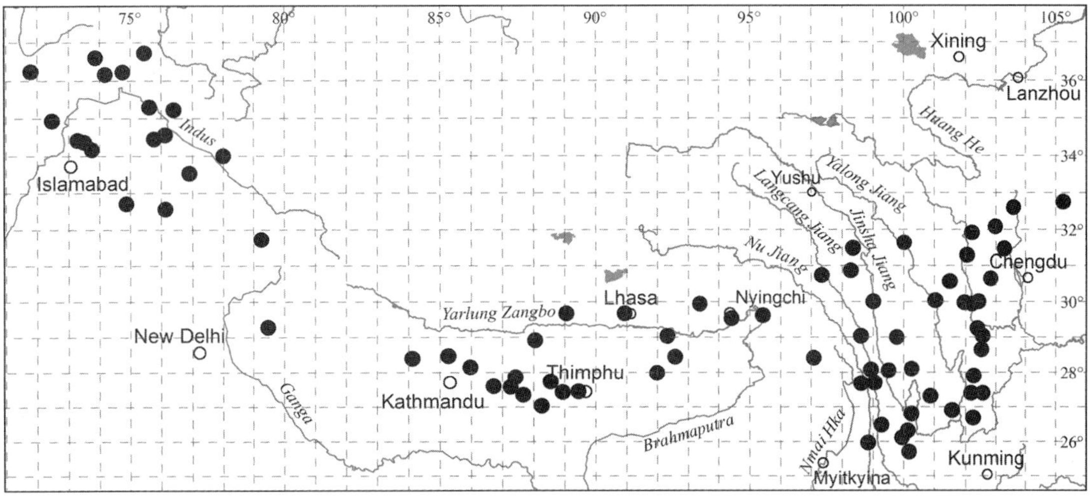

Map 47–129: **Campanula pallida** Wall.

Phenology: flowering and fruiting from May to September.

Habitats: grassy slopes, thickets, or open woods; 750–4300 m.

Distribution: all the regions except Vakhan, W and C Nepal, and U Irrawaddy (Map 47–129); also in the other parts of Afghanistan and SW China; in Laos and Thailand.

Notes: Both chasmogamous and cleistogamous flowers are found in this species. In the latter, development of the corolla is arrested at an early stage; as a result, the calyx lobes are deltoid (vs. narrowly triangular or subulate) and much shorter than those in the former. The capsule produced by cleistogamous flowers are globose (vs. obconic or obovoid) and much smaller than those from chasmogamous flowers. An individual may bear two types of flowers or, less commonly, only cleistogamous flowers; the type of *Campanula microcarpa C. Y. Wu* (1965), *P. H. Yü 645 (KUN),* is a specimen of the latter.

10. Campanula mekongensis Diels ex C. Y. Wu in Wu & Li, Yunnan Trop. Subtrop. Fl. Res. Rep. 1: 58. 1965. Type: China, Yunnan, Mengcang, 1 Jan. 1959, *Y. H. Li 576* (holotype: KUN).

Herbs perennial. Roots carrot-shaped, up to 5 mm in diameter. Stems usually single, erect or ascending, 15–30 cm tall, branched, white-villous. Leaves alternate, sessile, or with short petiole up to 2 mm long; leaf blade oblong or oblanceolate, 0.5–3 cm long, 0.3–1.2 cm broad, adaxially sparsely hirsute, abaxially hirsute entirely or only along veins, base cuneate, margin sparsely serrate or entire, apex obtuse. Flowers solitary and terminal or in a cyme of 2 or 3: pedicel up to 2 cm long. Calyx tube half-globose, hispid along 10 ribs; calyx lobes subulate, 2–4 mm long, up to 1 mm broad, hispid along rib, ciliate, with a pair of vcrrucose denticles. Corolla blue or purplish blue, campanulate, 6–10 mm long; lobed slightly over half, sparsely puberulent outside. Style shorter than corolla. Capsule obovoid, 3–5 mm long, 3–4 mm in diameter. Seeds yellow, ellipsoid, 0.5 mm long.

Phenology: flowering from January to May; fruiting from March to May.

Habitats: along river bank, sometimes on boulders; 500–1300 m.

Distribution: S Hengduan (Map 47–130); also in S Yunnan of China. **New records** for the Pan-Himalaya.

S HENGDUAN: **Fugong**, Lumadeng, W side of Nujiang, 1237 m, 2004, *Gaoligong Shan Biodiversity Survey 19756* (CAS); Fugong, Pihe, 1270 m, 2004, *Gaoligong Shan Biodiversity Survey 21043* (CAS).

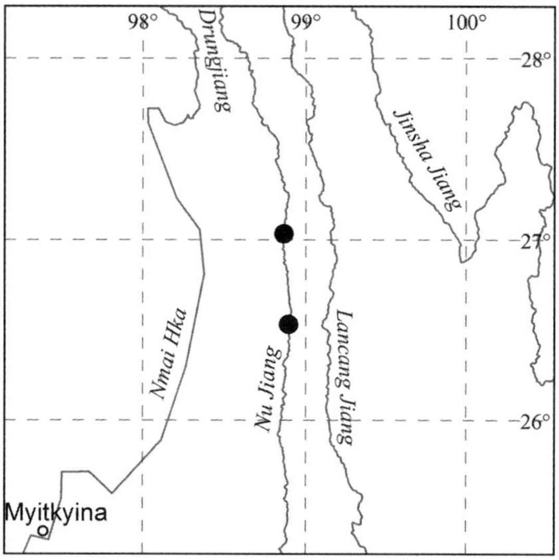

Map 47–130: **Campanula mekongensis** Diels ex C. Y. Wu

11. Campanula cana Wall. in Roxburgh, Fl. India 2: 101. 1824; Clarke in Hooker, Fl. Brit. India 3: 440. 1881; Hong, Fl. Reipubl. Popularis Sin. 73(2): 87. 1983. Type: India, "Himalaya near Gossain Than", *Wallich 1284* (holotype: K; isotype: BM).

Campanula pasumensis C. Marquand, J. Linn. Soc. Bot. 48 : 196. 1929. Type: China, Tibet, East Himalaya, Pasum Lake, 3050 m, 18 Aug. 1924. *F. Kingdon- Ward 6109* (holotype: K; isotype: BM).

Campanula aprica Nannf., Acta Horti Gothob. 5: 23 pl. 8. fig. a. 1930. Type: China, "Szechuan, reg. bor. – occid. Hsü-tsing", ca. 2200 m, 10 Oct. 1922, *H. Smith 4592* (holotype: GH, n. v.; isotype: BM).

Plants perennial, densely white-villous on all

parts except corolla. Tap root thickened, more or less carrot-shaped, up to 8 mm in diameter. Stems caespitose, usually diffuse, 15–30 cm long. Leaves sessile or lower ones with petiole up to 5 mm long; leaf blade ovate, elliptic, oblanceolate, or li-near-lanceolate, 4–25 mm long, 2–10 mm broad, base cuneate, margin nearly entire, crenulate or serrate, apex obtuse or acute. Flowers several in a terminal cyme; pedicel 0.5–3 cm long, ebracteolate. Calyx tube obconical to broadly obconical, 2–3 mm long at anthesis; calyx lobes subulate or narrowly triangular, 3–5 mm long, 1–2 mm broad, margin entire, crenulate, or sometimes dentate, with 1–2 pairs of teeth. Co-rolla blue, blue-purple, purple, or violet, campanulate, 1–1.8 cm long, outside villous; corolla lobes 1/2–2/3 as long as tube. Style slightly exserted. Capsule globose or ob-ovoid-globose, 3–5 mm long, 3–4 mm broad, 10-ribbed, 3-poricidal at base. Seeds brownish yellow, shiny, ellipsoid, 0.5–0.6 mm long, smooth.

Phenology: flowering and fruiting from May to November.

Habitats: rocky crevices, open rocky slopes, grassy slopes, or thickets; 1000–4300 m.

Ditribution: Jammu & Kashmir, U Ganga & Indus, U Yarlung Zangbo, W and C Nepal, Bhutan, M Yarlung Zangbo, N and S Hengduan, and U Irrawaddy (Map 47–131); also in Guizhou of China.

12. Campanula cashmeriana Royle, Ill. Bot. Himal. Mts. 1: 254. tab. 62. fig. 1. 1835; Clarke in Hooker, Fl. Brit. India 3: 441. 1881; Haridasan & Mukherjee in Hajra & Sanjappa, Fasc. Fl. India 22: 45. 1996. Type: Royle's tab. 62 cited.

Campanula evolvulacea Royle ex A.DC. in de Candolle, Prodr. 7: 473. 1839. ≡ *Campanula cashmeriana* var. *evolvulacea* (Royle ex A. DC.) C. B. Clarke in Hooker, Fl. Brit. India3: 441. 1881. Type: not seen.

Perennials caespitose. Roots attenuate, up to 6 mm in diameter. Stems many, ascending, 10–20 (–30) cm long, pubescent. Leaves nearly sessile or shortly petiolate; leaf blade ovate-lanceolate or obovate, (0.2–) 1–1.5 (–3.5) cm long, (0.1–) 0.7–1.2 (–1.5) cm broad, abaxially canescent, margin crenate or nearly entire. Inflorescences a lax corymb of 1- or 2-flowered cymes. Calyx canescent; calyx lobes triangular-lanceolate, 5–7 mm long, 1–2 mm broad, entire. Corolla white, pinkish white, pale blue to dark blue, or purple, campanulate, 1.3–2.8 cm long, 2 or 3 times as long as calyx lobes; corolla lobes ca. 1/2 as long as tube. Style included, much shorter than corolla. Capsule globose, 3–4 mm in diameter.

Phenology: flowering and fruiting from June to September.

Habitats: cliffs, rocks, or rocky slopes; 2130–4400 m.

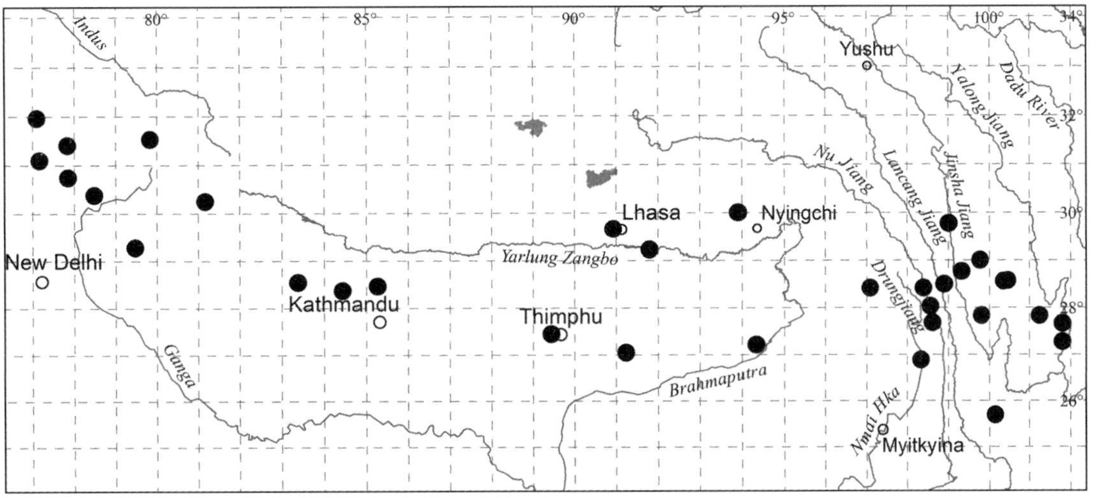

Map 47–131: **Campanula cana** Wall.

Distribution: Vakhan, N Pakistan, Jammu & Kashmir, U Ganga & Indus, and W Nepal (Map 47–132); also in W Pakistan (Baluchistan).

Notes: I have searched widely for the type specimen but I failed to find it. Thus I designate here Royle's figure as type.

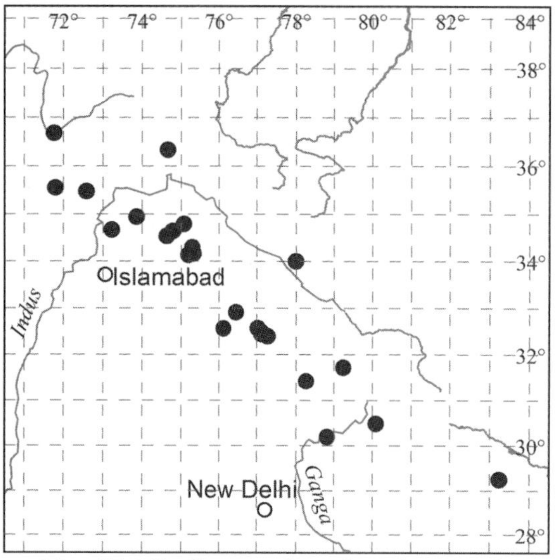

Map 47–132: **Campanula cashmeriana** Royle

VAKHAN: Bei Quazi-Deh, 2700 m, 1964, *H. Roemer 137* (W). **N PAKISTAN**: **Baltistan**, Cachemire, Minimarg, 2000 m, 1953, *F. Schmid 702* (E); **Chitral**, Drosh Gol, S of Drosh, 2130 m, 1958, *J. D. A. Stainton* (W); Drosh Gol, 2590 m, 1958, *S. A. Bowes Lyon 159* (E, W); **Gilgit**, Gilgit Kaghan Rd, near Burawai, 1972, *M. Qaiser & A. Ali 5323* (KUH); Hunza, Tokuchos Kasimabad, ca. 2700 m, 2005, *J. & N. Alam 3315* (KUH); **Hazara**, Buri Bandi, 1976, *Skankat & Nisar 1010* (ISL) ; Golimussla, 1899, *J. F. Duthie* s. n. (K); Hill Territory, Chailsar to Ganga Kandas, 2590 m, 1958, *B. L. Burtt & Makazmi 1238* (E); Setfal Lake-Muluh Road, 1978, *M. N. Chandh et al. 1753* (ISL); Siran Valley, 1896, *Inayat 19891* (K, W); Siran, 1976, *Shankat & Nisar 980* (ISL); **Swat**, Kalam, 2200 m, 1965, *K. H. Rechinger 1852* (E) & *K. H. Rechinger 30800* (W). **JAMMU & KASHMIR**: **Badwan**, 2290 m, 1929, *J. F. Duthie 1499* (K); **Bashahr state**, Labrang, 2900 m, 1890, *J. H. Lace 536* (E).; **Kolahoi Valley**, 3350–3660 m, 1893, *J. F. Duthie 13520* (K); **Ladak**, *M. L. Wathen s. n.* (K); **Liddar**, 3050 m, 1956, *O. Polunin 56–488* (E); **Rajparyan Sanctuary**, Upper Bringhi, 3050 m, 1943, *F. Ludlow & G. Sherriff 9328* (E); **Sekiwas**, 3350 m, 1956, *O. Polunin 56–590* (E); **Sonamarg**, 2740 m, 1919, *H. H. Rich 1159* (K); **Tragbal**, 3050 m, 1919, *R. R. Stewart 4822* (K); Tragbal, 3350 m, 1935, *F. Ludlow & G. Sherriff* 1536 (E) **U GANGA & INDUS**: **Chamba**, Wassa, 2130–3350 m, 1896, *G. A. Gammie 18563* (K); Chamba, Donai to Bindralani, 2440–2740 m, 1897, *J. H. Lace 1619* (E); **Garhwal**: Niti, 3350 m, *Strachey & Winterbottom 6* (K); **Kumaon**, Byans, Kulti Valley, 3960 m, 1884, *J. F. Duthie 3118* (K); **Milum Glacier**, 3810 m, 1848, *Strachey & Winterbotton 6* (K); **Kulu**, Khoksar to Sissu, 3350 m, 1935, *M. Nath 261* (KUH); **Kyelang**, 3140 m, 1941, *N. L. Bor 14693* (E, K); Kyelang, 3100 m, 1970, *E. Köckeis & W. Stangl* s. n. (W); **Shimla**, Baspa Valley, 3960 m, 1939, *G. Sherriff 7479* (E); **Upper Chenab** Valley, Road to Purti, 2440, 1881, *R. Sllis 1416* (K); **Zanda**, Diya Township, 3000 m, 1976, *Qinghai-Xizang Exped. 76–8158* (KUN, PE). **W NEPAL**: **Karnali**, Dolpa, ca. 4000 m, 1973, *Grey-Wilson & Phillips 386* (K); 2 km S of Mur (Koram Toppo River), ca. 4400 m, 1973, *Grey-Wilson & Phillips 570* (K).

13. Campanula alsinoides Hook. f. & Thomson, J. Proc. Linn. Soc. Bot. 2: 24. 1857; Clarke in Hooker, Fl. Brit. India 3: 441. 1881; Type: NW India, Jammu & Kashmir, Kishtwar, 2440–3050 m, 18 Jun. 1848, *T. Thomson* s. n. [**lectotype here designated**: K (K000814531)]; Kunawar, above Sungnam, 22 Aug. 1847, *T. Thomson* s. n. [paratype: K (K000814530)].

Herbs perennial. Roots attenuate, up to 5 mm in diameter. Stems many, caespitose, filiform, diffuse, up to 20 cm long, pubescent. Leaves alternate; petiole up to 3 mm long; leaf blade oblong-ovate, elliptic or spatulate, 5–20 mm long, 3–10 mm broad, both surfaces hirsute, base cuneate or rounded, margin entire to crenulate, apex rounded or acute. Flowers terminal, solitary or less often in a 2-flowered cyme. Calyx white-hirsute; calyx lobes triangular-lanceolate, 3–5 mm long, 1.5–2 mm broad. Corolla light purple or purple-red, tubular-campanulate, 8–10 mm long, 3–5 mm wide, sparsely hirsute outside; corolla lobes 1/2 as long as tube. Style included or as long as corolla. Capsule globose, 2–3 mm in diameter.

Phenology: flowering and fruiting from June to October.

Habitats: rocky places or cliffs; 900–3700 m.

Distribution: Jammu & Kashmir, U Ganga & Indus, and C Nepal (Map 47–133); also in Punjab of India.

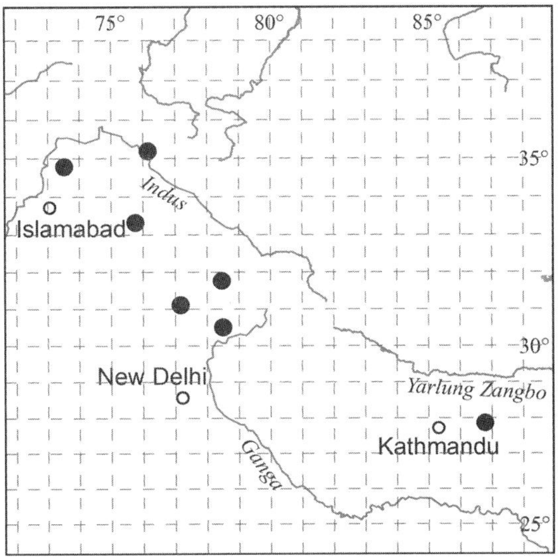

Map 47–133: **Campanula alsinoides** Hook. f. & Thomson

Notes: Cleistogamy is probably a prevailing reproductive strategy in this species. Among the specimens cited below including the type specimens only *J. S. Gamble 1519 A & B* (K) show normal corolla. The fruit produced by cleistogamous flowers are globose, with oblong calyx lobes.

N PAKISTAN: **Baltistan**, Karakoram, Nagar Village, 2380 m, 1960, *O. Polunin 6416* (E); **Hazara**, 3 km N of Hill, 2130 m, 1958, *B. L. Burtt & Makazmi 1271 A* (E). **U GANGA & INDUS**: **Garhwal**, *Falconer* s. n. (K); **Shimla**, s. loc., 910 m, 1878, *J. S. Gamble 6007* (K); near Shimla, Hattogach, 3170 m, 1876, *J. S. Gamble1579 A* (K) & *1579B* (K). **C NEPAL**: between Pangboche and Porche, 3660 m, 1876, *A. Schilling S2208* (K).

14. Campanula argyrotricha Wall. ex A. DC. in de Candolle, Prodr. 7: 473. 1839; Clarke in Hooker, Fl. Brit. India 3: 441. 1881; Type: India: Uttarakhand, Kumaon *Wallich 7138* (holotype: G)

Perennials. Roots 1.5 mm in diameter. Stems caespitose, slender, diffuse, 8–18 cm long, densely white villose. Leaves sessile, leaf blade ovate, oblong or elliptic, 5–23 mm long, 3–15 mm broad, mixed white hispidulous and sparsely white hispid, abaxially more densely so, base cuneate or rounded, margin dentate or crenulate, apex acute or rounded. Flowers terminal, solitary or often 2–several in a cyme; pedicel 0.5–3 cm long, white hispid. Calyx tube densely white or yellowish hispid; calyx lobes elliptic or lanceolate, widest at or below middle, above base, 4–7 mm long, 2–3 mm broad, white hispidulous and hispid mixed, entire. Corolla red or reddish blue, tubular-campanulate, 1.2–2 cm long, villose outside; corolla lobes ca. 1/2 as long as corolla tube, oblong, apex obtuse. Style included. **Chromosomes**: n=14.

Phenology: flowering from May to September; fruiting from September to November.

Habitats: rocks; 2400–4100 m.

Distribution: **endemic**; U Ganga & Indus, E Nepal, Sikkim & Darjiling, Bhutan, and M Yarlung Zangbo (Map 47–134).

U GANGA & INDUS: **Chachpur**, Raiengash, 2740 m, 1878, *A. Smythicer* s. n. (K); **Chamba**, 2590 m, *R. Strachey & J. E. Winterbottom 2* (K,P); 2440–2740 m, 1844, *M. P. Edgeworth* s. n. (K); **Mt Chur**, ca. 3350 m, 1884, *J. R. Drummond 2757* (E, K); **Garhwal**, near Mt Ku'ar Pass, 3350–3660 m, 1885, *J. F. Duthie 4171* (K); **Kulu** Valley, Bhahu Pass, 1941, *M. Nath 4334* (KUH); near **Shimla**, 2740 m, *J. R. Drummond 2755* (K). **U YARLUNG ZANGBO**: **Gyirong**, Gyirong Town, Zacun-Lado Pass, 3968 m, 2011, Y. S. Chen *et al. 395* (PE). **E NEPAL**: **Mechi**, Taplejung, 4120 m, 1991, *D. G. Long et al. 929* (E); **Sargamata**, Solukhumbu, 3800 m, 1997, *M. Staudinger N24/1a* (W). **SIKKIM & DARJILING: Sikkim**, Pheraothang, 3960 m, 1913, *R. E. Cooper 968* (E). **BHUTAN**: **Kumar Thang**, Paro Chu, 3810 m, 1949, *F. Ludlow et al. 16288* (E); Tzatogang to Dotanang, 2800 m, 1967, *H. Kanai et al. 1580* (E); **Paro Chu**, Mem

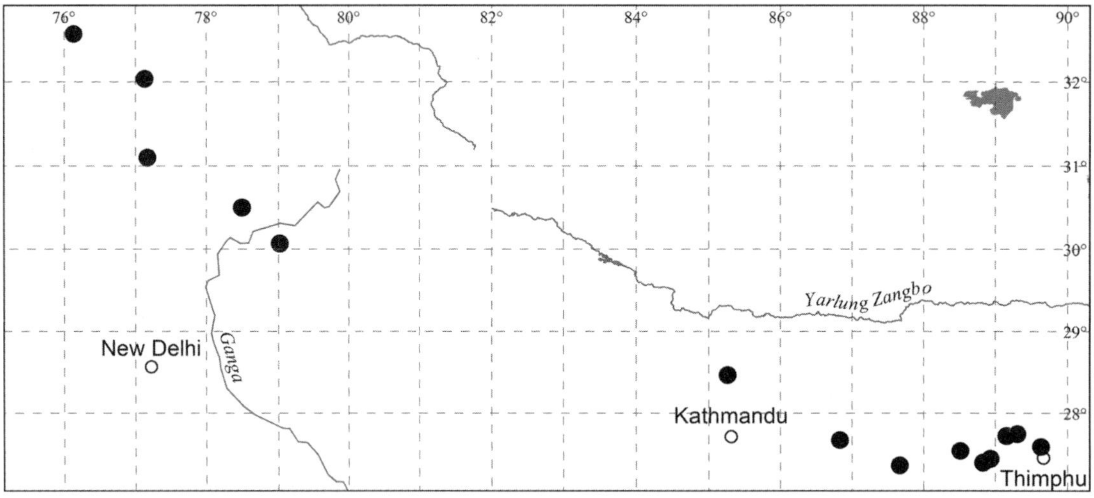

Map 47–134: **Campanula argyrotricha** Wall. ex A. DC.

La, 3506 m, 1949, *F. Ludlow et al. 17484* (E). **M YARLUNG ZANGBO**: **Yadong**, Upper Chumbi Valley, 3660 m, 1938, *B. J. Gould 1574* (K); Chumbi & Phari, 1879, *Dungboo* s. n. (K).

15. Campanula yunnanensis D. Y. Hong, Fl. Reipubl. Popularis Sin. 73(2): 184. pl. 14: 7–9. 1983; Hong, Lammers & Klein in Wu, Raven & Hong, Fl. China 19: 534. 2011. Type: China, Yunnan, Binchuan, Oct. 1946, *T. N. Liou 21570* (holotype: PE).

Plants perennial. Roots carrot-shaped, sometimes branched. Stems several from one caudex, almost entirely prostrate, 30–45 cm long. Lower leaves withered at anthesis; uppermost leaves sessile; leaf blade ovate to ovate-orbicular, 7–11 mm long, adaxially densely appressed hispid, abaxially densely tomentose, margin crenate. Flowers terminal, upright or somewhat declined. Calyx tube obconical, densely hirsute; calyx lobes triangular, densely tomentose, margin with a pair of teeth. Corolla purple or blue-purple, tubular-campanulate, ca. 8 mm long; corolla lobes as long as tube. Style ca. 5 mm long. Capsule obovoid-conical, 4–5 mm long, 3–4 mm in diameter. Seeds yellow-brown, ellipsoid, slightly compressed, ca. 0.5 mm long, smooth. Figure 47–62: c–e.

Phenology: flowering and fruiting from October to December.

Habitats: rocky slopes; 1900–2200 m.

Distribution: S Hengduan (Map 47–135); also in S & C Yunnan of China.

S HENGDUAN: **Yangbi**, Shangpingpo to Goutouqing, 1946, *T. N. Liou 22471* (PE).

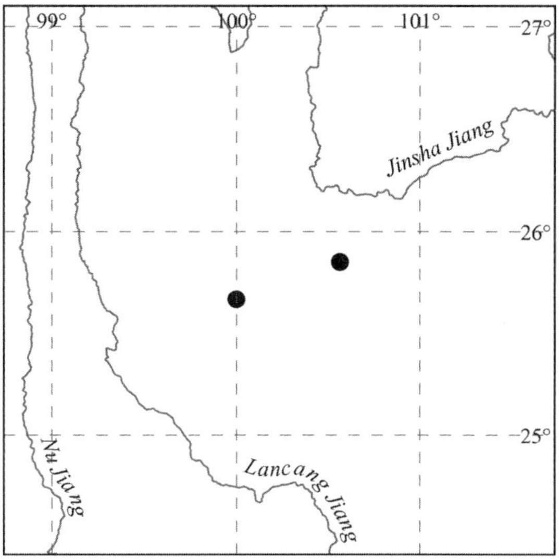

Map 47–135: **Campanula yunnanensis** D. Y. Hong

16. Campanula chinensis D. Y. Hong, Acta Phytotax. Sin. 18(2): 247. fig. 2: 6–7. 1980 & Fl. Reipubl. Popularis Sin. 73(2): 87. pl. 12: 5–6. 1983. Type: China, Yunnan, Gongshan, upper reaches of Drungjiang, 2400 m, *T. T. Yü 19610* (holotype: PE).

Plants perennial. Roots carrot-shaped. Stems caespitose, ascending, up to 35 cm long,

Figure 47–62: **a** & **b**, **Campanula pallida** Wall.: lower and upper parts of stem. **c–e**, **Campanula yunnanensis** D. Y. Hong: **c**, a part of plant; **d**, flower; **e**, abaxial surface of leaf blade. **a** & **b** redrawn by Ai-Li LI from *Icon. Cormophyt. Sin.* 4, 1975; **c–e** from *Fl. Reipubl. Popularis Sin.* 73(2), 1983.

branched, hispid. Leaves winged-petiolate; leaf blade elliptic or lanceolate, 1–3.5 cm long, 0.4–1.2 cm broad, adaxially appressed hispid, abaxially densely villous, base cuneate, margin entire or denticulate, apex obtuse, acute or acuminate. Flowers terminal on main stem and branches, solitary or several in a cyme. Calyx tube obconic, densely villous; calyx lobes triangular, margin entire. Corolla blue or blue-purple, campanulate but rotate above, 8–10 mm long; corolla lobes slightly longer than tube, linear-oblong. Stamens exserted; anthers all connivent into a tube, or connivent into a group of 2 and another of 3. Style 1.3–1.5 cm long, much exserted. Capsule globose, obovoid, or obconic. Seeds yellowish brown, ellipsoid, ca. 0.5 mm

Phenology: flowering from July to August; fruiting in September.

Habitats: rocks or poor pine forests; 2400–3800 m.

Distribution: **endemic**; Yarlung Zangbo-Brahmaputra, N Hengduan, and U Irrawaddy (Map 47–136).

YARLUNG ZANGBO-BRAHMAPUTRA: **Zayü** Town, suburb, 2450 m, 2012, *D. Y. Hong et al. H 12038* (PE); **Zhowagoin**, Sangjiu, 2009, *X. H. Jin et al. SET-ET 808* (PE). **U IRRAWADDY**: Sangjiu to **Ridong**, 2012, *FLPH Tibet Exped. 12-1508* (PE).

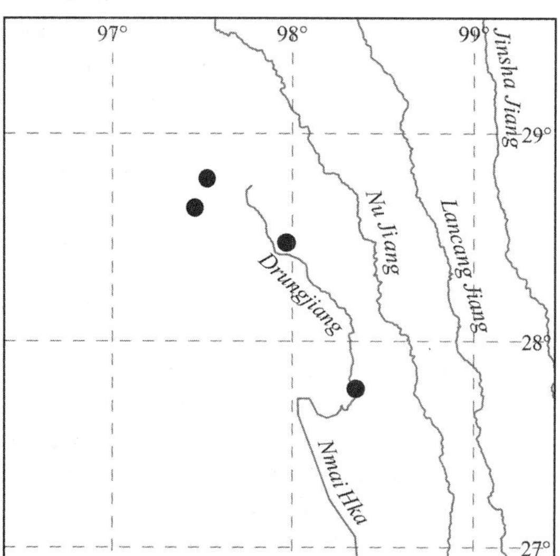

Map 47–136: **Campanula chinensis** D. Y. Hong

17. Campanula dimorphantha Schweinf., Beitr. Fl. Aethiop. 1:140. 1867; Thulin, Nord. J. Bot. 7: 419. 1987; Hong, Lammers & Klein in Wu, Raven & Hong, Fl. China 19: 534. 2011. ≡ *Campanula canescens* Wall. ex A. DC., Monogr. Camp. 292. 1830, non Roth (1827). ≡ *Campanula wallichii* Babu, J. Bombay Nat. Hist. Soc. 65: 808. 1969. Type: Nepal, Rampora, 1821, *Wallich 1289* (lectotype designated by Kitamura: K; isolectotype: BM).

Campanula benthamii Wall. ex Kitam., Fl. Afghan. 377. 1960. Type: Nepal, *Wallich 1290* (holotype: K)

Plants annual. Stems erect, up to 45 cm tall, simple, or branched from base, hispid. Basal leaves withered early, shortly petiolate; leaf blade spatulate or elliptic, base cuneate or attenuate, margin serrulate, apex acute; cauline leaves winged-petiolate; leaf blade spatulate, elliptic, or linear, 2–7 cm long, base attenuate, margin serrulate or nearly entire, apex acute or acuminate. Inflorescence a terminal panicle of many cymes; pedicel variable in length, terminal flower with longer pedicel. Calyx tube semiglobose to obconic, base acute; calyx lobes narrowly triangular, 3–5 mm long. Corolla purple or blue-purple, campanulate, hispid outside, glabrous inside; corolla lobes as long as tube. Style included. Capsule globose. Seeds yellow-brown, ellipsoid, compressed, ca. 0.3 mm long, smooth. Figure 47–61: e–h.

Phenology: flowering and fruiting in March and April.

Habitats: grassy slopes, roadsides, or river banks; below 2000 m.

Distribution: N Pakistan, Jammu & Kashmir, U Ganga & Indus, C Nepal, Sikkim & Darjiling, and Bhutan (Map 47–137); also in the other parts of Afghanistan, S and SW China (Guangdong, Guizhou, Shaanxi, Sichuan, Taiwan, and S Yunnan), E India (Assam, Meghalaya, and Madras), Sri Lanka, Laos, Vietnam; NE Africa.

Notes: Cleistogamy may be a predominant reproductive strategy of this species, as most specimens we examined have only cleistogamous flowers. The fruit produced by such flowers

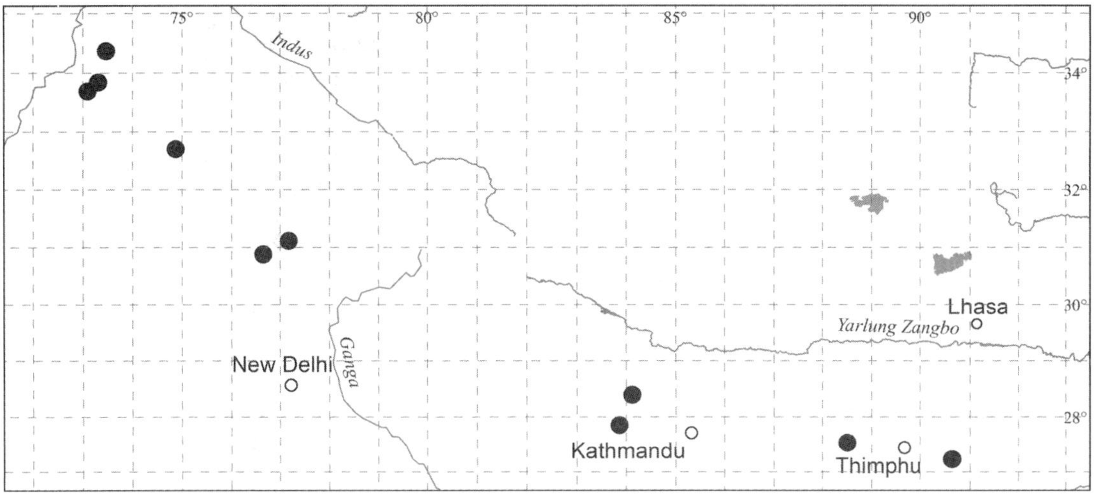

Map 47–137: **Campanula dimorphantha** Schweinf.

are of unequal sizes, the terminal flowers within a cyme producing larger fruit than the lower ones.

N PAKISTAN: **Azad Kashmir**, Muzaffarabad, ca. 760 m, 1976, *J. Mohammad 2033* (ISL); **Hazara**, Islamabad, Shakka Parian, 1975, *M. Hussain & J. Akhtar 885* (ISL); Kashmir Rd, 1976, *Shankat & Nisar 10* (ISL). **JAMMU & KASHMIR**: **Chenab Valley**, 1848, *s. coll. s. n.* (K); NW India, Khigzerabad, 1844, *M. P. Edgeworth s. n.* (K); **Kheri** Distr. (Upper Gangetic Plain), 1898, *Inayat* s. n. (K); **Sambalpur**, Surguja State 1070–1160 m, 1947, *H. F. Mooney 2851* (K). **U GANGA & INDUS**: **Shimla**: above Pagog, 1830 m, 18 7, *Gamble 4156A* (K). **SIKKIM & DARJILING**: **Sikkim**, Pawon Valley, 1220 m, *J. D. Hooker s. n.* (K). **BHUTAN**: **Tongsa**, 7 km S of Shamgong, 1850 m, 1982, *A. J. C. Grierson & D. G. Long 4202* (E, K).

18. Campanula punctata Lam., Encycl. 1: 586. 1785. Anonymous, Icon. Cormophyt. Sin. 4: 385. fig. 6184. 1975: Hong, Fl. Reipubl. Popularis Sin. 73(2): 80. 1983. Type: Gmelin's tab. 30 in *Fl. Sibirica* 3: 1767, cited by Lamarck.

Plants perennial, setose throughout or rarely nearly glabrous. Rhizomes long, horizontal. Stems erect, robust, 0.2–1 m tall, usually branched above. Basal leaves long-petiolate; leaf blade cordate or ovate; lower cauline leaves winged-petiolate, upper ones sessile; leaf blade ovate, lanceolate, or triangular, up to 7 cm long, base rounded or truncate, margin irregularly dentate, apex acute or acuminate. Flowers terminal on main stem and branches, pendent. Calyx lobes narrowly triangular; appendages reflexed, elliptic, triangular, ovate, or lanceolate. Corolla white, yellow, or pink, commonly purple- or red-punctate, tubular-campanulate, 3–6.5 cm long. Capsule semi-globose or obconic. Seeds gray-brown, ellipsoid, slightly compressed, ca. 1 mm long, smooth. **Chromosomes**: $2n = 34$. Fi-gure 47–60: b.

Phenology: flo-wering from June to September; fruiting in September and October.

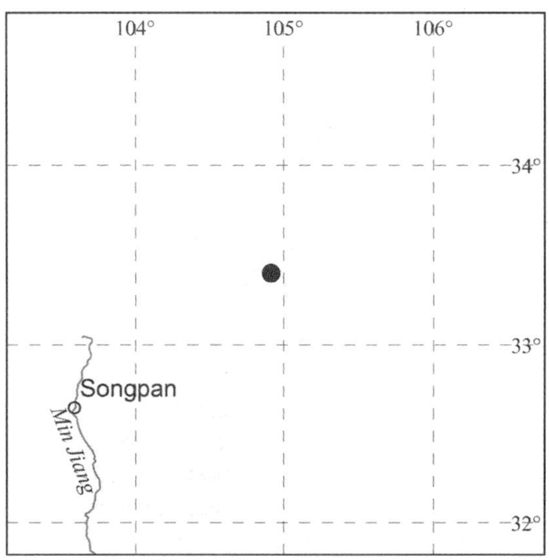

Map 47–138: **Campanula punctata** Lam.

Habitats: forests, thickets, or grassy places; below 2500 m.

Distribution: N Hengduan (Map 47–138); widely spread in E and NE Asia: N, NE and C China, Japan, Korea Peninsula, and Russia (Siberia and the Far East).

N HENGDUAN: Wudu, Nanshan, 2500 m, 1930, *Hopkinson 498* (PE).

19. Campanula leucoclada Boiss., Diagn. Pl. Or. Nov. ser. 23: 109. 1856; Rechinger & Schimann-Czeika in Rechinger, Fl. Iranica 13: 15. 1965; Nasir in Nasir & Ali, Fl. Pakistan 155: 10. fig. 2: D–E. 1984. Type: Pakistan, Baluchistan, *Stocks* s. n. (holotype: W; isotype: G).

Perennial herbs. Roots carrot-shaped, up to 1.5 cm in diameter, woody. Stems many, caespitose, woody at base, 20–45 cm tall, glabrous to sparsely hirsute. Leaves alternate, sessile, ovate or elliptic, 1–3 cm long, 0.5–1.3 cm broad, ada-xially glabrous to hispidulous, abaxially hirsute, margin nearly entire, serrate or dentate. Flowers in cymes, terminal on main stem and branches. Calyx setulose entirely or only along ribs and margin; calyx lobes triangular to lanceolate, 3–7 mm long, 1–3 mm broad; appendages smaller than calyx lobes, obtuse. Corolla pale violet, mauve to white, tubular-campanulate, 8–10 mm long; corolla lobes ca. as long as tube, puberulous. Style slightly exserted. Capsule nodding, globose, ca. 5 mm in diameter.

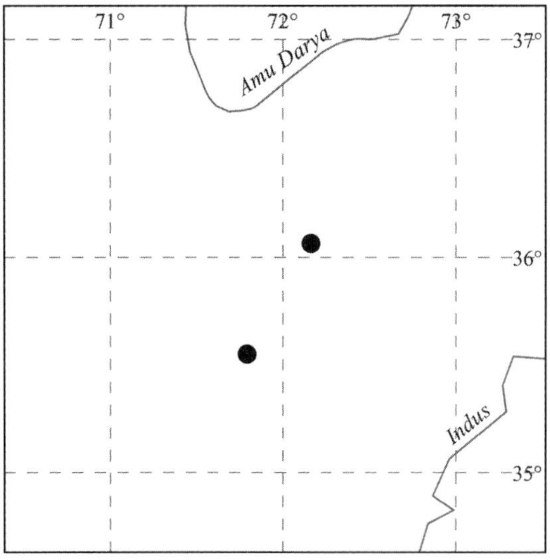

Map 47–139: **Campanula leucoclada** Boiss.

Phenology: flowering and fruiting from May to September.

Habitats: rocky hillsides, by cliffs, on limestones; 1100–3150 m.

Distribution: N Pakistan (Map 47–139); also in other parts of Afghanistan and W Pakistan.

N PAKISTAN: **Chitral**, Drosh, 1370 m, 1908, *R. A. Toppin 226* (K); Mastuj Track, Reshun, 1980 m, 1958, *J. D. A. Stainton 2526* (W).

20. Campanula polyclada Rech. f. & Schimann-Czeika in Rechinger, Fl. Iranica 13: 19. t. 3. fig. 2. 1965; Nasir in Nasir & Ali, Fl. Pakistan 155: 11. fig. 2: F, G. 1984. Type: Afghanistan, NW Panjao, 2900 m, 24 Jul. 1962, *K. H. Rechinger 18756* (holotype: W).

Herbs perennial. Roots attenuate, up to 1 cm in diameter, woody. Stems many from one caudex, caespitose, 10–15 cm tall, hirsute. Leaves alternate, sessile to shortly petioled; leaf blade ovate to orbicular, 5–15 mm long, 2–10 mm broad, abaxially densely pubescent, base broadly cuneate, margin nearly entire to denticulate, apex obtuse. Flowers terminal on main stem and branches, 2 to several in a cyme or solitary; pedicel as long as or slightly longer than flowers. Calyx glabrous or setulose along ribs and on margin; calyx lobes triangular, 3–4 mm long; appendages ca. 2 mm

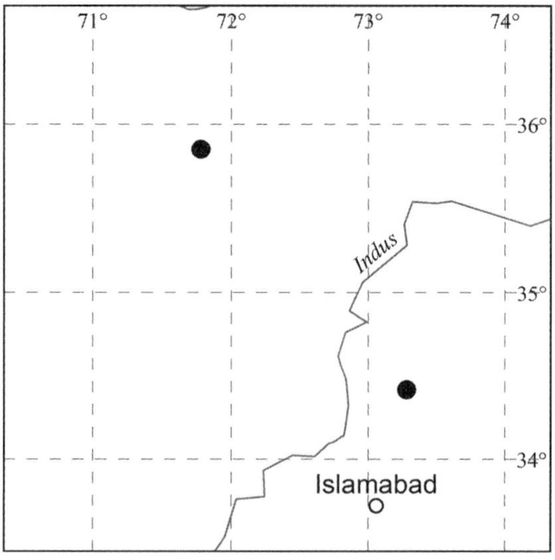

Map 47–140: **Campanula polyclada** Rech. f. & Schimann-Czeika

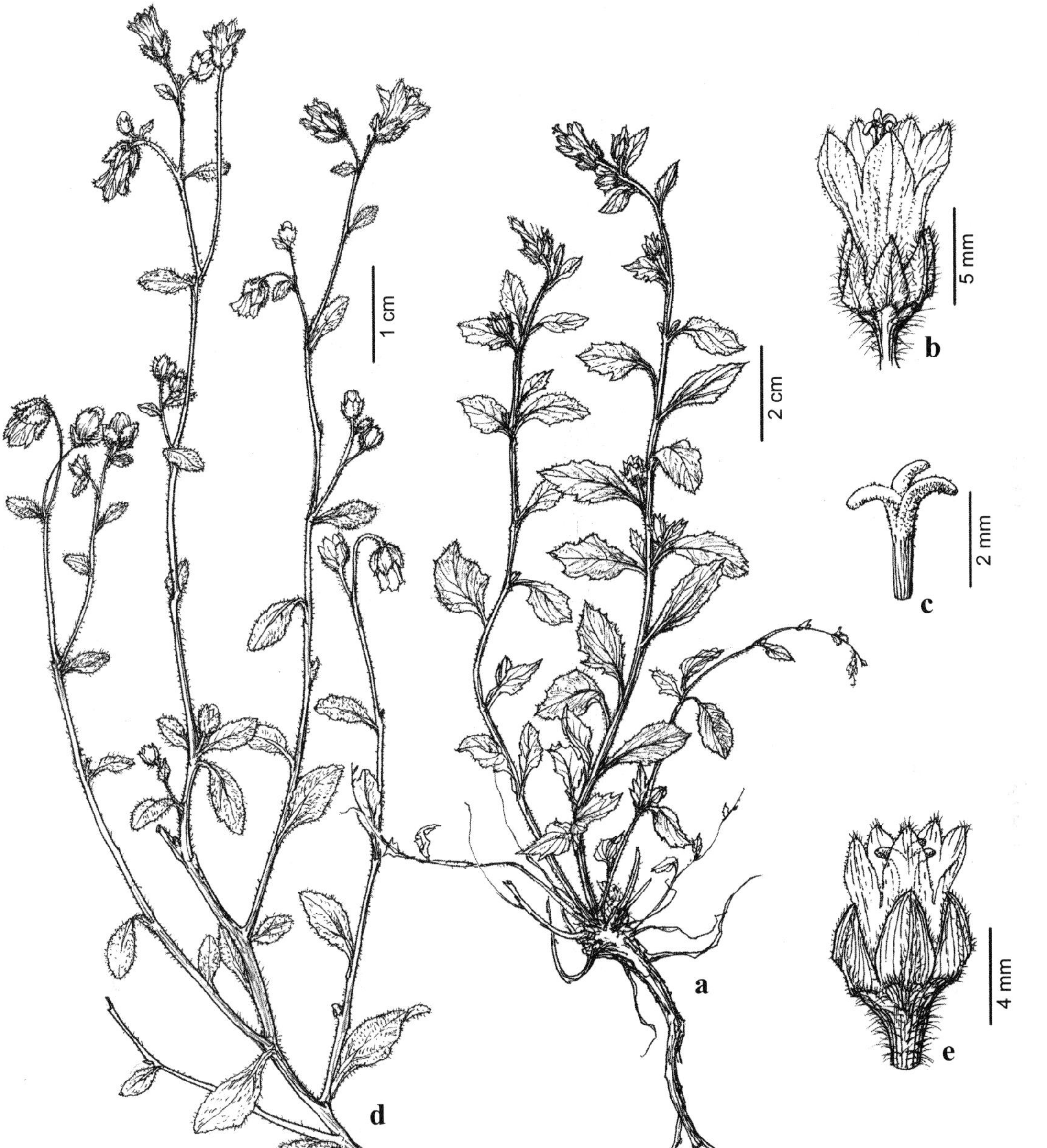

Figure 47–63: **a–c**, **Campanula polyclada** Rech. f. & Schimann: **a**, plant; **b**, flower; **c**, stigma. **d & e**, **Campanula leucantha** Gili: **d**, stem; **e**, flower. Drawn by Zhi-Juan CHEN.

long. Corolla tubular-campanulate, 8–10 mm long, pubescent outside; corolla lobes ca. 1/2 as long as tube. Style exserted. Seeds ellipsoid, ca. 0.6 mm long, 0.3 mm wide. Figure 47–63: a–c.

Phenology: flowering and fruiting in August.

Habitats: 1500–3200 m.

Distribution: N Pakistan (Map 47–140); also in other parts of Afghanistan.

N PAKISTAN: **Chitral**, Shehreshan towards Shoghore, 1584 m, 2007, *H. Ali 5781* (KUH); **Hazara**, Kagan Valley, 2012, *Z. D. Chen et al. 1061* (PE).

21. Campanula staintonii Rech. f. & Schimann-Czeika in Rechinger, Fl. Iranica 13: 16. tab. 3. fig. 1. 1965; Nasir in Nasir & Ali, Fl. Pakistan 155: 8. fig. 2: A, B. 1984. Type: Pakistan, Chitral, Mirkhani, SW of Drosh, 1370 m, 24 Apr. 1958, *J. D. A. Stainton 2288* (holotype: BM; isotypes:KUH, W).

Perennials erect, woody at base. Roots attenuate, ca. 1 cm in diameter. Stems many, caespitose, 20–30 cm tall, sparsely white-villous. Leaves alternate, sessile or shortly petiolate; leaf blade ovate-lanceolate, elliptic or narrowly oblong, 1.5–3 cm long, 0.4–0.6 cm broad, adaxially nearly glabrous, abaxially sparsely to densely hirsute, base cuneate, margin usually serrate to crenate, sometimes entire, apex acute. Flowers in a terminal cyme of 2–4; pedicel 0.5–1.5 cm long. Calyx glabrous; calyx lobes triangular-lanceolate, 5–6 mm long, 2–3 mm broad, hispid-ciliate, apex acute; appendages ca. 2 mm long, hispid. Corolla white with brown veins, campanulate, 1.5–2 cm long; corolla lobes ca. 1/2 as long as tube, glabrous. Style included.

Phenology: flowering in April.

Habitats: in crevices of shady cliffs; ca. 1370 m.

Distibution: **endemic**; confined to N Pakistan (Map 47–141).

Notes: Only the type collection with three duplicates has been seen so far respectively in the three herbaria.

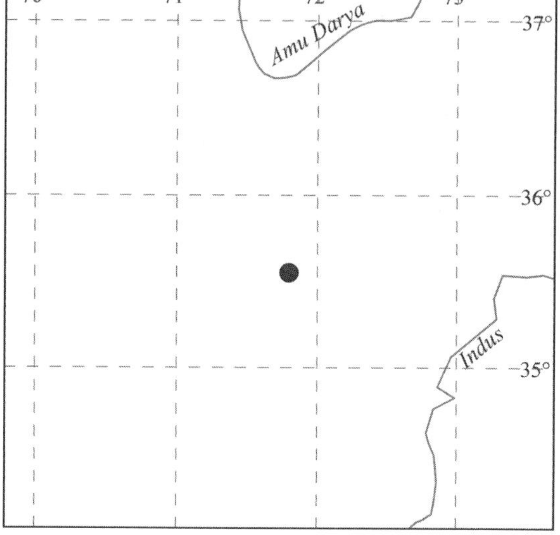

Map 47–141: **Campanula staintonii** Rech. f. & Schimann-Czeika

22. Campanula leucantha Gilli, Repert. Spec. Nov. Regni Veg. 68:87. 1963; Rechinger & Schimann-Czeika in Rechinger, Fl. Iranica 13: 16. tab. 2. 1965; Nasir in Nasir & Ali, Fl. Pakistan 155: 10. fig. 2: C. 1984.Type: Afghanistan, Palghan near Kabul. *Gilli 3873* (holotype: W).

Perennials, woody at base. Roots thick, elongated, woody. Stems 10–30 cm tall, branched, white hirsute. Leaves alternate, sessile or lower ones shortly petiolate; leaf blade ovate to ovate-lanceolate, or elliptic, 1.2–3 cm long, 0.5–1.5 cm broad, adaxially sparsely hirsute, abaxially fairly densely hirsute, base cuneate, margin serrate, apex acute. Inflorescens terminal, a corymb of 2-several cymes, each cyme 1–4-flowered; pedicel as long as flower. Calyx tube setose along ribs; calyx lobes lanceolate or subulate, 5–10 mm long, 2–3 mm broad, setose, appendages as long as or shorter than calyx tube. Corolla light pink, blue or white, tubular, 1.3–2 cm long, pubescent outside; corolla lobes 1/2 as long as tube. Style shorter than or as long as corolla. Figure 47–63: d & e.

Phenology: flowering from April to June; fruiting from July to September.

Habitats: 1500–2800 m.

Distribution: N Pakistan and U Ganga & Indus (Map 47–142); also in E and C Afghanistan.

N PAKISTAN: Baltistan, Skardu, Satpaea Village, 1978, *M. Ajas & M. Agani 835* (ISL); Skardu, Chaghmagar Shigal, 1978, M. *Ajas & M. Agani 1416* (ISL); **Chitral**, Ashreth, 1895, *Surg.-Lt. Harris 16316* (K); **Gilgit**, Hunza, Zurthgurban Nullah, Shimshal, 2007, *J. Alam et al. 4027* (KUH); **Hazara**, Babusar, 1975, *M. A. Siddiqi 23789* (ISL); Hazara, Kagan Valley, 2012, *Z. D. Chen et al. 469* (PE), *482* (PE) & *495* (PE). **U GANGA & INDUS**: **Lahul**, Mt Drilburi, 1867, *s. coll.* s. n. (WU).

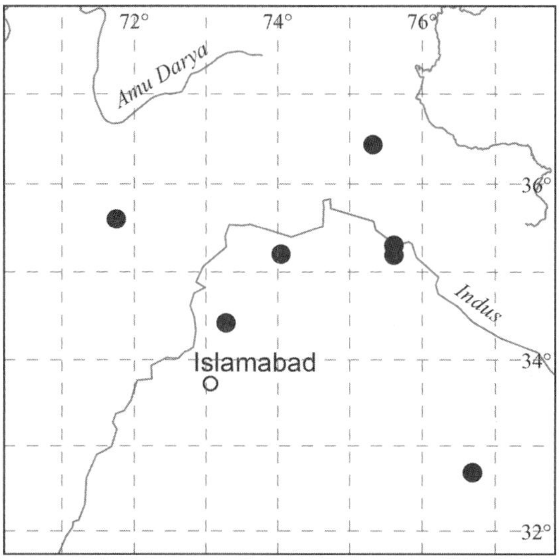

Map 47–142: **Campanula leucantha** Gilli

Subg. 2. Rapunculus (Dumort.) Kharadze, Syst. Geogr. Rast. 32: 54. 1976. ≡ *Campanula* sect. *Rapunculus* Dumort., Fl. Belg. 58. 1827.

Basal leaves rosulate, persistent at anthesis; cauline leaves mostly at base, upper ones sessile or nearly so, usually linear if present. Flowers solitary, terminal only on main stem, or also on branches. Capsule poricidal above middle.

About 140 species, throughout Arctic and N temperate regions, extending southwards to N Africa, westwards to N Mexico and SE United States (Florida); five species (four endemic) in the Pan-Himalaya.

23. Campanula crenulata Franch., J. Bot. (Morot) 9: 365. 1895; Nannfeldt in Handel-Mazzetti, Symb. Sin. 7: 1067. 1936; Anonymous, Icon. Cormophyt. Sin. 4: 386. fig. 6187. 1975. Type: China, Yunnan, Heqing (Ho-Kin), 26 Aug. 1884, *M. Delavay 60* [**lectotype here designated**: P (P00225841); isolectotype: P].

Campanula nephrophylla C. Y. Wu in Wu & Li, Yunnan Trop. Subtrop. Fl. Res. Rep. 1: 60. pl. 13: 5–7. 1965. Type: China, Yunnan, Zhongdian (Shangri-la), 3800 m, 17 Aug. 1937, *T. T. Yü 12754* (holotype: KUN; isotype: PE).

Herbs perennial. Roots often carrot-shaped, ca. 5 mm in diameter. Caudex enveloped by remains of old petioles. Stems 2–7, caespitose, ascending, 6–33 cm tall, simple or rarely branched, glabrous. Basal leaves several, often rosulate, long-petiolate, leaf blade orbicular, cordate, or ovate, 0.7–3.5 cm long, 0.6–2 cm broad, usually glabrous, sometimes adaxially sparsely hispid, margin crenulate; lower cauline leaves petiolate, leaf blade spatulate, elliptic, or ovate, 1–3 cm long, margin crenulate; upper cauline leaves nearly sessile, linear, margin entire or callose-denticulate. Flowers terminal, solitary, pendent or horizontal, glabrous throughout. Calyx tube obconic; calyx lobes subulate or narrowly triangular, 3–12 mm long, 1–2 mm broad, margin callose-denticulate. Corolla blue, blue-purple, or dark purple-red, campanulate, 1.2–2.6 cm long; corolla lobes ca. 2/3 as long as tube. Style included. Capsule cylindrical to narrowly obovoid, 1–1.8 cm long, ca. 5 mm in diameter. Seeds yellowish brown, ellipsoid, slightly compressed. Figure 47–64: a.

Phenology: flowering from July to September; fruiting from September to November.

Habitats: rocky places or grassy slopes; 2600–4200 m.

Distribution: **endemic**; confined to S Hengduan (Map 47–143).

S HENGDUAN: **Daocheng**, Gonggaling, 4100 m, 1981, *Qinghai-Xizang Exped. 5434* (KUN, PE); **Heqing**, Lianping, 3700 m, 1940, *R. C. Ching 24047* (PE); **Muli**, Baicaoping, 2600 m, 2011, *D. Y. Hong et al. H11025* (PE); Mt Gibu, 2900 m, 1937, *T. T. Yü 14171* (PE); **Ninglang**, Yongning, Mt Wuaha, 3810–3960 m, *J. F. Rock 24239* (B); **Panzhihua**, 1400 m, 1983, *Qinghai-Xizang Exped. 11217* (KUN); **Shangri-**

Figure 47–64: **a**, **Campanula crenulata** Franch. **b** & **c**, **Campanula calcicola** W. W. Smith: **b**, plant; **c**, a section of stem, showing abaxial surface of leaf. **a** redrawn by Ai-Li LI from *Icon. Cormophyt. Sin.* 4, 1975; **b** & **c** from *Fl. China* 19, 2011.

la (Zhongdian): Juatze, 3700 m, 1937, *T. T. Yü 12693* (PE); **Weixi**, Yezhi, 3600 m, 1935, *C. W. Wang 68647* (PE); **Yulong** (Lijiang), 3300–3800 m, 1942, *K. M. Feng 8961* (PE).

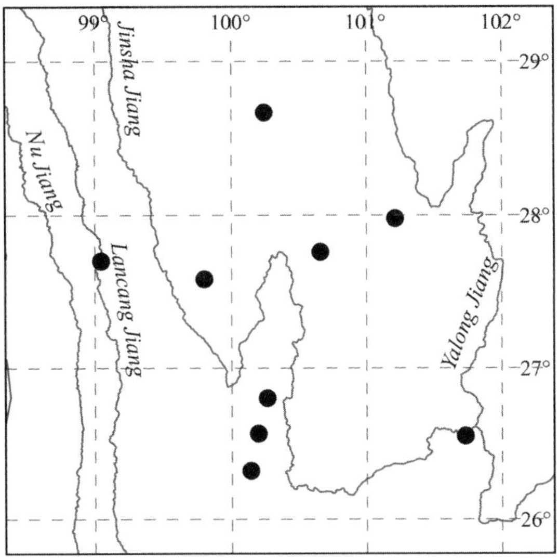

Map 47–143: **Campanula crenulata** Franch.

24. Campanula calcicola W. W. Smith, Notes Roy. Bot. Gard. Edinburgh 12: 196. 1920; Hong, Fl. Reipubl. Popularis Sin. 73(2): 91. pl. 13: 2. 1983; Hong, Lammers & Klein in Wu, Raven & Hong, Fl. China 19: 536. 2011. Type: China, Yunnan, Yungning-Yangtze divide, 27°40′N, 3050–3350 m, Jul. 1914, *G. Forrest 13081* (**lectotype here designated**: E); Shangri-la (Zhongdian), 3960 m, 1914, *G. Forrest 13108* (paratype: E).

Plants perennial. Roots carrot-shaped. Stems several, caespitose, ascending, 6–30 cm tall, usually branched, villous. Basal leaves long-petiolate; leaf blade ovate or orbicular, 0.5–2.4 cm long, 0.6–2.7 cm broad, adaxially villous, base cordate or truncate, margin crenate or callose-denticulate, apex obtuse or rounded; lower cauline leaves long-petiolate, leaf blade ovate or orbicular, 5–7 mm long, 4–6 mm broad, base cordate or truncate, apex obtuse or rounded; middle and upper cauline leaves shortly petiolate or sessile, leaf blade lanceolate or narrowly oblong, margin callose-dentate. Flowers terminal, often upright. Calyx tube obovoid or obconic, sparsely villous or sometimes glabrous, base rounded; calyx lobes narrowly oblong to subulate-linear, 2–6 mm long, margin with 1–3 pairs of teeth. Corolla blue, purple or blue-purple, broadly campanulate, 0.9–2 cm long; corolla lobes ca. 1/2 as long as tube. Capsule obovoid, ellipsoid, or subglobose, 4–8 mm long. Seeds brown-yellow, ellipsoid, ca. 0.5 mm long, 1-ribbed, smooth. Figure 47–64: b & c.

Phenology: flowering and fruiting from August to October.

Habitats: moist rocks or pine forests; 2300–3900 m.

Distribution: **endemic**; confined to S Hengduan (Map 47–144).

S HENGDUAN, **Dali**: Cangshan, 1946, *T. N. Liou 21091* (PE); **Daocheng**, Gonggaling, 3100 m, 1937, *T. T. Yü 12959* (PE); **Muli**, Ninglang Shan, 3400 m, 1983, *Qinghai-Xizang Exped. 14325* (PE); SW of Muli, 3660 m, 1922, *G. Forrest 22133* (W); **Xiangcheng**: No.4 Distr., 3600 m, 1973, *Sichuan Veget. Surv. (Xiang) 3261* (PE); **Yulong** (Lijiang), 3500 m, 1939, *K. M. Feng 2962* (PE).

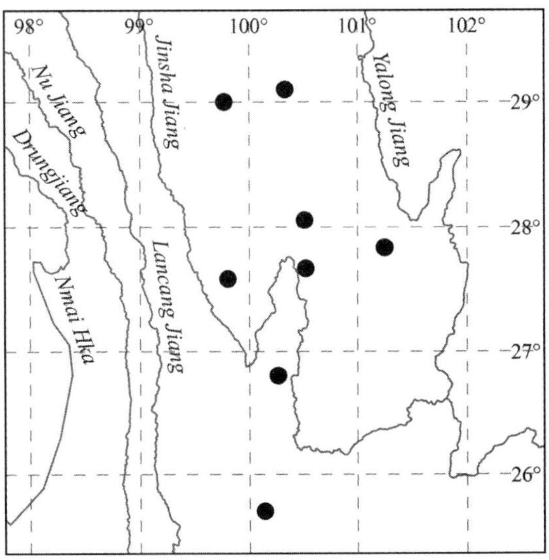

Map 47–144: **Campanula calcicola** W. W. Smith

25. Campanula chrysosplenifolia Franch., J. Bot. (Morot) 9: 364. 1895; Nannfeldt in Handel-Mazzetti, Symb. Sin. 7: 1067. 1936: Hong, Fl. Reipubl. Popularis Sin. 73(2): 91. 1983; Hong, Lammers & Klein in Wu Raven & Hong, Fl. China 19: 536. 2011. Type: China, Yunnan,

Heqing, Che-tcho-tze, Ta-Pin-tze, 3 Oct. 1882, *M. Delavay 430* [**lectotype here designated**: P (P00225602); isolectotype: P]

Campanula leucotricha C. Y. Wu in Wu & Li, Yunnan Trop. Subtrop. Fl. Res. Rep. 1: 58. pl. 13: 1–4. 1965. Type: China, Yunnan, Binchuan, Jizu Shan, 3300 m, 28 Sept. 1940, *R. C. Ching 24926* (holotype: KUN; isotype: PE).

Plants perennial. Roots carrot-shaped. Stems caespitose, filiform, erect, 7–22 cm tall, simple but often branched, glabrous. Basal leaves rosulate, long-petiolate, leaf blade ovate or orbicular, 6–17 mm long, 5–15 mm broad, adaxially villous, base cordate, margin crenulate; cauline leaves sessile or short-petiolate, leaf blade ovate to linear, 5–15 mm long, 0.2–15 mm broad, base cordate or truncate, margin crenate to serrate, sometimes callose-denticulate, apex acute or obtuse. Flowers terminal on main stem and branches, mostly solitary, sometimes 2 or 3 and, glabrous throughout. Calyx tube obovoid or narrowly obconic, base acute or obtuse; calyx lobes subulate or narrowly triangular, 4–9 mm long, 0.5–1.5 mm broad, margin with 1–3 pairs of verrucose teeth or entire. Corolla blue, campanulate, 8–17 mm long; corolla lobes ca. 1/2 as long as tube, ovate, apex acute to obtuse. Style included. Capsule narrowly obovoid, 9–19 mm long. Seeds yellowish brown to brown, oblong, slightly compressed, ca. 1 mm long, 1-ribbed.

Phenology: flowering from June to September.

Habitats: rocky places or meadows; 3000–4300 m.

Distribution: **endemic**; confined to S Hengduan (Map 47–145).

S HENGDUAN: **Dali**, W flank of Dali Range, 1913, *G. Forrest 11690* (PE); **Muli**, 3960–4270 m, 1918, *G. Forrest 16819* (W); **Shangri-la** (Zhongdian), s. loc., 1914. *G. Forrest. 13053* (PE); between Yongning and Shangri-la (Zhongdian), 3000 m, 1915, *Handel-Mazzetti 7647* (W, WU); **Yanbian**, Yankou, Dawan Village, 2657 m, 2011, *D. Y. Hong et al. H11044* (PE); **Yanyuan**, Bailing, 3700 m, 1983, *Qinghai-Xizang Exped. 12735* (KUN).

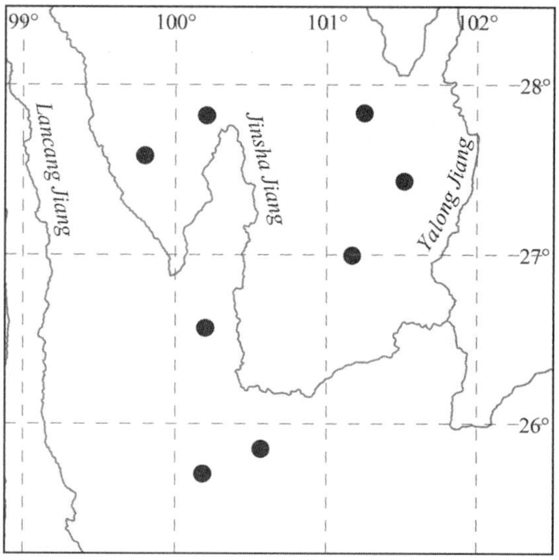

Map 47–145: **Campanula chrysosplenifolia** Franch.

26. Campanula aristata Wall. in Roxburgh, Fl. India 2: 98. 1824; Clarke in Hooker, Fl. Brit. India 3: 441. 1881; Hong, Fl. Reipubl. Popularis Sin. 73(2): 91. 1983. Type: Nepal, *Wallich 1291* (holotype: K).

Campanula aristata var. *longisepala* C. Marquand, J. Linn. Soc. Bot. 48: 196. 1929. Type: China, Tibet, Pan-Sum-Kye La, 4500 m, 25 Aug. 1924, *F. Kingdon-Ward 6133* (holotype: K).

Campanula cylindrica (Pax & K. Hoffm.) Nannf., Acta Horti Gothob. 5: 24. 1930. ≡ *Wablenbergia cylindrica* Pax & K. Hoffm., Repert. Spec. Nov. Regni Veg. 12: 501. 1922. Type: China, Sichuan, Dawu (Tatsien Lu-Dawo), 3800 m, *W. Limpricht 1884* (holotype: WRSL, n.v.; isotype: WU).

Plants perennial. Roots carrot-shaped. Stems usually caespitose, erect, 10–50 cm tall. Basal leaves long-petiolate, glabrous, leaf blade ovate or broadly elliptic; lower and middle cauline leaves long-petiolate, glabrous, leaf blade lanceolate, elliptic, or linear; middle and upper cauline leaves sessile, glabrous, linear, 1.5–7 cm long, glabrous, margin entire or sparsely denticulate. Calyx tube very narrowly oblong, 0.5–1.5 cm long; calyx lobes filiform, (3–) 7–18 (–25) mm long, usually longer than corolla. Corolla blue or blue-purple, 7–15 mm long; corolla lobes ca. 1/2 as long as tube. Capsule clavate, attenuate

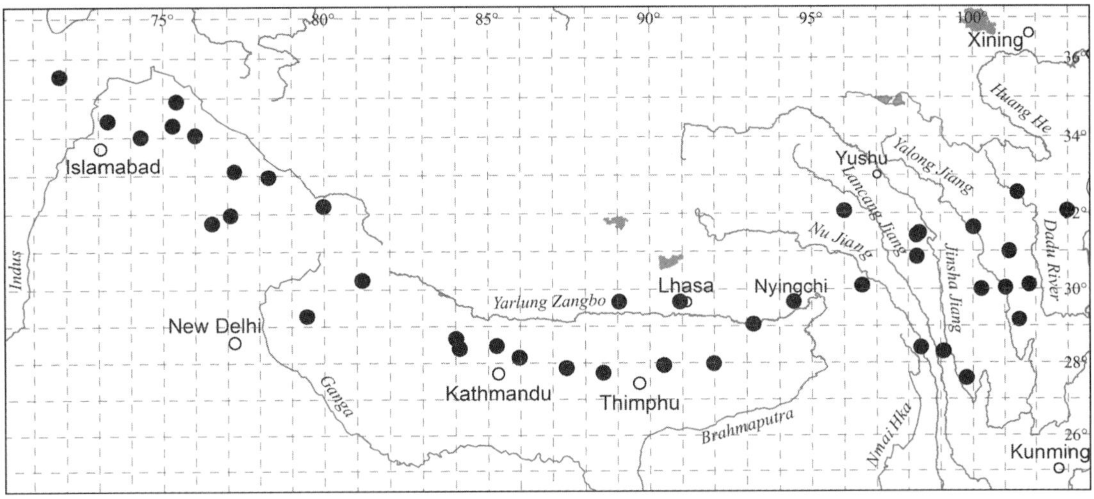

Map 47–146: **Campanula aristata** Wall.

towards base, 2–4 cm long. Seeds brown-yellow, ellipsoid, slightly compressed, ca. 0.7 mm long, smooth. Figure 47–65: a.

Phenology: flowering from June to August; fruiting in August and September.

Habitats: alpine meadows or thickets; 3500–5000 m.

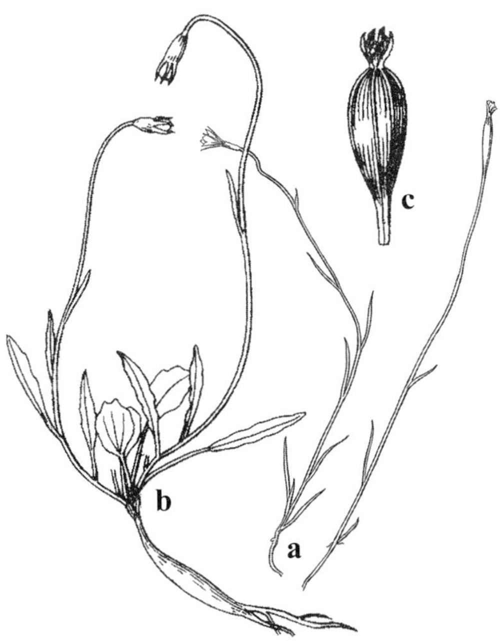

Figure 47–65: **a**, **Campanula aristata** Wall. **b** & **c**, **Campanula immodesta** Lammers: **b**, plant; **c**, fruit. **a** redrawn by Ai-Li LI from *Icon. Cormophyt. Sin.* 4, 1975; **b** & **c** from *Fl. Reipubl. Popularis Sin.* 73(2), 1983.

Distribution: N Pakistan, Jammu & Kashmir, U Ganga & Indus, U Yarlung Zangbo, C Nepal, Sikkim & Darjiling, Bhutan, M Yarlung Zango, L Yarlung Zangbo, Yarlung Zangbo-Brahmaputra, Tangut, N and S Hengduan (Map 47–146); also in other parts of Afghanistan, and in S Gansu (Xiahe), E Qinghai and Shaanxi (Mts. Taibai) of China.

27. Campanula immodesta Lammers, Novon 8: 34. 1998; Hong, Lammers & Klein in Wu, Raven & Hong, Fl. China 19: 536. 2011. ≡ *Campanula modesta* Hook. f. & Thomson, J. Proc. Linn. Soc. Bot. 2: 24. 1857, non Schott, Nyman & Kotschy (1854). Type: India, Sikkim, 3660–4270 m, *J. D. Hooker* s. n. (holotype: K).

Herbs perennial. Roots thickened, up to 7 mm in diameter. Stems single or 2 to several and caespitose, erect or ascending, 7–22 cm long. Basal leaves rosulate, petiole 5–45 mm long, leaf blade orbicular or ovate, 5–15 mm long, 4–8 mm broad, margin crenate or entire; lower cauline leaves long-petiolate, leaf blade linear or lanceolate; middle and upper cauline leaves 1–3, sessile, leaf blade linear or narrowly lanceolate, 5–20 mm long, margin crenate. Flowers terminal, solitary, more or less pendulous. Calyx tube obovoid, 2–10 mm long, glabrous; calyx lobes subulate to narrowly triangular, 1.8–8 mm long,

0.5–1 mm broad, usually shorter than corolla, entire. Corolla blue or purple-blue, campanulate, 4–11 mm long; corolla lobes deltoid, ca. 1/2 as long as tube. Capsule obovoid, 5.5–20 mm long, 2.5–6 mm in diameter. Figure 47–65: b & c.

Phenology: flowering in July and August; fruiting from August to October.

Habitats: alpine meadows; 3300–5180 m.

Distribution: **endemic**; U Ganga & Indus, U Yarlung Zangbo, Nepal, Sikkim & Darjiling, M and L Yarlung Zangbo, Yarlung Zangbo-Brahmaputra, and S Hengduan (Map 47–147).

Notes: The present author has not seen any specimen from Nepal, and thus readers may doubt the identity of *J. F. Duthie 602* from Tihri-Garhwal of India. I examined the collection in K, and found that it has calyx tube globose which must be the character state of *C. immodesta*, but not that of *C. aristata*.

U GANGA & INDUS: **Tihri-Garhwal**, Dudu Glacier, 4270–4570 m, 1883, *J. F. Duthie 602* (K). **U YARLUNG ZANGBO**: **Dinggyê**, Chengtang Gou, 4151 m, 2013, *PE-Tibet Exped. 3439* (PE). **SIKKIM & DARJILING**: **Sikkim**: Chapopla, 4880 m, 1911, *Ribu & Rhomoo 5260* (K); Chhortenima La, 5180 m, 1909, *Smith & Cave 2353* (B); Zemu Valley, 4420 m, 1909, *Smith & Cave 1375* (K); Zemu Vally, 4510 m, 1909, *Smith & Cave 1357* (B). **M YARLUNG ZANGBO**: **Yadong**, near Pagri, 1953. *P. C. Tsoong 5804* (PE). **L YARLUNG ZANGBO**: **Guxiang**, 4250 m, 1965, *J. S. Ying & D. Y. Hong 1300* (PE). **YARLUNG ZANGBO-BRAHMAPUTRA**: **Cona**, Boshan Pass, 4500 m, 1975, *C. Y. Wu et al. 75–1010* (PE); **Goyü** Pass, 4160 m, 2012, *D. Y. Hong et al. H12043* (PE); Demula Pass, 4700 m, 1973, *Qinghai-Xizang Exped. 73–1200* (PE). **S HENGDUAN**: **Cawarong**, Ngala, 3750 m, 1935, *C. W. Wang 66072* (PE); **Dêqên**, Baimashan, 4150 m, 1937, *T. T. Yü 9287* (PE); **Kangding**, Dongrergo, 4300–4400 m, 1922, *H. Smith 3921* (PE); **Muli**, Wa'erzhai, 3300 m, 1937, *T. T. Yü 14575* (PE); **Shangri-la** (Zhongdian), Xianrendong, 3400 m, 1937, *T. T. Yü 12119* (PE); **Weixi**, Dongquling, 3800 m, 1937, *T. T. Yü 9904* (PE).

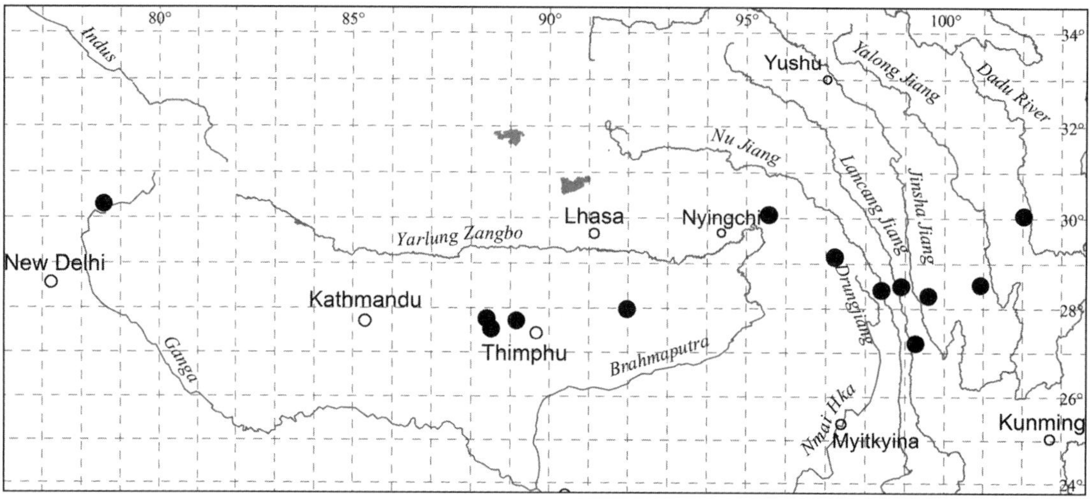

Map 47–147: **Campanula immodesta** Lammers

Imperfectly known species:

Campanula wattiana Nayar & Babu, India Bot. Soc. 49: 183. 1970; Haridasan & Mukherjee in Hajra & Sanjappa, Fasc. Fl. India 22: 52. 1996. Type: India, Himachal Pradesh, Bagi, *Watt 8609* (holotype: CAL, n. v.; isotype: CAL, n. v.). I have seen neither type specimens in CAL, nor the collections cited by Haridasan and Mukherjee (1996). According to the description provided by Haridasan and Mukherjee it seems to me that *C. wattiana* is quite possibly an element of *C. cana*, which is a rather common and variable species.

10. ASYNEUMA Griseb. & Schenk

Grisebach & Schenk, Arch. Naturgesch. 18: 335. 1852. Type: *Asyneuma canescens* (Waldst. & Kit.) Griseb. & Schenk (*Phyteuma canescens* Waldst. & Kit.).

Herbs perennial. Roots carrot-shaped. Leaves alternate. Cymes of several flowers, subtended by an involucral bract, clustered into a long and interrupted spike, spikes single or compound; pedicel short, with a pair of linear bracteoles at base. Calyx adnate to hypanthium up to top, 5-lobed; calyx lobes linear. Corolla 5-divided to near base, corolla lobes appearing free, linear. Stamens 5; filaments dilated at base, dilated part densely ciliate. Ovary inferior, 3-locular; style nearly as long as corolla, hairy above; stigma 3-fid, segments linear, recurved. Fruit a capsule, 3-poricidal slightly above middle. Seeds ovoid-ellipsoid or ovoid-oblong, angular or not. Plants protandrous.

About 30 species; temperate Eurasia, mainly in the Mediterranean; five species (one endemic) in the Pan-Himalaya.

1a. Stigma 2-fid; lower leaves with petiole 5–15 mm long, upper ones sessile, leaf blade ovate; flowers lax, not in a capitulum .... 5. *A. thomsonii*
1b. Stigma 3-fid; leaves all sessile or subsessile, leaf blade lanceolate or elliptic; flowers lax or in a capitulum.
 2a. Flowers lax, not in a cyme or capitulum; pedicel 5–10 mm long .... 4. *A. pakistanicum*
 2b. Flowers in a cyme or capitulum; pedicel much shorter.
  3a. Capsule obovoid-conical, base acute or acuminate; plants often glabrous .... 1. *A. fulgens*
  3b. Capsule globose, base truncate or rounded; plants always hairy.
   4a. Calyx-lobes often more or less reflexed at anthesis, 7–10 mm long; leaf blade white-hirsute on both surfaces .... 2. *A. chinense*
   4b. Calyx-lobes always upright, 3–7 mm long; leaf blade glabrous on adaxial surface .... 3. *A. strictum*

1. Asyneuma fulgens (Wall.) Briq., Candollea 4: 334. 1931; Hong, Fl. Reipubl. Popularis Sin. 73(2): 141. 1983; Hong, Lammers & Klein in Wu, Raven & Hong, Fl. China 19: 553. 2011. ≡ *Campanula fulgens* Wall. in Roxburgh, Fl. India 2: 99. 1824. Type: Nepal, *Wallich 1283* (holotype:K).

Roots fleshy. Stems often single, less frequently caespitose, erect, 30–100 cm tall, usually simple, usually glabrous, sometimes sparsely hirsute. Leaves all nearly sessile, or lower ones with petiole up to 2 cm long; leaf blade ovate, ovate-lanceolate, lanceolate or elliptic, 5–7 cm long, 1–3 cm broad, adaxially glabrous or sparsely hispidulous, abaxially usually glabrous, sometimes sparsely hispidulous along veins, margin serrate. Cymes clearly separated from each other, sometimes with gaps of up to 10 cm. Calyx glabrous; calyx tube obovoid-conical, longer than broad; calyx lobes 5–11 mm long. Corolla blue or purple; corolla lobes as long as or slightly longer than corolla tube. Style much shorter than corolla. Capsule obovoid-conical, 7–9 mm long, 3–4 mm in diameter.

Distribution: U Yarlung Zangbo, W and C Nepal, Sikkim & Darjiling, Bhutan, M Yarlung Zangbo, Yarlung Zangbo-Brahmaputra, and S Hengduan (Map 47–148); also in other parts of India and in Sri Lanka.

1a. Capsule glabrous; calyx lobes 5–10 mm long, less than 1 mm broad, not enlarged at fruiting; stems glabrous .... 1a. subsp. *fulgens*
1b. Capsule hispid along ribs; calyx 7–11 mm long at anthesis, 11–15 mm long, up to 2 mm broad at fruiting; stems villous .... 1b. subsp. *forrestii*

1a. Asyneuma fulgens subsp. **fulgens**

Stems glabrous. Leaf blade adaxially glabrous. Calyx lobes 5–10 mm long, less than 1 mm broad, not enlarged in fruit. Capsule glabrous. Figure 47–66: a-c.

Phenology: flowering from June to August; fruiting from August to October.

Habitats: grassy slopes, forests, or forest margins in ravine; 1800–3100 m.

Distribution: U Yarlung Zangbo, C Nepal, Sikkim & Darjiling, Bhutan, M Yarlung Zangbo, and Yarlung Zangbo-Brahmaputra; also in other parts of India and in Sri Lanka.

U YARLUNG ZANGBO: **Gyirong**: near Gyirong town, 2400 m, 1981, *C. C. Ni et al. 2296* (PE). **W NEPAL**: **Karnali**, E of Jumla, Luma, 2590 m, *Polunin, Sykes & Williams 4944* (E, P). **C NEPAL**: **Gandaki**, Gorkha, Sardu Khola, 2050 m, *M. Suzuki et al. 9485170* (E); below Curjahni, 2440 m, *Stainton, Sykes & Williams 3645* (E). **SIKKIM & DARJILING**: **Sikkim**, 1830–2130 m, 1849, *J. D. Hooker* s. n. (K); Lachong, 2740 m, 1849, *J. D. Hooker* s. n. (K). **BHUTAN**: **Byaka** Dzong, Bumtang Chu, 2 km N of Byaka, 2700 m, 1979, *A. J. C. Grierson & D. G. Long 2595* (K); **Thimphu**, Moti thang, 2434 m, 1971, *Ramesh Bedi 406* (K); Tsalimape to Simo Sampa, 2290–2350 m, 1938, *B. J. Gould 890* (K). **M YARLUNG ZANGBO**: **Yadong**, Chumbi, 1904, *C. H. Bell* s. n. (K); 3050 m, 1936, *F. S. Chapman 344* (K). **YARLUNG ZANGBO-BRAHMAPUTRA**: **Gyigang**, 2600 m, 1973, *Qinghai-Xizang Exped. 73–594* (KUN, PE).

1b. Asyneuma fulgens (Wall.) Briq. subsp. **forrestii** D. Y. Hong, **subsp. nov.** Type: Yulong, (Lijiang), Lijiang Range, 1910, *G. Forrest 6081* (holotype: PE).

Stems villous. Leaf blade hispidulous or villous on both surfaces. Calyx lobes 7–11 mm long at anthesis, 11–15 mm long, up to 2 mm broad in fruit. Capsule hispid along ribs.

Distribution: **endemic**; confined to S Hengduan.

Notes: *G. Forrest 6081* may represent a new species; its capsule is hispid along ribs; calyx lobes enlarged at fruiting, up to 11–15 mm long, ca. 2 mm broad; stems rather densely hirsute, and leaf blade hispidulous on both surfaces. I have seen only this specimen.

2. Asyneuma chinense D. Y. Hong, Fl. Reipubl. Popularis. Sin. 73(2): 188. 1983; Hong & Lammers in Wu, Raven & Hong, Fl. China, 19: 554. 2010. Type: China, Yunnan, NW Zhongdian [Shangri-la], Sept. 1939, *K M. Feng 2324* (holotype: PE).

Roots fleshy. Stems single, less frequently caespitose, erect, 40–100 cm tall, usually simple, hirsute. Leaves all nearly sessile, or lower cauline leaves with petiole up to 3 cm long; leaf

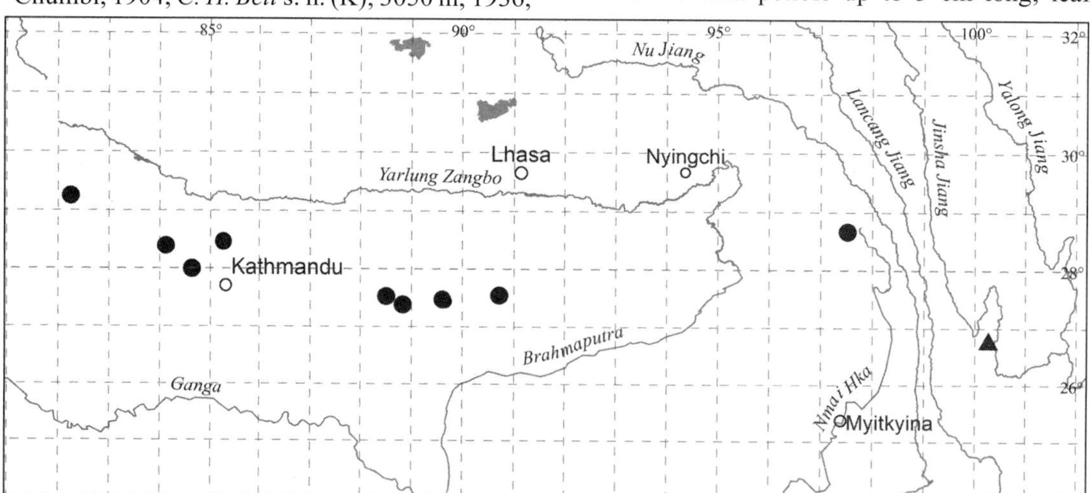

Map 47–148: **Asyneuma fulgens** (Wall.) Briq.
● subsp. **fulgens**, ▲ subsp. **forrestii**

Figure 47–66: **a–c**, **Asyneuma fulgens** (Wall.) Briq. subsp. **fulgens**: **a**, stem and inflorescence; **b**, flower; **c**, fruit. **d–f**, **Asyneuma chinense** D. Y. Hong: **d**, root and lower part of stem; **e**, flower; **f**, fruit. **a–c** from *Fl. Reipubl. Popularis Sin*. 73(2), 1983; **d–f** redrawn by Ai-Li LI from *Icon. Cormophyt. Sin.* 4, 1975.

blade ovate, ovate-lanceolate, lanceolate or elliptic, 2.5–8 cm long, 0.7–3.5 cm broad, both surfaces white hirsute, margin serrate. Spikes few-flowered; involucrate bracts sometimes pubescent, each subtending 1–4 flowers. Calyx usually glabrous, less frequently hispidulous; calyx tube globose; calyx lobes subulate, 7–10 mm long, slightly longer than corolla, often reflexed after anthesis. Corolla purple or blue. Style slightly shorter than corolla. Capsule globose, with 3 wide longitudinal furrows, 4–6 mm long, 4–5 mm in diameter, base truncate or even concave, broadest at base. Seeds ovoid-oblong, slightly compressed, with one rib, brown-yellow, ca. 0.5 mm long. Figure 47–66: d–f.

Phenology: flowering and fruiting from June to September.

Habitats: grassy slopes, forests, or forest margins; below 1500–3400 m.

Distribution: Yarlung Zangbo-Brahmaputra and S Hengduan (Map 47–149); also in other parts of China (Guangxi, Guizhou, S and E Yunnan).

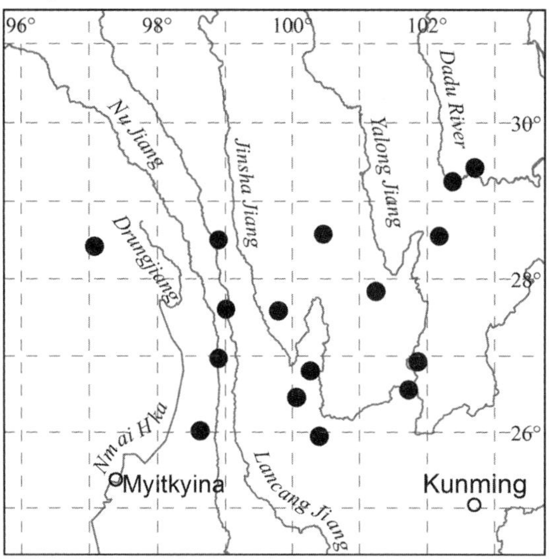

Map 47–149: **Asyneuma chinense** D. Y. Hong

Notes: *Asyneuma chinense* is rather common in SW China, but it was misidentified as *A. fulgens* by Nannfeldt (1936, in Handel-Mazetti, Symp. Sin. 7(4): 1076) who was followed by subsequent authors. It is distinct from the latter in having the capsule globose with truncate or even concave base, the leaf blade hairy on both surfaces and more or less reflexed calyx lobes.

YARLUNG ZANGBO-BRAHMAPUTRA: **Zayü**, 1980, *s. coll.* s. n. (PE). **S HENGDUAN**: **Binchuan**, Jizu Shan, 2200 m, 1965, *Tree Oil Surv. 65–0665* (KUN); **Daocheng**, Mengzi Commune, 2900 m, 1973, *Sichuan Plant Surv. 2505* (KUN, PE); **Dêqên**, Huanfuping, 3400 m, 1935, *C. W. Wang 69278* (PE); **Fugong**, Fengquan to Qiaomigulu, 1600–1800 m, 1982, *Qinghai-Xizang Exped. 7295* (PE); **Hanyuan**, Xiangling Shan, 2800 m, 1938, *T. P. Wang 9512* (PE); **Heqing**, Gualapo, forests, 1940, *R. C. Ching 23362* (KUN, PE); **Lushui**, Pianma Valley, 1900 m, 1981, *s. coll.* s. n. (PE); **Mianning**, Huangjia Shan, 2060 m, 1993, *Z. C. Zhao 4962* (PE); Xiaowang Gou, 2020 m, 1993, *Z. C. Zhao 4504* (PE); **Miyi**, Malong, 1500 m, 2002, *X. H. Jin et al. 437* (PE); **Muli**, Kumatian, 2500 m, 1937, *T. T. Yü 7675* (KUN, PE); Wanglang to Cabulang, 2500–2600 m, 1983, *Qinghai-Xizang Exped. 14132* (PE); Lea-Lang-Kang-din, 2800 m, 1937, *T. T. Yü 7482* (PE); **Panzihua**, Cycas Nature Reserve, 2203 m, 2010, *Y. Yang et al. 581* (PE); **Shangri-la** (Zhongdian), Tongpatze Shan, 1939, *K. M. Feng 2324* (PE); Lochihe, 2700 m, 1939, *K M. Feng 2743* (PE); **Shimian**, 1955, *C. C. Hsien 41456* (PE); **Weixi**, Kanpu, 2500 m, 1935, *C. W. Wang 64442* (KUN, PE); **Yulong** (Lijiang), Lishi, 2440 m, 1981. *Z. X. Tang et al. 969* (PE).

3. Asyneuma strictum Wendelbo, Nytt. Mag. Bot. 1: 60. 1952; Nasir in Nasir & Ali, Fl. Pakistan 155: 6. 1984. ≡ *Asyneuma thomsonii* (Hook. f. & Thomson) Bornm. var. *strictum* (Wendelbo) Kitam., Fl. Afgh. 377. 1960. Type: Pakistan, Chitral, Zapotili, western side of Main Barum Glacier, ca. 3500 m, 9 Jul. 1950, *P. Wendelbo* s. n. (holotype: BG, n. v.; isotype: K).

Herbs perennial. Roots thick, elongated. Stems erect, 30–50 cm tall, branched, hirsute. Leaves alternate, nearly sessile or petiole up to 7 mm long, hirsute; leaf blade lanceolate, 2–3.5 cm long, 0.6–1.5 cm broad, adaxially glabrous, abaxially hirsute, base cuneate to rounded, margin serrate, apex acuminate. Flowers in a cyme of 2–several; pedicel ca. 2 mm long, glabrous or puberulent, bracteoles at lower part. Calyx tube obovoid, glabrous or scaberulose; calyx lobes subulate-linear, 3–7 mm long, erect. Corolla blue or violet-blue, 7–10 mm long. Style nearly as long as corolla. Stigma 3-fid. Seeds oblong, ca. 1.5 mm long.

Phenology: flowering in June and July.

Habitats: grassy hillsides; 2500–3900 m.

Distribution: N Pakistan (Chitral, Gilgit) (Map 47–150); also in NE Afghanistan.

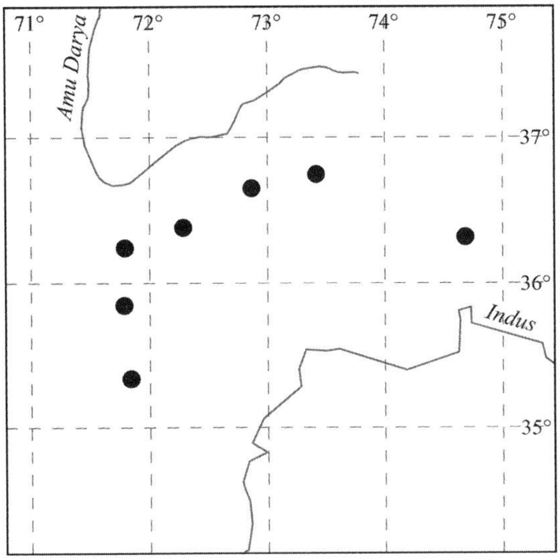

Map 47–150: **Asyneuma strictum** Wendelbo

N PAKISTAN: **Chitral**, Chitral, Pasti Rabat Ghari, 3839 m, 2007, *H. Ali 6533* (KUH); Darkot, 2960 m, 1976, *S. A.Bowes Lyon 8060* (K); Shogot, 2900 m, 1958, *J. D. A. Stainton 2657* (E); Yarkhum, 3050 m, 1958, *S. A. Bowes Lyon 1052* (E); Ziarat: 2740 m, 1895, *Surg.-Lt. Harriss 16326* (K, WU); **Gilgit**, Molikhoo, Ojhore top between Karimabad & Parsan, 3288 m, 2007, *H. Ali 5971* (KUH).

4. Asyneuma pakistanicum D. Y. Hong, Phytotaxa: 227(2): 196. fig.1. 2015. Type: Pakistan, Chitral, Lutkhoo, Buzur Hill, Garm Chashma, 3200 m, 27 Jun. 2007, *Haidar Ali 6294* (holotype: KUH).

Herbs perennial. Roots thickened, attenuate, 12 cm long, 3 mm thick. Stems caespitose, up to 23 cm long, glabrous. Leaves alternate, sessile or lower ones nearly sessile, ovate-lanceolate or lanceolate, 8–12 mm long, 4–6 mm broad, base cuneate or nearly rounded, margin entire, undulate, apex acute or shortly acuminate. Flowers 3–5 on each stem, solitary in axils of bracts; pedicel 5–10 mm long, without bracteole. Calyx tube rounded, glabrous; calyx lobes linear, 3–4 mm long, ca. 0.5 mm broad, glabrous. Corolla violet, glabrous, ca. 9 mm long, lobed nearly to base; corolla lobes linear, ca. 1 mm wide, style much longer than corolla, ca. 11 mm long; stigma 3-fid.

Phenology: flowering in June.

Habitats: ca. 3200 m.

Distribution: **endemic**; confined to N Pakistan (Map 47–151).

Notes: *Haidar Ali 6294* was identified as *Asyneuma thomsonii* (Hook. f.) Bornm. However, *A. thomsonii* has petiolate leaves, with the petiole 1–2 cm long, the leaf blade ovate or ovate-lanceolate, much larger, 3–6 cm long, 2–3 cm broad. The flowers in this newly described are in cymes of 1–3, and, particularly, the stigma is 3-fid, not 2-fid as in *A. thomsonii*. Therefore, this species is clearly different from *A. thomsonii*.

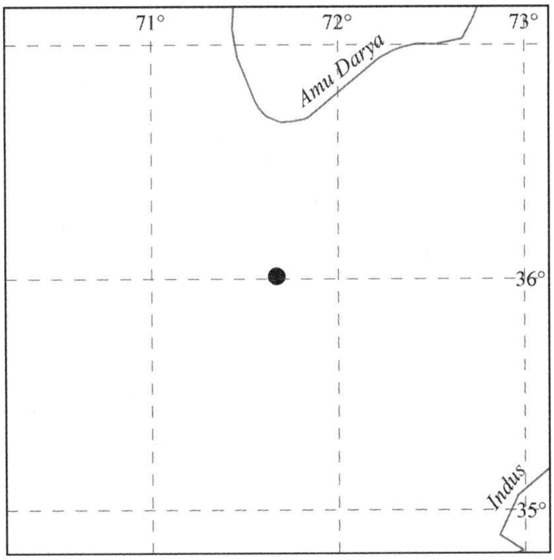

Map 47–151: **Asyneuma pakistanica** D. Y. Hong

5. Asyneuma thomsonii (Hook. f.) Bornm., Beih. Bot. Centrabl. 38, 2: 350. 1921; Nasir in Nasir & Ali S, Fl. Pakistan 155: 5. fig. 1: H. I. 1984. ≡ *Campanula thomsonii* Hook. f. in Hooker & Thomson, J. Proc. Linn. Soc. Bot. 2: 25. 1857. ≡ *Phyteuma thomsonii* (Hook. f.) C. B. Clarke in Hooker, Fl. Brit. India 3: 438. 1881. Type: India, Jammu, Banahal Pass, 1830–

2440 m, 12 May 1848, *T. Thomson* s. n. [**lectotype here designated**: K (K000814288)]; loc. eodem, 12 May 1848, *T. Thomson* s. n. (paratype: K).

Herbs perennial. Roots attenuate, up to 5 mm in diameter. Stems erect, 30–100 cm tall, glabrous to sparsely villous. Leaves alternate; petiole 1–2 cm long, glabrous or sparsely villous; leaf blade ovate, ovate-lanceolate, 3–6 cm long, 2–3 cm broad, glabrous on both surfaces or sparsely pubescent abaxially, base rounded or shallowly cordate, margin crenate to serrate, apex acute to acuminate. Flowers in 1–3-flowered cymes: pedicel 2–5 mm long, bracteoles at middle. Calyx tube ovoid-globose or globose, glabrous; calyx lobes filiform, 4–6 mm long, reflexed from anthesis. Corolla corncockle blue; corolla lobes linear to lanceolate, 6–8 mm long. Style exserted; stigma 2-fid. Capsule globose. Seeds oblong, ca. 1.5 mm long.

Phenology: flowering in May and June.

Habitats: 1800–3000 m.

Distribution: N Pakistan and Jammu & Kashmir (Map 47–152); also in NE Afghanistan.

N PAKISTAN: **Azad Kashmir**, Muzaffarabad, Kishan Ganga Valley, 1956, *S. M. A. Kazmi 325* (KUH); Poonch & Azad Kashmir, Hajipir, 2440 m, 1952, *R. R. Stewart 24061* (KUH); **Chitral**: 2130 m, 1895, *Surg. - Lt. Harriss 16324* (K, WU); Mansehra, Sathian Gali, ca. 2160 m, 1988, *B. U. Shra et al. 2174* (PMNH); **Swat**: Ushu Valley, Matiltan, 3000 m, 1987, *A. Ghefoor & S. Omer 3407*(KUH). **JAMMU & KASHIMIR**: **Aharzal**, 2440 m, 1977, *J. D. A.Stainton 7838* (E); **Ganha**, Bringhi Valley, 1980 m, 1940, *F. Ludlow & G. Sherriff 7602* (E); near **Gulmarg**, 2440–2740 m, 1892, *J. F. Duthie* s.n. (E, WU).

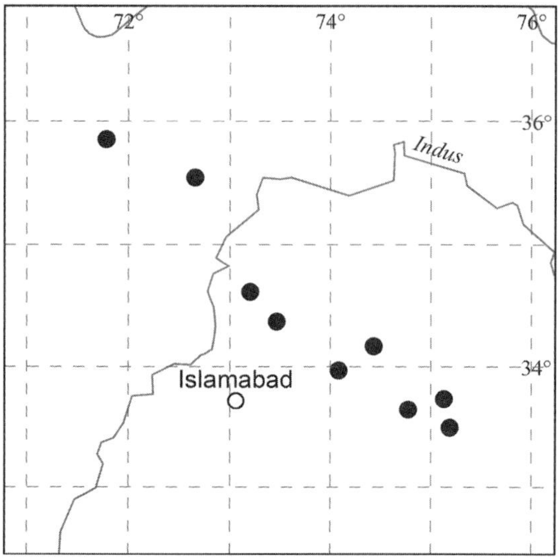

Map 47–152: **Asyneuma thomsonii** (Hook. f.) Bornm.

11. ADENOPHORA Fisch.

Fischer, Mém. Soc. Imp. Naturalistes Moscou 6: 165. 1823. Type: *Adenophora triphylla* (Thunb.) A. DC. (*Campanula triphylla* Thunb.).

Perennial herbs. Roots usually thickened, various in shape, mostly fleshy, carrot-shaped. Caudex often short, sometimes elongated and branched. Stems erect, less frequently ascending. Basal leaves often rosulate, usually long-petiolate, cordate; cauline leaves mostly alternate, rarely opposite or verticillate. Flowers in a cyme; cymes sometimes reduced to a branch with one terminal flower and one to several bracts, sometimes gathered into a panicle. Calyx epigynous, rarely semi-epigynous; calyx tube various in shape; calyx lobes 5, entire or dentate. Corolla always epigynous, campanulate, funnelform or tubular, 5-lobed or 5-fid to middle. Stamens 5; filaments much dilated into a lamella at base; lamellae densely ciliate-villous, valvate, enveloping disc; anthers elongate. Disc usually tubular, rarely annular. Ovary inferior, 3-locular; ovules many; style with pollen-receiving hairs; stigma 3-fid, stigma lobes narrow, recurved. Capsule 3-poricidal below persistent calyx lobes. Seeds many, ellipsoid, 1-angular.

The genus *Adenophora* is closely related to *Campanula*. The only diagnostic character distinguishing them is disc, present in the former, while absent in the latter.

About 60 species, in East, Central and Southeast Asia, southwards to India and Vietnam, and northwards to Siberia and the Far East; two species extending into Europe and one of them endemic to Crimea; 14 species (5 endemic) in the Pan-Himalaya.

1a. Cauline leaves (at least lower ones) petiolate, very rarely nearly sessile.

2a. Calyx lobes toothed; leaf blade lobed or with 2-several large teeth.

3a. Cauline leaves pinnatifid halfway to midrib, rarely only dentate-lobed; capsule long-ellipsoid, ca. 3 mm wide .... 12. *A. lobophylla*

3b. Cauline leaves with 2–several large marginal teeth; capsule ellipsoid, over 3 mm wide .... 9. *A. potaninii*

2b. Calyx lobes entire; leaf blade serrate or sinuate-crenate.

4a. Cauline leaves 2-several, aggregated at lower part of stem; leaf blade of lower one or two leaves cordate at base .... 1. *A. delavayi*

4b. Cauline leaves many, evenly distributed; leaf blade cuneate or rounded at base .... 2. *A. longipedicellata*

1b. Cauline leaves sessile, only some individuals of some species (e.g. *A. stricta*) with lower cauline leaves very shortly winged-petiolate.

5a. Corolla urceolate, less than 1.7 cm long; style strongly exserted, over 1.5 times longer than corolla; disc thinly tubular, 2–7 mm long, less than 1 mm wide.

6a. Calyx lobes filiform, (3–) 6–14(–20) mm long, sometimes with 1–2 pairs of verrucose teeth at lower part .... 14. *A. capillaris*

6b. Calyx lobes subulate, 2–6 mm long, entire .... 13. *A. liliifolioides*

5b. Corolla campanulate, usually over 1.7 cm long (1.3–1.6 cm long in *A. linearifolia*); style slightly exserted or included, less than 1.5 times longer than corolla; disc not much longer than broad.

7a. Calyx lobes entire.

8a. Calyx lobes 1.5–2.5 mm long, ca. 1 mm broad; disc ca. 0.5 mm long; leaves aggregated at lower part of stem, leaf blade linear .... 4. *A. dawuensis*

8b. Calyx lobes 5–10 mm long, 1–1.5 (–2); disc more than 1 mm long; leaves evenly distributed along stem, leaf blade ovate to linear.

9a. Calyx scaberulose; disc 1–2.5 mm long .... 3. *A. stricta*

9b. Calyx mostly glabrous; disc 3–8 mm long, 2–3 mm wide ...... 11. *A. himalayana*

7b. Calyx lobes toothed.

10a. Cauline leaves aggregated at middle part of stems, glabrous; leaf blade lanceolate or linear-elliptic .... 10. *A. wilsonii*

10b. Cauline leaves evenly distributed or aggregated at lower part, glabrous or hairy; leaf blade various in shape.

11a. Leaf blade with 2 to several large teeth on each side, hispidulous on both surfaces; disc 2–3 mm long, hairy at least at apex .... 9.*A. potaninii*

11b. Leaf blade entire or serrate, glabrous, hirsute or hispidulous beneath; disc less than 2 mm long, very rarely longer, glabrous, occasionally with some hairs.

12a. Leaf blade linear; stems and leaves mostly glabrous; calyx lobes linear-subulate, less than 1 mm broad; corolla 1.3–1.6 cm long .... 8. *A. linearifolia*

12b. Leaf blade ovate to lanceolate, very rarely linear-lanceolate; stems and leaves usually hairy; calyx lobes subulate, over 1 mm broad; corolla (1.3–)1.6–3 (–4) cm long.

13a. Corolla bowl-shaped, lobes more than 1/2 as long as corolla tube; calyx lobes erect; leaf blade mostly entire .... 6. *A. jasionifolia*

13b. Corolla campanulate, lobes 1/3–1/2 as long as corolla tube; calyx lobes mostly reflexed since anthesis, less often spreading; leaf blade mostly serrate.

14a. Leaves evenly distributed on stems; calyx lobes 3–7 (–10) mm long; disc less than 1 mm long; style as long as or longer than corolla .... 5. *A. khasiana*

14b. Leaves mostly aggregated in lower half of stem; calyx lobes 8–20 mm long; disc 1.2–3.5 mm long; style shorter than corolla .... 7. *A. coelestis*

1. Adenophora delavayi (Franch.) D. Y. Hong, **comb. nov.** Basionym: *Campanula delavayi* Franch., J. Bot. (Morot) 9: 364. 1895. Type: China, Yunnan, Lan Kong (Heqing), 2500 m, 19 Sept. 1883, *M. Delavayi 1709* [**lectotype here designated**: P (P00225912); isolectotypes: P].

Adenophora brevidiscifera D. Y. Hong, Fl. Reipubl. Popularis Sin. 73(2): 184. pl. 16: 1–3. 1983, **syn nov**. Type : China, Sichuan: Yanyuan, Yousuo Township, 2850 m, pine forests, 16 Aug. 1963, *Q. L. Zhang 1765* (holotype: PE).

Stems ascending, ca. 35 cm tall, simple, densely hirsute below, glabrous above. Basal leaves long-petiolate, hirsute, petiole 2.5–4.5 cm long, leaf blade ovate, ca. 3 cm wide, adaxially hirsute, abaxially sparsely hirsute along veins, base cordate with sinus up to 4 mm deep, margin sinuate-crenate or serrate; lowermost cauline leaves most similar to basal ones but slightly smaller, petiole 0.5–2 cm long; upper cauline leaves 2–4, sessile, less than 1 cm long, linear. Cymes 2–3, lax,one-flowered; bracts filiform, minute; pedicel slender. Calyx tube obovoid, glabrous; calyx lobes subulate, 2–3 mm long, ca. 0.5 mm broad, margin entire or with a pair of denticles. Corolla blue, blue-purple, or pale purple, narrowly campanulate, 1.8–2.5 cm long; corolla lobes narrowly ovate-elliptic, as long as tube, ca. 4 mm wide, apex acute. Disc annular, less than 0.5 mm high. Style included, ca. 12 mm long. Capsule obovoid, 5–7 mm long, 3–4 mm in diameter. Figure 47–67: a–c.

Phenology: flowering in August and September; fruiting in September.

Habitats: dry *Pinus yunnanensis* forests; 2700–3000 m.

Distribution: **endemic**; confined to S Hengduan (Map 47–153).

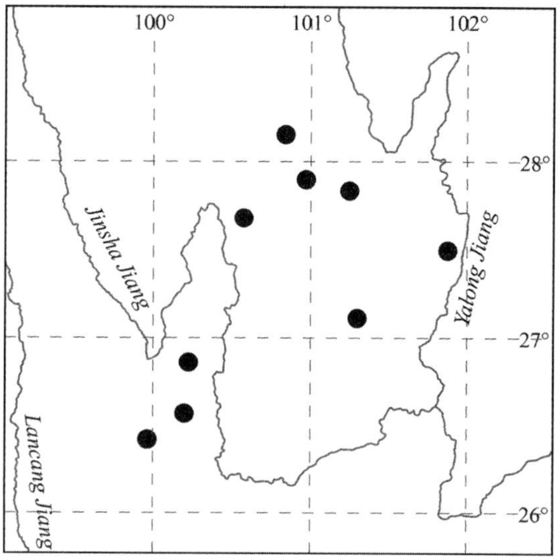

Map 47–153: **Adenophora delavayi** (Franch.) D. Y. Hong

Notes: *Adenophora delavayi* (Franch.) D. Y. Hong resembles *A. jasionifolia*, from which it differs in having entire calyx lobes, long-petiolate and cordate lower cauline leaves, and a shorter disc.

S HENGDUAN: **Jianchuan**, Yuhua Township, 2680 m, 2010, *D. Y. Hong et al. H10014* (PE); **Muli**, SE of Muli, 3660 m, 1922, *G. Forrest 22129* (W); near Wachang, 3000 m, 1983, *Qinghai-Xizang Exped. 14180* (PE); **Yanbian**, Yankou, 2700 m, 2002, *X. F. Gao et al. 3670* (PE); **Yanyuan**, Gongmu Shan, 1986, *Z. T. Wang 869–395* (PE); Yongning (Ninglang)-Muli,

3100–3500 m, 1915, *Handel-Mazzetti 7266* (W); **Yulong** (Lijiang), Longshan, 2500 m, 1983, *Qinghai-Xizang Exped. 15000* (PE); Lijiang Range, 2740–3050 m, 1910, *G. Forrest 6473* (P, PE, W); Ganhaizi, 2900 m, 2010, *D. Y. Hong et al. H10030* (PE).

2. Adenophora longipedicellata D. Y. Hong, Fl. Reipubl. Popularis Sin. 73(2): 185. pl. 19: 6–7. 1983; Hong et al. in Wu, Raven & Hong, Fl. China 19: 540. 2011. Type: China, Sichuan, Ya'an, Jinfeng-si, 750 m, 20 Aug. 1963, *K. C. Kuan et al. 2503* (holotype: PE).

Stems 1–3 m tall, simple or with slender branches up to 70 cm long, glabrous. Basal leaves ovate, base cordate. Cauline leaves (at least lower ones) short-petiolate; leaf blade ovate-elliptic or lanceolate, 7–12 cm long, 2–5 cm broad, glabrous or abaxially sparsely hispid along veins, base cuneate or obtuse, margin serrate or double serrate, apex acuminate. Flowers in cymes forming a large lax panicle with long branches, glabrous or puberulent; pedicel 1–3 cm long. Calyx tube subglobose or obovoid, glabrous; calyx lobes subulate-lanceolate, 8–14 mm long, entire. Corolla white, purple, or pale blue, funnelform, 1.9–2.1 cm long; corolla lobes deltoid, 5–6 mm long. Disc annular, ca. 1 mm long or shorter, glabrous. Style ca. 2.1 cm long, almost as long as or slightly longer than corolla. Figure 47–67: d & e.

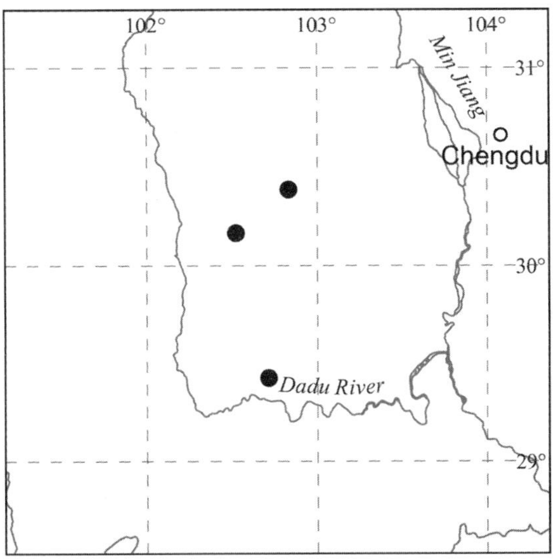

Map 47–154: **Adenophora longipedicellata** D. Y. Hong

Phenology: flowering and fruiting from August to October.

Habitats: grassy places, scrubs, or cliffs; 750–2400 m.

Distribution: S Hengduan (Map 47–154); also in Chongqing, Guizhou, Hubei, C Sichuan, and NE Yunnan of China.

S HENGDUAN: **Baoxing,** *K. L. Chu 3842* (PE); **Hanyuan**, Xiangling Shan, 2400 m, 1938, *T. P. Wang 8857* (KUN, PE); **Tianquan**: Erlang Shan, Huangzhuping, 1951, *W. K. Hu & Z. He 11738* (PE).

3. Adenophora stricta Miq., Ann. Mus. Bot. Lugduno-Batavum 2: 192. 1866; Tsoong, Contr. Inst. Bot. Nat. Acad. Peiping 3(3): 81. 1935; Hong, Fl. Reipubl. Popularis Sin. 73(2): 104. 1983; Hong *et al.* in Wu, Raven & Hong, Fl. China 19: 541. 2011. Type: Japan, collections by *Buerger*, *Siebold*, & *Textor* (syntypes, n.v.)

Stems 40–80 cm tall, simple, often hispidulous or villous, rarely glabrous. Cauline leaves sessile but lower ones sometimes with extremely short and winged petiole; leaf blade elliptic or narrowly ovate, 3–11 cm long, 1.5–5 cm broad, sparsely hispidulous, hirsute, or nearly glabrous, base cuneate. Flowers often in a pseudo-raceme or narrow panicle; pedicel less than 5 mm long. Calyx tube obovoid or obconical, puberulent, scaberulose, or less often glabrous; calyx lobes typically subulate, less often linear-lanceolate, 6–8 mm long, 1–1.8 mm broad, entire. Corolla blue or purple, broadly campanulate, 1.5–2.5 cm long, glabrous or hirsute at least along veins outside; corolla lobes deltoid-ovate, ca. 1/2 as long as tube. Disc shortly tubular, 1–2.5 mm long, glabrous. Style usually slightly longer (rarely shorter) than corolla. Capsule ellipsoid-globose, very rarely ellipsoid, 6–10 mm in diameter. Seeds brown-yellow, slightly compressed, ca. 1.5 mm long. **Chromosomes**: 2n=34.

Distribution: N and S Hengduan (Map 47–155); widely distributed in E China and Japan.

Four subspecies; three in this region, but the typical one in East China and Japan.

Figure 47–67: **a–c**, **Adenophora delavayi** (Franch.) D. Y. Hong: **a**, lower part of plant; **b**, upper part of stem; **c**, a flower with corolla and stamens removed, showing annular disc. **d** & **e**, **Adenophora longipedicellata** D. Y. Hong: **d**, a section of stem, showing a lateral branch and inflorescence; **e**, a flower with corolla and stamens removed, showing disc. **a–c** from *Fl. Reipubl. Popularis Sin.* 73(2), 1983, with modification; **d** & **e** from *Fl. Reipubl. Popularis Sin.* 73(2), 1983.

1a. Leaves glabrous or sparsely villous .... 3b. subsp. *confusa*
1b. Leaves hispidulous.
  2a. Corolla 2–2.5 cm long; disc 1.8–2.5 mm long .... 3a. subsp. *aurita*
  2b. Corolla 1.5–2.3 cm long; disc 1–1.8 mm long .... 3c. subsp. *sessilifolia*

3a. Adenophora stricta subsp. **aurita** (Franch.) D. Y. Hong & S. Ge, Novon 20: 428. 2010. ≡ *Adenophora aurita* Franch., J. Bot. (Morot) 9: 366. 1895. Type: China, Sichuan, Ta-tsien-lou (Kangding), 1893, *J. A. Soulie 501* [**lectotype here designated**: P (P00242602); isolectotypes: G, K, P].

Leaves hispidulous. Calyx mostly hairy, rarely nearly glabrous; calyx lobes 1–1.8 mm wide. Corolla 2–2.5 cm long. Disc 1.8–2.5 mm long. Figure 47–68.

Phenology: flowering and fruiting from late July to September.

Habitats: grassy places, waste fields, or forest margins; 2100–3900 m.

Distribution: **endemic**; confined to N & S Hengduan.

N HENGDUAN: **Barkam**, near Dangba, 2800 m, 1957, *X. Li 71830* **(PE)**; **Dawu,** Haitze Shan, 3900 m, 1934. *H. Smith 11375* (PE); **Jinchuan**: Anning, Muma Shan, 3200 m, 1958, *X. Li 78567* (PE); **Lixian**: 1952, *Z. He & Z. L. Zhou 13607* **(PE)**; **Jiuzhaigou** (Nanping), 2350 m, 1986, *Z. Y. Wu et al. 86–1066* (KUN); **Pingwu**, Pingwu to Nanping, 2100 m, 1986, *Z. Y. Wu et al. 86–469* (KUN); **Wenchuan**, Wolong, 1982, *K. Y. Lang et al. 1370* (KUN); **Xiaojin:** Meiwogou, 2900 m, 1958, *S. S. Chang* **&** *Y. X. Ren 6452* **(PE). S HENGDUAN**: **Kangding,** South Valley, 2600 m, 1934, *H. Smith 10797* (PE); **Luding**: E'rlang Shan, above Machang, 2500 m, 1953, *H. L. Tsiang 37847* (PE); **Tianquan**: summit of Erlang Shan, 1959, *Sichuan Econ. Plant Surv. (Ya) 890* (PE).

3b. Adenophora stricta subsp. **confusa** (Nannf.) D. Y. Hong, Fl. Reipubl. Popularis Sin. 73(2): 105. 1983. ≡ *Adenophora confusa* Nannf. in Handel - Mazzetti, Symb. Sin. 7: 1068. 1936. Type: China, Yunnan, Lidjiang (Lijiang), Yülung-schan (Yulong Snow Range), s. d. *G. Forrest 6282* (holotype: E; isotype: K).

Stems usually hispidulous: leaves glabrous or sparsely villous only along veins. Calyx usually glabrous.

Phenology: flowering and fruiting from August to October.

Habitats: open slopes or forests: 1000–3200 m.

Distribution: S Hengduan; also in C and S Yunnan of China.

S HENGDUAN: **Bijiang**, 2500 m, 1939, *H. T. Tsai 54174* (KUN); **Dali**, Longquan Shan, 1941, *H. C. Wang 1378* (PE); **Eryuan**, 1975, *B. Y. Qiu 61067* (KUN); **Lanping**, 2100 m, 1964, *S. G. Wu 9018* (KUN); **Weixi**, 2800 m, 1935, *C. W. Wang 67604* (PE); **Yulong** (Lijiang), 1939, *R. C. Ching 21863* (KUN).

3c. Adenophora stricta subsp. **sessilifolia** D. Y. Hong, Fl. Reipubl. Popularis Sin. 73(2): 185.

Figure 47–68: **Adenophora stricta** Miq. subsp. **aurita** (Franch.) D. Y. Hong. Redrawn by Ai-Li LI from *Icon. Cormophyt. Sin.* 4, 1975.

1983; Hong *et al.* in Wu, Raven & Hong, Fl. China 19: 542. 2011. Type: China, Hubei, Lichuan, Moudaoxi, 23 Sept. 1957, *G. X. Fu & Z. S. Zhang 1580* (holotype: PE).

Stems and leaves hispidulous. Calyx mostly hispidulous or scaberulose, less frequently glabrous. Corolla 1.5–2.3 cm long, usually glabrous or only with a few hairs along veins outside; disc 1–1.8 mm long.

Phenology: flowering and fruiting from July to September.

Habitats: grassy places at forest margins; 600–2700 m.

Distribution: N & S Hengduan; also in C China and NE Yunnan (Zhenxiong).

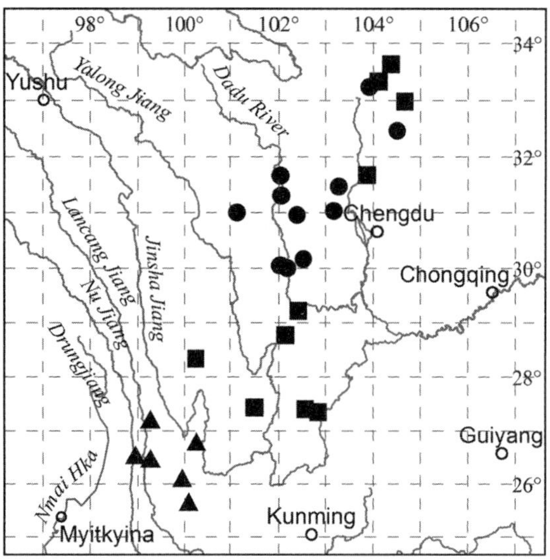

Map 47–155: **Adenophora stricta** Miq.
● subsp. **aurita,** ▲ subsp. **confusa,** ■ subsp. **sessilifolia**

N HENGDUAN: Maoxian, Huilong Shan 2000 m, 1975, *Y. C. Tang et al. 86* (PE); **Jiuzhaigou** (Nanping), Baihe, 2650 m, 1964, *Q. X. Li & X. C. Zhao 2895* (PE); **Wenxian**, 2200 m, 1984, *W. L. Chen et al. 8757* (PE); **Zhugqu**, Jiao'erqiao, 1800 m, 1958, *Y. Q. He 693* (PE). **S HENGDUAN**: **Butuo**, Jiaojihe, 2500 m, 1959, *Sichuan Econ. Plant Surv. (Liang 59) 5892* (PE); **Hanyuan,** 2700 m, 1938, *T. P. Wang 9586* (PE); **Meigu**, 1959, *Sichuan Econ. Plant Surv. (Liang 59) 5981* (PE); **Mianning**, Xiaogeda, 2000 m, 1959, *Sichuan Econ. Plant Surv. (Xi 59) 2161* (PE); **Puge**, S of the city, 1900 m, 1959, *Sichuan Econ. Plant Surv. (Liang 59) 4620* (PE); **Yanyuan**, 3000–3640 m, 1914, *Handel-Mazzetti 5384* (WU).

4. Adenophora dawuensis D. Y. Hong, **sp. nov.** Type: China, Sikang (Sichuan), Taofu (Dawu), 3100 m, 18 Sept. 1934, *H. Smith 12215* (holotype: PE).

Roots attenuated, ca. 7 mm in diameter. Stems up to 12, caespitose, erect, 15–40 cm tall, glabrous. Leaves alternate, mainly aggregated at lower part of stem, only 1–3 at middle and upper parts, all sessile, leaf blade linear, 0.5–5 cm long, 1–1.5 mm broad, glabrous, margin entire. Flowers 1–4, terminal on main stem and short branches, pendulous; pedicel 2–15 mm long. Calyx entirely glabrous, calyx tube obovoid, ca. 2–3 mm long, 1.5–2 mm wide; calyx lobes subulate, 1.5–3 mm long, ca. 1 mm broad, entire. Corolla campanulate, 1.5–1.8 cm long, glabrous; corolla lobes deltoid, 4–5 mm long. Filaments ca. 5 mm long, much dilated at base, dilated part ca. 2.6 mm long, densely villous-ciliate; anthers ca. 6 mm long. Disc less than 0.5 mm long. Style exserted. Capsule obovoid, ca. 8 mm long, 4 mm in diameter. Figure 47–69: a & b.

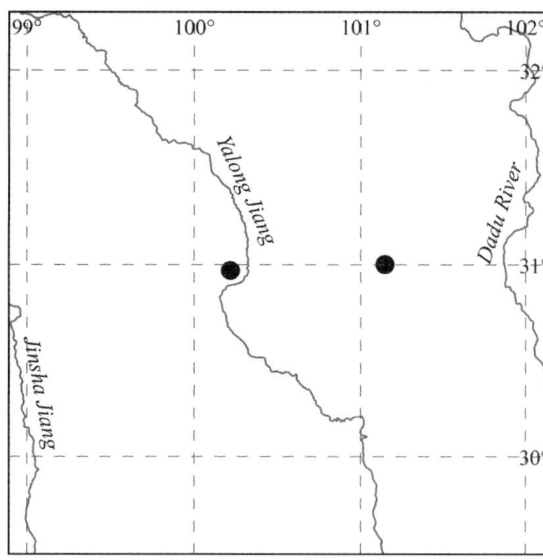

Map 47–156: **Adenophora dawuensis** D. Y. Hong

Phenology: flowering in August and September.

Habitas: 3100–3200 m.

Figure 47–69: **a** & **b**, **Adenophora dawuensis** D. Y. Hong: **a**, stem; **b**, a flower with corolla opened, showing stamens, disc, style, and stigma. **c** & **d**, **Adenophora linearifolia** D. Y. Hong: **c**, upper part of stem; **d**, a flower with corolla opened, showing stamens, disc, style, and stigma. Both drawn by Ai-Li LI.

Distribution: **endemic**; confined to N Hengduan (Map 47–156).

Notes: The new species is very isolated from all other species in having needle-like leaves, the whole plant entirely glabrous except for stamens, tiny calyx lobes with entire margin, and very short disc. It resembles *A. delavayi* (Franch.) D. Y. Hong and *A. liliifolioidea* Pax & K. Hoffm., but *A. delavayi* has long-petiolate and cordate or ovate leaves, hairy stems, and included style, while *A. liliifolioidea* has urceolate corolla, much longer disc, and strongly exserted style.

N HENGDUAN: Xinlong, Garzê to Xinlong, 3200 m, 2006, *D. E. Boufford et al. 37291* (PE).

5. Adenophora khasiana (Hook. f. & Thomson) Oliv. ex Collett & Hemsl., J. Linn. Soc. Bot. 28: 80. 1890; Tsoong, Contr. Inst. Bot. Nat. Acad. Peiping. 3(3): 73. 1935; Clement, Fl. Bhutan 2(3): 1380. 2001; Hong *et al.* in Wu, Raven & Hong, Fl. China 19: 546. 2011. ≡ *Campanula khasiana* Hook. f. & Thomson, J. Proc. Linn. Soc. Bot. 2: 25. 1857. ≡ *Adenophora khasiana* (Hook. f. & Thomson) Feer, Bot. Jahrb. Syst. 12: 617. 1890. Type: India, Khasia, 1520 m, *Lobb* s. n. (**lectotype here designated**: K); loc. eodem, *Griffith* s. n. (paratype: K); loc. eodem, 1520–1830 m, *J. D. Hooker & Thomson* s. n. (paratype: K).

Adenophora bulleyana Diels, Notes Roy. Bot. Gard. Edinburgh 5: 175. 1912. Type: China, Yunnan, Lichiang (Lijiang), 27°12′ N, 3050 m, Aug. 1906, *G. Forrest 2653* (**lectotype here designated**: E; isolectotype: K).

Adenophora diplodonta Diels, Notes Roy. Bot. Gard. Edinburgh 5: 176, 1912. Type: China, Yunnan, "Mekong and Yangtze Valleys, dry hillsides", 1520–2740 m, Sept. 1904, *G. Forrest 396* (holotype: E).

Adenophora albescens C. Y. Wu in Wu & Li, Yunnan Trop. Subtrop. Fl. Res. Rep. 1: 63, pl. 27: 3. 1965. Type: Shangri-la (Zhongdian), pine forests, 16 Aug. 1939, *K. M. Feng 2045* (**lectotype here designated**: KUN; isolectotype: PE).

Adenophora dimorphophylla C. Y. Wu in Wu & Li, Yunnan Trop. Subtrop. Fl. Res. Rep. 1: 67. pl. 15. 1965, p. p., excl. isotype in PE (01230331 = *A. coelestis*). Type: China, Yunnan, Shangri-la [Zhongdian], Haba, 2900 m, glades in pine forests, 17 Sept. 1962. *Zhongdian Exped. 2129* [holotype: KUN; isotype: PE(01230904)].

Adenophora likiangensis C. Y. Wu in Wu & Li, Yunnan Trop. Subtrop. Fl. Res. Rep. 1: 69. pl. 29: 1. 1965, Type: China, Yunnan, Lijiang, Snow Range, Heibaishui, 2500 m, mixed forests, 18 Aug. 1942, *K. M. Feng 9076* (**lectotype here designated**: KUN; isolectotype: PE).

Roots carrot-like, 7–8 cm long, 1.5–1.6 cm in diameter. Stems often single or sometimes 2 from one caudex, simple, (30–) 50–100 cm tall, usually hirsute. Cauline leaves evenly distributed along stems, sessile or petiole up to 2 cm long; leaf blade ovate-orbicular, ovate, obovate, or elliptic, 3–9 cm long, 1.5–6.5 cm broad, adaxially scaberulose, abaxially rather densely hirsute (sometimes only along veins), base cuneate, margin serrate or irregularly biserrate, apex acute or acuminate. Inflorescences a narrow panicle or a pseudo-raceme of several flowers; pedicel short. Calyx tube campanulate, glabrous, less often rather densely hispidulous; calyx lobes narrowly triangular, 3–7 (–10) mm long, 1–1.5 mm broad, margin with 1–4 pairs of denticles. Corolla blue or pale blue, infundibular-campanulate, 1–2.4 cm long, lobes to ca. 1/2 as long as tube. Disc shortly tubular, less than 1 mm long. Style slightly longer than corolla to obviously exserted. **Chromosomes**: 2n=34. Figure 47–70: a–d.

Phenology: flowe-ring from August to October; fruiting from September to November.

Habitats: forests, scrubs, or grassy places: 1000–3100 m.

Distrbution: Bhutan, L Yarlung Zangbo, and S Hengduan (Map 47–157); also in Meghalaya of India, other parts of Yunnan in China, and C Myanmar.

BHUTAN: **Tashiling**, 2590 m, 1914, *R. E. Cooper 2064* (E); Tobrang near **Tashi Yangtsi**, Upper Kulong Chu 2200 m, 2000, *S. A. Bowes Lyon 13187* (E); between Denchang and Rib La, 2290 m, 1949, *F. Ludlow et al. 21406* (E); Ruku-

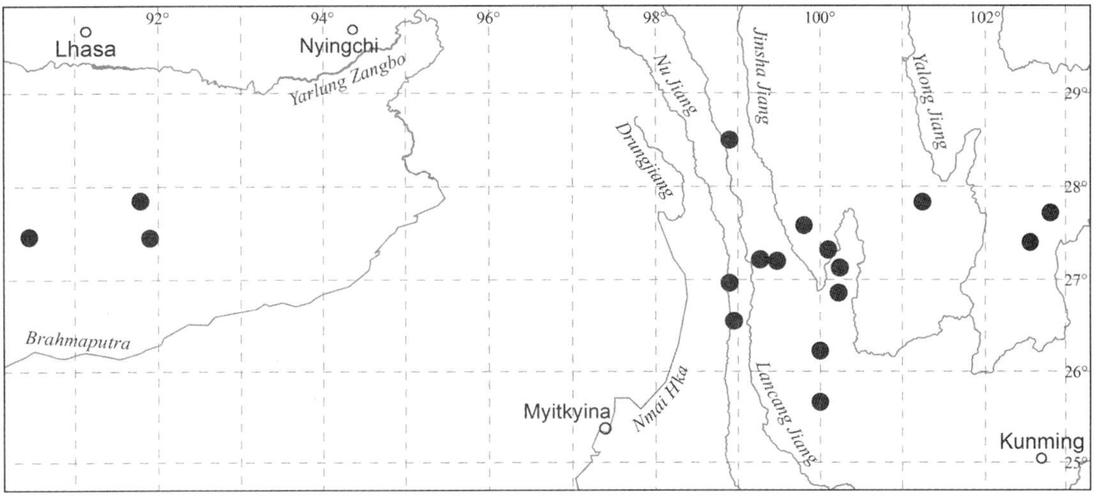

Map 47–157: **Adenophora khasiana** (Hook. f. & Thomson) Oliv. ex Collett & Hemsl.

leje, 1914, *R. E. Cooper 2323* (E). **YARLUNG ZANGBO-BRAHMAPUTRA**: **Mama–Tea Farm**, 2600 m, 1975. *Qinghai-Xizang Suppl. Exped. 751639* (KUN, PE). **S HENGDUAN**: **Bijiang,** Gaoligong Shan, 2500 m, 1979, *Z. G. He & Y. M. Zhan 831* (KUN); **Butuo**, 1600 m, 1959, *Sichuan Econ. Plant Surv.* (Liang) *5892* (PE); **Dêqên**, 2500 m, 1935, *C. W. Wang 69350* (PE); **Eryuan**, Sanying, 1975, *B. Y. Qiu 60905* (KUN); **Fugong**, 1850 m, 1933, *H. T. Tsai 54854* (KUN); **Muli**, Rangtzantze, 2800 m, 1937, *T. T. Yü 14032* (KUN, PE); **Puge**, Tuogu, 2200 m, 1959, *Sichuan Econ. Plant Surv.* (Liang) *5550* (PE); **Shangri-la** (Zhongdian), 2350–2700 m, 1939, *K. M. Feng 2730* (PE); **Weixi**, 2800 m, 1934, *H. T. Tsai 59513* (KUN, PE); **Yangbi**, W slope of Cangshan, 2300 m, 1980, *K. M. Feng et al. 80–72* (KUN).

6. Adenophora jasionifolia Franch., J. Bot. (Morot) 9: 365. 1895; Nannfeldt in Handel-Mazzetti, Symb. Sin. 7: 1073. 1936; Hong, Fl. Reipubl. Popularis Sin. 73(2): 101. 1983. Type: China, W Sichuan, Tongolo, 1893, *J. A. Soulie 242* [holotype: P (P00242408); isotypes: K, P].

Adenophora forrestii Diels, Notes Roy. Bot. Gard. Edinburgh 5: 174. 1912. Type: China, Yunnan, Hoching (Heqing) and Lichiang (Lijiang) Valley, 2130–2440 m, Sept. 1904, *G. Forrest 394* (holotype: E; isotype: K).

Adenophora forrestii Diels var. *minor* W. W. Smith, Notes Roy. Bot. Gard. Edinburgh 8: 330. 1915. Type: China, Yunnan, Atuntsi (Dêqên), 3960 m, *F. Kingdom-Ward 1077* (holotype: E; isotype: E).

Adenophora forrestii Diels var. *handeliana* Nannf. in Handel-Mazzetti, Symb. Sin. 7: 1070. 1936. Type: Yunnan, Shangri-la (Dschungdien), 3550–3650 m, 10 Aug. 1914, *Handel-Mazzetti 13100* (holotype: Wu).

Adenophora pumila P. C. Tsoong, Contr. Inst. Bot. Nat. Acad. Peiping 3(3): 76. fig. 2. 1935. Type: China, Sichuan (Hsikang), Kangding, 18 Jul. 1934, *C. S. Liou 787* (holotype: PE)

Adenophora atuntzensis C. Y. Wu in Wu & Li, Yunnan Trop. Subtrop. Fl. Res. Rep. 1: 64. pl. 14. 1965. Type: China, Yunnan, Atuntze [Dêqên], Baima Shan, 3900 m, 12 Aug. 1937, *T. T. Yü 9599* (**lectotype here designated**: KUN; isolectotype: PE).

Adenophora microcodon C. Y. Wu in Wu & Li, Yunnan Trop. Subtrop. Fl. Res. Rep. 1: 70. pl. 29: 2. 1965. Type: Shangri-la (Zhongdian), grassy places, 17 Jul. 1939, *K. M. Feng 1679* (**lectotype here designated**: KUN; isolectotype: PE).

Roots carrot-shaped, up to 1 cm in diameter. Caudex sometimes with horizontal branches. Stems (1) 2-several, simple, less often branched, 15–55 cm tall, leafy mostly along lower half, sparsely villous, rarely glabrous. Cauline leaves

Figure 47–70: **a–d**, **Adenophora khasiana** (Hook. f. & Thomson) Collett & Hemsl.: **a**, middle part of stem; **b** & **c**, upper parts of stem; **d**, a flower with corolla and stamens removed, showing short disc. **e–g**, **Adenophora coelestis** Diels: **e**, a section of stem; **f**, flower; **g**, a flower with corolla and stamens removed, showing toothed calyx lobes and disc. **a–d** redrawn by Ai-Li LI from *Icon. Cormophyt. Sin.* 4, 1975; **e–g** from *Fl. Reipubl. Popularis Sin.* 73(2), 1983.

sessile or shortly winged-petiolate; leaf blade ovate-orbicular, elliptic, lanceolate, or linear-lanceolate, 2–8 cm long, 0.2–1.8 cm broad, usually puberulent or rarely glabrous on both surfaces, base cuneate, margin entire, less often crenulate or serrate, apex acute or acuminate. Flowers terminal and solitary, or several in a pseudo-raceme; pedicel 2–10 mm long. Calyx tube obconical, base acute or sometimes rounded, glabrous; calyx lobes narrowly triangular, 5–10 mm long, 1.5–1.7 mm broad, margin with several verrucose denticles, sometimes sparsely ciliate. Corolla blue or purple-blue, bowl-shaped, 1.3–2.2 cm; corolla lobes as long as, rarely up to 1/2 as long as, tube, deltoid-orbicular. Disc annular, 0.5–1 mm high. Style 1.3–1.5 cm, included. Capsule ellipsoid, 8–11 mm long, 5–6.5 mm in diameter. Seeds yellow-brown, ca. 1 mm long. **Chromosomes**: 2n=34.

Phenology: flowe-ring from July to September; fruiting in September and October.

Habitats: meadows, thickets, or grassy places at forest margins; 2900–4700 m .

Distribution: **endemic**; Tangut, N and S Hengduan (Map 47–158).

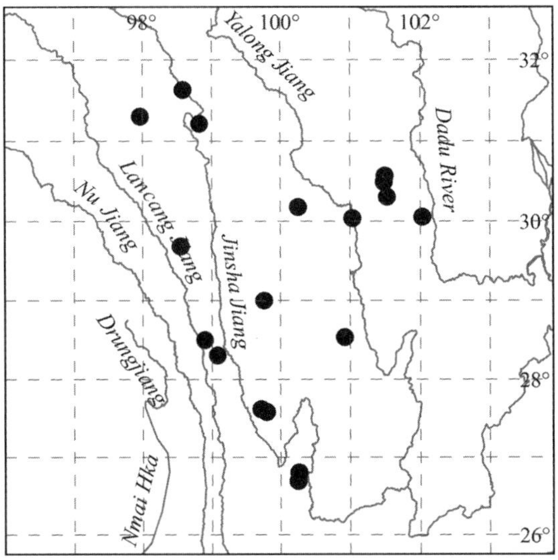

Map 47–158: **Adenophora jasionifolia** Franch.

TANGUT: Jomda, Gamtog, 1952, *P. C. Tsoong 5190* (PE). **N HENGDUAN: Baiyü**, Changtai, 4150 m, 1982, *Z. X. Tang et al. 744* (PE); **Dawu**: Taining, 3700 m, 1934, *H. Smith 11735* (PE); Qianning, near the town, 3600 m, 1961, *J. S. Ying et al. 9806* (PE); **Gonjo**, Harjia-Zhongsha, 4012 m, 2010, *Kham Exped. 10-1818* (PE); **Litang**: 13 km N of the city, 4400 m, 1982, *Tang & Xia 466* (PE); **Markam**, Angduo-Luoni, 4043 m, 2010, *Kham-Exped. 10-2065* (PE); **Yajiang**: No. 4 Distr., 3720 m, 1974, *Q. Q. Wang 5008* (PE). **S HENGDUAN: Daocheng,** Mowu, 3800 m, 1937, *T. T. Yü 12866* (PE); **Dêqên**, Baima Shan, 1935, *C. W. Wang 69717* (PE); **Kangding**, Yülinggong, 3500 m, 1934, *H. Smith 10821* (PE); between Xinduqiao and Bamei, 3400 m, 2012, *D. Y. Hong et al. H12064* (PE); **Muli,** near Wa'erzhai, 3100 m, 1937, *T. T. Yü 14433* (PE); **Shangri-la** (Zhongdian), Bigu Forest Farm, 3550 m, 1981, *Y. F. Han et al. 81–1500* (PE); **Xiangcheng**: No. 1 Distr., 4200 m, 1973, *Sichuan Veget. Surv. 3081* (PE); **Yulong** (Lijiang), ca. 2900 m, 1918, *C. Schneider 1900* (B).

7. Adenophora coelestis Diels, Notes Roy. Bot. Gard. Edinburgh 5: 173. 1912; Tsoong, Contr. Inst. Bot. Nat. Acad. Peiping 3(3): 69. 1935; Hong, Fl. Reipubl. Popularis Sin. 73(2): 101. pl. 14: 7–9. 1983. Type: China, Yunnan, E flank of Lichiang (Lijiang) Range, 27°12′N, 3050–3350 m, limestone cliffs, Aug. 1906, *G. Forrest 2718* (**lectotype here designated**: E; isolectotype: K); loc. eodem, 27°15′N, 3200–3350 m, Aug. 1906, *G. Forrest 2810* (paratype: E).

Adenophora megalantha Diels, Notes Roy. Bot. Gard. Edinburgh 5: 175. 1912. Type: China, Yunnan, Lichiang (Lijiang), 27°12′N, 3050–3350 m, Aug. 1906, *G. Forrest 2661* (**lectotype here designated**: E; isolectotype: K).

Adenophora ornata Diels, Notes Roy. Bot. Gard. Edinburgh 5: 174. 1912. Type: China, Yunnan, Lichiang (Lijiang), 27°15′N, 3350–3660 m, Aug. 1906, *G. Forrest 2801* (**lectotype here designated**: E; isolectotype: K); Yunnan, Dali, 25°40′N, 3350–3660 m, Sept. 1906, *G. Forrest 3861* (paratype: E); Dali, *G Forrest 3862* (paratype: P).

Adenophora pachyrhiza Diels, Notes Roy. Bot. Gard. Edinburgh 5: 176. 1912. Type: China, Yunnan, Tali (Dali) Range, 25°40′N, 3350 m, Sept. 1906, *G. Forrest 3860* (holotype: E).

Adenophora dimorphophylla C. Y. Wu in Wu & Li, Yunnan Trop. Subtrop. Fl. Res. Rep. 1: 67. pl. 15. 1965, p. p., quoad isotype in PE (01230331).

Caudex sometimes with horizontal branches. Stems 1 or 2 from each caudex, simple, usually 50–80 cm tall, glabrous to rather densely hispid, usually densely leafy in lower half. Cauline leaves sessile or sometimes lower ones short-petiolate; leaf blade ovate-rhombic, obovate, ovate, or linear-lanceolate, 2–10 cm long, 0.5–3 cm broad, adaxially sparsely hispidulous, abaxially often hispidulous along veins, margin irregularly serrate. Flowers several in a pseudo-raceme or narrow panicle; pedicel short. Calyx tube globose-obovoid or obovoid-conical, glabrous; calyx lobes narrowly triangular, 8–20 mm long, 1–1.5 (–2) mm broad, margin with 1-several pairs of denticles. Corolla blue or blue-purple, broad-campanulate, 1.5–4 cm long; corolla lobes ovate-deltoid, 6–11 mm long, ca. 1/2 as long as tube. Disc (1.2–) 2–3 (–3.5) mm long, glabrous or sometimes hairy. Style included. Capsule ellipsoid, 10–14 mm long, 5–7 mm in diameter. **Chromosomes**: 2n=102. Figure 47–70: e–g.

Phenology: flo-wering and fruiting from July to October.

Habitats: grassy slopes, forests, or pine forest openings; 2000–4000 m.

Distribution: N and S Hengduan (Map 47–159); also in C & N Yunnan of China.

N HENGDUAN: **Baiyü**, Acha-Garzê, 4197 m, 2010, *Kham Exped. 10-0806* (PE); **Gonjo**, Shadong, 4000 m, 2010, *Kham Exped. 10-1942* (PE). **S HENGDUAN**: **Dali**, Cangshan, Zhonghe-si, 1946, *T. N. Liou 21040* (PE); **Dechang,** Cihuan Township, 2200 m, 1959, *S. F. Zhu 20140* (PE); **Heqing**: Lianping, 2650 m, 1940, *R. C. Ching 23942* (KUN, PE); **Huili**, 2900 m, 1932, *T. T. Yü 1536* (PE); **Liziping**, Baicaoping Village, 2750 m, 2011, *D. Y. Hong et al. H11033* (PE); **Muli**, Gibo Shan, 2800 m, 1937, *T. T. Yü 7624* (KUN, PE); **Puge**, 1959, *Sichuan Econ. Plant. Surv. (Liang) 5438* (PE); **Shangri-la** (Zhongdian), Haba, 3800 m, 1962, *Zhongdian Exped. 1322* (PE); **Xiangcheng**, Ma'an Shan, 3873 m, 2010, *Kham Exped. 10-0219* (PE); **Xichang**, 2700–3000 m, 1932, *T. T. Yü 1270* (PE); **Yulong** (Lijiang), Ganhaizi, 2900 m, 2010, *D. Y. Hong et al. H10026* (PE); Yulong Snow Range, 2900–3000 m, 1942, *K. M. Feng 8908* (KUN, PE).

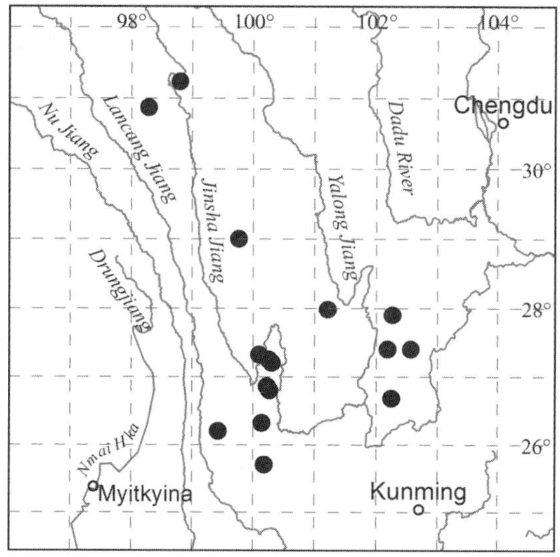

Map 47–159: **Adenophora coelestis** Diels

8. Adenophora linearifolia D. Y. Hong, **sp. nov.** Type: China, Sichuan "Thibet oriental: (Province de Batang), Yargong: bois, pelouses," Aug. 1903, *M. A. Soulié 3. 292* [holotype: P (P00235706); isotypes: P].

Roots unknown. Stems erect, simple, 20–40 cm tall, glabrous. Basal leaves unknown; cauline leaves all alternate, mostly at lower half of stem, lowermost one or two long-petiolate, petiole 0.3–1.5 cm long, leaf blade elliptic or linear-elliptic, 0.8–3 cm long, 0.4–0.6 cm broad, margin sparsely serrate or entire; all others sessile, linear, 3–6 cm long, 0.2–0.4 cm broad, glabrous on both surfaces, margin entire. Flowers in a terminal pseudo-raceme of 2–4, rarely solitary and terminal; bracts leaf-like, but much smaller; pedicel 1–2 cm long. Calyx glabrous, calyx tube ovoid, base rounded; calyx lobes linear-subulate, 3–12 mm long, less than 1 mm broad, margin with 1–2 pairs of verrucose denticles. Corolla campanulate, 1.3–1.6 cm long; corolla lobes ca. 1/2 as long as tube. Stamens free; filaments ca. 5 mm long, glabrous, base much dilated; anthers ca. 5 mm long. Disc short, less than 1 mm long, glabrous. Style slightly longer than corolla. Figure 47–69: c & d.

Phenology: flowering in August and September.

Habitats: meadows or pine forests; 2700–3600 m.

Distribution: **endemic**; confined to S Hengduan (Map 47–160).

Notes: This newly described species mostly resembles *A. jasionifolia* Franch., but differs from the latter in having mostly linear leaves, calyx lobes with only 1 or 2 pairs of denticles, corolla 1.3–1.6 cm long, lobed to 1/3 at most, and style slightly longer than corolla.

S HENGDUAN: **Daocheng**, Mengzi Commune, 3600 m, 1973, *Sichuan Veget. Surv. 2544* (PE): **Kangding** (?),Yargong, 1904, *M. A. Soulié 3132* (P); **Muli**, Donglang, 2800 m, 1983, *Qinghai-Xizang Exped. 14665* (PE); on border of **Shangri-la** (Zhongdian), 2700 m, 1939, *K. M. Feng 2936* (PE).

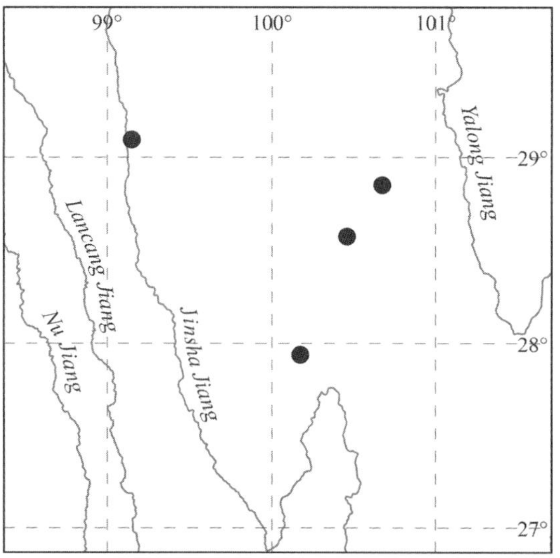

Map 47–160: **Adenophora linearifolia** D. Y. Hong

9. Adenophora potaninii Korsh., Mém. Acad. Imp. Sci. Saint Pétersbourg (sér. 7) 42(3): 39. 1894; Ge & Hong, J. Syst. Evol. 48 (6): 452. 2010. Type: China, Sichuan "China borealis prov. Tschili, Schansi, Kansu occidentali et orientali et Szetschuan septentrionali", China, Sichuan, 9 Aug. 1885, *G. N. Potanin* s. n. (**lectotype here designated**: LE); loc. eodem, 22 Aug. 1884, *G. N. Potanin* s. n. (paratype: LE).

Two subspecies, but only the typical one in our region, while the other, *A. potaninii* subsp. *wawreana* (Zahlbr.) S. Ge & D. Y. Hong, in N and NE China (Hebei, Henan, Liaoning, SE Nei Mongol, and Shanxi).

Adenophora potaninii subsp. **potaninii**

Adenophora bockiana Diels, Bot. Jahrb. Syst. 29: 605. 1901. ≡ *Adenophora potaninii* Korsh. var. *bockiana* (Diels) S. W. Liu, Fl. Qinghai. 3: 317. 1996. Type: China, Sichuan, "W Tsa Ku lao" (Lixian), Weikuan, s. d., *H. A. von Rosthorn 2531* (B).

Stems often single, simple, 30–100 cm tall, usually densely, less often sparsely retrorse-hispidulous. Cauline leaves sessile or short-petiolate; leaf blade obovate, ovate-elliptic, oblong, occasionally linear-elliptic or linear, 2–7 cm long, 0.5–4 cm broad, both surfaces hispidulous, base obtuse or cuneate, margin with 2–several large teeth, apex obtuse, acute, or shortly acuminate. Flowers usually several glomerated into a pseudo-raceme or panicle; pedicel less than 1 cm long. Calyx-tube obovoid or globose-obovoid, base rounded or obtuse, glabrous; calyx lobes triangular-subulate, 3–9 mm long, margin with 1 or 2 pairs of long denticles. Corolla purple, blue, or blue-purple, rarely white, campanulate, 1.5–2.5 cm long; corolla lobes ovate-deltoid, 5–8 mm. Disc tubular, 2–2.6 (–3) mm, hairy at least at apex. Style nearly as long as or slightly longer than corolla. Capsule globose-ellipsoid or ellipsoid, 7–11 mm long, 4–6 mm in diameter. Seeds brown-yellow, ellipsoid, 1-ribbed.

Phenology: flowering from July to October; fruiting in September and October.

Habitats: sunny grassy slopes, less often scrubs or forests; 1000–3300 m.

Distribution: N and S Hengduan (Map 47–161); also in C and NW China (E Gansu, Ningxia, E. Qinghai, Shaanxi, SW Shanxi, and W Sichuan).

N HENGDUAN: **Barkam**, Dalangjia Gou, 2640 m, 1957. *X. Li 72454* (PE); **Dawu**, Qianning, 2900 m, 1974, *Sichuan Plant Surv. 5732* (PE); **Heishui**, Caoshiba Gou, 2900 m, 1957, *X. Li 73233* (PE); **Jinchuan**, 2200 m, 1957, *X. Li.*

& *G. Z. Zhu 75380* (PE); **Lixian**, 1952, *Z. He & Z. L. Zhou 14054* (PE); **Jiuzhaigou** (Nanping), Baihe, 3000 m, 1964, *Q. X. Li & X. C. Zhao 3134* (PE); **Songpan**, 3300 m, 1937, *T. P. Wang 7743* (PE); **Wenchuan**, 2800 m, 1959, *Sichuan Econ. Plant Surv. (A 59) 2445* (KUN, PE); **Xiaojin**, Mongguqiao, 2400 m, 1975, *Sichuan Plant Surv. 9830* (PE). **S HENGDUAN: Baoxing**, Donghe, 1958, *S. S. Chang & Y. X. Ren 7626* (PE); **Tianquan**, 1936, *K. L. Chu 4113* (PE).

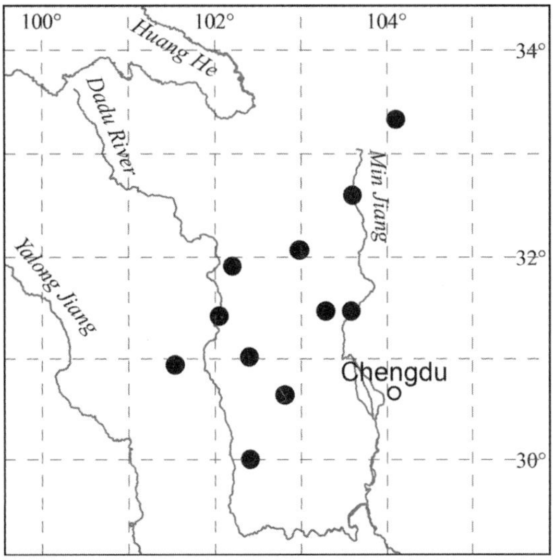

Map 47–161: **Adenophora potaninii** Korsh. subsp. **potaninii**

10. Adenophora wilsonii Nannf. in Handel-Mazzetti, Symb. Sin. 7: 1075. 1936; Hong, Fl. Reipubl. Popularis Sin. 73(2): 123. pl. 20. fig. 1–3.1983. Type: C China (W Hubei), Nov. 1907, *E. H. Wilson 2605* (holotype: K).

Stems often 2–several from one caudex, erect, simple or branched above, 25–80 cm tall, glabrous; basal leaves withered at anthesis, middle part of stem densely leafy. Leaves shortly petiolate; leaf blade linear-elliptic or lanceolate, 4–10 cm long, 0.5–1.2 cm broad, both surfaces glabrous, base attenuate into winged petiole, margin serrate or crenate. Flowers in a panicle; pedicel up to 1 cm long. Calyx tube obovoid or obconical, sometimes subglobose, glabrous; calyx lobes subulate or linear-lanceolate, 5–7 mm long, ca. 1 mm broad, with 1 or 2 pairs of verrucose denticles. Corolla purple or blue-purple, funnelform-campanulate, 1.5–2 cm long; corolla lobes ovate-deltoid, ca. 1/2 as long as tube. Disc annular or shortly tubular, less than 1.2 mm long, glabrous. Style 2–2.5 cm long, ca. 5 mm longer than corolla. Capsule globose-ellipsoid, 7–8 mm long, 4–5 mm in diameter. Figure 47–71: a-c.

Phenology: flowering from August to October; fruiting in September and October.

Habitats: thickets, by rocks in ravine; 800–1600 m.

Distribution: S Hengduan (Map 47–162); also in C China (Chongqing, W Hubei, N Guizhou, S Shaanxi, and E Sichuan).

S HENGDUAN: **Tianquan**, Shuitaping, 1400 m, 1953, *H. L. Tsiang 37931* (PE).

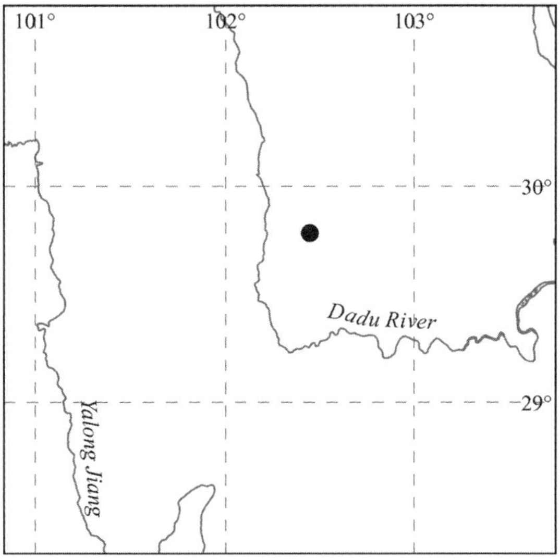

Map 47–162: **Adenophora wilsonii** Nannf.

11. Adenophora himalayana Feer, Bot. Jahrb. Syst. 14: 618. 1890; Hong, Fl. Reipubl. Popularis Sin. 73(2): 132. 1983 & in C. Y. Wu, Fl. Xizang. 4: 591. fig. 257, 3 & 4. 1985. Type: India, Kumaon, Byans, above Napalcha, rocks in valley, 3960 m, 13 Sept. 1884, *J. F. Duthie 3122* (lectotype designated by Turner, Edinburgh J. Bot. 70: 2. 2013; K); Kumaon, Kutti Valley in Byans, 3960 m, 09 Sept. 1884, *J. F. Duthie 3121* (paratype: K); loc. eodem, 3050–3350 m, 29 Jul. 1881, *J. F. Duthie 5735* (paratype: K).

Roots often thickened, up to 1 cm in diameter. Stems often several from one caudex, sim-

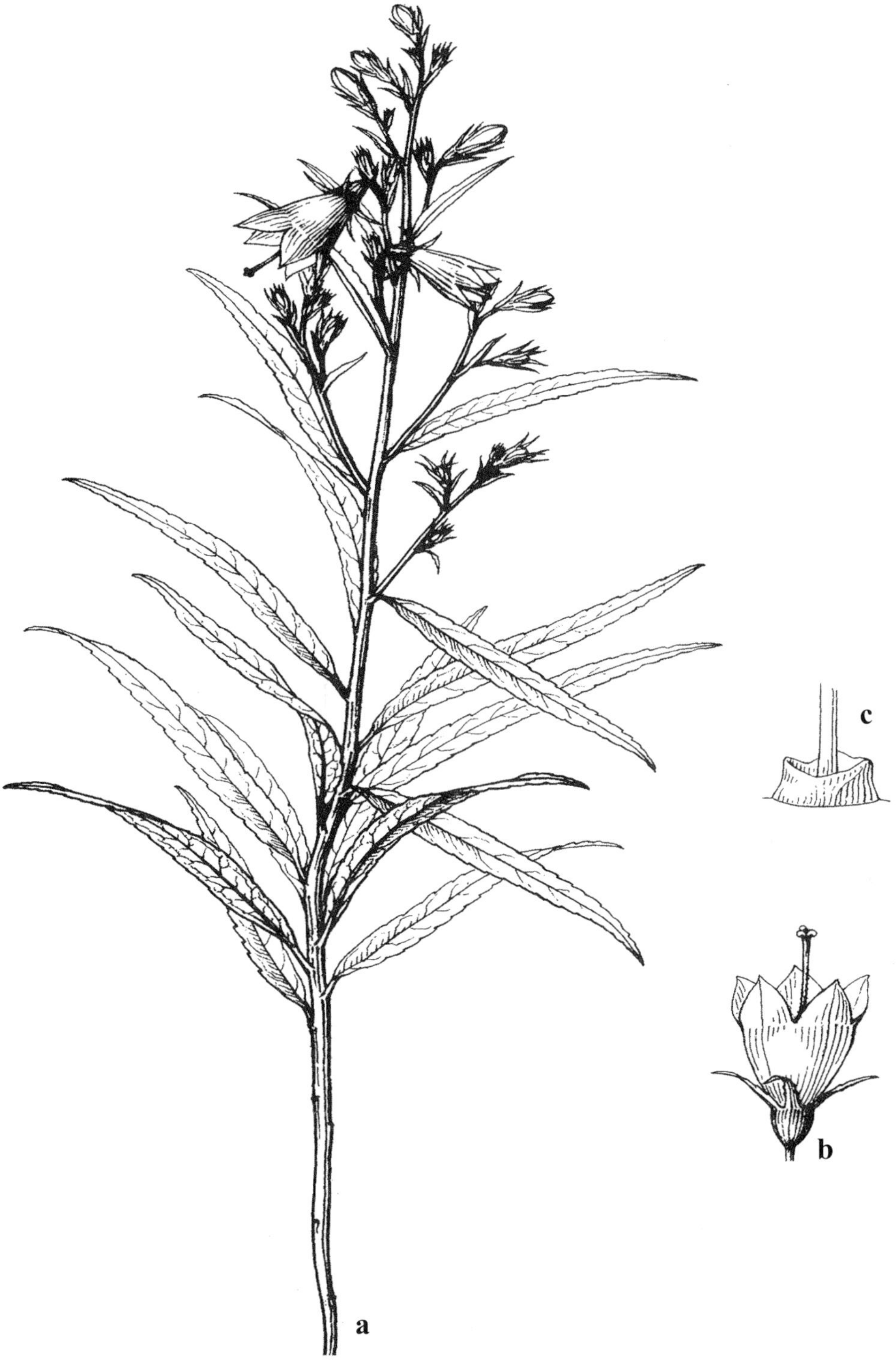

Figure 47–71: **Adenophora wilsonii** Nannf.: **a**, stem; **b**, flower; **c**, disc and lower part of style. From *Fl. Reipubl. Popularis Sin.* 73(2), 1983.

ple, 15–60 cm tall, glabrous, less often retrorse-hispidulous or sparsely retrorse-hirsute. Cauline leaves sessile or lower ones short-petiolate; leaf blade ovate, lanceolate, narrowly elliptic, or linear, 3–14 cm long, 0.1–2.5 cm broad, glabrous, very rarely hairy, margin entire or sparsely serrate. Flowers solitary, or several in a pseudo-raceme, never in a panicle. Calyx tube obconical or obovoid, glabrous; calyx lobes subulate, 5–10 mm long, 1–1.5 (–2) mm broad, entire, less often with verrucose denticles. Corolla blue or blue-purple, campanulate, 1.7–2.2 cm long; corolla lobes ovate-deltoid, 4–7 mm long. Disc thickly tubular, 3–8 mm long, 2–3 mm wide. Style nearly as long as or slightly longer than corolla. Capsule obovoid, 5–10 mm long, 4–7 mm in diameter.

Distribution: widely distributed in the Pan-Himalaya (Map 47–163) and its adjacent regions.

1a. Calyx lobes usually entire, very occasionally with verrucose denticles on margin; leaf blade usually broadly linear, less often narrowly elliptic to ovate-lanceolate .... 9a. subsp. *himalayana*

1b. Calyx lobes usually with verrucose denticles on margin, rarely entire; leaf blade mostly ovate to ovate-lanceolate, up to 2.5 cm broad, rarely broadly linear .... 9b. subsp. *alpina*

11a. Adenophora himalayana subsp. **himalayana**

Campanula lacei Duthie ex Bedd., J. Roy. Hort. Soc. 32: 207. 1907. Type: India, Chamba State, Darwas, 2440 m, 23 Jun. 1897, *J. H. Lace 1642* (holotype: E).

Leaf blade usually broad-linear, rarely narrow-elliptic to ovate-lanceolate. Calyx lobes entire, very occasionally with verrucose denticles on margin. Disc 2–3 mm in diameter. Style usually slightly exserted. Figure 47–72: a & b.

Phenology: flowering from July to September; fruiting in September and October.

Habitats: alpine meadows, scrubs, or forests; 2800–4700 m.

Distribution: N Pakistan, U Ganga & Indus, M and L Yarlung Zangbo, Yarlung Zangbo-Brahmaputra, Tangut, N and S Hengduan; also in other parts of China (E Gansu, NE Qinghai and Xinjiang), E Kazakhstan, Kyrgyzstan, and Tajikistan.

Notes: *Campanula younghusbandi* Prain, based on a collection by *H. J. Walton* (Hills above Lhasa, Aug. 1904: K) is an invalid name that belongs here.

N PAKISTAN: **Gilgit**, Nitar Valley, 3050–3350 m, 1892, *J. F. Duthie 12419* (K); Naltar, 3050 m, 1960, *D. McVean* s. n. (E). **U GANGA & INDUS**: **Burang**: Koja, 3800 m, 1976, *Qinghai-Xizang Exped. 8464* (KUN, PE); W. Himalaya, Kharmalak, 4880 m, 1899, *J. F. Duthie* s. n. (K). **M YARLUNG ZANGBO**: **Gyaca**, Nada Township, 3700 m, 1982, *Xizang Chinese Mater. Medic. Herbs Exped. 4384* (PE); **Lhasa**, hills above Lhasa, 1904, *H. J. Walton* s. n. (K); **Lhozhag**, Serxiang, Lapu Spring, 2013, *Y. S. Chen et al. 13-1686* (PE); **Nêdong**, 3650 m, 2013, *Y. S. Chen et al. 13-0108* (PE). **L YARLUNG ZANGBO**: **Gongbo'gyamda**, 4000 m, 1976, *Wu 4829* (KUN). **YARLUNG ZANGBO-BRAHMAPUTRA**: **Lhünze**, Gema, 3800 m, 1960, *G. X. Fu 1181* (PE). **TANGUT: Aba:** 3300 m, 1961, *S. Jiang (Aba) 6815* (PE); **Baxoi:** Demula Pass, 4500 m, 1980. *C. C. Ni et al. 1218* (PE); **Garzê**: Chuosijia (Guanyinqiao), 3800 m, 1958, *P. X. Li 10474* (PE); W of Garzê, 3520–3550 m, 2005, *D. E. Boufford et al. 33987* (PE); **Nangqên**: Dongba to Ziqu River, 4300 m, 1965, *Y. C. Yang 1292* (PE); **Qamdo**: Toba, 3900 m, 1976, *Qinghai-Xizang Exped. 12643* (PE); **Riwoqê**: Sangdo, 3800 m, 1976, *Qinghai-Xizang Exped. 12955* (PE); **Yushu**: Longbao, 4300 m, 1964. *s. coll. 489* (PE); **Zadoi**: Sulu Township, 1965, *S. W. Liu 373* (PE); **Zamtang**, NNE of the city, 3140 m, 2007, *D. E. Boufford et al. 38979* (PE). **N HENGDUAN**: **Dawu**, 3100 m, 1934, *H. Smith 12150* (PE). **Gongjo**: 3500 m, 1976, *Qinghai-Xizang Exped.12612* (KUN, PE); **Xinlong**, Xinlong to Garzê, 3125 m, 2006, *D. E. Boufford et al. 36264* (PE); **S HENGDUAN**: **Daocheng**, 3200 m, 1981, *Qinghai-Xizang Exped. 4294* (KUN); **Dêngqên**: Shagong, 3600 m, 1991, *J. S. Yang 91-634* (PE); **Dêrong**, 3400 m, 1981, *Qinghai-Xizang Exped. 3313* (KUN); **Kangding**: Xinduqiao, 3500 m, 1963. *K. C. Kuan et al. 1351* (PE); **Muli**, 3400–3600 m, 1983, *Qinghai-*

Figure 47–72: **a** & **b**, **Adenophora himalayana** Feer subsp. **himalayana**: **a**, upper part of stem; **b**, a flower with corolla and stamens removed, showing long and thick disc. **c–f**, **Adenophora lobophylla** D. Y. Hong: **c**, middle part of stem; **d**, upper part of stem; **e**, a flower with corolla and stamens removed, showing toothed calyx lobes and disc; **f**, fruit. From *Fl. Reipubl. Pepularis Sin*. 73(2), 1983.

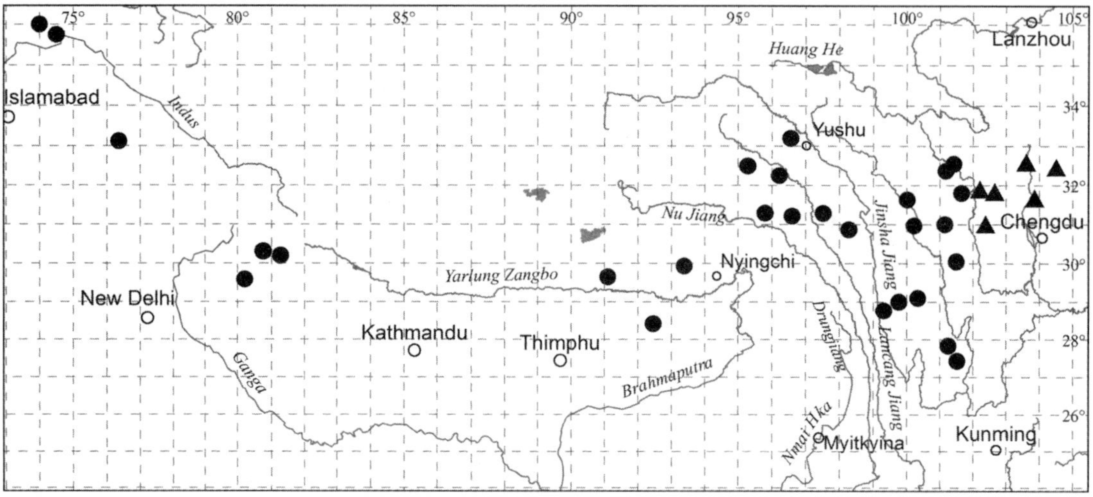

Map 47–163: **Adenophora himalayana** Feer

● subsp. **himalayana,** ▲ subsp. **alpina**

Xizang Exped. 12963 (KUN); **Xiangcheng**, 3700 m, 1981, *Qinghai-Xizang Exped. 4659* (KUN); **Yanyuan**, 3400–3600 m, 1983, *Qinghai-Xizang Exped. 12430* (KUN).

11b. Adenophora himalayana subsp. **alpina** (Nannf.) D. Y. Hong, Fl. Reipubl. Popularis Sin. 73(2): 132. 1983; Hong *et al.* in Wu, Raven & Hong, Fl. China 19: 544. 2011. ≡ *Adenophora alpina* Nannf., Acta Hort. Gothob. 5: 14. pl. 3. 1930. Type: China, Sichuan, Songpan, Dongrergo, 4000 m, 1922, *H. Smith 3508* (**lectotype here designated**: GH, n. v.; isolectotype: E).

Adenophora smithii Nannf., Acta Horti Gothob. 5: 21. pl. 11. fig. a-c. 1930. Type: China, Sichuan Sung-pan (Songpan), 3200–3400 m, 1922, *H. Smith 2520* (**lectotype here designated**: GH, n. v.; isolectotypes: E, PE).

Adenophora smithii Nannf. f. *crispa* Nannf., Acta Hort. Gothob. 5: 22. pl. 11. d & e. 1930. Type: China, Sichuan, Sung-Pan (Songpan), 3300–3400 m, 1922, *H. Smith 2515* (holotype: GH, n. v.).

Leaf blade mostly ovate to ovate-lanceolate, less often broadly linear, up to 2.5 cm wide, abaxially often sparsely hispidulous. Calyx lobes often with verrucose denticles, occasionally entire. Disc 1.5–2.5 mm in diameter. Style often included.

Phenology: flowering in July.

Habitats: meadows or grassy places at forest margins; 2500–4200 m.

Distribution: **endemic**: N Hengduan; also in other parts of Gansu and S Shaanxi of China.

Notes: *Adenophora alpina* was recorded in U Ganga & Indus (Himachal Pradesh) by Haridasan & Mukherjee (1996: Fasc. Fl. India 22). This record was based on misidentification of *A. himalayana* subsp. *himalayana*; *A. alpina* (= *A. himalayana* subsp. *alpina*) is confined to NW Sichuan of China.

N HENGDUAN: **Barkam**, Shuanjingsi, 4200 m, 1957, *X. Li 74112* (PE); **Hongyuan**, summit of Zhegu Shan, 4200 m, 1957, *X. Li 23784* (KUN); **Maoxian**, E Buji Shan, 1900 m, 1961, *S. Jiang 6815* (KUN); **Pingwu**, Wangba, Chubaihe, 2510 m, 1958, *s. coll. 10793* (PE); **Songpan**: Chelong, 2800 m, 1961, *S. Jiang (Aba)7309* (PE); Dongrergo, *H. Smith 3623* (PE); **Xiaojin**, 4230–4350 m, 2007, *D. E. Boufford et al. 38608* (KUN).

12. Adenophora lobophylla D. Y. Hong, Fl. Reipubl. Popularis Sin. 73(2): 186. 1983; Ge & Hong, J. Syst. Evol. 48 (6): 453. 2010; Hong *et al.* in Wu, Raven & Hong, Fl. China 19: 549. 2011. Type: China, Sichuan, Jinchuan, Anning, Baguadiaogou, 2400 m, by stream, 17 Aug. 1958, *X. Li 78384* (holotype: PE).

Stems erect, simple, 20–100 cm tall, sparsely retrorse-hirsute. Cauline leaves all or only lower and middle ones petiolate; petiole 0.5–2.5 cm long; leaf blade ovate or ovate-lanceolate, 2–5 cm long, 1–2 cm broad, adaxially sparsely puberulent, abaxially hirsute along veins, margin pinnatifid, rarely only dentate-lobed. Flowers in a panicle, or several in a pseudo-raceme, rarely solitary and terminal; pedicel 5–20 mm long. Calyx tube long-obovoid, base obtuse or rounded, glabrous; calyx lobes subulate, 3–4 mm long, less than 1 mm broad, margin with 1 or 2 pairs of verrucose denticles, rarely ciliate. Corolla blue or purple-blue, campanulate, 11–17 mm long; corolla lobes ovate-deltoid, 4–6 mm long. Disc tubular, 2–3 mm long, sparsely puberulent apically. Style nearly as long as or slightly longer than corolla. Capsule long-ellipsoid, 6–8 mm long, ca. 3 mm in diameter, base rounded. Seeds yellow, ellipsoid, ca. 0.8 mm long. **Chromosomes**: 2n=34, 34+2B. Figure 47–72: c–f.

Phenology: flowering from July to October; fruiting from August to October.

Habitats: thickets, open woods, or margins of forests; 2000–3400 m.

Distribution: **endemic**; confined to N Hengduan (Map 47–164).

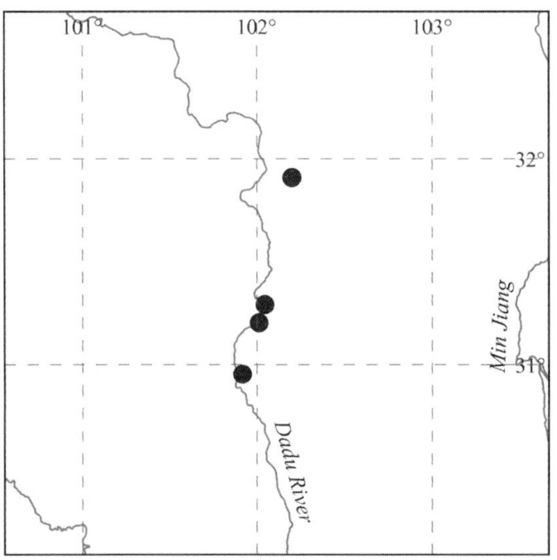

Map 47–164: **Adenophora lobophylla** D. Y. Hong

N HENGDUAN: **Barkam**, Dalangjiao Valley, 2650 m, 1957. *X. Li 229833* (PE); **Danba**, Dongguhe, 2500 m, 1995, *D. Y. Hong et al. 95051* (PE); **Jinchuan**, Anning, Baguadiaogou Valley, 2450 m, 1991, *S. Ge Population 22* (PE); Jinchuan, Zengda, 2300 m, 1995, *D. Y. Hong et al. 95048* (PE).

13. Adenophora liliifolioides Pax & K. Hoffm., Repert. Spec. Nov. Regni Veg. 12: 499. 1922; Hong, Fl. Reipubl. Popularis Sin. 73(2): 134. 1983; Hong & Ge, Novon 20: 426. 2010. Type: China, Sichuan, "Ost.-Tibet: Dege (Dêgê)", 3900 m, *W. Limpricht 2112* (lectotype designated by Hong & Ge, 2010: WU).

Adenophora gracilis Nannf., Acta Horti Gothob. 5: 17. pl. 5. fig. b. 1930. Type: China, Sichuan, Sung-pan (Songpan), ca. 3200 m, 14 Jul. 1922, *H. Smith 2747* (holotype: GH, n. v.; isotypes: E, PE).

Stems often single, simple, 30–100 cm tall, usually hirsute, less often glabrous. Cauline leaves sessile; leaf blade very variable in shape, from ovate to linear, 2–11 cm long, 0.4–3 cm broad, adaxially glabrous or hispidulous, abaxially hispidulous with hirsute along major veins or, less often, totally glabrous, base cuneate or attenuate, margin entire or sparsely serrate, apex acuminate. Flowers in a narrow panicle, or only several in a pseudo-raceme. Calyx tube ellipsoid, glabrous; calyx lobes narrowly triangular, 2–6 mm long, 0.6–1 mm broad, margin entire or very rarely with verrucose denticles. Corolla blue, purple-blue, or pale purple, rarely white, urceolate, 8–12 mm long; corolla lobes ca. 1/5 as long as tube. Disc narrowly tubular, 3–6.5 mm long, usually glabrous. Style 1.5–2 cm long, almost 2 times as long as corolla. Capsule ovoid or ellipsoid, 6–9 mm long, 3–4 mm in diameter. Seeds brownish yellow, ellipsoid, 1-ribbed, 1.2–1.5 mm long, finely striate. Figure 47–73.

Phenology: flowering from July to September; fruiting from August to October.

Habitats: meadows, scrubs, or among debris; 2400–4600 m.

Distribution: Bhutan, M and L Yarlung Zangbo, Yarlung Zangbo-Brahmaputra, Tangut, N and S Hengduan (Map 47–165); also in other parts of Gansu and S Shaanxi of China.

BHUTAN: **Thimphu**, Paro Chu, 2800 m, 2003, *S. A. Bowes Lyon 15002* (E). **M YARLUNG ZANGBO**: **Lhasa**, mountains near Kalasi, 4600 m, 1965, *Y. T. Chang & K. Y. Lang 1996* (PE); Lhasa, 3660 m, 1904, *L. A. Waddell.* s. n. (K); **Nêdong**, Jeba Township, 3700–3864 m, 2012, *FLPH Tibet Exped. 12–0741* (PE); **Zhag'yab**, Jitang, 3600 m, 1976, *Qinghai-Xizang Exped. 12318* (KUN, PE). **L YARLUNG ZANGBO**: **Sumzom**, 3500 m, 1965, *J. S. Ying & D. Y. Hong 651135* (PE); **Gongbo'gyamda**, south side of Niyang Chu, 3670–3800 m, 2010, *D. Y. Hong & Q. Wang H10004* (PE); ca. 3 km from Bahe, 3275 m, 2000, *D. E. Boufford et al. 30056* (PE); **Gyaca**, Zhoinda, 3300 m, *Qinghai-Xizang Suppl. Exped. 750660* (PE); **Mainling**, near Jaga, 3000 m, 1982, *C. C. Ni et al. 2770* (PE);**Nyingchi**, Naixi, 1965, *Y. T. Chang & K. Y. Lang 1238* (KUN); **Rawu**, 3900 m, 1980, *C. C. Ni 1261* (PE); Midui Glacier, 3500 m, 2012, *D. Y. Hong et al. 12025* (PE). **YARLUNG ZANGBO-BRAHMAPUTRA**: **Tumbatse**, 3350 m, 1924, *F. Kingdon-Ward 6052* (E, K). **TANGUT**: **Aba**, Mêruma, by Yellow River, 3400 m, 1984, *W. L. Chen et al. 7780* (PE); **Biru**, Mt Xiala, 4000 m, 1976, *D. D. Tao 11290* (PE); **Dêgê**, Manigango, 3900 m, 1982, *Z. X. Tang & al. 646* (PE); **Garzê**, Baimai, 3500 m, 1960. *J. S. Ying 4537* (PE); **Jomda**, Gamtog, 3200 m, 1976, *Qinghai-Xizang Exped. Veget. Group 9921* (PE); **Qamdo**, near the city, 1951, *Y. W. Tsui 5572* (PE); **Sêrtar**, Wungda, 3400 m, 1974, *Z. S. Qing 6574* (PE); **Sêrxü**, Dainkog, near the town, 3450 m, 1974, *s. coll.7661* (PE); **Sogxian**, Rongbo, 3800–4000 m, 1976, *D. D. Tao 10938* (KUN, PE); **Zamtang**, N of the city, 3075–3525 m, 2007, *D. E. Boufford et al. 38967* (PE). **N HENGDUAN**: **Baiyü**, Zhangdu, 3300 m, 1983, *J. T. Peng 22113* (PE); by Jinsha Jiang, 2983 m, 2010, *Kham Exped. 10–0678* (PE); **Barkam**, Sha'er, 2750 m, 1975, *Sichuan Plant Surv. 9543* (PE); **Batang**, between Batang and Shama, 3660 m, 2010, *Kham Exped. 10-0503* (PE); **Dawu**, Lhamo Mondeh La, 4000 m, 1934, *H. Smith 13210* (PE); Qianning, S of the city, 3450 m, 1959, *S. Jiang & C. L. Jin 2133* (PE); **Gongjo**, 3510–3550 m, 2009, *D. E. Boufford et al. 41531* (KUN); **Heishui**, Luhua, 2440 m, 1957, *X. Li & J. X. Zhou 73648* (PE); **Hongyuan**, Waqên Pasture, 3500 m, 1975. *Y. C. Tang et al. 50* (PE); **Jinchuan**, Wanli Township, 3150 m, 1958, *X. Li 78129* (PE) ; **Lixian**, s. loc., 1952, *Z. He & Z. L. Zhou 13569* (PE); **Songpan**, Huanglongsi, 3000 m, 1959, *Sichuan Econ. Plant Surv. (Aba) 1614* (PE); **Xiaojin**, Lianhe, 3300 m, 1959, *Z. G. Liu 428* (PE); **Xinlong**, Xinlong to Garzê, 3141 m, 2006, *D. E. Boufford 36282* (PE); **Yajiang**, Jianziwan Shan, 3100–3200 m, 1983, *K. Y. Lang et al. 2984* (KUN). **S HENGDUAN**: **Daocheng**, Riwa-Gongga, 3400 m, 1981, *Qinghai-Xizang Exped. 4365* (PE); **Kangding**, Zhonggu Village, 3230 m, 1963, *K. C. Kuan, et al. 213* (PE); Xinduqiao-Bamai, 3400 m, 2012, *D. Y. Hong et al. H12065* (PE); **Luding**, Ganhaizi, 2980 m, 1963, *K. C. Kuan,et al. 1803* (PE); **Luhuo**, Zhêhor, 3750 m, 1961, *S. Jiang 9351* (PE).

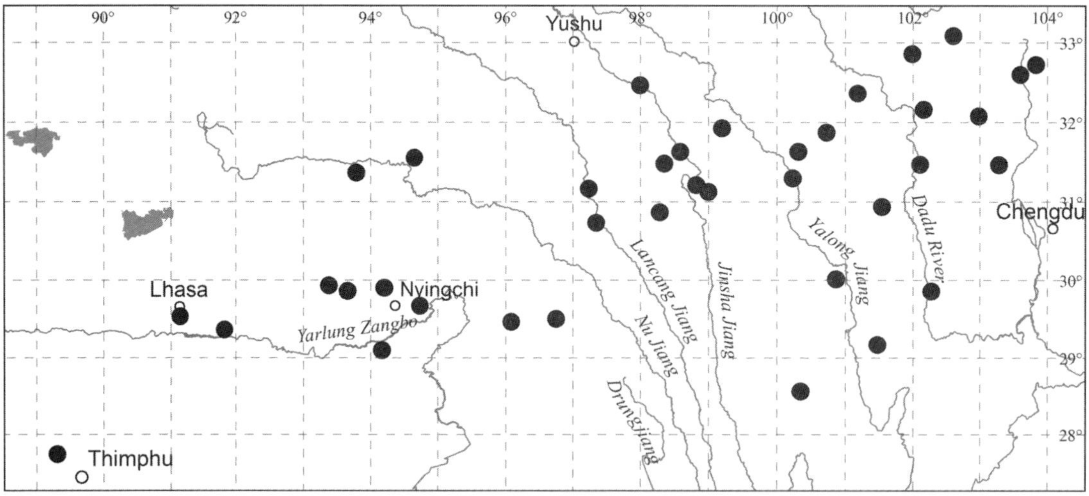

Map 47–165: **Adenophora liliifolioides** Pax & K. Hoffm.

Figure 47–73: **Adenophora liliifolioides** Pax & K. Hoffm.: **a**, middle part of stem; **b**, upper part of stem; **c**, a section of stem, showing a different form of leaves; **d**, flower; **e**, calyx tube and disc. Redrawn by Ai-Li LI from *Icon. Cormophyt. Sin*. 4, 1975.

14. Adenophora capillaris Hemsl., J. Linn. Soc. Bot. 26: 10. 1889; Tsoong, Contr. Inst. Bot. Nat. Acad. Peiping 3(3): 69. 1935; Anonymous, Icon. Cormophyt. Sin. 4: 394. fig. 6201. 1975; Hong, Fl. Reipubl. Popularis Sin. 73(2): 135. 1983. Type: China, Hubei, Patung Distr. (Badong CO), s. d., *A. Henry 954* (**lectotype here designated**: K); Hubei, Ichang, May 1888, *A. Henry 4799* (paratype: K).

Stems single, 0.5–1 m tall, glabrous or hirsute. Cauline leaves usually sessile; leaf blade ovate-lanceolate, elliptic-lanceolate, or linear, 3–19 cm long, 0.2–5 cm broad, glabrous or hirsute, margin entire or serrate (denticulate when narrow), apex acuminate. Flowers in a large and lax panicle, or in a narrow panicle, rarely several in a pseudo-raceme; main axis and branches filiform. Calyx-tube ellipsoid or sometimes ovoid; calyx lobes filiform, (3–) 6–14 (–20) mm long, spreading or reflexed, margin entire or sometimes with 1 to several verrucose denticles near their base. Corolla pale blue, pale purple, or white, urceolate, nearly tubular, or tubular-funnelform, 1–1.7 cm long; corolla lobes narrowly triangular, 3–4 mm long. Disc narrowly tubular, 2–5 mm long, often glabrous. Style 1.5–2.5 cm long, strongly exserted. Capsule globose, ellipsoid, or ovoid, 4–9 mm long, 4–5 mm in diameter. Seeds brown, ellipsoid, 1-ribbed, 1–1.6 mm long.

Three subspecies, but subsp. *paniculata* (Nannf.) D. Y. Hong & S. Ge in central and northern China.

Distribution: N and S Hengduan, and U Irrawaddy (Map 47–166); also in C and SW China.

1a. Calyx lobes (3–) 6–9 (–20) mm, usually entire; corolla 1–1.4 (–1.7) cm long; stems and leaves mostly glabrous .... 14a. subsp. *capillaris*

1b. Calyx lobes (4–) 9–14 (–20) mm long, usually denticulate; corolla 1.3–1.7 cm long; stems and leaves mostly more or less hirsute .... 14b. subsp. *leptosepala*

14a. Adenophora capillaris subsp. **capillaris**

Adenophora longisepala P. C. Tsoong, Contr. Inst. Bot. Nat. Acad. Peiping 3(3): 74. 1935. Type: China, Sichuan, Mt Omei (Emei), 15 Aug. 1928, *W. P. Fang 2999* (holotype: PE; isotype: K).

Stems and leaves usually glabrous. Inflorescence branches up to 27 cm long. Calyx lobes usually (3–) 6–9 (–20) mm long. Corolla 1–1.4 (–17) cm long. Capsule globose, ellipsoid, or ovoid. Figure 47–74

Phenology: flowering and fruiting from July to October.

Habitats: forests, forest margins, or grasslands; 1400–2800 m.

Distribution: N and S Hengduan; also in Chongqing, Guizhou, W Hubei, S Shaanxi, and other parts of Sichuan in China.

N HENGDUAN: **Heishui**, Luhua, 2380 m, 1957, *X. Li 73679* (PE); **Lixian**, 1952, *Z. He & Z. L. Zhou 13180* (PE); **Pingwu**, Wangba, 2000 m, 1958, *H. L. Tsiang 10629* (PE); **Wenchuan**, Wolong, 2750 m, 1975, *Sichuan Econ. Plant Surv. 8289* (PE); **Wenxian**, Baishuijiang Nature Reserve, 2000 m, 2006, *Baishuijiang Exped. 0747* (PE); **Xiaojin**, W slope of Balong Shan, 2600–2700 m, *K. Y. Lang et al. 1515* (KUN). **S HENGDUAN**: **Baoxing**, Guobayan, 2080 m, 1933. *T. H. Tu 4362* (PE); **Butuo**, Jiaojihe, 2300 m, 1959, *Sichuan Econ. Plant Surv. 5841* (KUN, PE); **Ganluo**, Pingba, 2400 m, 1959, *Sichuan Econ. Plant Surv. 4357* (KUN, PE); **Kangding**, between Togola and Bamei Town, 2012, *D. Y. Hong et al. H12066* (PE); Yala, 1951, *W. K. Hu & Z. He 10228* (PE); **Luding**, Erlang Shan, 2500 m, 1953, *H. L. Tsiang & J. H. Xiong 34469* (PE); **Xuexi**, Bao'an, 2700 m, 1959, *Sichuan Econ. Plant Surv. (Liang 59) 3820* (KUN).

14b. Adenophora capillaris subsp. **leptosepala** (Diels) D. Y. Hong, Fl. Reipubl. Populuaris Sin. 73(2): 136. 1983; Hong *et al.* in Wu, Raven & Hong, Fl. China 19: 550. 2011. ≡ *Adenophora leptosepala* Diels, Notes Roy. Bot. Gard. Edinburgh 5: 175. 1912. Type: China, Yunnan, Dali Range, E flank, Sept. 1906, *G. Forrest 3857* (**lectotype here designated**: E; isolectotype: K); Dali, 3350 m, Sept. 1906, *G. Forrest 3858* (paratypes: E, K).

Adenophora leptosepala Diels var *linearifolia* C. Y. Wu in Wu & Li, Yunnan Trop. Subtrop. Fl. Res. Rep. 1: 69. pl. 28: 4: 1965. Type: China, Yunnan, Heqing, 9 Aug. 1940, *R. C. Ching 23611* (holotype: KUN).

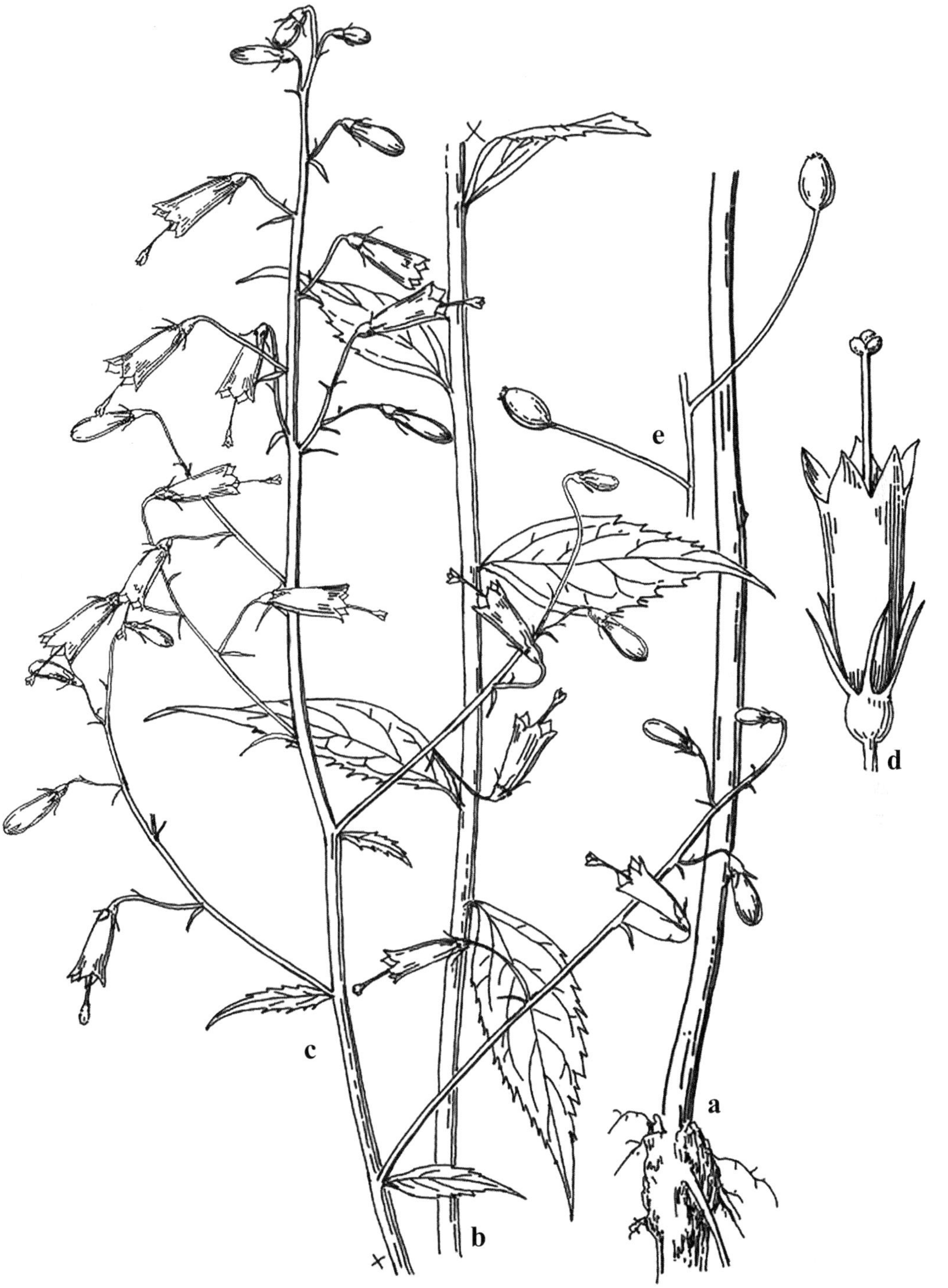

Figure 47–74: **Adenophora capillaris** Hemsl. subsp. **capillaris**: **a**, basal part of plant; **b**, middle part of stem; **c**, upper part of stem; **d**, flower; **e**, fruit. Redrawn by Ai-Li LI from *Icon cormophyt. Sin.* 4, 1975.

Adenophora urceolata C. Y. Wu, in Wu & Li, Yunnan Trop. Subtrop. Fl. Res. Rep. 1: 74. pl. 29: 4. 1965. Type: China, Yunnan, Binchuan, Jizu Shan, 27 Aug. 1938, *W. C. Wu, C. Y. Yang & C. Y. Wu 11633* (holotype: KUN; isotype: PE).

Stems and leaves usually hirsute. Calyx lobes (4–) 9–14 (–20) mm long; margin usually toothed. Corolla 1.3–1.7 cm long. Capsule globose or ovoid.

Phenology: flowering and fruiting from July to September.

Habitats: grassy places, forest margins, or forests; 2000–3700 m.

Distribution: S Hengduan and U Irrawaddy; also in NE Yunnan of China.

S HENGDUAN: **Dali**, E side of Diancang Shan, Yinglofeng, 2750–3200 m, 1984, *1984 Sino-Amer. Bot. Exped. 820* (B, KUN, PE); **Eryuan**, Sanying, 2950 m, 1963, *NW Yunnan Jinshajiang Team 6193* (KUN, PE); **Gongshan**, Cham-pu-tung, 3200 m, 1935, *C. W. Wang 66706* (KUN, PE); **Heqing**, Lianping, 3500 m, 1940, *R. C. Ching 24015* (KUN, PE); **Lanping**, 3200 m, 1933, *H. T. Tsai 53747* (KUN, PE); **Lushui**, Luobenzhuo, 2130 m, 2005, *Gaoligong Shan Biodiversity Survey 25759* (CAS); **Muli**, Kumatian, 2900 m, 1937, *T. T. Yü 7723* (KUN, PE); **Shangri-la** (Zhongdian), outside Alpine Bot. Gard., 3300 m, 2010, *D. Y. Hong et al. H10016* (PE); Tehgoh, 3200 m, 1937. *T. T. Yü 12383* (KUN, PE); Napohai, 3300 m, 2008, *Q. E. Yang & Q. Yuan (Yangqe) 1997* (PE); **Weixi**, Yezhi, 3600 m, 1935, *C. W. Wang 68669* (PE); **Yulong**, (Lijiang), Snow Range, 3200 m, 1942, *K. M. Feng 9153* (KUN, PE). **U IRRAWADDY**: **Chimili Valley**, 3050 m, 1919, *F. Kingdon-Ward 1231* (E); NE Upper Burma, 3050 m, 1924–1925, *G. Forrest 24942* (E, K); 3660–3960 m, 1925, *G Forrest 27216* (E, K); **Senighku Valley**, 3050–3350 m, 1926, *G Forrest 7502* (K).

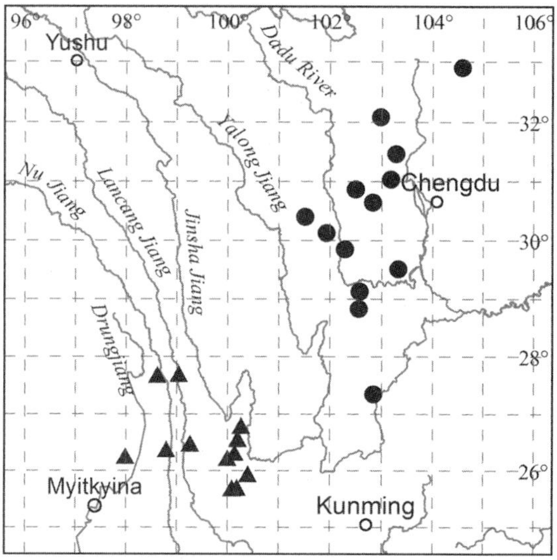

Map 47–166: **Adenophora capillaris** Hemsl.
● subsp. **capillaris**, ▲ subsp. **leptosepala**

Imperfectly known species

Adenophora watsonii W. W. Smith, Notes Roy. Bot. Gard. Edinburgh 8: 175. 1914. Type:?

Smith's protologue states that this taxon was based on the plant raised from seeds collected by Mr. C. M. Watson in 1908 from near Tatsienlu (Kangding in W Sichuan Prov.). No specimen was cited by the author. According to the protologue it does not belong to any species here recognized, but I have not seen any specimen so far, nor wild plants, which look like the plant described by Smith, though I have visited so many herbaria, and I have made three trips to Kangding area.

12. PERACARPA Hook. f. & Thomson

Hooker & Thomson, J. Proc. Linn. Soc., Bot. 2: 26. 1857. Type: *Peracarpa carnosa* (Wall.) Hook. f. & Thomson (*Campanula carnosa* Wall.).

Perennial herbs. Rhizomes elongate, with scales and buds, usually with a tuber at end. Leaves alternate. Flowers solitary and axillary, long-pedicellate. Calyx entirely epigynous, 5-lobed. Corolla infundibular-campanulate, 5-cleft to middle or slightly over middle. Stamens 5, free from corolla;

filaments ciliate, base dilated into narrow triangle; anthers elongated. Ovary inferior, 3-locular; style villous above; stigma 3-lobed, lobes elongated, recurved. Fruit dry, 3-locular or one locule reduced and thus becoming 2-locular, indehiscent, or sometimes irregularly dehiscent. Seeds several to many, ellipsoid, smooth.

One to three species according to different accounts, from the Himalaya eastwards to Philippines and Russia (the Far East); one species in our region.

1. Peracarpa carnosa (Wall.) Hook. f. & Thomson, J. Proc. Linn. Soc., Bot. 2: 26. 1857; Nannfeldt in Handel-Mazzetti, Symb. Sin. 7(4): 1076. 1936: Anonymous, Icon. Cormophyt. Sin. 4: 395. fig. 6203. 1975; Hara, Fl. East Himal. 2 Report: 129. 1971. ≡ *Campanula carnosa* Wall. in Roxburgh, Fl. India 2: 102. 1824. Type: *Wallich 1282* (holotype: K; isotype: BM).

Perennials slender, forming small mats. Stems fleshy, 5–15 cm long, up to 1 mm in diameter, glabrous. Leaves mostly at upper part; petiole 3–15 mm long; leaf blade membranous or thinnly papery, ovate-orbicular or orbicular, 0.8–2.5 cm long, 0.7–2 cm broad, both surfaces glabrous or adaxially sparsely appressed-hispidulous, base truncate or shallowly cordate, margin sinuate, but with a spinule at sinus, apex rounded or acute; lower cauline leaves distant and smaller. Pedicel slender but straight, 1–6 cm long. Calyx glabrous; calyx tube obovoid-conical; calyx lobes triangular to linear-lanceolate. Corolla white or blue-purple; corolla lobes as long as tube or slightly longer, linear-elliptic. Fruit ob-

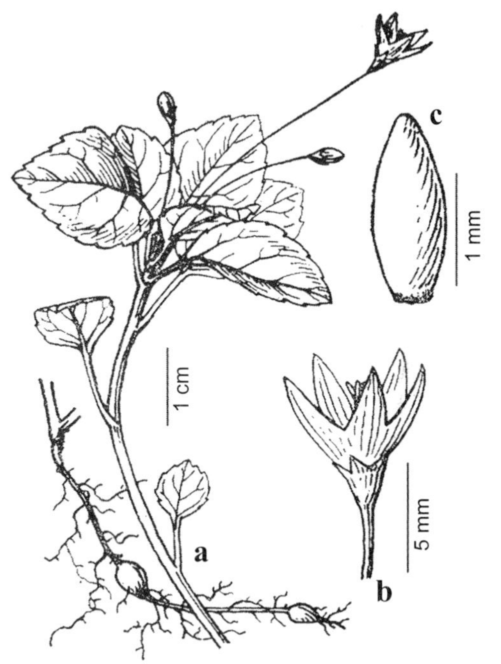

Figure 47–75: **Peracarpa carnosa** (Wall.) Hook. f. & Thomson: **a**, root and stem; **b**, flower; **c**, seed. From *Acta Phytotax. Sin.* 18(4), 1980.

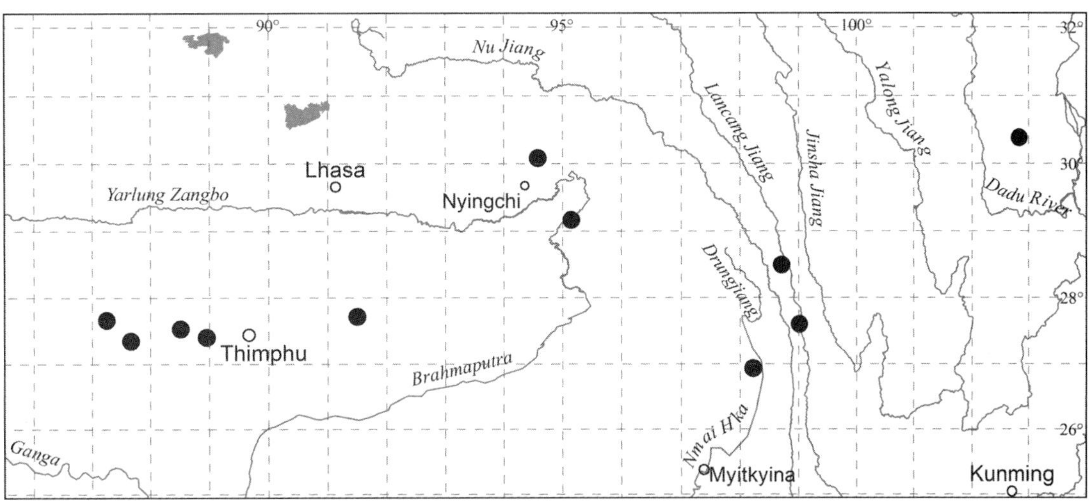

Map 47–167: **Peracarpa carnosa** (Wall.) Hook. f. & Thomson

ovoid, 4–5 mm long. Seeds brown, ca. 1.7 mm long. **Chromosomes**: 2n=30. Figure 47–75

Phenology: flowe-ring from March to May; fruiting from April to November.

Habitats: forests or moist rocks by stream; below 3000 m.

Distribution: E Nepal, Sikkim & Darjiling, Bhutan, M and L Yarlung Zangbo, Yarlung Zangbo-Brahmaputra, S Hengduan, and U Irrawaddy (Map 47–167); widely distributed, extending eastwards to Philippines, Japan, and Russia (the Far East).

E NEPAL: **Kosi**, Sankhuwasabha Distr., 1988, *M. Suzuki et al. 8820290* (BM); **Mechi**, Taplejung, 1969, *L. H. J. Williams 670* (BM). **SIKKIM & DARJILING**: **Sikkim**, 1830–2740 m, *J. D. Hooker* s. n. (BM). **BHUTAN**: **Ritang** (2400 m) – Ratsoo (1850 m), 1967, *Kanai et al. 9986* (BM); **Tobrang** (Trashiyangse Chu), 2130 m, 1949, *F. Ludlow et al. 20224* (BM). **M YARLUNG ZANGBO**: **Yadong,** Asangcun, 2840 m, 1975, *Qinghai- Xizang Suppl. Exped. 750152* (PE). **L YARLUNG ZANGBO**: **Yi'ong**, 1965, *J. S. Ying & D. Y. Hong 1271* (PE). **YARLUNG ZANGBO-BRAHMAPUTRA: Baibung**, 1690 m, 1983, *S. Z. Cheng et al. 3687*(PE). **S HENGDUAN**: **Baoxing**, 1954, *T. P. Soong* s. n. (PE); **Dêqên**, Kaakerpu, 3000 m, 1937, *T. T. Yü 8474* (PE); **Weixi**, Kangpu, 3000 m, 1935, *C. W. Wang 64577* (PE). **U IRRAWADDY**: **North Triangle**, above Ahkail, 2440 m, 1953, *F. Kingdom- Ward 20903* (BM).

13. HOMOCODON D. Y. Hong

Hong, Acta Phytotax. Sin. 18: 473. 1980 & Fl. Reipubl. Popularis Sin. 73(2): 143. 1983; Lammers, World Checklist Bibliography-Campanulaceae. 354. 2007. Type: *Homocodon brevipes* (Hemsl.) D. Y. Hong (*Wahlenbergia brevipes* Hemsl.).

Annuals prostrate. Leaves alternate. Flowers 1 or 2 on extremely short branches, sessile or short-pedicelled. Calyx epigynous, 5-lobed; calyx lobes dentate. Corolla epigynous, tubular-campanulate, 5-lobed. Stamens 5, free from corolla and each other; filaments slightly dilated at base, dilated part sparsely ciliate. Ovary inferior, 3-locular; stigma 3-fid, lobes linear, recurved. Fruit dry, lacerate or irregularly poricidal in lower part. Seeds ellipsoid, not angular, shallowly reticulate.

This group was treated as congeneric with *Heterocodon* Nutt., but its flowers are on short branches, not solitary and opposite to leaves on main stem, and seeds ellipsoid and reticulate, not angular and smooth.

Two species; both (one endemic) in the Pan-Himalaya.

1a. Pedicel absent; corolla ca. 3.5 mm long; style as long as corolla .... 1. *H. brevipes*
1b. Pedicel present, up to 6 mm long; corolla 7–9 mm long; style obviously exserted .... 2. *H. pedicellatus*

1. Homocodon brevipes (Hemsl.) D. Y. Hong, Acta Phytotax. Sin. 18: 474. fig. 1: 7–10. 1980 & Fl. Reipubl. Popularis Sin. 73(2): 144. 1983; Hong & Lammers in Wu, Raven & Hong, Fl. China 19: 552. 2011. ≡ *Wahlenbergia brevipes* Hemsl., Hooker's Icon. Pl. 8: t. 2768. 1903. ≡ *Heterocodon brevipes* (Hemsl.) Nannf. in Handel-Mazzetti, Symb. Sin. 7(4): 1075. 1936. Type: China, Yunnan, Mengzi (Mengtse), 1520 m, *A. Henry 10941* (holotype: K).

Wahlenbergia monantha H. J. P. Winkl., Repert. Spec. Nov. Regni Veg. 12: 501. 1922. Type: China, Sichuan, Kangding (Tatsienlu), 3700 m, *W. Limpricht 2338* (holotype: WRSL, n.v.)

Annuals prostate, glabrous throughout. Stems up to 50 cm long, 3-winged, branches extremely short, with one to several leaves appearing fascicled. Leaves alternate; petiole 2–9 mm long; leaf blade deltoid-orbicular or ovate-orbicular, 7–12 mm long, 7–12 mm broad, base nearly truncate, margin serrate, apex acute. Flowers sessile, ca. 5 mm long. Calyx tube ovoid, 1–1.5 mm long; calyx lobes narrowly triangular, ca. 2 mm long, margin with a pair of teeth. Corolla white, pale blue or pale purple, tubular-campanulate, ca. 3.5 mm long, 5-cleft; corolla lobes slightly longer than tube, linear-oblong. Filaments ca. 1 mm long; anthers elongated. Style as long as corolla. Fruit ovoid-globose, ca. 2.5 mm long, pericarp shin. Seeds several, brown, ca. 0.7 mm long. Figure 47–76: a–d.

Phenology: flowering and fruiting from April to August.

Habitats: forests, thickets, or grassy slopes, often on rocks; 1000–2900 m.

Distribution: Bhutan and S Hengduan (Map 47–168); also in other parts of China (W Guangxi, SW Guizhou, and S Yunnan).

Notes: According to specimens I have examined, the species is disjunct between NE Bhutan and SW China (W Guangxi, SW Guizhou, SW Sichuan, and Yunnan). I wonder whether this distribution really reflects the distribution pattern of the species considering that it is a small, inconspicuous herb, often creeping on rocks.

BHUTAN: **Upper Kuru Chu**, 2134 m, 1949, *F. Ludlow et al. 20914* (BM). **S HENGDUAN**: **Dali**, Qingbixu, 2100 m, 1981, *Sino-British Exped. to Cangshan 1981 1095* (K); Dali, 2040–2440 m, *G. Forrest 4787* (P); **Ganluo**, Haitang, 2200 m, *s. coll. 4314* (PE); **Hanyuan**, Daxiangling, ca. 1300 m, 1934, *H Smith 10086* (PE).

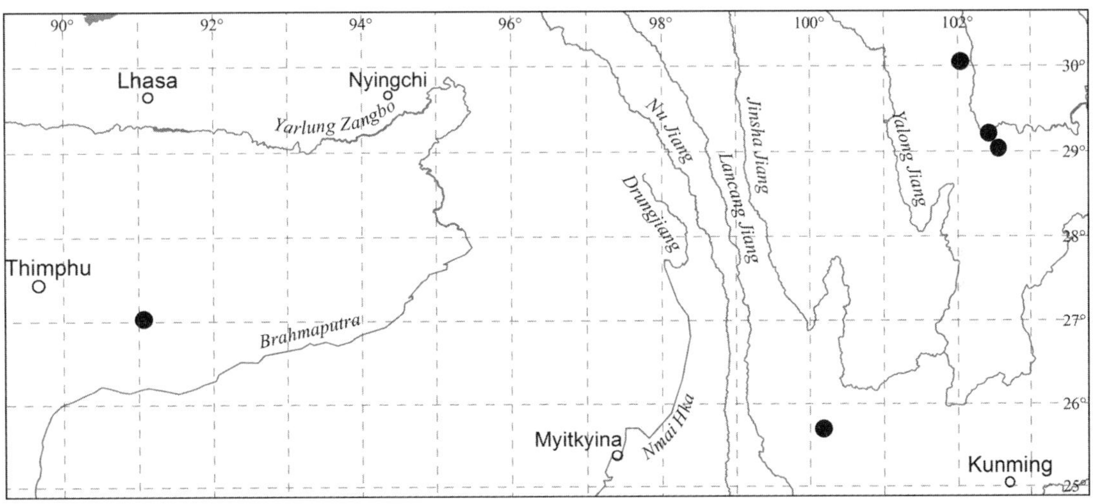

Map 47–168: **Homocodon brevipes** (Hemsl.) D. Y. Hong

2. Homocodon pedicellatus D. Y. Hong & L. M. Ma, Acta Phytotax. Sin. 29: 268. fig. 1, 1991; Hong & Lammers in Wu, Raven & Hong, Fl. China 19: 552. 2011. Type: China, Sichuan: Luding, Hailuogou, 1400–1600 m, 12 Jul. 1982. *K. Y. Lang et al. 569* (holotype: PE).

Annuals prostrate. Stems slender, glabrous or very sparsely villous, 3-winged, almost all axils on main stem with a short branch, branches with several leaves. Leaves alternate, glabrous; petiole 2–4 mm long; leaf blade fan-shaped, 4–9 mm long, 3–12 mm broad, leaves on branches obviously smaller, margin dentate. Flowers solitary in axils of leaves on branches, 1–3 on each branch, entirely glabrous; pedicel conspicuous, up to 6

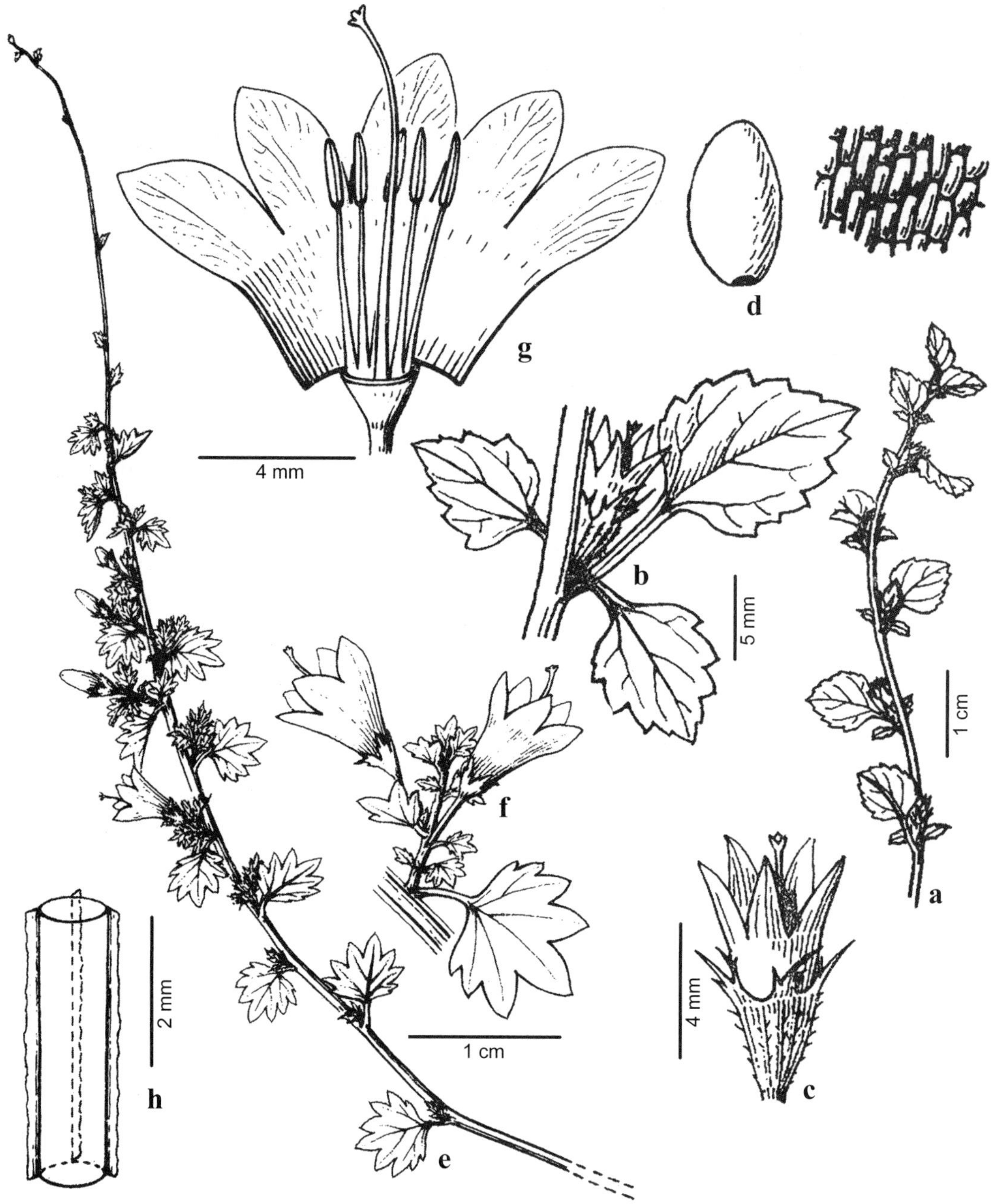

Figure 47–76: **a–d**, **Homocodon brevipes** (Hemsl.) D. Y. Hong: **a**, upper part of stem; **b**, a section of stem, showing lateral branch and terminal flower; **c**, flower; **d**, seed and a part of seed-coat. **e–h**, **Homocodon pedicellatus** D. Y. Hong & L. M. Ma: **e**, upper part of stem; **f**, a section of stem, showing lateral branch and axillary flowers; **g**, flower with calyx lobes removed and corolla opened, showing stamens, style, and stigma; **h**, a section of winged stem. **a–d** from *Acta Phytotax. Sin.* 18(4), 1980; **e–h** from *Acta Phytotax. Sin.* 29(3): 269. fig. 1. 1991, with permission.

mm long. Calyx epigynous; calyx tube obconical, ca. 1.5 mm long; calyx lobes triangular, ca. 2 mm long, margin with a pair of teeth. Corolla blue, tubular-campanulate, 7–9 mm long, 5-cleft; corolla lobes as long as tube, oblong. Filaments ca. 4.5 mm long; anthers terete, ca. 2 mm long. Style obviously exserted; stigma 3-fid, lobes linear, recurved at anthesis. Figure 47–76: e–h.

Phenology: flowering in July.

Habitats: ravines; 1400–1600 m.

Distribution: **endemic**; confined to S Hengduan (Map 47–169). No any specimen has been added after the type collection.

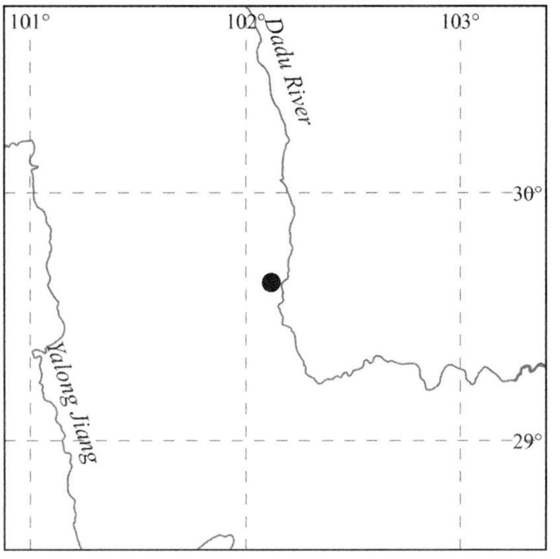

Map 47–169: **Homocodon pedicellatus** D. Y. Hong & L. M. Ma

LOBELIACEAE

Annuals, perennials, subshrubs, or shrubs, rarely trees (not in our region), lactiferous, mostly poisonous. Leaves alternate, rarely opposite or verticillate, pinnatifid or undivided, estipulate. Inflorescences racemose, sometimes paniculate or flowers solitary in leaf axils. Flowers perfect, rarely unisexual (plants dioecious), zygomorphic, 5-merous, often pedicel twisted so flower is resupinate; calyx epigynous, calyx lobes equal or unequal in length, persistent. Corolla epigynous, cleft to base at back, limb 2-lipped or nearly 2-lipped, occasionally all five corolla lobes spread into a plane (*Lobelia chinensis*); corolla lobes monomorphic or dimorphic. Stamens connate; filaments united into a tube; anthers longitudinally dehiscent, upper three longer than lower two, lower two or all barbate at top. Ovary inferior, half-inferior, rarely superior, (1- or) 2-locular, placentation axile, rarely parietal; ovules anatropous, generally many; style with a ring of collecting hairs just below stigma; stigma 2-fid, occasionally entire. Fruit a capsule or a berry, or dry and indehiscent. Seeds ellipsoid, oblong or 3-angular; embryo straight, imbedded in copious endosperm.

The Lobeliaceae is included in the Campanulaceae s. l. in APG III. The Campanulaceae s. l. comprises five monophyletic groups, one of which is the Lobelioideae. Therefore, it is a matter of taste to recognize the group as a family or a subfamily. However, it is distinguished from the Campanulaceae s. s. by a number of diagnostic characters: morphologically by the rusupinate flowers with a zygomorphic corolla, the connate filaments and anthers, and by the predominantly 2-locular ovary; palynologically by the non-spinuliferous exine of pollen, and chemically by the production of chelidonic acid and pyridine alkaloids but not caffeic acid.

The Lobeliaceae encompasses 30 genera and about 1200 species, and is nearly cosmopolitan, but is mostly in the tropics and subtropics (Lammers, 1992 in Peng, Inst. Bot. Acad. Sin. Monogr. Ser. 12: 43–61); only one genus in our region.

LOBELIA L.

Linnaeus, Sp. Pl. 1: 929. 1753; Lammers, Ann. Missouri Bot. Gard. 98: 42. 2011. Type: *Lobelia cardinalis* L.

Mostly herbs, annual, monocarpic or perennial, some species woody at base, and some in Africa and Hawaii Islands with woody trunks. Leaves alternate, in two rows or spiral, sometimes in basal rosette. Inflorescences a terminal raceme, sometimes a panicle, or flowers solitary in axils of leaves. Flowers bisexual, rarely unisexual (some Australian species dioecious); bracteoles present or absent. Calyx tube ovoid, semiglobose or shallowly campanulate; calyx lobes equal or nearly equal in length, rarely labiate, entire or serrate, persistent. Corolla zygomorphic; upper side often longitudinally divided to base or nearly so; limb 2-lipped or nearly so, a very few species with all corolla lobes spread in one plane on lower side; upper lip 2-lobed, lower lip 3-lobed. Stamens connate, stamen-tube enveloping style, exserted from corolla on upper side; anther tube mostly dark gray, all anthers or only lower 2 barbate at top. Ovary inferior, semi-inferior or very rarely superior, 2-locular; stigma 2-fid, villous on receptive surface; ovules many. Fruit a capsule or berry. Capsule 2-valved at apex. Seeds many, oblong or 3-angled, sometimes winged, smooth, foveolate, striate, or verrucose.

Recent reference: Lammer, 2011, Revision of the infrageneric classification of *Lobelia* L. (Campanulaceae: Lobeloideae). Ann. Missuri Bot. Gard. 98: 37–62.

About 415 species, mainly in tropics and subtropics, particularly tropical Africa and America, a few species extending to temperate regions; 19 species (9 endemic) in the Pan-Himalaya.

According to Lammers (2011) the genus comprises 18 sections, of which five occur in our region.

Key to sections

1a. Anthers with a single bristle at top of each of lower 2, very rarely glabrous; seeds reticulate .... sect. ***Hypsela*** (C. Presl) Lammers (1 sp., 19)

1b. Anthers witha tuft of filiform hairs at top of lower 2 or all 5; seeds striate or striate-reticulate.

2a. Plants annual or perennial, up to 50 cm tall; stems gracile; corolla bilabiate, lobes strongly dimorphic, upper 2 much shorter and narrower than lower 3.

3a. Pedicel usually lacking bracteoles; only lower 2 anthers barbate at top .... sect. ***Stenotium*** (C. Presl) Lammers (1 sp., 4).

3b. Pedicel with 2 bracteoles; all 5 anthers barbate at top .... sect. ***Delostemon*** (E. Wimm.) J. Murata (3 spp., 1–3)

2b. Plants perennial (biannual in *L. nudigena*), usually 1–9 m tall; stems erect, robust; corolla unilabiate or nearly bilabiate, lobes monomorphic or scarcely dimorphic, all similar in shape and size.

4a. Flowers solitary in axils of upper leaves, rarely in a terminal raceme; leaves petiolate, rarely sessile; fruit a berry .... sect. ***Speirema*** (Hook. f. & Thomson) Lammers (1 sp., 18).

4b. Flowers in a many-flowered terminal raceme or panicle; leaves sessile, rarely petiolate; fruit a capsule .... sect. ***Rhynchopetalum*** (Fresen.) Benth. (13 spp., 5–17).

Key to species.

1a. Fruit baccate, nearly plane at top.

2a. Stems prostrate; leaf blade ovate or broad-ovate, 0.8–2.5 cm long; flowers solitary and axillary .... 19. *L. angulata*

2b. Stems erect; leaf blade elliptic to lanceolate, 7–15 cm long; flowers axillary or in a terminal raceme .... 18. *L. montana*

1b. Fruit capsular, more or less conical at top.

3a. Dwarf herbs; stems erect or prostrate; corolla lobes nearly equal in width.

4a. Stems winged; pedicel usually lacking bracteoles; anthers only lower 2 barbate at top .... 4. *L. heyneana*

4b. Stems terete or angular, wingless; pedicel with 2 bracteoles; all anthers barbate at top.

5a. Stems prostrate, rooted at nodes; leaf blade 1–5.4 cm long, 0.8–3.2 cm broad; petiole 3–12 mm long .... 1. *L. zeylanica*

5b. Stems erect, not rooted or rooted only at basal nodes; leaf blade smaller, nearly sessile.

6a. Stems and leaves villous; bracteoles absent or minute .... 3. *L. terminalis*

6b. Stems and leaves glabrous; bracteoles linear-lanceolate .... 2. *L. alsinoides*

3b. Robust herbs or subshrubs; corolla nearly labiate, corolla lobes of upper lip linear, those of lower lip orbicular to ovate-lanceolate.

7a. Inflorescences with numerous and densely arranged flowers; bracts pendulous, wooly, covering entirely flowers .... 17. *L. nubigena*

7b. Inflorescences with limited and usually lax-arranged flowers; bracts upright, glabrous or with other types of hairs, never covering flowers.

8a. Rhzomes present; stems simple, rarely branched, less than 1 m tall; leaf blade less than 13 cm long.

9a. Leaves sessile, densely arranged at upper part of stems, lanceolate to linear-lanceolate, 3–16 mm wide .... 16. *L. sessilifolia*

9b. Leaves petiolate, evenly arranged at stems.

10a. Calyx tube densely white-hirsute, calyx lobes reflexed in fruiting .... 14. *L. pleotricha*

10b. Calyx tube glabrous or with other types of hairs, calyx lobes upright, rarely spread.

11a. Stems densely villous, branched .... 15. *L. drungjiangensis*

11b. Stems glabrous or covered with other types of hairs, usually simple.

12a. Stems glabrous; leaf blade less than 8 cm long, glabrous .... 12. *L. taliensis*

12b. Stems hairy; leaf blade usually over 8 cm long, hairy on both surfaces.

13a. Pedicel 8–12 mm long; corolla densely hispid along ribs outside, glabrous inside .... 11. *L. gaoligongshanica*

13b. Pedicel 3–5 mm long; corolla glabrous outside, villose at throat .... 13. *L. erectiuscula*

8b. Rhizomes lacking; stems usually multibranched, usually 1–2 m tall; leaf blade over 12 cm long, rarely smaller.

14a. Stems, leaves, pedicel and calyx all densely pannose; calyx lobes entire, densely pannose, 11–15 mm long .... 10. *L. rosea*

14b. Stems, leaves, pedicel and calyx glabrous or hispidulous; calyx lobes denticulate or entire, glabrous, hirsute or scaberulose, longer or shorter.

15a. Calyx lobes shorter than 11 mm, occasionally longer, denticulate; leaf blade 7–17 cm long.

16a. Pedicel 3–4 mm long; corolla glabrous outside, villous below throat, 1.5–1.8 cm long .... 9. *L. sichuanensis*

16b. Pedicel 8–12 mm long; corolla densely hispid along ribs outside, glabrous inside, 1.8–2 cm long .... 11. *L. gaoligongshanica*

15b. Calyx lobes longer than 11 mm, entire or denticulate; leaf blade usually 12–33 cm long.

17a. Capsule globose; anther tube white-hirsute at back; seed-coat striate .... 5. *L. pyramidalis*

17b. Capsule ellipsoid; anther tube glabrous; seed-coat foveolate.

18a. Bracteoles just below hypanthium, leafy, dentate, 6–12 mm long, 1–2 mm broad .... 6. *L. foliiformis*

18b. Bracteoles at middle part of pedicel, subulate, 1–2 mm long, ca. 0.2 mm broad.

19a. Calyx lobes entire; bracteoles near top of pedicel, ca. 1 mm below calyx tube .... 7. *L. seguinii*

19b. Calyx lobes verrucose-denticulate; bracteoles minute, usually at middle part of pedicel, sometimes absent ...... 8. *L. doniana*

1. Lobelia zeylanica L., Sp. Pl. 2: 932. 1753; Clarke in Hooker, Fl. Brit. India 3: 424. 1881, pro syn.; Tsoong, Contr. Inst. Bot. Nat. Acad. Peiping 3(3): 115. 1935; Hong & Lammers in Wu, Raven & Hong, Fl. China, 19: 557. 2011. Type: Sri Lanka, not seen.

Lobelia affinis Wall. ex A. DC. in de Candolle, Prodr. 7: 360. 1839; Wall., Cat. 1311. 1828, nom, nud. Type: *Wallich 1311* [holotype: K (K000781058); isotypes: BM, E].

Lobelia lobbiana Hook. f. & Thomson, J Proc. Bot. Linn. Soc. 2: 28. 1857. *Lobelia zeylanica* L. var. *lobbiana* (Hook. f. & Thomson) Y. S. Lian. Type: India, Khasia Mts. *Lobb* s. n. (holotype: K).

Herbs annual, succulent, rooting from lower nodes. Stems prostrate, 10–60 cm long, laxly branched, 4-angular, glabrous or puberulent. Leaves spirally alternate; petiole 3–15 mm long, puberulent; leaf blade broadly deltoid-ovate or ovate, 5.5 cm long, 0.8–4 cm broad, adaxially glabrescent, abaxially sparsely scaberulose along veins, base truncate, shallowly cordate or broadly cuneate, margin serrate, apex acute or obtuse. Flowers solitary and axillary; pedicel 1–1.5 cm long, sparsely puberulent, bracteoles 2 at base, 1–2 mm long, sometimes caducous. Calyx tube campanulate, 2–5 mm long, puberulent; calyx lobes linear-lanceolate, 3–7 mm long, ca. 1 mm broad, ciliate. Corolla purple, pale purple or white, 5–15 mm long, 2-lipped, lobes of upper lip obovate-oblong, those of lower lip broadly elliptic, abaxially sparsely villous along middle ribs. Filaments connate into a tube above 2/3; anther tube 1–1.5 mm long, abaxially puberulent, anthers all barbate at top. Capsule obconical to oblong, 5–7 mm long, 2–4 mm in diameter, obviously veined. Seeds 3-angular, red-brown. Figure 47–77: a–c.

Phenology: flowering and fruiting all the year round.

Habitats: by water and stream in ravines; below 1500 (–2000) m.

Distribution: C Nepal, Sikkim & Darjiling, Yarlung Zangbo-Brahmaputra, and U Irrawaddy (Map 47–170); also in other parts of China (Fujian, Guangdong, Guangxi, Taiwan, S Yunnan), other parts of India and Myanmar, in Indonesia, Malaysia, Phi-lippines, Sri Lanka, and Thailand; Papua New Guinea.

C NEPAL, **Narayani**, *DNEP2 B84* (E). **SIKKIM & DARJILING**: **Darjiling**, 1923, *J. M. Cowan s. n.* (K); Phansidowa, 1876, *C. B. Clarke 31797A* (K); **Sikkim**, Terai, 150 m, 1884, *C. B. Clarke 36625B* (K); Chakoong, 914–1830 m, 1845, *J. D. Hooker* s. n. (BM, K). **YARLUNG ZANGBO-BRAHMAPUTRA**: **Mêdog**, Xiran, 600 m, 1992, *H. Sun et al. 1591* (KUN); **West Siang** Distr., 850 m, 1993, *K. Shawe S. 18* (E). **U IRRAWADDY**: **Ngawchang Valley**, Mankyi, 914–1220 m, 1918, *F. Kingdon-Ward 3027* (E).

2. Lobelia alsinoides Lam., Dict. Bot. 3(2): 588. 1792; Lian in Hong, Fl. Reipubl. Popularis Sin. 73 (2): 145. 1983; Hong & Lammers in Wu, Raven & Hong, Fl. China 19: 557. 2011. Type: s. loc., *Poiret* s. n. (holotype: P).

Lobelia hancei H. Hara, J. Jap. Bot. 17: 23. 1941. ≡ *Lobelia alsinoides* subsp. *hancei* (H. Hara) Lammers, Bot. Bull. Acad. Sin. 33: 286. 1992, **syn. nov.** Type: China, near Guangzhou (circa Canton), *Hance 634* (collected by Samson, Aug. 1869) (holotype: A).

Lobelia trigona Roxb., Fl. India 2: 111. 1824. ≡ *Dortmannia trigona* (Roxb.) Kuntze, Revis. Gen. Pl. 2: 380. 1891. Type: East India, *Rox-*

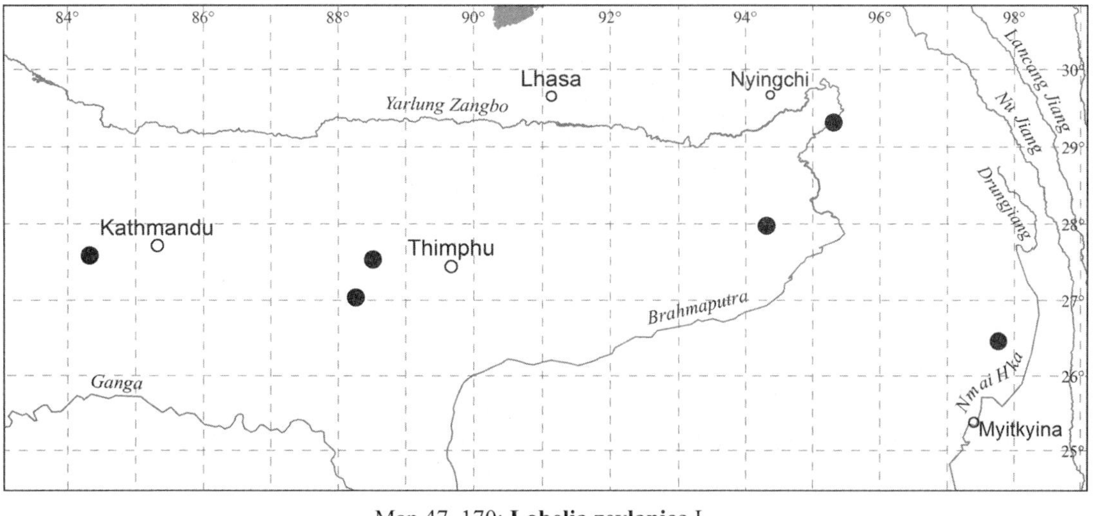

Map 47–170: **Lobelia zeylanica** L.

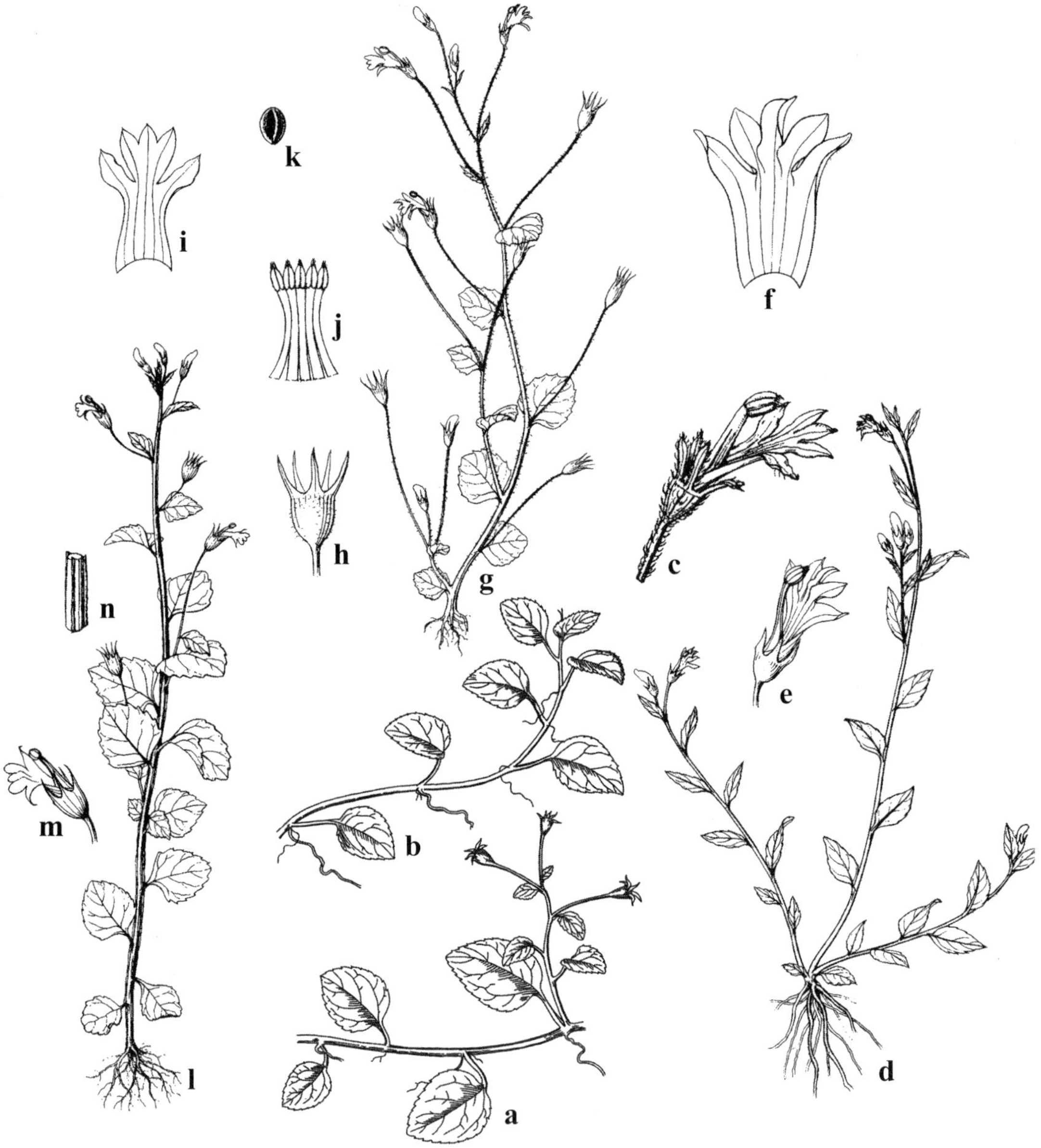

Figure 47–77: **a–c**, **Lobelia zeylanica** L.: **a** & **b**, plant, **c**, flower. **d–f**, **Lobelia alsinoides** Lam.: **d**, plant; **e**, flower; **f**, opened corolla. **g–k**, **Lobelia terminalis** C. B. Clarke: **g**, plant; **h**, fruit; **i**, opened corolla; **j**, opened tube of connate stamens; **k**, seed. **l–n**, **Lobelia heyneana** Roem. & Schult.: **l**, plant; **m**, flower; **n**, a section of winged stem. **a** & **b** redrawn by Ai-Li LI from *Icon. Cormophyt. Sin.* 4, 1975; **c** from *Fl. China* 19, 2011; **d–n**, from *Fl. Reipubl. Popularis Sin.* 73(2), 1983.

burgh s. n. [holotype: K (K000781062)].

Herbs annual, succulent. Stems prostrate to ascending, 20–30 cm tall long, robust, angular, laxly branched, glabrous. Leaves spirally alternate, sparse; petiole 1–3 mm long, glabrous; leaf blade suborbicular to ovate-lanceolate, 1–1.8 cm long, 0.8–1.8 cm broad, both surfaces scabrous but glabrous, base shallowly cordate, margin serrate or obscurely toothed, apex rounded or acute. Flowers in a lax raceme, subtended by leaf-like bracts; pedicel (1.5–) 2–2.7 cm long, slender, glabrous, base with 2 lanceolate bracteoles ca. 3 mm long. Calyx tube funnelform-campanulate, glabrous, 2–3 mm long; calyx lobes linear-lanceolate, entire, glabrous, slightly longer than tube at anthesis, but shorter than tube in fruit. Corolla pale blue or pink, 4–5 mm long, 2-lipped, upper lobes oblong-oblanceolate, upright, lower lobes oblong-elliptic, patent. Stamens connate above middle of filaments; filament-tube glabrous; anther tube ca. 1 mm long, dorsally glabrous, all anthers barbate at top. Capsule ellipsoid, ca. 5 mm long, 3–4 mm in diameter. Seeds many, 3-angular, dark brown. Figure 47–77: d–f.

Phenology: flowering and fruiting all the year round.

Habitats: wet places, paddy fields, by water, or among wet grasses in forests; 250–2400 m.

Distribution: W, C and E Nepal, and Sikkim & Darjiling (Map 47–171); also in other parts of China (Guangdong, Guangxi, Hainan, Taiwan, and S Yunnan), India and Myanmar; in Bangladesh, Indonesia, S Japan (RyuKyu) Malaysia, Philippines, Sri Lanka, Thailand, and Vietnam; Papua New Guinea.

Notes: Lian (1983) recognized two species, *L. alsinoides* and *L. hancei*, while Hong & Lammers (2011) treated them as two subspecies under *L. alsioides*. However, I found that the differences between the two taxa as Hong & Lammers indicated are obscure, and their distribution are sympatric, with no ecological or geographical differentiation. Thus I here merge them into a single species without infraspecific division.

W NEPAL: **Rapti**, Rukum, 2440 m, 1954. *J. D. A. Stainton et al. 4773* (E). **C NEPAL**: **Lumbini**, Nawalparasi, 250 m, 1996, *D. R. Dangol 12190* (W). **E NEPAL**: **Mechi**, Taplejung, Dumhan, by Tamur River, 700 m, 1963, *G. Murata et al. 6302811* (K). **SIKKIM & DARJILING**: **Darjiling**, 1923, *J. M. Cowan* s. n. (K).

3. Lobelia terminalis C. B. Clarke in Hooker, Fl. Brit. India 3: 424. 1881. ≡ *Dortmannia trigona* (Roxb.) Kuntze var. *terminalis* (C. B. Clarke) Kuntze, Rev. Gen. 2: 380. 1891. Type: Bengal, Bhyadanga, W Mymensingh, 20 Nov. 1868, *C. B. Clarke 8044* (holotype: K).

Lobelia thorelii E. Wimm., Repert. Spec. Nov. Regni Veg. 26: 3. tab. 71, fig. 3. 1929. Type:

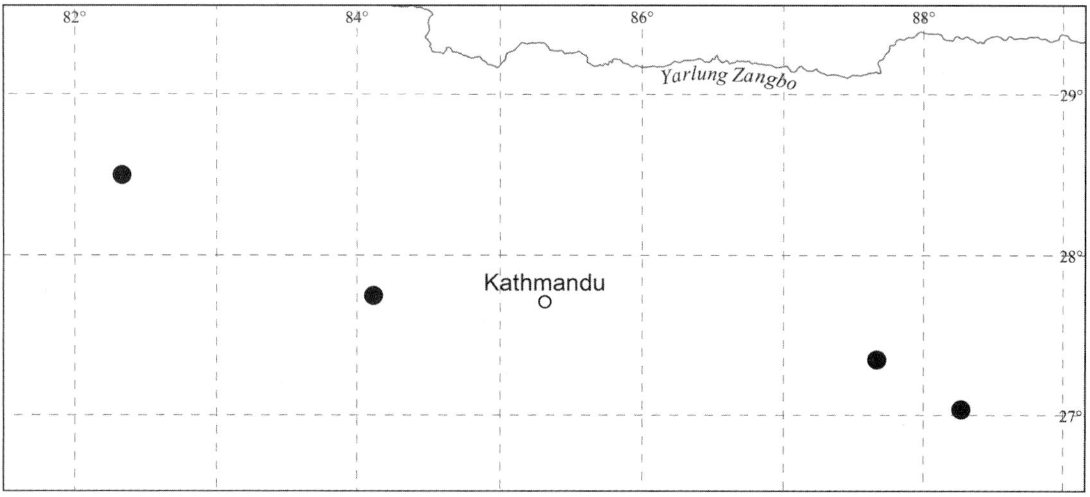

Map 47–171: **Lobelia alsinoides** Lam.

"Expedition du ME-KONG", 1866–1868, *Thorel* s. n. (holotype: P).

Lobelia terminalis var. *minuta* C. B. Clarke in Hooker, Fl. Brit. India 3: 424. 1881. Type: India, Darjiling, Terai, Phansidowa, 150 m, 17 Dec. 1876, *C. B. Clarke 31778* (holotype: K).

Herbs annual, slender. Stems ascending or erect, 10–20 (–40) cm tall, branched, angular but not winged, puberulent. Leaves spirally and laxly arranged; petiole 1–3 mm long, hairy; leaf blade nearly orbicular to elliptic, 6–10 (–15) mm long, 6–10 mm broad, both surfaces puberulent, base truncate, subcordate or rounded, margin nearly entire at lower part, but obviously serrate at upper part, apex rounded. Flowers in a lax terminal raceme, subtended by leaf-like bracts, lower ones by normal leaves; bracts ovate-lanceolate, margin serrate, hairy; pedicel slender, 1.5–4 cm long, sparsely puberulent; bracteoles 2, persistent or caducous. Calyx tube obovoid or semiglobose, ca. 2 mm long, puberulent; calyx lobes linear-subulate, 2–3 mm long, ca. 0.5 mm broad, entire. Corolla pale purple, 2-lipped, 4–5 (–8) mm long, glabrous; upper lobes oblong-linear, upright, lower lobes oblong, slightly shorter than upper ones. Stamens connate above middle of filaments into a tube; filament-tube glabrous; anther tube ca. 1 mm long, abaxially sparsely villous, all anthers shortly barbate at apex. Capsule obovoid, 4–5 mm long, ca. 3 mm in diameter, sparsely puberulent. Seeds many, brown-yellow, 3-angular. Figure 47–77: g–k.

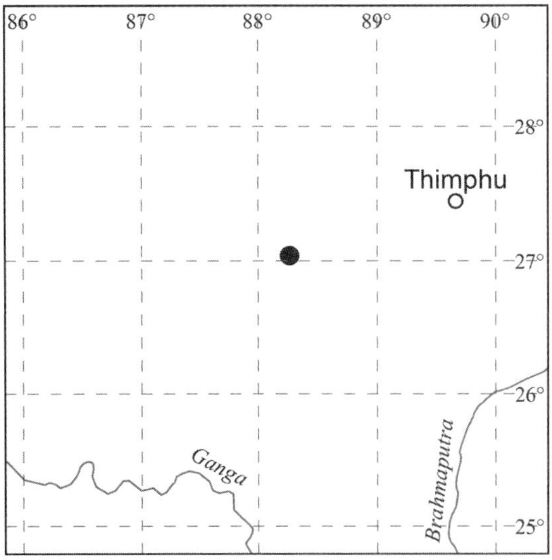

Map 47–172: **Lobelia terminalis** C. B. Clarke

Phenology: flowering in November.

Habitats: wet places in forests; 150–900 m.

Distribution: Sikkim & Darjiling (Map 47–172); also in S Yunnan of China, Laos, Thailand, and Vietnam.

Note: There is only one record in our region and that is the type of *Lobelia terminalis* var. *minuta* C. B. Clarke.

4. Lobelia heyneana Roth **ex** Roem. & Schult., Syst. Veg. 5: 50. 1819. Type: not seen.

Lobelia trialata Ham. ex D. Don, Prodr. Fl. Nepal. 157. 1825. Type: s. loc., *Hamiltan* s. n. (holotype: BM)

Herbs annual. Stems erect, 10–50 (–60) cm tall, branched, 3-winged, glabrous; Leaves alternate; petiole 1–4 mm long, winged; leaf blade broadly deltoid-ovate, 0.8–2.5 cm long, 0.4–2.5 cm broad; upper ones smaller and narrower, adaxially sparsely puberulent, abaxially glabrous, base truncate or shallowly cordate, often decurrent, margin dentate, apex rounded or acute. Flowers in a poorly defined raceme, subtended by leaf-like bracts; upper bracts narrowly elliptic, shorter than pedicel; pedicel 1–1.5 (–2) cm long, slender, with 2 bracteoles at base; bracteoles subulate, less than 1 mm long, often caducous. Calyx tube campanulate, 1–2 mm long, glabrous; calyx lobes longer than calyx tube at anthesis, but shorter than tube after anthesis, subulate, 2–3 mm long, entire. Corolla pale purple, 3–5 mm long, sparsely puberulent inside; limb 2-lipped, upper lobes linear, uptight, less than 1 mm long; lower lobes elliptic to nearly orbicular, spread, ca. 2 mm long. Stamens connate above middle of filaments into tube; anther tube ca. 1 mm long, lower two anthers barbate at apex. Capsule oblong or obovoid-oblong, 4–5 mm long, 2–4 mm in diameter. Seeds pale red-brown, slightly compressed, ellipsoid, smooth. Figure 47–77: l–n.

Phenology: flowering and fruiting all year around.

Habitats: Moist places; 500–2700 m.

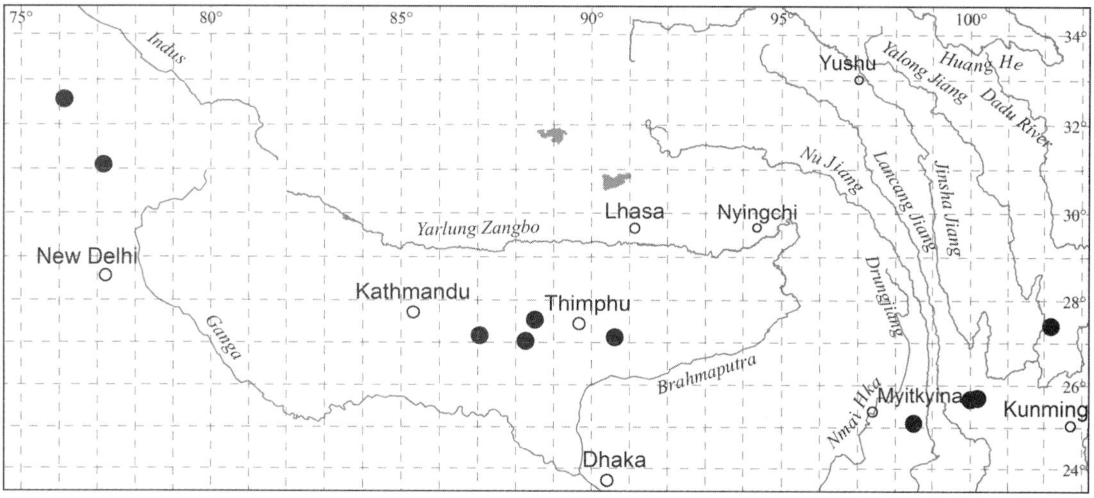

Map 47–173: **Lobelia heyneana** Roth ex Roem. & Schult.

Distribution: U Ganga & Indus, E Nepal, Sikkim & Darjiling, Bhutan, S Hengduan, and U Irrawaddy (Map 47–173); also in the area from Sri Lanka to SE Asia, Papua New Guinea, and tropical Africa.

U GANGA & INDUS: **Chamba**, 400 m, 1880, *J. R. Drummond 1332* (K); 914 m, 1874, *C. B. Clarke 23733A* (K); **Shimla**, 1520 m, 1884, *J. R. Drummond 2750* (K); Shimla, 1520 m, 1885, *H. Collett 640* (K). **E NEPAL**: **Kosi**, Bhojpur, 1580 m, 1995, *M. Mikage et al. 9558261* (E). **SIKKIM & DARJILING**: **Darjiling**, Sumbung, 1876, *J. S. Gamble 2037A* (K); **Sikkim**, 915–1520 m, *J. D. Hooker* s. n. (W); Sikkim, 1520 m, 1849, *J. D. Hooker* s. n. (K). **BHUTAN**, **Kabiuhung** (Kalimhung?), 1372 m, 1879, *J. S. Gamble 7311* (K). **S HENGDUAN**: **Dali**, 1800 m, 1915, *Handel-Mazzetti 8989* (E); **Dechang**, 1300 m, 1914, *Handel-Mazzetti 5623* (W); **Yangbi**, 1400–1800 m, 1916, *Handel-Mazzetti 13028* (W). **U IRRAWADDY**: **Tengchong**, 1830–2130 m, 1903, *G. Forrest 1009* (E).

5. Lobelia pyramidalis Wall., Acta Soc. Asiat. 13: 376. 1820; Lian in Hong, Fl. Reipubl. Popularis Sin. 73(2): 159, 1983. Type: Nepal, *Wallich 1302* (**lectotype here designated**: BM; isolectotypes: P).

Rapuntium wallichiana C. Presl, Prodr. Monogr. Lobel. 23, 24. 1836. ≡ *Lobelia pyramidalis* var. *wallichiana* (C. Presl) Steud., Nomencl. Bot. ed. 2, 2: 62. 1841. ≡ *Lobelia wallichiana* (C. Presl) Hook. f. & Thomson, J. Proc. Linn. Soc. Bot. 2: 29. 1857. Type: India, Khasia, *Wallich,* not seen.

Herbs, 1–2.5 m tall, much branched above. Stems glabrous except for setose inflorescence raches. Leaves alternate, nearly leathery, sessile or shortly petiolate; basal leaves spatulate, lower cauline leaves oblong, up to 25 cm long, middle and upper cauline leaves narrowly lanceolate, 13–25 cm long, 2.5–4 cm broad, both surfaces glabrous, base broadly cuneate, margin densely minute-serrate, apex acuminate. Flowers in a terminal panicle of racemes; racemes one-sided, rachis glabrous or minutely setose; flowers dense; pedicel 1–2.4 cm long, glabrous or setulose, arched; bracts linear, entire, often shorter than flowers; bracteoles usually absent, but sometimes 1 or 2, various in position,. Calyx tube shortly cylindrical, 5–7 mm long, 4–6 mm wide, glabrous or puberulent; calyx lobes lanceolate-linear, 12–30 mm long, 1–3 mm broad, margin entire. Corolla white, rose or bluish, 2.5–3 cm long, glabrous outside, densely villous inside corolla tube, nearly 2-lipped; lobes of upper lip linear, 2/3 as long as total length of corolla; those of lower lip ovate-lanceolate, 1/3 as long as corolla. Stamens connate above base into tube; filament-tube glabrous; anther tube ca. 5 mm long, densely villous along sutures, lower two anthers barbate at apex. Capsule nearly globose, 6–8 mm in diame-

Figure 47–78: **a–c**, **Lobelia pyramidalis** Wall.: **a**, a section of stem; **b**, opened corolla; **c**, a flower with corolla removed, showing calyx, stamens, and stigma. **d–h**, **Lobelia foliiformis** T. S. Zhang & D. Y. Hong: **d**, upper part of stem; **e**, leaf; **f**, opened corolla; **g**, a flower with corolla removed, showing calyx, stamens, and stigma; **h**, a calyx lobe. **a–c** from *Fl. Reipubl. Popularis Sin.* 73(2), 1983; d–i from *Acta Phytotax. Sin.* 30(2): 156. fig. 19. 1992, with permission.

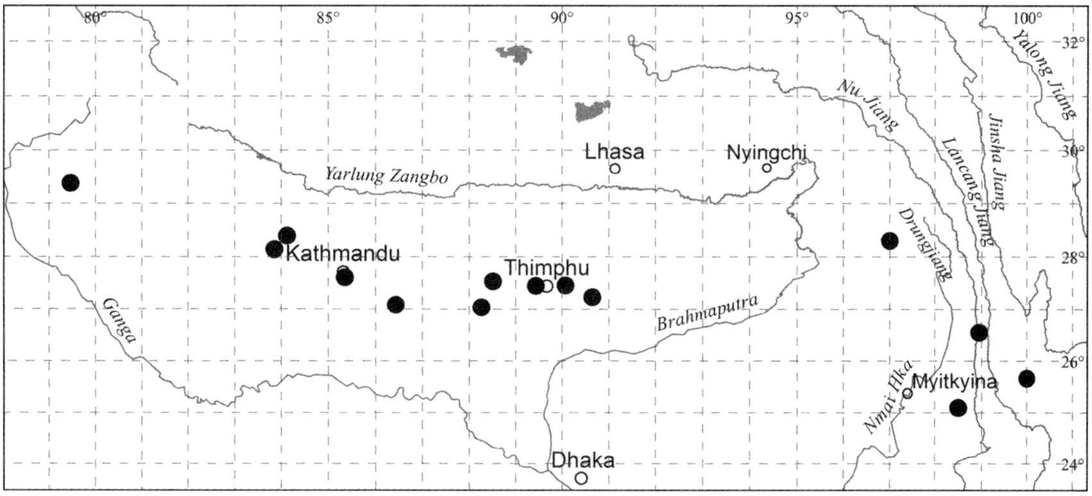

Map 47–174: **Lobelia pyramidalis** Wall.

ter, glabrous, often pendulous. Seeds with pale margin, oblong, obviously compressed. Figure 47–78: a–c.

Phenology: flowering from January to May; fruiting from June to October.

Habitats: grassy slopes, scrubs, or roadsides; below 1900 m.

Distribution: U Ganga & Indus, C Nepal, Sikkim & Darjiling, Bhutan, Yarlung Zangbo-Brahmaputra, S Hengduan, and U Irrawaddy (Map 47–174); also in SW China, Cambodia, Laos, Thailand, and Vietnam.

U GANGA & INDUS: **Nainital**, 1520 m, *R. Strachey & J. Winterbottom 1* (K, P); **C NEPAL**: **Bagmati**, Rasuwa, 2500 m, 2001, *ENEP 67* (E); **Kathmandu**, Godavari, 1680 m, 1974, *J. D. A. Stainton 7315* (E); Naudanda, 1800 m, 1980, *W. Schwabe* s. n. (B). **SIKKIM & DARJILING**: **Darjiling**, 2130 m, 1875, *C. B. Clarke 26751* (K); **Sikkim**, s. loc., 1520–2130 m, *J. D. Hooker* s. n. (K). **BHUTAN**: **Tongsa**, Shamgong, 1980 m, 1982, *A. J. C. Grierson & D. G. Long 4237* (E, K); **Thimphu-Dochu La**, 1967, *H. Hara et al. 10324* (E); **Wangdu**, Phodrang-Pele La, 23 Km E of Wangdu, 2000 m, 1975, *A. J. C. Grierson & D. G. Long 528* (E). **YARLUNG ZANGBO-BRAHMAPUTRA**: **Xiazayü**, Chibagou, 1500 m, 2012, *FLPH Tibet Exped. 12–1305* (PE). **S HENGDUAN: Bijiang**, 1942, *H. C. Wang 1867* (KUN); **Yangbi**, Ziyang, 2300 m, 1981, *Sino-Brit. Cangshan Pl. Exped. (S. B. E. C.) 0453* (KUN). **U IRRAWADDY**: **Tengchong**, S of city, 1580 m, 1981, *S. W. Yu & Q. T. Zhang 556* (KUN).

6. Lobelia foliiformis T. J. Zhang & D. Y. Hong, Acta Phytotax. Sin. 30(2): 155. 1992; Hong & Lammers in Wu, Raven & Hong, Fl. China 19: 560. 2011. Type: China, Yunnan, Dali, 2300–3000 m, 29 Aug. 1989, *T. J. Zhang 0026* (holotype: PE).

Herbs or subshrubs, 1–2 m tall, branched, glabrous. Cauline leaves sessile or shortly petiolate; leaf blade oblong to broad-lanceolate, papery, up to 25 cm long, 3.5 cm broad, adaxially glabrous or sparsely puberulent, abaxially often puberulent, base cuneate, margin glandular-serrate, apex acuminate. Racemes terminal, densely flowered, 5–20 cm long, often pendulous; bracts linear-lanceolate, usually entire; pedicel ca. 5 mm long, with 2 bracteoles at top; bracteoles leafy, 6–12 mm long, ca. 1 mm broad, margin glandular-denticulate. Calyx tube obconical, ca. 6 mm long, 4 mm wide, glabrous to puberulent; calyx lobes linear, upright or spreading, 12–16 mm long, usually denticulate. Corolla blue-purple or purple-red, 2–3 cm long, puberulent; upper two lobes ca. 1.7 cm long; lower three lobes ovate-lanceolate, ca. 8 mm long. Filaments 12–15 mm long, puberulent below; anther tube gray-blue, ca. 5 mm long, dorsally densely setulose; lower two anthers sparsely barbate at apex. Capsule ellipsoid, ca. 1.2 cm long, 0.8–1 cm in diameter. Seeds dark brown, ellipsoid. Figure 47–78: d–h.

Phenology: flowering and fruiting in August.

Habitats: sunny slopes; 2300–3000 m.

Distribution: **endemic**; confined to S Hengduan (Map 47–175). No additional collection has been seen.

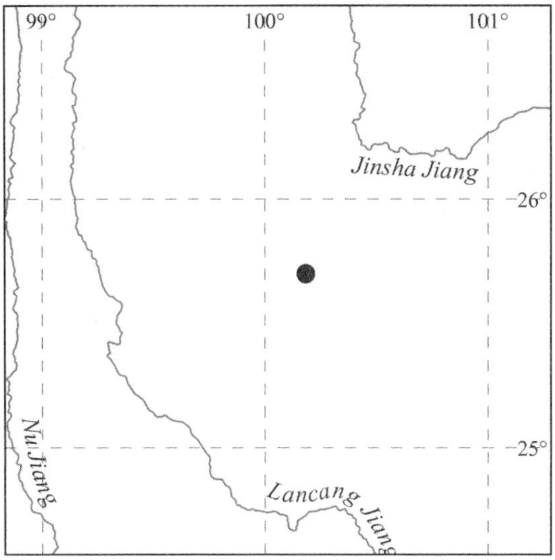

Map 47–175: **Lobelia foliiformis** T. J. Zhang & D. Y. Hong

7. Lobelia seguinii H. Lév. & Vaniot, Repert. Spec. Nov. Regni Veg. 12: 186. 1913. Type: China, Guizhou (Kouy-Tchéou), near Hang-Ko-Chou, 8 Oct. 1898, *J. Séguin* s. n. (**lectotype here designated**: E); s. loc., 31 Mar. 1904, *J. Esquirol* s. n. (paratype: E).

Herbs, annual or perennial, 1–2.5 (–5) m tall, much branched. Stems glabrous. Leaves alternate, papery, lower leaves long-petiolate, leaf blade long-oblong, up to 25 cm long; middle and upper cauline leaves sessile or short-petiolate, lanceolate, 6–28 cm long, 1.2–5.5 cm broad, both surfaces glabrous, base attenuate, margin serrate or double-serrate, apex acuminate. Racemes terminal, densely flowered, one-sided. Lower bracts linear-lanceolate, longer than flowers, margin denticulate; upper ones linear, shorter than flowers, margin entire; pedicel 5–8 mm long, slightly compressed, curved backwards, with 2 linear bracteoles ca. 1 mm from top. Calyx tube obovoid-oblong to obconical, 5–8 mm long, glabrous; calyx lobes linear-lanceolate, (8–) 16–20 (–25) mm long, 1.5–2 mm broad, glabrous, entire. Corolla purple red, purple-blue or pale blue, 2.5–3 (–3.5) cm long, densely villous below throat inside; upper lobes linear, 2/3 as long as total length of corolla, ca. 1 mm broad, upright or spreading; lower ones lanceolate, 1/2 as long as total length of corolla, spreading. Filament-tube nearly as long as corolla tube, glabrous except base; anther tube 5–7 mm long, base with several tufts of short hairs, dorsally glabrous, lower 2 anthers barbate at apex. Capsule cylindrical, 10–12 mm long, 5–7 mm in diameter, glabrous, pendent. Seeds oblong, foveolate.

Phenology: flowering and fruiting from August to October.

Habitats: grassy slopes, forest margins, or roadsides; 500–3000 m.

Distribution: S Hengduan and U Irrawaddy (Map 47–176); also in C and SW China (Guangxi, Guizhou, W Hubei, and other parts of Sichuan and Yunnan).

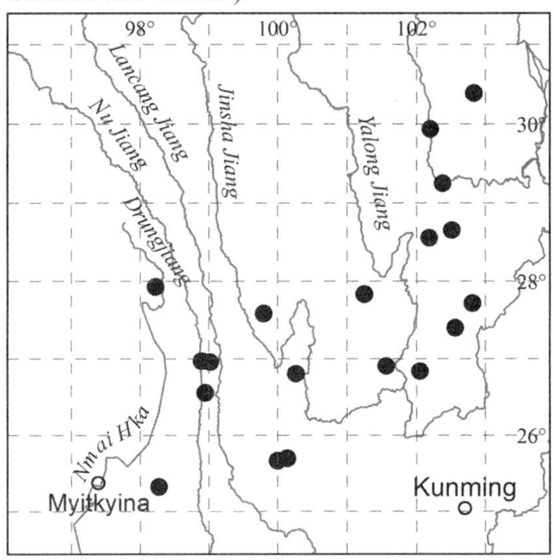

Map 47–176: **Lobelia seguinii** H. Lév. & Vaniot

S HENGDUAN: **Baoxing**, s. loc., 1936, *K. L. Chu 3786* (PE); Gaotizi, 1000 m, 1933, *T. H. Tu 4923* (PE); **Bijiang**, 1500 m, 1933, *H. T. Tsai 54684* (KUN); **Butuo**, 1200 m, 1959, *Sichuan Econ. Plant. Surv. (Liang) 5715* (KUN, PE); **Dali**, Cangshan, 2000 m, 1983, *Sino-Amer.*

Exped. 0007 (KUN); Cangshan, 1982, *H. Koyama et al. 876* (KUN); **Fugong**, 2800 m, 1934, *H. T. Tsai 58659* (KUN); Fugong, 2000 m, 1933, *H. T. Tsai 54337* (PE); **Lanping**, Biluoxue Shan, 2400–3000 m, 1964, *S. K. Wu 9030* (KUN); **Luding**, Luding to Qingbapo, 1951, *W. K. Hu & Z. He 11511* (PE); **Mianning**, Jinkuan, 1400 m, 1960, *J. S. Ying 4277* (PE); **Miyi**, Puwei, 1934 m, 2007, *J. Liu & M. Yuan My-198* (KUN); **Muli**, Shawan, 3000 m, 1937, *T. T. Yü 14191* (KUN, PE); Mosogou, 1200–2280 m, 1959, *S. K. Wu 2397* (KUN); **Puge**, 1800 m, 1959, *Sichuan Econ. Plant Surv. (Liang) 5597* (KUN, PE); **Shangri-la** (Zhongdian), 1939, *K. M. Feng 2249* (KUN, PE); **Shimian**, Wajiao, 1400 m, 1993, *Z. C. Zhao 2986* (PE); Tianba, 1600 m, 1993, *Z. C. Zhao 2859* (PE); **Yanbian**, 1380 m, 1959, *S. K. Wu 140* (KUN); **Yangbi**, Madi, W slope of Cangshan, 2860 m, 1980, *K. M. Feng 80–51* (KUN); Yangbi Farm, 1550 m, 1984, *H. Hara & K. Y. Guan* s. n. (KUN); **Yuexi**, Lizhou, 1930, *Z. P. Huang 2037* (PE); **Yulong** (Lijiang), Dadong Valley, 1550 m, 1985, *H. Y. Yu & H. Sun 77/008* (KUN); Yangtze Bend, 2440–2740 m, 1913, *G. Forrest 11171* (PE). **U IRRAWADDY**: **Dandanghe**, 2200 m, 1991, *Drungjiang Exped. 3278* (KUN); **Tengchong**, Guyong, 1850 m, s. d., *S. K. Wu 6675* (KUN).

8. Lobelia doniana Skottsberg, Acta Horti. Gothob. 4: 19. 1928; Lian in Hong, Fl. Reipubl. Popularis Sin. 73(2): 159. 1983. ≡ *Lobelia seguinii* H. Lév. & Vaniot var. *doniana* (Skottsberg) E. Wimm., Pfl. R. Heft. 107: 651. 1953, Type: Nepal, not seen.

Herbs perennial, up to 1.5 m tall, branched above. Stems erect, up to 1.5 cm in diameter, mostly glabrous, rarely hispidulous above. Cauline leaves alternate, at least lower ones petiolate, petiole cuneate-winged, up to 3 cm long; leaf blade elliptic to linear-elliptic, 8–33 cm long, 1–6 cm broad, adaxially glabrous, abaxially glabrous or sparsely hispidulous, base cuneate, margin minute-serrate, apex acuminate. Racemes terminal on main stem and branches; lower bracts leaf-like, gradually smaller upwards along stem; pedicel 5–15 mm long, usually densely hispidulous; bracteoles 2, at middle part of pedicel, filiform, minute. Calyx tube obconical, densely or sparsely scaberulose; calyx lobes subulate, 1.1–2.2 cm long, with 1 or 2 minute-verrucose teeth on each side. Corolla 2–2.5 cm long, purple, purple-red or blue-purple, scaberulose below outside, villous inside, 2-lipped; upper 2 lobes filiform-linear, slightly longer than lower 3. Filaments ciliate, connate above base, filament-tube glabrous; anther tube white hirsute only dorsally; lower two anthers barbate at top. Capsule obovoid, 8–12 mm long, 5–7 mm in diameter. Seeds ovoid, short-striate.

Phenology: flowering in September and October; fruiting in October and November.

Habitats: grassy slopes, forest margins, or glades; (800–) 1400–3200 m.

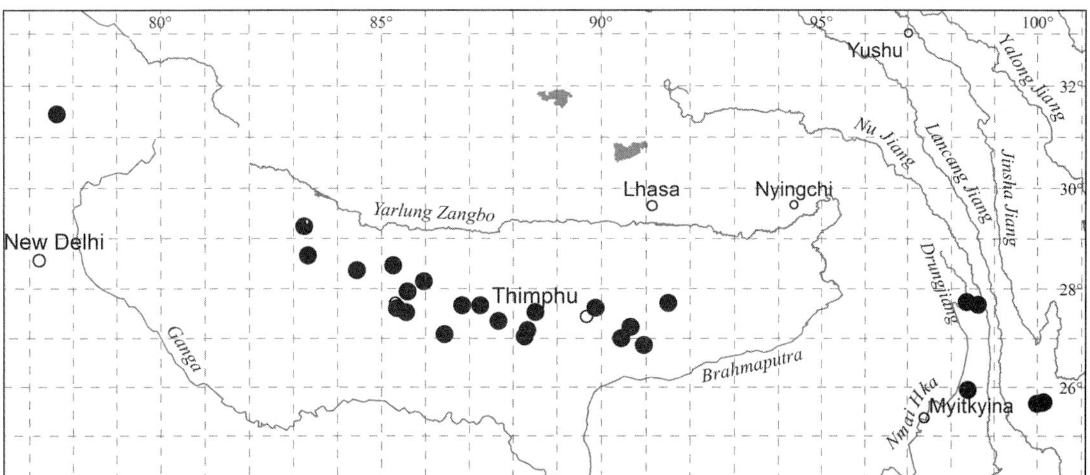

Map 47–177: **Lobelia doniana** Skottsberg

Distribution: **endemic**; U Ganga & Indus, U Yarlung Zangbo, W, C, and E Nepal, Sikkim & Darjiling, Bhutan, S Hengduan, and U Irrawaddy (Map 47–177).

U GANGA & INDUS: N Garhwal, Rampur, 1938, *K. Rain 8846* (E). **U YARLUNG ZANGBO**: **Gyirong**, Salu Township, 2400–3100 m, 1990, *B. S. Li et al. 13375* (PE); **Nyalam**, Zham, 2500 m, 1972, *Xizang Chinese Medic. Mater. & Herbs Exped. 1476* (PE); loc. eodem, 2200 m, 1981, *C. C. Ni et al. 1987* (PE). **W NEPAL**: **Karnali**, Dolpa, 2990 m, 1991, *M. Minaki et al. 9109220* (E). **C NEPAL**: **Bagmati**, Shermathang, 1979, *W. Schwabe* s. n. (B); Lalitpur & Kabhrepalanchok, 2320 m, 1995, *M. Mikage et al. 9552718* (E) ; Rasuwa, *D. G. Long et al. ENEP 167* (E) ; **Dhawalagiri**, Myagdi, 2430 m, 1996, *M. Mikage et al. 9682553* (E); **Gandaki**, Madi Khola, 2100 m, 1990, *Chadwell & McKelvie 242* (E) Godavari, S of **Kathmandu**, 1620 m, 1965, *A. D. Schilling 687* (K); **E NEPAL**: **Kosi**, Sankhuwasabha, Arun Valley, 2134 m, 1956, *J. D. A. Stainton 1480* (E); **Mechi**, Taplejung, 2350 m, 1989, *KEKE 862* (E); **Sagarmatha**, Solukhumbu, 1995, *F. Miyamoto et al. 9596601* (E). **SIKKIM & DARJILING**, **Darjiling**, 2130 m, 1873, *J. S. Gamble 2315A* (K); **Sikkim**, 2130 m, 1879, *J. S. Gamble 7110* (K). **BHUTAN**: **Denchong**, Kuomachu, 2134 m, 1949, *F. Ludlow et al. 21464* (BM); **Mo chu**, 2377 m, 1967, *S. Bowes Lyon 5182* (BM); **Punakha**, Gichha, 2440 m, 1914, *R. E. Cooper 3144* (E); **Shamgong**, 1985, *L. Broad* s. n. (E); **Tongsa**, 2440 m, 1914, *R. E. Cooper 2291* (E); **Trashi Yangtsi** Valley, Tobrang, 2440 m, 1934, *F. Ludlow & G. Sherriff 930* (E). **S HENGDUAN**: **Dali**, Cangshan, Zhonghe Peak, 2004, *Y. Liu* s. n. (PE); Cangshan, 2300 m, 1989, *T. J. Zhang 0023* (PE); **Gongshan**, Sijitong, Champu-tung, 2800 m, 1935, *C. W. Wang 67432* (PE); **Yangbi**, Halamo, 1946, *T. N. Liou 22859* (PE). **U IRRAWADDY**: **Drungjiang**, Kiukiang Valley, 1500 m, 1938, *T. T. Yü 20417* (PE); **Htawgaw**, 1830 m, 1924, *G. Forrest 25415* (E)

9. Lobelia sichuanensis (Y. S. Lian) D. Y. Hong, **stat. nov.** Basionym: *Lobelia davidii* Franch. var. *sichuanensis* Y. S. Lian in Hong, Fl. Reipubl. Popularis Sin. 73(2): 188. 1983. Type: China, Sichuan, Puge, Luoji Shan, 3000 m, 24 Aug. 1959, *Sichuan Econ. Plant Surv. (Liang 59) 5486* [**lectotype here designated**: PE; isolectotype: PE].

Herbs perennial, 40–100 cm tall or taller, lacking rhizomes. Tap roots single or two, slightly thickened, 8–10 cm long, or longer, 0.5–1.2 cm in diameter, lateral roots fibrous, up to 1 mm wide. Stems yellow-green or purple, erect, simple or branched, white-hispidulous or scaberulose. Leaves sessile, or winged-petiolate; leaf blade elliptic, 7–12 (–25) cm long, 2–4 cm broad, both surfaces white-hispidulous, base cuneate, margin serrate, apex acute or acuminate. Racemes terminal, 15–50 cm long, lower flowers shorter than bracts, middle and upper ones longer than bracts; pedicel 3–4 mm long, always densely white-hispidulous. Calyx tube globose, scaberulose or hispidulous entirely or only along ribs; calyx lobes subulate, 6–11 mm long, 1–1.5 mm broad, hispidulous, ciliate, with 1–5 verrucae or denticles on each side. Corolla 1.5–1.8 cm long, villous below throat, lobes of upper lip linear; those of lower lip elliptic or elliptic-lanceolate. Stamens connate from base, filament tube glabrous or villous near top; anther tube white-setose entirely or only dorsally; lower two anthers barbate at apex. Capsule globose, 6–10 mm in diameter. Seeds yellow-brown, ellipsoid, slightly compressed. Figure 47–79: a–c.

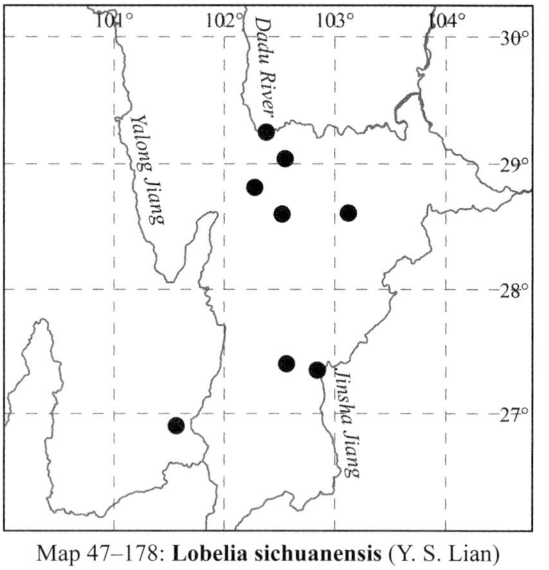

Map 47–178: **Lobelia sichuanensis** (Y. S. Lian) D. Y. Hong

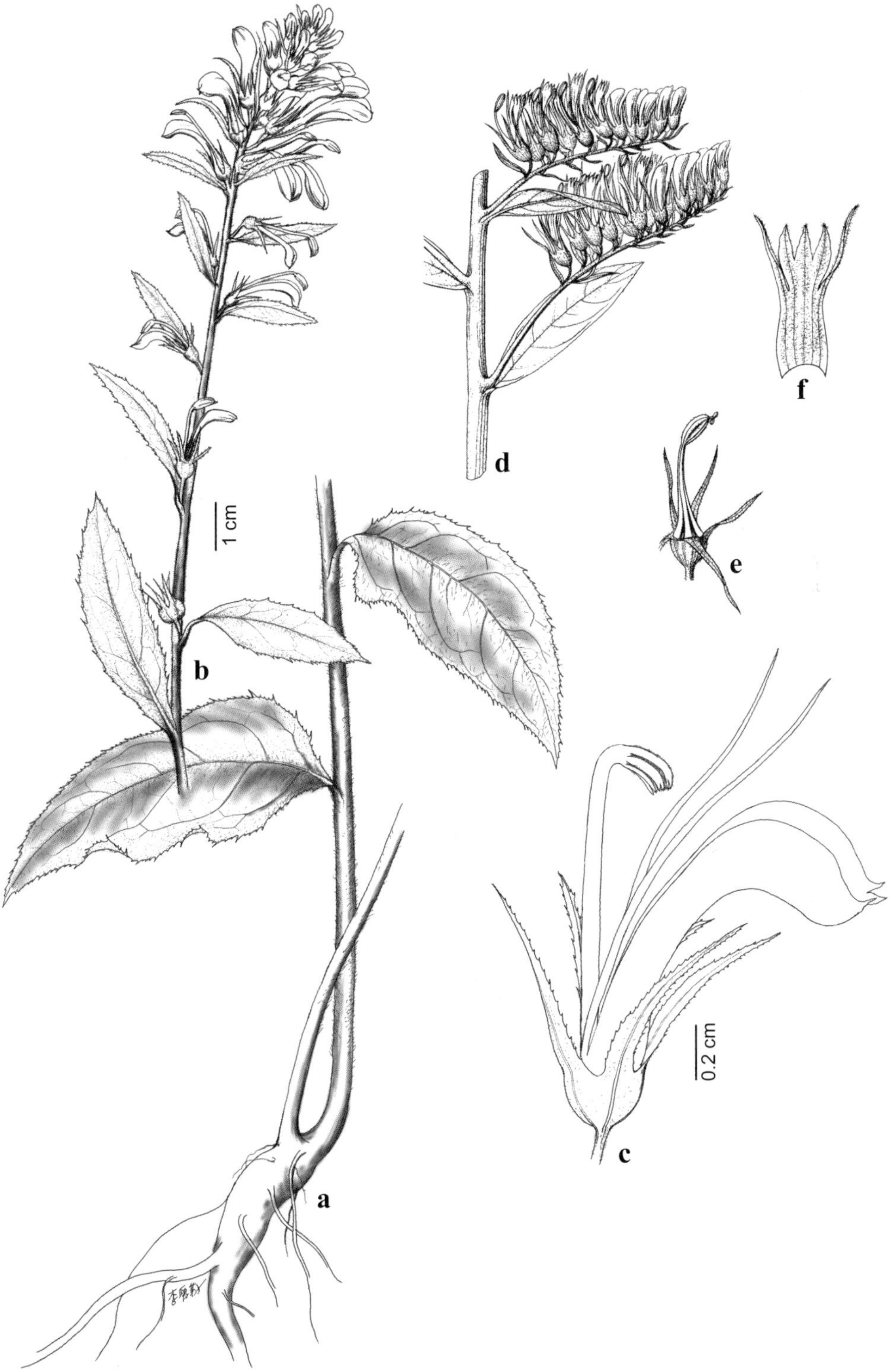

Figure 47–79: **a–c**, **Lobelia sichuanensis** (Y. S. Lian) D. Y. Hong: **a**, root and lower part of stem; **b**, upper part of stem; **c**, flower. **d–f**, **Lobelia rosea** Wall.: **d**, a section of stem; **e**, a flower with corolla removed, showing calyx, stamens and stigma; **f**, opened corolla. **a–c** drawn by Ai-Li LI; **d–f** from *Fl. Reipubl. Popularis Sin.* 73(2), 1983.

Phenology: flowering and fruiting from August to October.

Habitats: edges of forests, forests, or sunny slopes; 2000–3000 m.

Distribution: **endemic**; confined to S Hengduan (Map 47–178).

Notes: 1) Although the author of *Lobelia davidii* var. *sichuanensis*, Lian, designated *Sichuan Econ. Plant Surv.* (*Liang*) *5846* as type which has two sheets in PE, he did not indicate a particular sheet. Thus I here designate lectotype and isolectotype. 2) *Lobelia sichuanensis* is distinct from *L. davidii* in having tap roots but no rhizomes, leaf blade elliptic to long-lanceolate, and anther tube densely villose on back.

S HENGDUAN: **Butuo**, Jiaojihe, 2500 m, 1959, *Sichuan Econ. Plant Surv. (Liang 59) 5891* (PE); **Decang**, Xiaogaoqiao, 2700 m, 1959, *S. F. Zhu 20314* (PE); **Ganluo**, Haitang, 2000 m, 1959, *Sichuan Econ. Plant Surv. (Liang 59)4231* (KUN, PE); **Meigu**, Hongxi, 2500 m, 1976, *Sichuan Plant Surv. 13159* (PE); 1959, *Sichuan Econ. Plant Surv. 1682* (PE); **Mianning**, Tuowu, 2400 m, 1959, *S. F. Zhu 20319* (PE); **Shimian**, Gongyiqiao, 1600 m, 1993, *Z. C. Zhao 2360* (PE); **Yanbian**, Zhemulong, 2700–3000 m, 2002, *X. F. Gao et al. 3357* (PE). **Yuexi**, Dongshan, 1900 m, 1959, *Sichuan Econ. Plant Surv. (Liang) 3636* (KUN).

10. Lobelia rosea Wall., Pl. Asiat. Rar. 2: 42, t. 152. 1831; Clarke in Hooker, Fl. Brit. India 3: 427. 1881. Type: Wallich's tab. 152 of protologue **(lectotype here designated).**

Lobelia clavata E. Wimm., Repert. Spec. Nov. Regni Veg. 38: 78. 1935, **syn. nov**. Type: China, Yunnan, Simao, *A. Henry 12663* [**lectotype here designated**: BM (BM000996386); isolectotype: BM].

Herbs 1.5–3 m tall, much branched. Stems terete, up to 1 cm in diameter, tomentose. Cauline leaves alternate, sessile; leaf blade oblong or elliptic, 15–28 cm long, 3–8 cm broad, thickly papery, both surfaces densely hispidulous, base broadly cuneate to nearly rounded, margin densely denticulate, short-ciliate, apex acute. Racemes one-sided, glomerate into a panicle; flowers dense, oriented towards one side, upright; involucral bracts leaf-like, 5–7 cm long, 2–3 cm broad, margin double-serrate, ciliate; bracts lanceolate-linear, 10–15 mm long, 1–3 mm broad, shortly pannose; pedicel 5–10 mm long, terete, pannose, twisted towards posterior side; bracteoles often absent or 1, minute, near base. Calyx densely pannose; calyx tube semiglobose, base rounded; calyx lobes lanceolate-linear, 11–15 mm long, 1–1.5 mm broad, margin entire, sometimes ciliate. Corolla white or pink, or tube white but limb pale mauve, 2–3.2 cm long, pannose outside, villous inside, rarely 2-lipped; upper lobes linear, ca. 2/3 as long as total length of corolla; lower

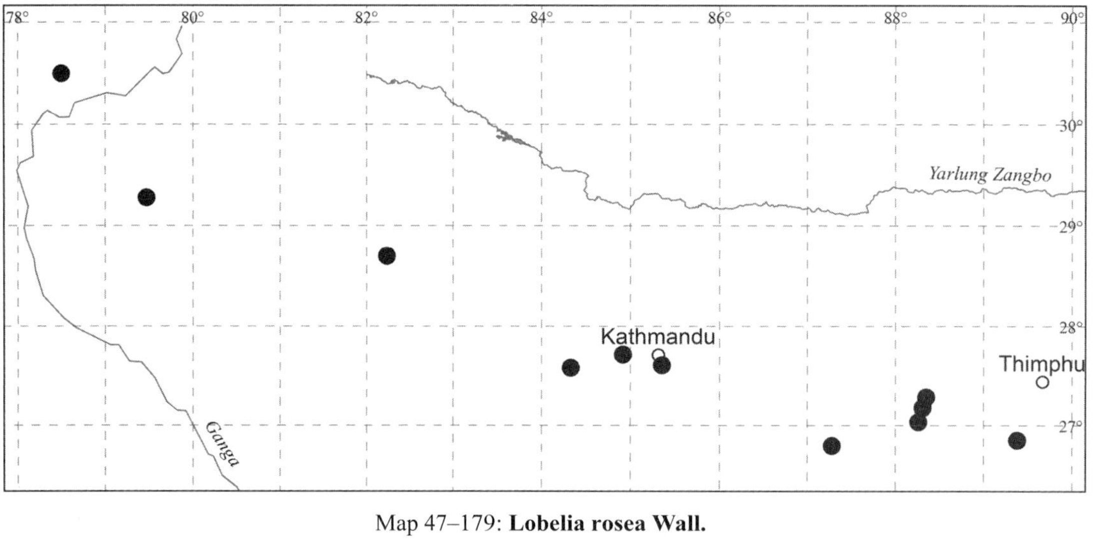

Map 47–179: **Lobelia rosea Wall.**

lobes ovate-lanceolate, 1/3 as long as total length of corolla. Stamens connate into tube above base; filament-tube densely puberulent; anther tube 5–6 mm long, densely villous along sutures; lower two anthers barbate at apex. Capsule pendulous, nearly globose or shortly cylindrical, 6–9 mm long, 5–7 mm in diameter, densely puberulent. Seeds ellipsoid, slightly compressed, smooth, margin pale in colour. Figure 47–79: d–f.

Phenology: flowering and fruiting from December to April.

Habitats: grassy slopes, fo-rests, or roadsides; below 1900 m.

Distribution: U Ganga & Indus, W, C, and E Nepal, Sikkim & Darjiling, and Bhutan (Map 47–179); also in other parts of Myanmar (Shan States) and China (SW Guizhou and S Yunnan).

Notes: Wallich's collections are mostly preserved in K or BM, rarely in CAL and C. I visited the former three herbaria and examined all the collections there but I failed to find the type specimen. Thus I designate here Wallich's illustration as type.

U GANGA & INDUS, Garwhal, Nagpur, 1520–2130 m, 1844, *M. P. Edgeworth 238* (K); **Kumaon**, W Nanirk, 2290 m, *R. Strachy & J. E. Winterbottom 2* (K). **W NEPAL**: **Bheri**, Jajarkot, 1220 m, 1952, *O. Polunin et al. 3810* (E). **C NEPAL**: **Narayani**, Chitwan, ca. 200 m, 1996, *M. Mikage et al. 9614227* (E); **Kathmandu,** 45 km NE of Kathmandu, ca. 1600 m, 1968, *S. W. U. V. Breckle J64* (W); S of Kathmandu, Godavari, 1620 m, 1965, *A. D. Schilling 687* (K). **E NEPAL**: **Kosi**, Dharan, 610 m, 1969, *L. H. J. Williams 100* (E). **SIKKIM & DARJILING**: **Darjiling**, 2130 m, 1873, *J. S. Gamble 2315A* (K); 2130 m, 1879, *J. S. Gamble 7110* (K); Rungeet, 1876, *C. B. Clarke 27264C* (K); **Sikkim**, Yoksum Bakhim, 1980 m, 1983, *AGSES 329* (K); 1520–2440 m, *J. D. Hooker* s. n. (K); 915–1520 m, *J. D. Hooker* s. n. (K). **BHUTAN**: **Phuntsholing**, 780 m, 1982, *A. J. C. Grierson & D. G. Long 2966* (E).

11. Lobelia gaoligongshanica D. Y. Hong, **sp. nov**. Figure 47–80. Type: China, Yunnan, Longyang Distr., Nujiang Township, 24°56′55″N, 98°45′15″E, 2240–2320 m, 4 Nov. 1998, *H. Li et al. 11518* (holotype: CAS).

Herbs perennial. Underground part unknown. Stems erect, terete, 0.5–2 m tall, densely puberulent, with short branches. Leaves alternate, sessile or shortly winged-petiolate; petiole 2–4 mm long; leaf blade elliptic, 5–11 cm long, 1.5–3 cm broad, those on branches conformable but much smaller, both surfaces hispidulous along veins, base cuneate, margin serrate, apex acuminate. Flowers solitary, axillary on main stem and branches; pedicel 8–12 mm long, densely hispidulous, without bracteoles. Calyx tube obovoid or half-globose, puberulent; calyx lobes 6–8 mm long, ca. 1 mm broad, 3-veined,hispidulous along veins, margin with several pairs of verrucose denticles, hispidulous-ciliate. Corolla lavender, 1.8–2 cm long, densely hispid along ribs outside, hispid along ribs at lower and midde parts inside, two lobes of upper lip linear, ca. 10 mm long, 1.5 mm broad; three of lower lip ca. 7 mm long, 3 mm broad. Stamens connate into tube from base, filament tube ca. 13 mm long, hispidulous along ridges; anther tube 4–5 mm long, sparsely hispid on back, lower two anthers barbate at top. Fruit a capsule, subglobose, 6–8 mm long, 5–7 mm in diameter. Seed yellow, ellipsoid, ca. 0.5 mm long. Figure 47–80.

Phenology: flowering in September; fruiting in November.

Habitats: forested slopes or grassy area, on decomposed granites; 2300–2400 m.

Distribution: **endemic**; S Hengduan and U Irrawaddy (Map 47–180).

Notes: This new species is distinct from its relatives in having stems and both surfaces of leaves hispidulous or puberulent, calyx puberulent and verrucose-denticulate, corolla hispid along ribs outside, and fruit subglobose.

U IRRAWADDY: **Tengchong**, Sanyun, Gaoligong Shan State Nature Reserve, 2300 m, 2003, *Gaoligong Shan Biodiversity Survey 18708* (CAS); Gaoligong Shan State Nature Reserve, 2400 m, 2003, *Gaoligong Shan Biodiversity Survey 18749* (CAS).

Figure 47–80: **Lobelia gaoligongshanica** D. Y. Hong: **a**, middle part of stem; **b**, a flower, showing the indumentum of corolla outside and inside; **c**, capsule. Drawn by Ai-Li LI.

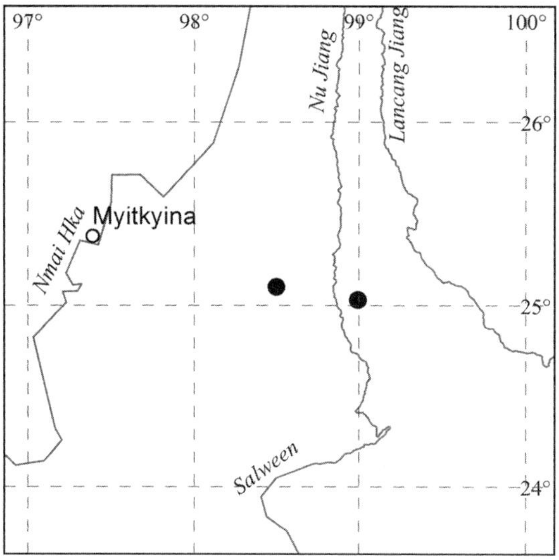

Map 47–180: **Lobelia gaoligongshanica** D. Y. Hong

12. Lobelia taliensis Diels, Notes Roy. Bot. Gard. Edinburgh 5: 170. 1912. Type: China, Yunnan, Dali, 25°40′N, 2900 m, Oct. 1906, *G. Forrest 3868* (**lectotype here designated**: E); Yunnan, Lichiang (Lijiang), 27°20′N, 2743–3050 m, Sept. 1906, *G. Forrest 3000* (paratypes: B, BM, E, K, P).

Lobelia hybrida C. Y. Wu in Wu & Li, Yunnan Trop. Subtrop. Fl. Res. Rep. 1: 92. pl. 32–3. 1965. Type: China, Yunnan, Longchuan, 1000 m, 30 Oct. 1958, *S. K. Wen 580518* (holotype: KUN).

Herbs perennial, 0.5–1.2 cm tall. Rhizomes short. Stems often purplish, erect, simple or few-branched, terete, glabrous. Basal leaves petiolate; petiole narrowly winged; leaf blade spatulate, up to 8 cm long; cauline leaves sessile or shortly petiolate, leaf blade obovate-oblong to obovate-lanceolate or elliptic, 3.5–7 cm long, 1.5–2 (–3) cm broad, papery, both surfaces glabrous, base cuneate, margin with glandular denticles, apex obtuse or acute. Racemes terminal, lax, one-sided; bracts leaflike, sessile; pedicel compressed, 4–7 mm long, puberulent; bracteoles 2, at middle of pedicel. Calyx tube obovoid, 3–4 mm long, puberulent, base rounded; calyx lobes linear or linear-lanceolate, 9–12 mm long, 1–1.5 mm broad, margin glandular-denticulate. Corolla pale blue or rose, 2.4–3 cm long, lobes of upper lip linear, slightly upright, ca. 1.5 cm long, abaxially sparsely villous along mid-veins; those of lower lip oblong-lanceolate, ca. 8 mm long, patent. Filament-tube ca. 1.1 cm long, glabrous; anther tube ca. 6 mm long, dorsally sparsely villous at apex, lower two anthers barbate at apex. Capsule oblong, 4–6 mm long, 3–5 mm in diameter. Seeds ellipsoid. Figure 47–81: a–d.

Phenology: flowering and fruiting from August to October.

Habitats: grassy slopes, forests, or edges of forests; 1600–2600 m.

Distribution: **endemic**; S Hengduan and U Irrawaddy (Map 47–181).

S HENGDUAN: **Dali**, Cangshan, Zhonghesi, 1946, *T. N. Liou 21048* (PE) & 21045 (PE); Cangshan, 2200 m, 1983, *Sino-Amer. Exped. 0033* (KUN); Shimeng Village, 1940, *R. C. Ching 22885* (KUN); **Heqing**: Xiangshuihe, 2600 m, 1940, *R. C. Ching 24354* (KUN, PE); **Yangbi**, Nanshan, E of the city, 2100 m, 1940, *R. C. Ching 25243* (PE); **Yulong** (Lijiang), 27°20′N, 2740–3050 m, 1906, *G. Forrest 3000* (paratypes: BM, K, P). **U IRRAWADDY**: **Tengchong**, No. 4 Distr., 1960, *W. Q. Yin 60–1081* (KUN) ; Mazhang, 1930 m, 1998, *H. Li et al. 10951* (CAS).

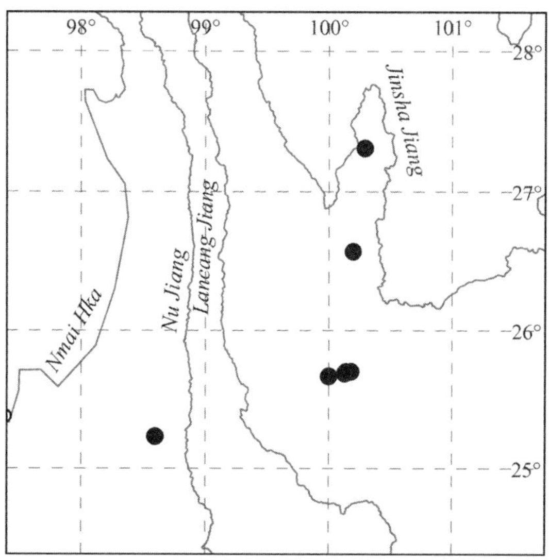

Map 47–181: **Lobelia taliensis** Diels

Figure 47–81: **a–d**, **Lobelia taliensis** Diels: **a**, lower part of plant; **b**, upper part of stem; **c**, flower; **d**, fruit. **e–i**, **Lobelia pleotricha** Diels: **e**, upper part of stem; **f**, flower; **g**, opened corolla; **h**, fruit; **i**, upper part of enlarged leaf. All from *Fl. Reipubl. Popularis Sin.* 73(2), 1983.

13. Lobelia erectiuscula H. Hara, J. Jap. Bot. 40: 328. 1965. ≡ *Lobelia erecta* Hook. f. & Thomson, Proc. J. Linn. Soc. Bot. 2: 28. 1857, non De Vriese (1845). Type: India, Sikkim, Yakla, 3050 m, 16 Oct. 1869, *C. B. Clarke 10025 A* (**neotype here designated**: K).

Lobelia tibetica W. L. Zheng, Acta Phytotax. Sin. 36: 549. 1998, p. p., quoad paratype, *W. L. Zheng 1390* (PE).

Herbs perennial, 60–80 cm tall. Stems erect, simple, retrorse-puberulent. Leaves spirally arranged, petiole of lower leaves up to 6 cm long, leaf blade narrowly elliptic or elliptic-lanceolate, 5–13 cm long, 3–4 cm broad, hispidulous along veins on both surfaces or glabrous, base narrowly cuneate, margin crenate or irregularly double serrate, apex acuminate; upper leaves with winged petiole 0.5–1.5 cm long, leaf blade elliptic or ovate, 4–8 cm long, 3–4 cm broad, base broad-cuneate or nearly rounded, margin obscurely serrate, apex acute. Racemes 10–15 cm long; pedicel 3–5 mm long, densely scaberulose; bracts leaf-like, narrowly elliptic to linear-elliptic, from lower ones much longer than corolla to upper ones shorter than corolla. Calyx tube nearly globose, ca. 5 mm long, puberulose; calyx lobes lanceolate, 6–8 mm long, 1.5–2 mm broad, ciliate, denticulate. Corolla purple-blue, 1.3–1.9 cm long, villose at throat, upper lobes linear, lower lobes deltoid-ovate, stamens connate above base, hairy at conjuctures; anther-tube 4 mm long, lower two anthers barbate at top. Capsule globose, 1–1.2 cm in diameter. Seeds ellipsoid, slightly compressed, pale brown.

Phenology: flowering and fruiting in August and September.

Habitats: thickets at edges of forests or glades; 3000–4000 m.

Distribution: **endemic**; C and E Nepal, Sikkim & Darjiling, Bhutan, M Yarlung Zangbo, Yarlung Zangbo-Brahmaputra and U Irrawaddy (Map 47–182).

Notes: Hooker and Thomson enumerated in the protologue the collection, *J. D. Hooker* s. n. from India, Sikkim, 2740–3960 m. J. D. Hooker's collections are all in K, but I failed to find the collection. Thus I designate here Clarke's collection also from Sikkim as neotype.

C NEPAL: **Bagmati**, Chautara, 1979, *W. Schwabe* s. n. (B); Helambu, 1982, *W. Schwabe* s. n. (B); **Gandaki**, Annapurna Himal. Seti Khola, 3350 m, 1954, *Stainton et al. 6681* (BM, E); **Janakpur**, Dolakha, 1983, *H. Ohba et al. 8351484* (E). **E NEPAL**: **Kosi**, Sankhuwasabha, 3300 m, 1988, *M Suzuki et al. 8850745* (E, KUN) & *8821108* (BM); Sankhuwasabha, Arun Valley, 3050 m, 1956, *J. D. A. Stainton 1344* (BM, E). **SIKKIM & DARJILING**: **Darjiling**, Phalut-Singalila, 3650 m, 1972, *H. Kanai et al. 723178* (KUN); **Sikkim**, Lachung, 1892, *G. A. Gammie* (BM); Changu, 3960 m, 1913, *R. E. Coo-*

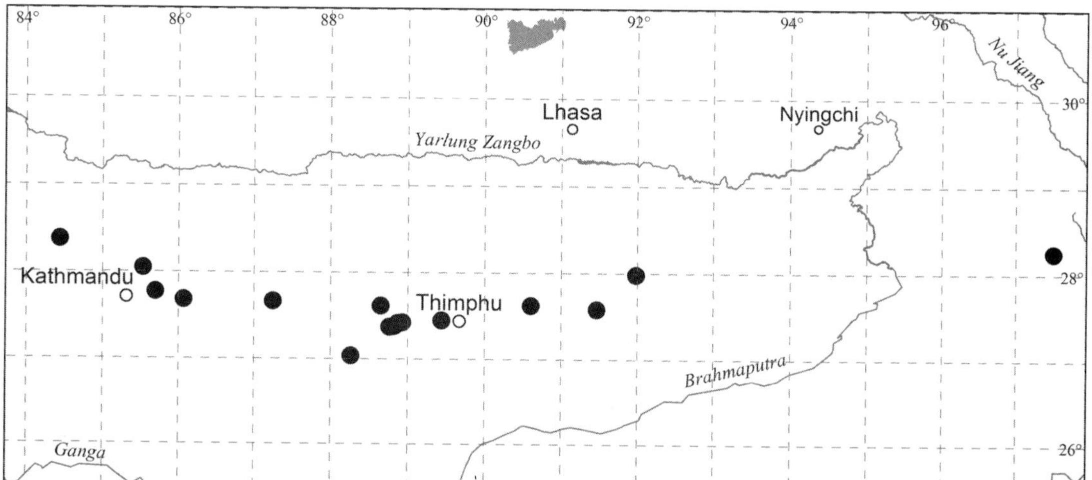

Map 47–182: **Lobelia erectiuscula** H. Hara

per 961 (E). **BHUTAN**, **Pemiltauka** to Bele-la, 2545–3660 m, 1938, *B. J. Gould 953* (K); **Phoji Ding**, 3720 m, 1971, *R. Bedi 594* (K); **Ridang**, Aujduphnang, 2700 m, 1914, *R. E. Cooper 2035* (E); **Thimphu**, 3320 m, 1979, *A. J. C. Grierson & D. G. Long 2737* (K); **Tongsa**, Yuto La 3350 m, 1949. *F. Ludlow et al. 19551* (BM, E); **Trashi Yangtsi**, Dib La, 3050–3960 m, 1934, *F. Ludlow & G. Sherriff 954* (BM). **M YARLUNG ZANGBO**: **Yadong**, Champitang, 3660 m, 1936, *F. S. Chapman 452* (K); below Champitang, 1953, *P. C. Tsoong 5965* (PE); Chumpi & Phari, Nattung, 1879, *Dungboo* s. n. (K). **YARLUNG ZANGBO-BRAHMAPUTRA**: **Boshan**, 3800 m, 1975, *Qinghai-Xizang Suppl. Exped. 751869* (KUN, PE); **Cona**, Mama, 3000 m, 1975, *Qinghai-Xizang Suppl. Exped. 751806* (KUN, PE); **Sêmgou**, Jiba Village, 3500 m, 1975, *s. coll. 75–1752* (PE). **U IRRAWADDY**: **Nam Tamai Valley** (Adung Wang-Gamlang Wang), 3660 m, 1937, *F. Kingdon-Ward 13356* (BM).

14. Lobelia pleotricha Diels, Notes Roy. Bot. Gard. Edinburgh 5: 170. 1912. ≡ *L. davidii* var. *pleotricha* (Diels) E. Wimm., Pfl. R. Heft. 107: 568. 1953. Type: China, Yunnan, E flank of the Dali Range, 25°40′ N, 2440-2740 m, Oct. 1906, *G. Forrest 3869* (**lectotype here designated**: E; isolectotype: K).

Lobelia handelii E. Wimm., Sitz. Akad. Wiss. Wien 61: 109. 1924. ≡ *Lobelia davidii* Franch. var. *handelii* (E. Wimm.) E. Wimm., Pfl. R. Heft 107: 658. 1953. ≡ *Lobelia pleotricha* Diels var. *handelii* (E. Wimm.) C. Y. Wu in Wu & Li, Yunn. Trop. Subtrop. Fl. Res. Rep. 1: 95. 1965. Type: China, Yunnan, Salween-Drungjiang Divide, Champutung, 3900 m, 15–17 Aug. 1916, *Handel-Mazzetti 9901* (holotype, n. v.; isotype: K).

Lobelia tibetica W. L. Zheng, Acta Phytotax. Sin. 36(6): 549. 1998. **syn. nov.** Type: China, Tibet, Mainling, 3200 m, 2 Sept. 1992, *W. L. Zheng 0472* (holotype: XZE; isotype: PE).

Herbs perennial, 60–80 cm tall. Stems dark red, sparsely pubescent. Leaves alternate; lower leaves petiolate, petiole 2–3 cm long, narrowly winged; upper leaves sessile or shortly petiolate, leaf blade elliptic-lanceolate, 6–12 cm long, 2–3.5 cm broad, both surfaces densely white hirsute, base cuneate , margin sinuate or irregularly crenate, apex acuminate. Racemes terminal, 10–30 cm long; bracts leaf-like, lower ones much longer, but upper ones sometimes shorter than flowers; pedicel 5–13 mm long, densely white hirsute; bracteoles 2, near base. Calyx tube nearly globose, 4–6 mm long, 4–6 mm wide, densely, rarely sparsely, white-hirsute; calyx lobes linear-lanceolate, 6–14 mm long, 1–1.5 mm broad, reflexed in fruit, densely white-hirsute, margin glandular-denticulate. Corolla purple-red to blue-purple, 1.8 –2.2 cm long, corolla lobes hairy along midrib and ciliate; upper lobes linear, upright; lower lobes ovate-lanceolate, spreading. Filament-tube glabrous or sparsely puberulent; anther tube ca. 5 mm long, dorsally sparsely pubescent; lower 2 anthers barbate at apex. Capsule shortly columnar, 7–8 mm long, 5–6 mm in diameter, base truncate, concave. Seeds ellipsoid, slightly compressed. Figure 47–81: e–i.

Phenology: flowering and fruiting from August to October.

Habitats: grassy slopes, thickets, or margins of bamboo forests; 2000–4300 m.

Distribution: **endemic**; Yarlung Zangbo-Brahmaputra, S Hengduan and U Irrawaddy (Map 47–183).

YARLUNG ZANGBO-BRAHMAPUTRA: **Mêdog**, 80 K, 2200 m, 2012, *D. Y. Hong et al. H12008* (PE); S of Zhaolong Shan Pass, 1982, *S. Z. Cheng & B. S. Li 280* (PE). **S HENGDUAN**: **Bijiang**, 4350 m, 1964, *S. K. Wu 8792* (KUN); **Dali**, Cangshan, Zhonghesi, 1946, *T. N. Liou 21152* (PE); Gantongsi, 1944, *H. C. Wang 4524* (PE); Malongfeng, 1944, *H. C. Wang 4434* (PE); **Gongshan**, Cikai, Heipushan, 1943, *K. M. Feng 8239* (PE); Geputong, 3600–3800 m, 1940, *K. M. Feng 7957* (PE); Salween-Drungjiang Divide, 3600 m, 1938, *T. T. Yü 20713* (PE); Bingzhongluo, 2000–3400 m, 2013, *X. H. Jin et al. ST2063* (PE); **Lushui**, Pianma, 2600–3000 m, 1989, *H. Sun 8086* (KUN). **U IRRAWADDY**: **Chawchi Pass**, 4000 m, 1920, *R. F. 1866* (E); **Drungjiang**, 3300 m, 2013, *X. H. Jin et al. ST0719* (PE); Salween-Kiukiang Divide, Haipuh, 3100 m, 1938, *T. T. Yü 20315* (E) ;

Tengchong, Danza, 1983, *Lin Qin 770686* (KUN).

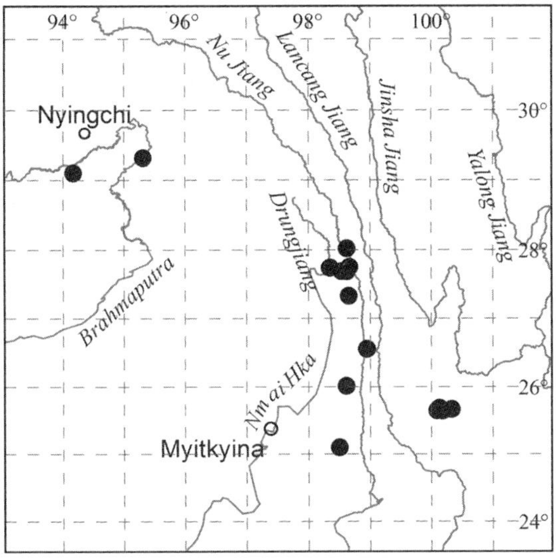

Map 47–183: **Lobelia pleotricha** Diels

15. Lobelia drungjiangensis D. Y. Hong, **sp. nov**. Figure 47–82. Type: China, Yunnan, Gongshan, Cikai, vicinity of Dimupo, 27°45′46.9″N, 98 °34′43″E, 2700 m, deciduous broadleaf forest, 30 Aug. 2006, *Gaoligong Shan Biodiversity Survey 34363* (holotype: CAS).

Herbs perennial. Rhizomes horizontal, ca. 5 mm in diameter. Roots fibrous or thickened and attenuate, up to 4 mm across. Stems erect, 50–80 cm tall, white-villous, with short branches up to 8 cm long. Leaves alternate, lower cauline ones wilted at anthesis, middle and upper ones short-petiolate, petiole up to 1.5 cm; leaf blade elliptic, 4–7 cm long, 1.5–3 cm broad, gradually becoming smaller up stem, and becoming bracts at upper part of stem, both surfaces white-hirsute entirely or along veins, base cuneate, margin dentate, apex acute. Flowers solitary, axillary at axils of leaves or bracts, bracts conformable with leaves; pedicel 5–7 mm long, densely hirsute. Calyx tube globose, white-hirsute; calyx lobes subulate, 3–4 mm long, ca. 1 mm broad, white-hirsute along rib and on margin. Corolla 1–1.2 cm long, all corolla lobes in a plane, two lobes of upper lip linear, less than 1 mm broad, hispid along ribs outside three of lower lip band-shaped, ca. 1.2 mm broad, hispid along ribs outside, villous inside. Stamens connate into tube from base; filament tube villous, more densely at top; anther tube white-hispid along conjuncture sutures, lower two anthers barbate at top. Fruit a capsule, globose, 5–6 mm in diameter. Seeds yellow, ellipsoid, ca. 0.5 mm long. Figure–82.

Phenology: flowering in August; fruiting in November.

Habitats: forests; 1600–2700 m.

Distribution: **endemic**; S Hengduan and U Irrawaddy (Map 47–184).

U IRRAWADDY: **Drungjiang** Township, vicinity of Meiliwang, 1600 m, 23 Nov. 1990, *Dulong Jiang Investigation Team 806* (CAS).

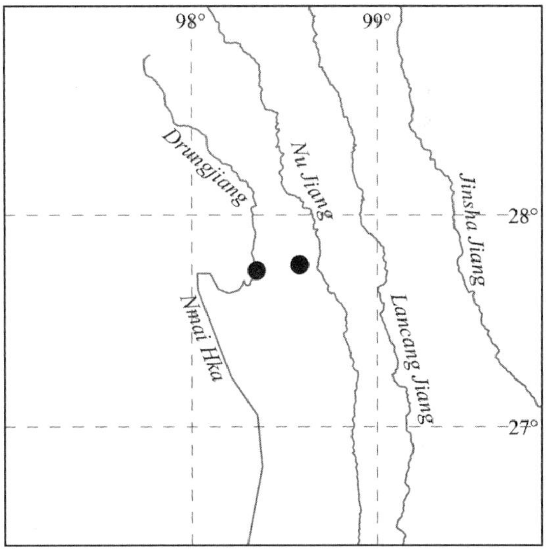

Map 47–184: **Lobelia drungjiangensis** D. Y. Hong

16. Lobelia sessilifolia Lamb., Trans. Linn. Soc. London, Bot. 10: 260. 1811; Hong & Lammers in Wu, Raven & Hong, Fl. China 19: 561. 2011. Type: not seen.

Herbs perennial, 0.6–1.2 m tall. Stems terete, usually simple, glabrous. Leaves alternate, densely aggregated at middle part of stem, sessile; leaf blade thickly papery, ovate-lanceolate to linear-lanceolate, 2.5–5.5 (–7) cm long, 0.3–1.6 cm broad, both surfaces glabrous, base nearly rounded to broadly cuneate, margin serrate, apex acuminate. Racemes terminal, 8–35 cm

Figure 47–82: **Lobelia drungjiangensis** D. Y. Hong: **a**, lower part of plant, showing rhizome; **b**, middle and upper parts of stem; **c**, a flower showing the indumentum of corolla outside and inside. Drawn by Ai-Li LI.

long, glabrous; bracts leafy, narrowly lanceolate, shorter than flowers; pedicel 5–12 mm long. Calyx glabrous, tube funnelform-campanulate, ca. 4 mm long; calyx lobes deltoid-lanceolate, 5–7 mm long, ca. 2 mm broad, margin entire. Corolla blue-purple, 2.5–3 (–3.5) cm long, nearly 2-lipped, glabrous outside, villous inside; upper 2 lobes spatulate, 15–20 mm long, 3–4 mm broad, longer than corolla tube, upright; lower lobes elliptic, ca. 15 mm long, 4–5 mm broad, nearly as long as corolla tube, densely ciliate. Stamens connate above base; filament-tube glabrous; anther tube 3–4 mm long, densely villous along conjunctive suture; lower two anthers barbate at apex. Capsule obovoid, 8–10 mm long, 5–7 mm in diameter. Seeds brown-red, semiglobose, one side thick, other thin, ca. 1.5 mm long, smooth.

Phenology: flowering and fruiting from July to September.

Habitats: wet meadows; 900–3000 m.

Distribution: S Hengduan (Map 47–185); also in E Asia: E China, Japan, Korea Peninsula, and Russia (the Far East and Siberia).

S HENGDUAN: **Weixi**, 2600 m, 1934, *H. T. Tsai 57885* (KUN); **Yulong** (Lijiang), Snow Range, 2940 m, 1923, *J. F. Rock 10717* (W); Xuesongcun, 2800 m, 1942, *K. M. Feng 8883* (KUN); between Yangtze and Mekong, 2800 m, 1915, *Handel-Mazzetti 7907* (W).

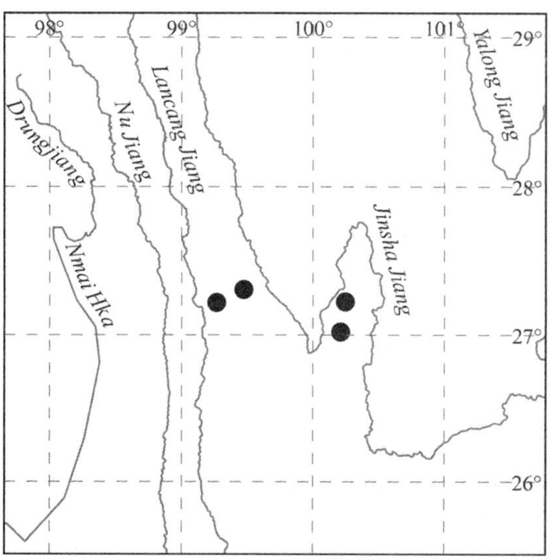

Map 47–185: **Lobelia sessilifolia** Lamb.

17. Lobelia nubigena J. Anthony, Notes Roy. Bot. Gard. Edinburgh 19: 175. 1936. Type: Bhutan, Tibdah La, 27°20′N, 90°40′E, 3960 m, Jul. 1915, *R. E. Cooper 4099* (holotype: E; isotype: E).

Herbs biennial, monocarpic 1–1.5 m tall, cylindrical in overall shape. Stems up to 2.8 cm in diameter, simple. Basal leaves rosulate, oblanceolate, up to 30 cm long, 5 cm broad, glabrescent; cauline leaves sessile, linear- or oblong-lanceolate, up to 12 cm long, 2 cm broad, dark purplish-green above, green beneath, base decurrent. Inflorescence terminal, racemose, densely flowered, conspicuously bracteate; bracts recurved, covering flowers lanceolate, 4–6 cm long, 1.5–2 cm broad, densely white-wooly, apex acuminate; pedicel villose, 0.5–1 cm long. Calyx longer than corolla, shortly campanulate; calyx lobes narrow-triangular, 1.5–2 cm long, 0.2–0.4

Figure 47–83: **Lobelia nubigena** J. Anthony: plant. Drawn by Ai-Li LI.

cm broad, densely white-villous. Corolla pale purple-blue, irregularly 5-lobed; corolla lobes ca. 1 cm long, 0.2–0.3 cm broad at base, upper 2 slightly longer than lower 3. Filaments ca. 1 cm long, sparsely villous, bases dilated, ca. 2 mm broad, upper parts connivent; anthers 7 mm long, upper and larger 3 glabrous, lower 2 barbate. Style ca. 1.3 cm long; stigma 2-fid. Seeds ellipsoid, 1–1.5 mm long. Figure 47–83.

Phenology: flowering in July and August.

Habitats: open alpine peaty turf; 3960–4300 m.

Distribution: endemic; confined to Bhutan (Map 47–186).

BHUTAN, **Bumthang**, Pegula, Yato La Ridge, 4300 m, 1994, *R. Pradhan 36* (E); **Tibdah La**, Yuto La, 4115 m, 1937, *F. Ludlow & G. Sherriff 3489* (E); **Tongsa**, Yato La Ridge, 3960 m, 1915, *R. E. Cooper 4766* (E).

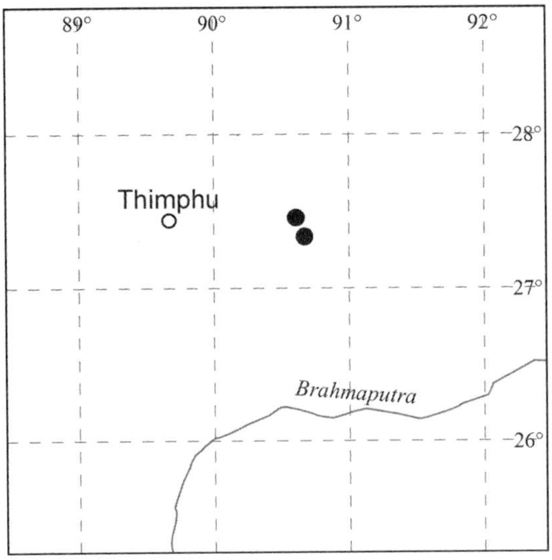

Map 47–186: **Lobelia nubigena** J. Anthony

18. Lobelia montana Reinw. ex Bl., Bijdr. 728. 1825; Hong & Lammers in Wu, Raven & Hong, Fl. China 19: 562. 2010. ≡ *Pratia montana* (Reinw. ex Bl.) Hassk., Flora 25: 2. 1842. Type: not seen.

Lobelia reflexisepala Lammers, Novon 8: 34. 1998. ≡ *Pratia reflexa* Y. S. Lian, Acta Phytotax. Sin. 17(3): 123, 1979, non *Lobelia reflexa* Stokes (1812). Type: China, Tibet, Mêdog, Nage-Hanmi, 2700 m, 1974, *Qinghai-Xizang Exped. 74–3970* (holotype: PE).

Herbs perennial. Stems up to 1.3 m long, erect or arching, terete, simple or branched, upper internodes zigzagged, glabrous. Leaves alternate, petiolate; petiole 0.6–2.5 cm long, often narrowly winged, glabrous; leaf blade elliptic or ovate-elliptic, 7–15 cm long, 2.5–4 cm broad, papery, both surfaces glabrous, base broad-cuneate to nearly rounded, margin crenulate, glandular-denticulate, apex caudate-acuminate. Flowers solitary and axillary; pedicel slender, 2–4 cm long, glabrous. Calyx tube semiglobose, 4–5 mm long, ca. 4 mm wide, glabrous; calyx lobes linear, 7–20 mm long, entire, glabrous, recurved but not reflexed. Corolla purple, or purple-striate, 1.3–1.5 cm long, glabrous outside, villous inside; upper 2 lobes linear-lanceolate; lower 3 long-ovate, apex with rostrum 2–3 mm long. Filament-tube villous; anther tube ca. 5 mm long, back sparsely villous; lower two anthers barbate at apex. Berry globose, purple or black-purple, 0.7–1 cm in diameter, base concave. Seeds ovoid-globose, slightly compressed, verruculose. Figure 47–84: a–e.

Phenology: flowering and fruiting from June to November.

Habitats: moist ravines, glades, or forest margins; 1100–2800 m.

Distribution: Sikkim & Darjiling, Bhutan, Yarlung ZangboBrahmaputra, S Hengduan, and U Irrawaddy (Map 47–187); also in SE Asia (Cambodia, Malaysia, Thailand, and Vietnam) and S Yunnan of China.

SIKKIM & DARJILING: **Darjiling**, Goompahar, 2440 m, 1875, *C. B. Clarke 27472 B* (K); Palmojua, 2200 m, 1972, *H. Kanai et al. 723180* (KUN); **Sikkim**, 1520–2130 m, *J. D. Hooker s. n* (K); Dholijhna, 1830 m, 1882, *J. S. Gamble 10538* (K). **BHUTAN**: **Chukka**, 2170 m, 1982, *A. J. C. Grierson & D. G. Long 3205* (E); Lota La, 2440 m, 1915, *R. E. Cooper 4682* (E); **Mongar**, Sengor (Senghor), 2200 m, 2000, *L. S. Bowes 13195* (E). **S HENGDUAN**: **Fugong**, 2600 m, 1933, *H. T. Tsai 54382* (KUN); **Gongshan**, W of Champutung, 2800 m, 1935, *C. W. Wang 67373* (PE); Gongshan to Drungjiang, 1900–2000 m, 1982, *Qinghai-*

Figure 47–84: **a–e**, **Lobelia montana** Reinw. ex Bl.: **a**, lower part of plant; **b**, a section of stem, showing a lateral inflorescence; **c**, opened corolla; **d**, a flower with corolla removed, showing filament tube and anther tube; **e**, berry. **f–i**, **Lobelia angulata** G. Forster: **f** & **g**, plant; **h**, opened corolla; **i**, opened filament and anther tube. **a** & **e** drawn by Zhi-Juan CHEN; **b–d** from *Fl. Reipubl. Popularis Sin*. 73(2), 1983; **f–i** redrawn by Ai-Li LI from *Icon. Comophyt. Sin*. 4, 1975 and *Fl. China* 19, 2011.

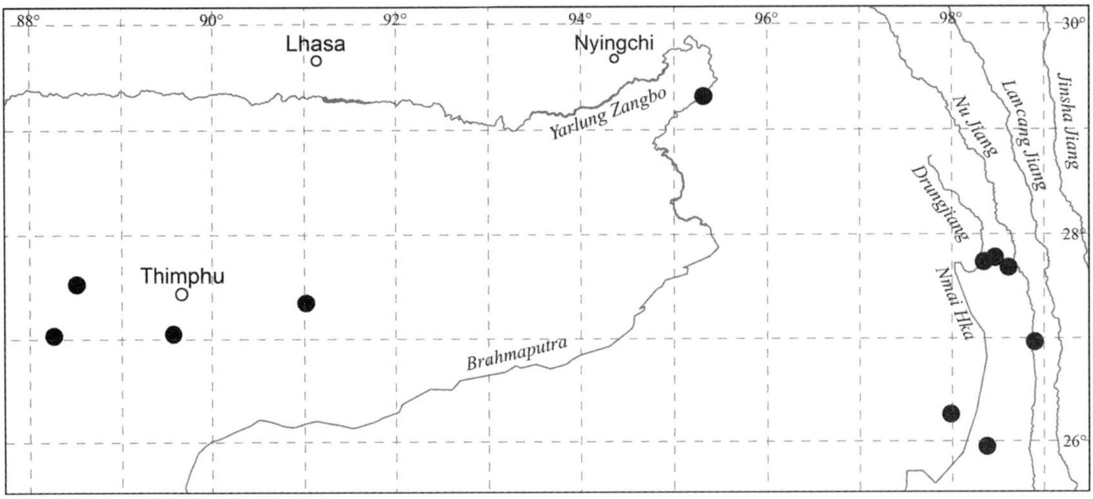

Map 47–187: **Lobelia montana** Reinw. ex Bl.

Xizang Exped. 8123 (KUN). **U IRRAWADDY**: **Chawnghka**, 610–915 m, 1919, *F. Kingdon-Ward 3519* (E); **Drungjiang**, Township, Bapo, 1350 m, 2002, *H. Li et al. 15064* (CAS); **Htawgaw**, 2130 m, 1924, *G. Forrest 24854* (E); **North Triangle** (Hkinlum), 1220 m, Aug. 1953, *F. Kingdon-Ward 21228* (BM); North Triangle (Tawa Bum), 2134–2440 m, Oct. 1953, *F. Kingdon-Ward 21442* (BM); E Upper Burma, 2134 m, Aug. 1924, *G. Forrest 24890* (E, K).

19. Lobelia angulata G. Forster, Fl. Ins. Austr. 58. 1786 [non (R. Brown) Kuntze, Revis. Gen. Pl. 2: 378. 1891]. ≡ *Pratia angulata* (G. Forster) Hook. f., Fl. Antarct. i. 43. 1844. Type: not seen.

Lobelia begoniifolia Wall., Asiat. Res. 13: 377. 1820. Type: Nepal, *Wallich 1306* (holotype: K).

Lobelia nummularia Lam., Dict. 3: 589. 1789. Type: Java, *Commerson* s. n. (holotype: P).

Herbs perennial. Stems prostrate, 12–55 cm long, villous, simple or branched at base, rooting at nodes. Leaves alternate, petiolate; petiole 1–16 mm long, villous; leaf blade ovate-orbicular, cordate or ovate, 0.8–2.5 cm long, 0.6–2.5 cm broad, both surfaces puberulent, base obliquely cordate, margin serrate, apex acute or obtuse. Flowers solitary, axillary; pedicel 0.7–3.5 cm long, glabrous. Calyx tube urceolate, 3–4 mm long, 2–3 mm wide, glabrous; calyx lobes linear-lanceolate, 2.5–6 mm long, straight, margin with 2–3 pairs of denticles. Corolla purple-red, pale purple, green or yellow-white, 6–10 mm long; corolla tube glabrous outside, villous inside; limb 2-lipped; upper two lobes linear-lanceolate; lower three lobes lanceolate. Stamens connate above middle; filament-tube glabrous; anther tube ca. 1 mm long, back villous; lower two anthers each with a single seta at apex. Fruit a berry, purple-red, ellipsoid-globose, 1–1.3 cm long. Seeds nearly globose, slightly compressed, verruculose. Figure 47–84: f–i.

Phenology: flowering and fruiting all year around.

Habitats: very common by fields, roadsides, wet places, grassy slopes, or open woods; below 2000 m.

Distribution: E Nepal, Sikkim & Darjiling, Bhutan, Yarlung Zangbo-Brahmaputra, S Hengduan, and U Irrawaddy (Map 47–188); widely distributed in E and SE Asia; Papua New Guinea.

E NEPAL: **Kosi**, Sankhuwasabha, 1988, *M Susuki 8820284* (KUN). **SIKKIM & DARJILING**: **Darjiling**, 915 m, 1880, *J. S. Gamble 8330* (K); 2133 m, 1872, *J. S. Gamble 2321A* (K); **Sikkim**, Gangtok, 1981, *P. Chakraborty 1620* (K); Leebong, 1524 m, 1874, *s. coll.* s. n. (K). **YARLUNG ZANGBO-BRAHMAPUTRA**: **Baibung**, 1000 m, 1992, *H. Sun et al. 0299* (KUN);

Dexing, 850 m, 1993, *H. Sun et al. 5898* (KUN). **S HENGDUAN**: **Bijiang**, 1400 m, 1964, *S. K. Wu 8525* (KUN); **Dali**, 1946, *N. T. Liou 14038* (KUN); **Fugong**, 1600 m, 1933, *H. T. Tsai 54743* (KUN). **U IRRAWADDY**: Kiukiang **(Drungjiang)** Valley, Michiaoton, 1600 m, 1938, *T. T. Yü 19927* (KUN); **Tengchong**, No. 9 Distr., 1800–2000 m, 1960, *W. Q. Yin 60–1335* (KUN); E Upper Burma, 2130 m, 1924, *G. Forrest 24864* (K).

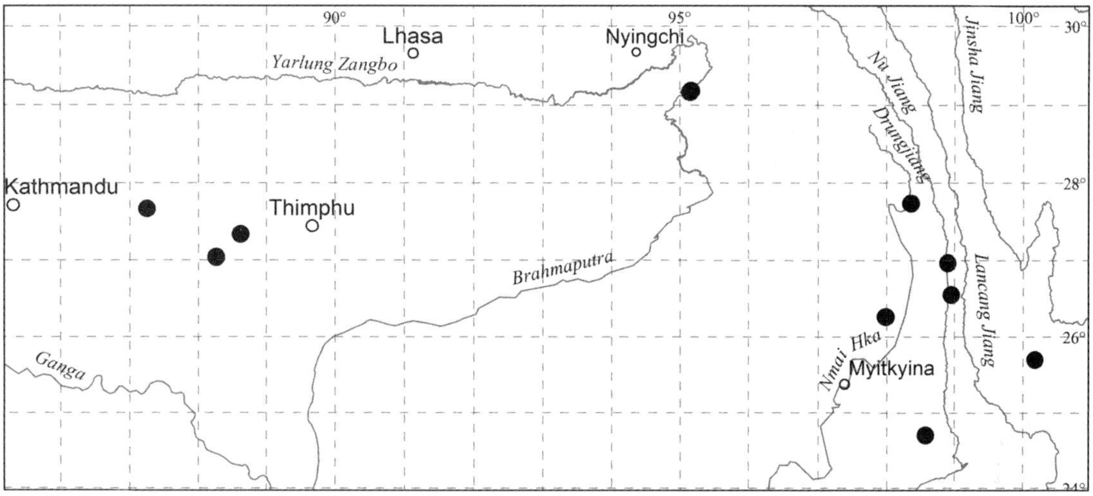

Map 47–188: **Lobelia angulata** G. Forster

MENYANTHACEAE

Herbs perennial, aquatic or nearly so, with rhizomes. Leaves alternate, rarely opposite, simple or trifoliolate; stipules absent. Inflorescences racemose, or flowers clustered at nodes of stem. Flowers (4-) 5-merous, perfect, regular, often distylous. Sepals free or connate at base. Corolla sympetalous; corolla lobes valvate in bud. Stamens 5, filaments attached to corolla tube, alternate with petals; anthers mostly sagittate, tetrasporangiate, dithecal, longitudinally dehiscent. Ovary of 2 carpels, superior, 1-locular; style linear; stigma 2-lobed; ovules many. Fruit a capsule, dehiscent or indehiscent. Seeds few to many, winged or wingless; endosperm copious.

Five genera and about 60 species, cosmopolitan; two genera with four species in the Pan-Himalaya.

1a. Leaves 3-foliolate, emergent; flowers in an emergent raceme; fruit dehiscent .... 1. *Menyanthes*

1b. Leaves entire, floating on water surface; flowers in a sessile cluster at water level; fruit indehiscent .... 2. *Nymphoides*

1. MENYANTHES L.

Linnaeus, Sp. pl. 1: 145. 1753.

Herbs perennial, aquatic or nearly so. Rhizomes long, prostrate, nodes with roots and scale-like leaves. Leaves basal, emergent from water; petiole base sheathing stem; leaf blade 3-foliolate. Inflorescences from rhizome nodes, a raceme, scapose, held clear of water. Flowers 5-merous, distylous. Calyx lobed almost to base. Corolla lobed just over half. Stamens inserted on corolla tube. Capsule dehiscent by 2 valves. Seeds smooth.

One species with circumboreal distribution.

1. Menyanthes trifoliata L., Sp. Pl. 1: 145. 1753; Ho & Ornduff in Wu & Raven, Fl. China 16: 140. 1995. Type: "Habitat in Europae paludibus". *Herb. Linn. No. 203.4* (lectotype designated by Qaiser in Nasir & Ali, *Fl. W. Pakistan* 111: 3 .1977: LINN, n. v.).

Rhizomes over 20 cm long, ca. 2 cm in diameter, usually in mud but sometimes floating. Petiole straight, erect, 10–20 (–30) cm long, base vaginate; leaflets elliptic 2.5–5 (–8) cm long, 1.2–2 (–4) cm broad, base cuneate, margin entire or crenulate, apex obtuse, midrib distinct. Racemes many-flowered, 3–14 cm long; peduncle (scape) 9–18 cm long; bracts ovate, 5–7 mm long, entire; pedicel 1–1.8 cm long. Calyx 4–5 mm long. Corolla white, tubular, 1.4–1.7 cm long; corolla lobes elliptic- lanceolate, 7.5–10 mm long, apex obtuse, fimbriate-pilose inside, glabrous outside. Filaments linear, 5.5–6.5 mm long; anthers 1.8–2 mm long. Short style 6–7 mm long; long style 10–12 mm long. Capsule globose, 6–7 mm in diameter. Seeds rounded, 2–2.5 mm in diameter. Figures 47– 85.

Phenology: flowering and fruiting from May to August.

Habitats: shallow water or swamp; 2500–3900 m.

Distribution: U Ganga & Indus, Yarlung Zangbo-Brahmaputra, and S Hengduan (Map 47–189); also in other N temperate regions.

U GANGA & INDUS: s. loc. *Falconer* s. n. (K): **YARLUNG ZANGBO- BRAHMAPUTRA**: **Mêdog**, Nage, 3200 m, 1974, *Qinghai-Xizang Exped. 74–3789* (PE); S of Nage, 3300 m, 1980, *Ecol. Exped. Alpine Team 10333* (PE); Hami to Nage, 3000– 3200 m, 1983, *B. S. Li & S. Z. Cheng 5267* (PE). **S HENGDUAN**: **Muli**, Kulu, 3600 m, 1937. *T. T. Yü 7112* (A, PE); Kulu, 3850 m, 1929, *J. F. Rock 18017* (A, K); **Weixi**, Lidiping, 3100 m, 1981, *Acad. Sin. Inst. Bot. Exped. to Hengduan Mts 1736* (PE); **Yanyuan**, Luguhu, 2500 m, 1937, *T. T. Yü 5289* (PE); **Yulong** (Lijiang), Ludian, 3000 m, 1956, *P. I. Mao 161* (PE); NW Lijiang, Ah-S-Chi, 1939, *R. C. Ching 20843* (A).

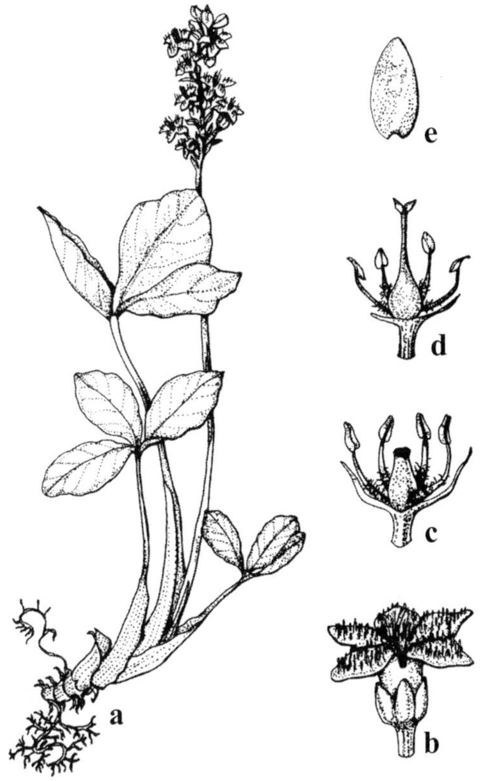

Figure 47–85: **a–e**, **Menyanthes trifoliata** L.: **a**, plant; **b**, short-styled flower; **c**, short-styled flower with calyx and corolla removed, showing stamens and pistil. **d**, long-styled flower with calyx and corolla removed, showing stamens and pistil; **e**, a calyx lobe. From *Fl. China* 16, 1999.

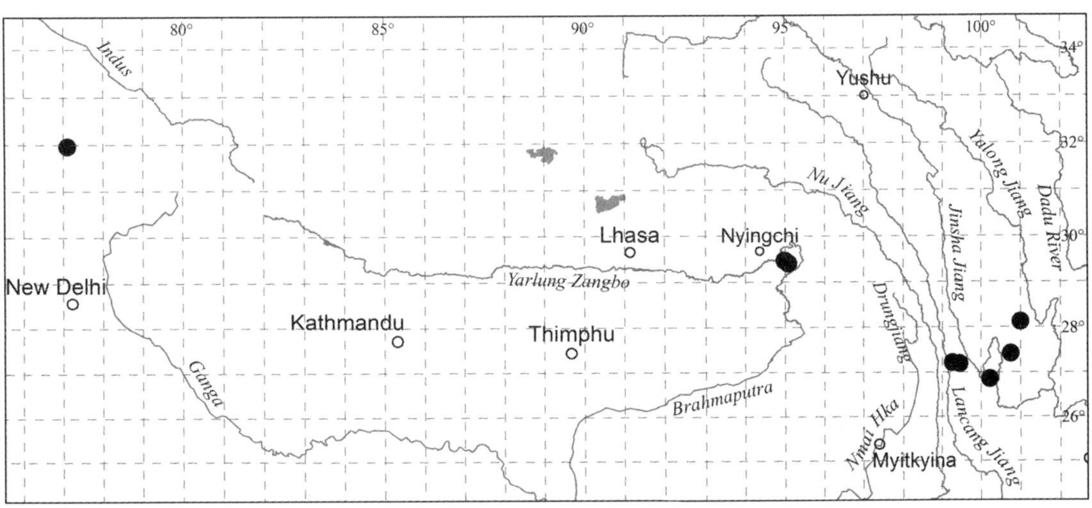

Map 47–189: **Menyanthes trifoliata** L.

2. NYMPHOIDES Séguir

Séguier, Pl. Veron. 3: 121. 1754. Type: *Nymphoides peltata* (S. G. Gmelin) O. Kuntze (*Limnanthemum peltatum* S. G. Gmelin).

Limnanthemum S. G. Gmelin, Novi Comment. Acad. Sci. Imp. Petrop. 14 (1): 527. 1770.

Herbs perennial, rarely annual, aquatic, often carpeting surface of lakes or ponds, with short rhizomes producing many slender, petiole-like stolons. Stems usually long, floating, sometimes producing fibrous roots from nodes. Leaves alternate, rarely opposite; leaf blade floating, veins palmate. Flowers clustered at nodes, (4) 5(–8)-merous, distylous, rarely homostylous. Calyx cleft to near base. Corolla rotate, cleft to near base, rarely less deeply cleft and campanulate, each corolla lobe with a tuft of long fimbriae at throat. Stamens inserted on corolla tube. Nectaries 5, attached at ovary base. Capsule indehiscent or rupturing irregularly, few-seeded. Seeds compressed or globose, smooth or ornamented.

Recent reference: Sivarajan & Joseph, 1993. The genus *Nymphoides* Séguier (Menyanthaceae) in India. *Aquatic Bot.* 45: 145–170.

About 35–40 species, cosmopolitan; three species in the Pan-Himalaya.

1a. Stems unbranched; lower leaves alternate, upper ones opposite; corolla 2.5–3 cm across, yellow, corolla lobes broadly membranous on margin; capsule 1.7–2.5 cm long; seeds compressed, 4–5 mm long, ciliate .... 1. *N. peltata*
1b. Stems branched; leaves solitary and terminal, sometimes also 1–3 leaves at stem nodes; corolla 0.5–1.5 cm across, white with yellow centre, corolla lobes not membranous on margin; capsule 2–6 mm long; seeds ellipsoid, 1–1.5 mm long, glabrous.
2a. Corolla lobes densely long-pilose inside, without a longitudinal fold; capsule ellipsoid .... 2. *N. indica*
2b. Corolla lobes glabrous inside, with a longitudinal fold; capsule subglobose ..... 3. *N. cristata*

1. Nymphoides peltata (S. G. Gmelin) O. Kuntze, Revis. Gen. Pl. 2: 429.1891; Ho & Ornduff in Wu & Raven, Fl. China 16: 141. 1995. ≡ *Limnanthemum peltatum* S. G. Gmelin, Novi Comment. Acad. Sci. Imp. Petrop. 14(1): 527. 1770. Type: not seen.

Rhizomes horizontal. Stems branched, sometimes producing fibrous roots from nodes. Leaves alternate at lower part of stem, opposite at upper part; petiole 5–10 cm long, base widened into an amplexicaul sheath; leaf blade ovate-orbicular to orbicular, 1.5–8 cm across, subleathery, adaxially glabrous, abaxially densely glandular, base cordate, margin entire, veins indistinct. Flowers usually densely clustered at nodes, 5-merous, distylous; pedicel 3–7 cm long. Calyx 7–9 mm long; calyx lobes elliplic-lanceolate to elliptic, apex obtuse. Corolla golden-yellow, 2.5–3 cm long, rotate; corolla lobes obovate, margin broadly membranous and laciniate, apex rounded to emarginate. Filaments sparsely pilose. Short-styled flowers: filaments 3–4 mm long; anthers curved, 4–6 mm long; ovary 5–7 mm long; style 1–2 mm long; stigma small. Long-styled flowers: filaments 1–2 mm long; anthers 2–3.5 mm long; ovary 0.7–1.7 mm long; style up to 1 cm long; stigma large, 2-lobed; lobes suborbicular; glands golden yellow. Capsule ellipsoid, 1.7–2.5 cm long, 0.8–1.1 cm in diameter. Seeds brown, compressed, elliptic, 4–5 mm long, densely ciliate. **Chromosomes**: 2n=54. Figures 47– 86.

Phenology: flowe-ring and fruiting from April to October.

Habitats: standing water; 100–1800 m.

Distribution: Jammu & Kashmir and S

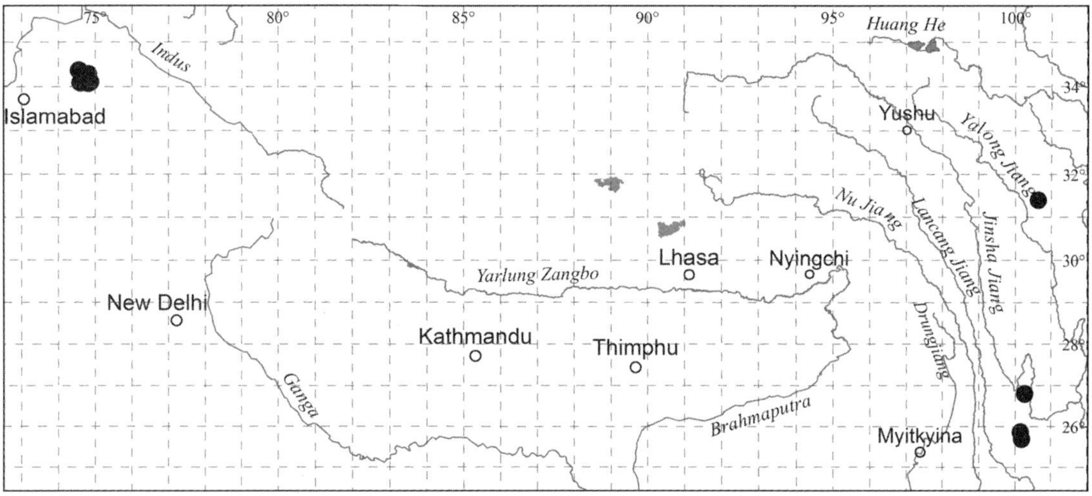

Map 47–190: **Nymphoides peltata** (S. G. Gmelin) O. Kuntze

Hengduan (Map 47–190); also in C and SW Asia and Europe; naturalized in N America.

JAMMU & KASHMIR: **Dal Lake**, 1918, *E. M. Saunders* s. n. (K); **Kangan**, Sind Valley, 1901, *J. F. Duthie 25692* (K); **Kashmir**, Sumbul 1520 m, 1928, *B. B. Osmaston 64* (K); **Srinagar**, 1700 m, 1976, *F. Billiet & J. Leonard 6698* (K); Srinagar to tangmarg, 1680 m, 1931, *R. R. Stewart 13139a* (K); **Woolar Lake**, 1520–1830 m, 1892, *J. F. Duthie 10857* (K). **S HENGDUAN**: **Dali**, Erhai, 1914, *C. Schneider 2744* (A); **Eryuan**, Chiwen Island, 1984, *Sino-Amer. Bot. Exped. (1984) 1085* (A); **Luhuo**, 2006, *D. E. Boufford et al. 36384* (KUN); **Yulong** (Lijiang), Likiang Valley, 1939, *R. C. Ching 21024* (A).

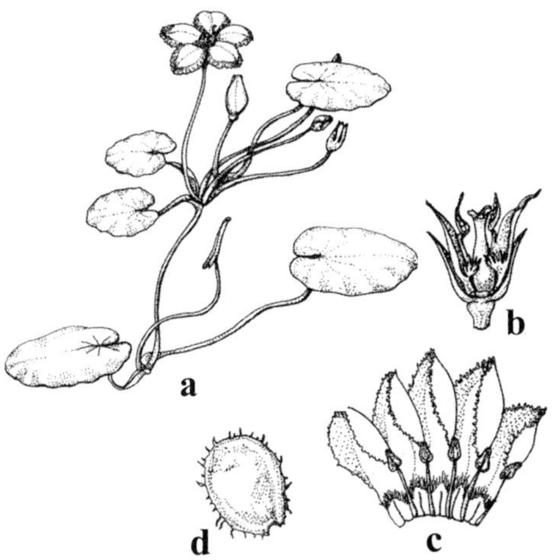

Figure 47–86: **Nymphoides peltata** (S. G. Gmelin) Kuntze: **a**, flowering and fruiting branch; **b**, longitudinal section of flower, showing stamens and pistil. **c**, opend corolla, showing stamens; **d**, seed. From *Fl. China* 16, 1999.

2. Nymphoides indica (L.) O. Kuntze, Revis. Gen. Pl. 2: 429. 1891; Sivarajan & Joseph, Aquatic Bot. 45: 156. 1993. ≡ *Menyanthes indica* L., Sp. Pl. 1: 145. 1753. ≡ *Limnanthemum indicum* (L.) Griseb., Gen. Sp. Gent. 343. 1838. Type: S India/Sri Lanka, *Herb. Linn. 203.2* (lectotype designated by A. Raynal, Adansonia, sér. 2, 14: 416. 1974: LINN, n. v.).

Limnanthemum esquirolii H. Lév., Repert. Spec. Nov. Regni Veg. 13: 259. 1915. Type: China, Kouy-Tchéou, Aug. 1909, *J. Esquirol 2532* (holotype: E; isotype: K).

Rhizomes horizontal. Stems branched. Petiole 1–2 cm long; leaf blade broad-ovate to subcordate, 3–18 cm long, subleathery, abaxially densely glandular, base cordate, margin entire; veins indistinct, palmate. Flowers many, 5-merous, distylous; pedicel 3–5 cm long. Calyx 3–6 mm long, calyx lobes lanceolate to narrow-elliptic, apex obtuse. Corolla white with a yellow centre, 7–12 mm long; corolla lobes ovate-elliptic, apex obtuse. Filaments 1.5–1.7 mm long; anthers 2–2.2 mm long. Stigma lobes triangular. Capsule ellipsoid, 3–5 mm long, few-seeded. Seeds brown, globose, 1.2–1.5 mm in diameter, smooth. **Chromosomes**: 2n=18, 36.

Phenology: flowering and fruiting from August to September.

Habitats: standing water; 100–1600 m.

Distribution: U Ganga & Indus and Nepal (Map 47–191); also in other parts of India and Myanmar; in Cambodia, E and S China, Indonesia, Japan, Korea Peninsula, Malaysia, Sri Lanka, and Vietnam; Australia and Pacific Islands.

U GANGA & INDUS: ca. 10 km N of **Roorkee**, 490 m, 1972, *Pertti Uotila 17660* (K); **Upper Gangetic Plain**, Gorakpur, 1898, *J. S. Gamble 22313* (K).

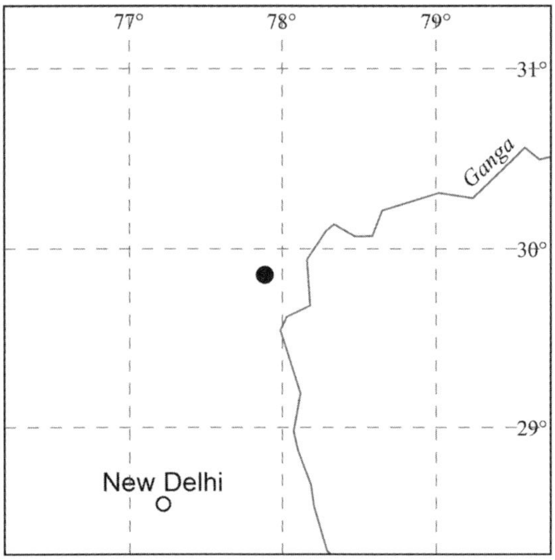

Map 47–191: **Nymphoides indica** (L.) O. Kuntze

3. Nymphoides cristata (Roxb.) O. Kuntze, Revis. Gen. Pl. 2: 429. 1891; Ho & Ornduff in Wu & Raven, Fl. China 16: 141. 1995. ≡ *Menyanthes cristata* Roxb., Pl. Coromandel 2 : 3. 1798. Type: *Roxburgh* s. n. (holotype: K)

Rhizomes horizontal. Stems branched, with a terminal leaf. Petiole 1–3 cm long; leaf blade ovate-orbicular to suborbicular, 3–10 cm long, subleathery, abaxially densely glandular, base cordate, margin entire, veins indistinct. Flowers many, 5-merous, distylous; pedicel 3–4.5 cm long. Calyx 3–5.5 mm long; calyx lobes ovate-elliptic to ovate, apex obtuse. Corolla white with a yellow centre, 4–8 mm long; corolla lobes ovate with a keeled fold from base to apex, apex obtuse. Filaments 1–2 mm long; anthers 0.8–1 mm long. Stigma lobes triangular. Capsule subglobose, 3–5 mm in diameter, few-seeded. Seeds pale brown, globose, 1.3–1.5 mm in diameter, smooth or scabrous. **Chromosomes**: 2n=18.

Phenology: flowering and fruiting from April to December.

Habitats: standing water.

Distribution: U Ganga & Indus and E Nepal (Map 47–192); also in S China (Fujian, Guangdong, and Hainan Hubei, Hunan, Jiangsu, other parts of Sichuan, and Taiwan).

U GANGA & INDUS: **Upper Gangetic Plain**, Kheri (Oudh), 1998, *Inaqat* s. n. (K); loc. eodem, 1998, *Inaqat 22314* (K). **E NEPAL**: Sibganja-Gauriganja, 1963, *H. Hara et al. 6302622* (K).

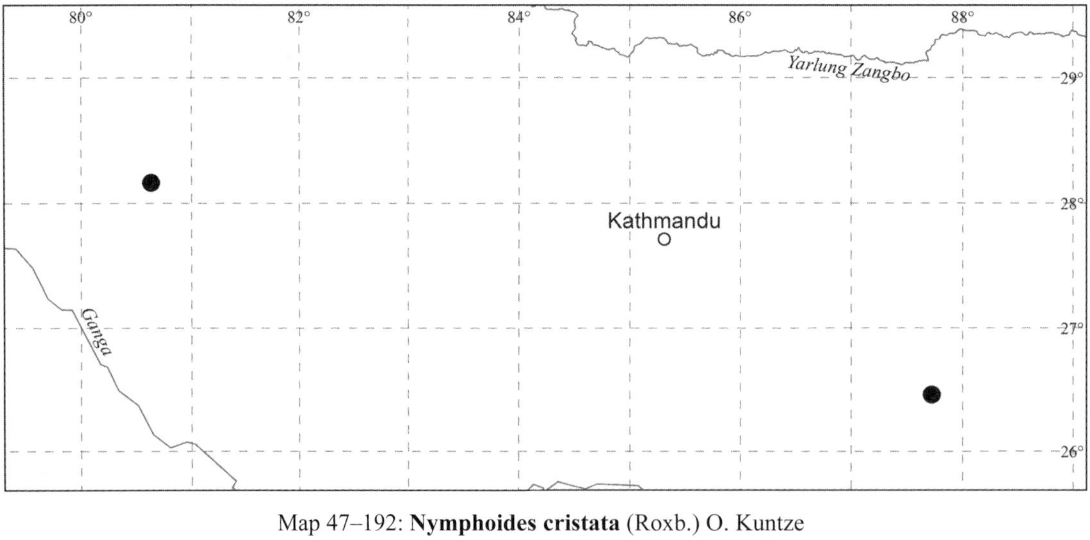

Map 47–192: **Nymphoides cristata** (Roxb.) O. Kuntze

List of nomenclatural novelties published in the present volume

Adenophora albescens C. Y. Wu, **lectotype designation**

Adenophora atuntzensis C. Y. Wu, **lectotype designation**

Adenophora brevidiscifera D. Y. Hong, **syn. nov.** = **Adenophora delavayi** (Franch.) D. Y. Hong

Adenophora bulleyana Diels, **lectotype designation**

Adenophora capillaris Hemsl., **lectotype designation**

Adenophora capillaris Hemsl., subsp. **leptosepala** (Diels) D. Y. Hong, **lectotype designation**

Adenophora coelestis Diels, **lectotype designation**

Adenophora dawuensis D. Y. Hong, **sp. nov.**

Adenophora delavayi (Franch.) D. Y. Hong, **comb. nov. & lectotype designation**

Adenophora himalayana Feer subsp. **alpina** (Nannf.) D. Y. Hong, **lectotype designation**

Adenophora khasiana (Hook. f. & Thomson) Oliv. ex Collett & Hemsl., **lectotype designation**

Adenophora likiangensis C. Y. Wu, **lectotype designation**

Adenophora linearifolia D. Y. Hong, **sp. nov.**

Adenophora megalantha Diels, **lectotype designation**

Adenophora microcodon C. Y. Wu, **lectotype designation**

Adenophora ornata Diels, **lectotype designation**

Adenophora potaninii Korsh, **lectotype designation**

Adenophora smithii Nannf., **lectotype designation**

Adenophora stricta Miq. subsp. **aurita** (Franch.) D. Y. Hong, **lectotype designation**

Asyneuma fulgens (Wall.) Briq. subsp. **forrestii** D. Y. Hong, **subsp. nov.**

Asyneuma thomsonii (Hook. f.) Bornm., **lectotype designation**

Campanula alsinoides Hook. f. & Thomson, **lectotype designation**

Campanula calcicola W. W. Smith, **lectotype designation**

Campanula chrysospleniifolia Franch, **lectotype designation**

Campanula colorata Wall., **lectotype designation**

Campanula crenulata Franch., **lectotype designation**

Campanula mekongensis Diels ex C. Y. Wu, **new record** for the Pan-Himalaya

Campanula sylvatica Wall., **lectotype designation**

Cyananthus subg. *Micranthus* K. K. Shrestha, **syn. nov.** = **Cyananthus** sect. **Anuui** (Y. S. Lian) K. K. Shrestha

Cyananthus sect. Stenolobi (Franch.) Y. S. Lian, **lectotype designation**

Cyananthus hayanus ('hayana') C. Marquand, **syn. nov.** = **Cyananthus lobatus** Wall. ex Benth.

Cyananthus himalaicus K. K. Shrestha, **syn. nov.** = **Cyananthus pedunculatus** C. B. Clarke

Cyananthus macrocalyx Franch. subsp. *spathulifolius* (Nannf.) K. K. Shrestha, **syn. nov.** = **Cyananthus macrocalyx** Franch.

Cyananthus microphyllus Edgew. subsp. *williamsonii* K. K. Shrestha, **syn. nov.** = **Cyananthus microphyllus** Edgew.

Cyclocodon parviflorus (Wall. ex A. DC.) Hook. F. Thomson, **lectotype designation**

Cyclocodon truncata Wall. ex A. DC., **lectotype designation**

Helwingia himalaica Hook. F. & Thomson, **lectotype designation**

Ilex atrata W. W. Smith, **lectotype designation**

Ilex dipyrena Wall., **lectotype designation**

Ilex fragilis Hook. f., **lectotype designation**

Ilex hookeri King, **lectotype designation**

Ilex hypotricha Loes., **lectotype designation**

Ilex intermedia Loes., **lectotype designation**

Ilex intricata Hook. f., **lectotype designation**

Ilex kingiana Cockerell, **lectotype designation**

Ilex longecaudata H. F. Comber var. *glabra* S. Y. Hu, **syn. nov.** = **Ilex longecaudata** H. F. Comber

Ilex macrostigma C. Y. Wu ex Y. R. Li, **syn. nov.** = **Ilex corallina** Franch.

Ilex pernyi Franch., **lectotype designation**

Ilex subodorata S. Y. Hu, **lectotype designation**

Ilex szechwanensis Loes., **lectotype designation**

Ilex umbellulata (Wall.) Loes., **lectotype designation**

Ilex venulosa Hook. f. var. *simplicifrons* S. Y. Hu, **syn. nov.** = **Ilex venulosa** Hook. f.

Ilex wattii Loes., **lectotype designation**

Lobelia clavata E. Wimm., **leclotype designation & syn. nov.** = **Lobelia rosea** Wall.

Lobelia drungjiangensis D. Y. Hong, **sp. nov.**

Lobelia erectiuscula H. Hara, **neotype designation**

Lobelia gaoligongshanica D. Y. Hong, **sp. nov**.

Lobelia hancei H. Hara, **syn. nov.**= **Lobelia alsinoides** Lam.

Lobelia pleotricha Diels, **lectotype designation**

Lobelia pyramidalis Wall., **lectotype designation**

Lobelia rosea Wall., **lectotype designation**

Lobelia seguinii H. Lév. & Vaniot, **lectotype designation**

Lobelia sichuanensis (Y. S. Lian) D. Y. Hong, **stat. nov**. Basionym: *L. davidii* Franch. var. *sichuanensis* Y. S. Lian

Lobelia taliensis Diels, **lectotype designation**

Lobelia tibetica W. L. Zheng, **syn. nov.** = **Lobelia pleotricha** Diels

Platycodon grandiflorus (Jacq.) A. DC., **lectotype designation**

Wahlenbergia agrestis (Wall.) A. DC., **lectotype designation**

Wahlenbergia hirsuta (Edgew.) Tuyn, **lectotype designation**

Index of Scientific Names

ISBN 978-1-107-15897-9

9 781107 158979

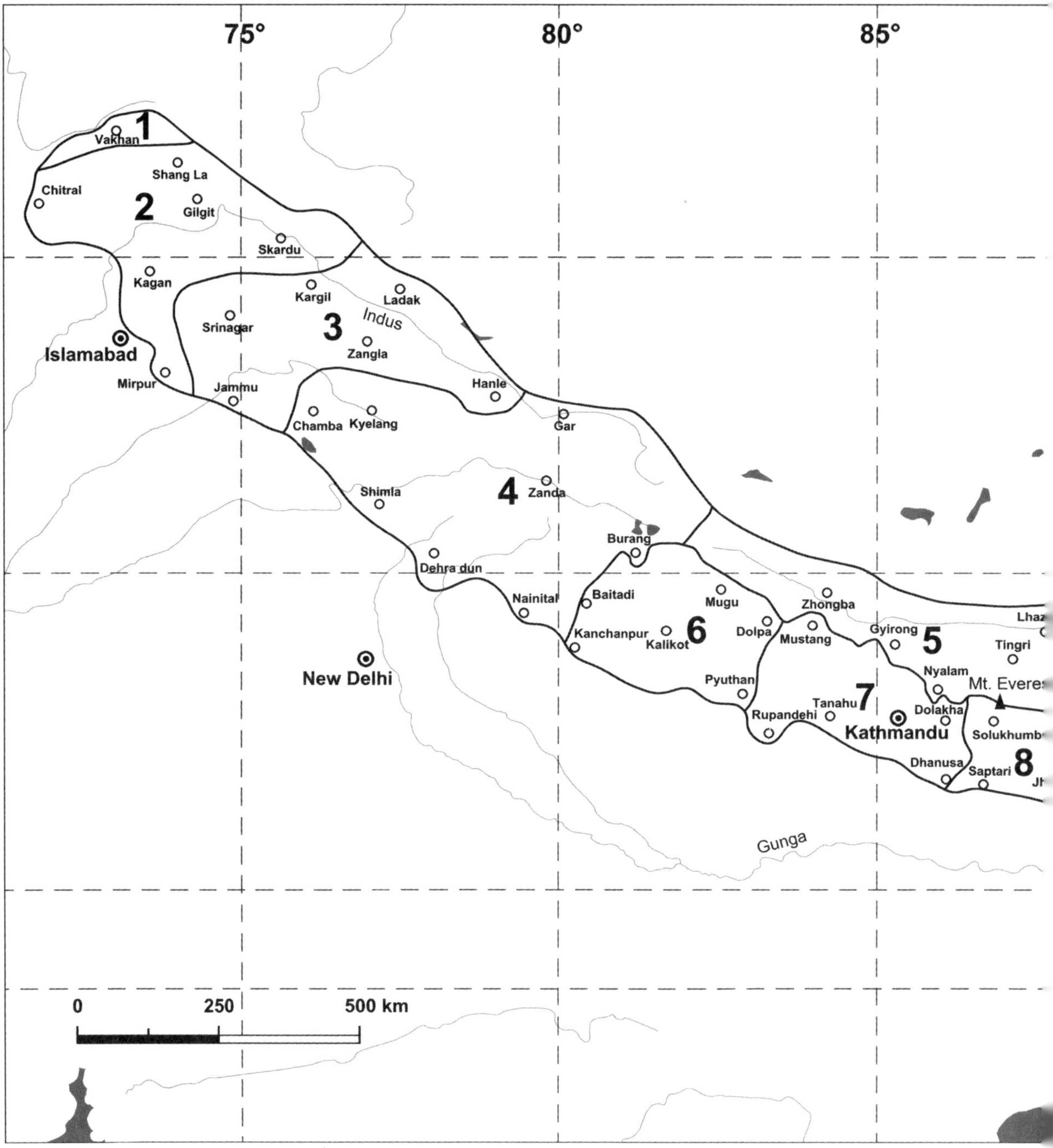

Subdivisions c

1.Vakhan; 2.N Pakistan; 3. Jammu & Kashmir; 4. U Ganga & Indus; 5
10. Bhutan; 11. M Yarlung Zangbo; 12. L Yarlung Zangbo; 13. Yarlung Za